Teaching General Chemistry:
A Materials Science Companion

Tom -
Best regards.
Art

Teaching General Chemistry: A Materials Science Companion

Arthur B. Ellis

Margret J. Geselbracht

Brian J. Johnson

George C. Lisensky

William R. Robinson

American Chemical Society, Washington, DC

1993

Library of Congress Cataloging-in-Publication Data

Teaching general chemistry: a materials science companion / Arthur B. Ellis ... [et al.].

p. cm.

Includes bibliographical references and index.

ISBN 0–8412–2725–X

1. Solid state chemistry. 2. Chemistry—Study and teaching (Secondary).

I. Ellis, Arthur B., 1951– .

QD478.T43 1993
541'.0421—dc20

93–29739
CIP

Cover: Magnetic spiking phenomenon observed when a cow magnet is placed beneath a Petri dish containing a ferrofluid. Chapter 2 discusses this phenomenon.

About the binding of this book: This book is bound with a lay-flat binding process. When the book is opened, the pages will lay flat, leaving the reader's hands free.

PRINTED IN THE UNITED STATES OF AMERICA

Second printing 1995

1993 Advisory Board

The Authors

From left to right: Margret J. Geselbracht, Assistant Professor, Department of Chemistry, Reed College, Portland, OR; Brian J. Johnson, Associate Professor, Department of Chemistry, St. John's University, Collegeville, MN; William R. Robinson, Professor, Department of Chemistry, Purdue University, West Lafayette, IN; Arthur B. Ellis, Meloche-Bascom Professor, Department of Chemistry, University of Wisconsin–Madison, Madison, WI; and George C. Lisensky, Professor, Department of Chemistry, Beloit College, Beloit, WI. This is a composite photograph. To the best of the authors' knowledge, the five authors were never all at the same place at the same time!

Contents

Laboratory Experiments

Appendices

Demonstrations

Topic Matrix

Topic	*Chapter*										*Experiment*
Atoms	1	2									1
Acids and Bases								8			15
Bands							7	8			7, 8
Batteries		2	3					8			
Bohr model for hydrogen atom								8			
Bonding			3		5	6	7				2
Conductivity, thermal and electrical							7				8, 11, 12
Coordination numbers/geometry					5						2
Crystal structure		2	3		5		7		9		2, 4, 5
Defects						6		8	9		6
Diffusion										10	13
Dipoles		2									
Electrochemistry								8			
Electromagnetic radiation				4		6	7	8			4,9
Electronegativity							7				7
Electrons		2					7	8			7,8
Entropy								8	9		
Equilibrium								8	9		
Free energy								8	9		
Heat capacity		2									1
Intermolecular forces		2									2
Ionic solids					5						2
Ionization								8			
Kinetics						6				10	
Lasers				4				8			4,9
Le Chatelier's principle									9		10
Magnetism		2							9		11, 14
Metals			3		5	6	7				2,10
Molecular orbital theory							7				
Nuclear chemistry						6					6
pH								8			15
Periodic properties			3			6	7				7
Phase changes						6			9		10, 11, 12
Quantum mechanics						6		8			
Redox		2							9	10	8,11
Semiconductors			3		5		7	8			2, 7, 9
Smart materials	1	2							9		10
Solid solutions			3			6	7		9	10	3, 7, 14
Spectroscopy - Beer's Law							7	8			7, 9
Stoichiometry			3		5				9	10	2, 11
Structures of solids			3		5	6	7		9		2, 5, 10
Superconductivity	1								9		11
Thermochemistry		2							9		
Thermodynamics						6			9	10	
Transition metals		2								10	3,10,11,12
VSEPR				4	5						2
X-ray diffraction				4							4, 5

Preface

The Spirit of the Companion

With the arrival of the 90s, it is an appropriate time to update and broaden the career advice Dustin Hoffman received in the 1967 movie, "The Graduate." Instead of urging readers to work with plastics, the intent of this book is to show, more generally, that materials are both a source of great opportunities for chemists in the coming decade and a natural vehicle for introducing chemical principles. Whether the solids are semiconductors, metals, superconductors, polymers, or composites, our society is increasingly dependent on advanced materials and devices that deliver performance within the constraints set by our limited energy and environmental resources. The introductory chemistry course can provide a firm foundation for understanding the burgeoning field of materials chemistry.

Despite the key role that chemistry plays in what is often called "materials science," there has been relatively little materials chemistry in introductory chemistry courses. In fact, a 1989 National Science Foundation *(1)* report concluded that "the historic bias of chemistry curricula toward small molecule chemistry, generally in the gaseous and liquid states, is out of touch with current opportunities for chemists in research, education, and technology." Furthermore, the report notes that "the attractiveness of chemistry and physics for undergraduate majors could be enhanced by greater emphasis on materials-related topics which would help students better relate their studies to the 'real world'."

Our experience has been, however, that many chemistry teachers are uncomfortable teaching materials chemistry, which has a strongly interdisciplinary flavor: Teachers may be less confident describing solid-state synthesis, processing, structure and bonding, and those physical and chemical properties of solids which have a language that is largely rooted in physics, engineering, and materials science. The extended, three-dimensional structures of solids, which are often hard to visualize, may represent another formidable obstacle.

The goal of this volume, *Teaching General Chemistry: A Materials Science Companion*, and its supporting instructional materials, is to demystify materials chemistry so that its essence—the interrelationship of synthesis, processing, structure, bonding, and physicochemical properties—can be readily brought into introductory chemistry courses.

A second goal of the Companion is to refresh and enliven general chemistry. The topics presented here enable instructors to put a new spin on the material that is traditionally covered in general chemistry. The examples from cutting-edge research, as well as everyday life, will serve to maintain student interest while illustrating the basic ideas that are important to an understanding of chemistry.

The Ad Hoc Committee

An ad hoc committee was formed in 1990, in recognition of the fact that both the interdisciplinary nature of materials chemistry and a lack of appropriate instructional tools were preventing many teachers from discussing the subject. The committee's composition reflects both technical and pedagogical expertise: Research areas spanning much of the breadth of modern materials chemistry are represented, as is skill at bringing chemical concepts into the classroom and laboratory *(2)*. The Ad Hoc Committee for Solid-State Instructional Materials comprises Aaron Bertrand, Georgia Institute of Technology; Abraham Clearfield, Texas A&M University; Denice Denton, University of Wisconsin-Madison; John Droske, University of Wisconsin-Stevens Point; Arthur B. Ellis, University of Wisconsin—Madison (Chair); Paul Gaus, The College of Wooster; Margret Geselbracht, Reed College; Martha Greenblatt, Rutgers University; Roald Hoffmann, Cornell University; Allan Jacobson, University of Houston; Brian Johnson, St. John's University; David Johnson, University of Oregon; Edward Kostiner, University of Connecticut; Nathan S. Lewis, California Institute of Technology; George Lisensky, Beloit College; Thomas E. Mallouk, University of Texas at Austin; Robert E. McCarley, Iowa State University; Ludwig Mayer, San Jose State University; Joel Miller, University of Utah; Donald W. Murphy, AT&T Bell Laboratories; William R. Robinson, Purdue University; Don Showalter, University of Wisconsin—Stevens Point; Duward F. Shriver, Northwestern University; Albert N. Thompson, Jr., Spelman College; M. Stanley Whittingham, SUNY at Binghamton; Gary Wnek, Rensselaer Polytechnic Institute; and Aaron Wold, Brown University.

Objectives

The principal objectives of the Ad Hoc Committee in producing the Companion are to revitalize the introductory chemistry course for all students and to increase the number and diversity of technically able students who will pursue careers as chemists, chemistry teachers, scientists, and engineers. To accomplish these objectives, the committee collected and created instructional materials and developed pedagogical strategies for the introduction of these materials into the curriculum.

The philosophy underpinning the committee's effort to mainstream materials chemistry into the curriculum is that *virtually every topic typically discussed in a general chemistry course can be illustrated with examples from materials chemistry*. The Companion is intended to collect, in one place, a critical mass of text, demonstrations, laboratory experiments, and leads to supporting instructional materials (software and kits, for example) that illustrate how materials chemistry fits into the traditional introductory chemistry course. At the same time, the Companion needs to be sufficiently flexible that it can accommodate, and even help to define, new introductory course structures.

The Companion is presented in a ready-to-use format that will enable teachers to integrate materials chemistry into their courses with minimal effort. Instructors will then have solid-state examples to complement molecular examples. By presenting both, the interconnectedness and universality of scientific thinking can be emphasized.

The Companion assumes essentially no background in materials chemistry but builds on a foundation of molecular chemistry shared by many college and pre-college chemistry teachers. Because the Companion is written for teachers, extra depth is sometimes included beyond the level one might present in an introductory chemistry course; most teachers are more comfortable when they have extra material available for support. The Companion contains enough information in the text, footnotes, and leading references to permit more quantitative and sophisticated treatments for upper-level undergraduate and even graduate courses.

As will be evident throughout the Companion, materials chemistry is intimately connected to other traditional disciplines like physics and engineering. Many colleagues in related scientific and engineering disciplines, who recognize the role of chemistry as a foundation course for students pursuing technical careers, have graciously contributed their expertise to this project *(3)*. Some of the Companion's contents will also be useful in their courses, which span the undergraduate science and engineering curriculum.

We encourage our readers to use and adapt materials from the Companion and to send us constructive comments that can be passed on to other users.

The authors' royalties from the book are being donated to the Institute for Chemical Education (ICE) to assist with dissemination of solid-state instructional materials.

Selection Criteria for Inclusion of Topics

In developing the Companion, we have omitted a great deal of information that would have provided a more comprehensive view of materials chemistry in favor of a more focused, less daunting offering. Furthermore, with a few exceptions, polymers are conspicuously absent from the Companion. This omission is deliberate: A related project, headed by John Droske and also funded by the National Science Foundation (*see* reference 4), will soon be releasing instructional materials describing how polymers can be integrated into the curriculum. John has graciously shared his group's work with us so that we could best complement rather than duplicate one another's efforts.

Key criteria for including materials in the Companion were that they be thought-provoking, illustrative of core chemical concepts, and easily incorporated into existing course structures. We particularly sought to include high-tech materials and advanced devices that have or will become an important part of our environment. Student recognition that these advanced materials and devices derive from chemical principles is a major pedagogical objective. We feel, too, that the incorporation of state-of-the-art examples from materials chemistry will provide a strong sense of relevance that will help revitalize the introductory chemistry course for all students, whether or not they pursue technical careers.

Two other criteria were cost and safety considerations. Most of the demonstration and laboratory experiments described herein require relatively little time or money to set up and dismantle, and instructions for doing so safely are provided throughout the Companion. In field tests of these instructional materials, we have found that the ease with which many of the solids can be safely handled and transported and the ability to use them repeatedly, minimizing expense and waste disposal, are particularly appealing to teachers.

How the Companion is Organized

An Introduction to Solids

The first ten chapters of the Companion represent a general introduction to solids that can either be read sequentially, for an overview; or topically, if particular kinds of solid-state illustrations are sought. The first chapter is meant to provide a context for this volume: the chapter describes materials chemistry, provides examples of cutting-edge research and applications that illustrate why materials chemistry is expected to be one of the most rapidly moving scientific and technological frontiers in the next decade, and highlights the role that chemists are playing in advancing this field.

To facilitate integration into the existing course structure at many institutions, the remainder of the first part of the Companion is organized under traditional general chemistry textbook chapter headings, such as "Stoichiometry" and "Equilibrium." Teachers will recognize some of the Companion's content as being simply a new twist on well-established and widely used material. In other cases, we believe that teachers will discover, as we have, exciting new approaches to the presentation of fundamental chemical concepts.

Demonstrations are an integral part of the Companion and are listed at the front of the Companion for ease of location. In field tests and workshops we have found demonstrations to be a particularly effective means for stimulating interest in materials chemistry.

Lists of additional reading materials have been included at the end of each chapter. More advanced texts that can be drawn upon are A.R. West's *Basic Solid State Chemistry* (Wiley, NY, 1988), P. A. Cox's *The Electronic Structure and Chemistry of Solids* (Oxford University Press,

Oxford, 1987), and L. Smart and E. Moore's *Solid State Chemistry* (Chapman and Hall, London, 1992). *The Materials Research Society Bulletin* (MRS Bull.), published monthly (Pittsburgh, PA), is recommended as a source of current information on all aspects of materials chemistry. Appendix 3 lists articles relevant to materials chemistry that have appeared in the *Journal of Chemical Education* since 1982, arranged according to topic.

A variety of representative exercises of varying levels of difficulty have been included, along with answers.

Laboratory Experiments

The second part of the Companion consists of laboratory experiments provided by members of the Ad Hoc Committee, whose authorship is noted thereon. Most of these experiments have been field-tested in introductory college chemistry courses.

Topic Matrix and Glossary

For much of the material in the Companion, a topic, demonstration experiment, or laboratory experiment could be used under any of several traditional chapter headings, and an arbitrary choice was made. Other possible choices, intended to give teachers maximum flexibility, are identified by using the matrix that follows the table of contents. A glossary has been included (Appendix 1) for quickly identifying definitions of key terms used in the Companion.

Supporting Instructional Materials

Sources of materials needed for demonstration and laboratory experiments are given in the Supplier Information, Appendix 2. A current list of suppliers will be maintained by ICE and can be obtained by writing ICE at the Department of Chemistry, University of Wisconsin—Madison, Madison, WI 53706. If you find alternate suppliers, please send us this information to add to the list.

Among the supporting instructional materials, the ICE Solid-State Model Kit (SSMK) is particularly noteworthy. Because of the periodic three-dimensional nature of many of the solids discussed in the Companion, it is critical that teachers and students have a good model with which to view such structures. The ICE SSMK has been designed to permit ready assembly and viewing of common structures in conjunction with the Companion.

Electronic mail can be used to obtain answers to questions that may arise from the use and adaptation of materials from the Companion. Questions submitted to ellis@chem.wisc.edu (INTERNET) will be answered as expeditiously as possible.

We hope there aren't any errors. However, given the breadth of material covered and our efforts to make this volume available to the community quickly, there may well be some errors, for which we apologize in advance. Please contact the authors to inform us of any mistakes.

The "Big Picture"

At the start of this project, one of the Ad Hoc Committee members, Joel Miller, noted that our objective should be to try to make it possible for solids to be used for at least ten percent of the examples in general chemistry courses. The Companion makes it possible, at the teacher's discretion, for solids to comprise as much as fifty percent of the examples. Lisensky and Ellis have incorporated solids for several consecutive semesters in small (25 students) and large (250 to 350 students) introductory chemistry courses at Beloit College and UW—Madison, respectively. Their findings, based on subsequent course performance and exit surveys, is that the solids-enriched course provides equivalent preparation for further study and gives students, particularly nonmajors, a perspective on chemistry that the students themselves describe as broad and relevant.

James Trefil notes in several of his books that science is like an interconnected web of ideas that can be entered from virtually any point *(5)*. The matrix linking topics in the Companion with traditional chemical concepts and the connections noted throughout with other disciplines is an illustration of this philosophy. What we hope to have shown with the Companion is that materials chemistry is an excellent launching point for exploring the web of chemically based ideas.

At the same time, we would be disappointed if this volume were treated as an end point. The cutting-edge examples presented herein show that chemistry, and science in general for that matter, is a living discipline. Like our research enterprise, the introductory course should be treated as a "moving target" with many opportunities for innovation. The Companion is intended to show that there is a natural synergism between research and teaching that can be used to suffuse introductory chemistry courses with relevance and vitality.

Acknowledgments

Many individuals and organizations contributed to the Companion. It is a pleasure to acknowledge ACS Books staff, Cheryl Shanks, Robin Giroux, Janet Dodd, and the production team, who handled the publication of this volume. We thank Sylvia Ware and Glenn Crosby, of the ACS Society Committee on Education, and the Committee, which provided seed money for our efforts; Patricia Daniels, Susan Hixson, James Harris, David Nelson, Robert Reynik, Curtis Sears, Robert Watson, and Gene Wubbels of the National Science Foundation, which provided major funding for this project through grants from the Education Directorate (USE-9150464) and from the Division of Materials Research; Robert Lichter of the Camille and Henry Dreyfus Foundation and the Foundation, which provided support for two Dreyfus Conferences that were crucial for planning this effort; Frank Di Salvo and Lynn Schneemeyer, who hosted our second Dreyfus Conference in conjunction with the 1992 Solid-State Gordon Research Conference; Robert Nowak, Tom Kennedy, Ted Tabor, and Warren Knox of Dow Chemical Company and the Dow Chemical Company Foundation, and Pat Takemoto of the UW—Madison Office of Outreach Development, which provided seed money for the ICE Solid-State Model Kit.

We thank John and Betty Moore and their co-workers, who helped develop supporting instructional materials through the Institute for Chemical Education and Project Seraphim; Joe Casey of Milwaukee Area Technical College, who served as a consultant for the Model Kit; other colleagues in the UW—Madison Department of Chemistry and College of Engineering's Materials Science Program: Jill Banfield, Kevin Bray, Judith Burstyn, Reid Cooper, Fleming Crim, Larry Dahl, Art Dodd, Bob Hamers, Eric Hellstrom, Tom Kelly, Dan Klingenberg, Tom Kuech, Max Lagally, Clark Landis, David Larbalestier, Rich Mayti, Leon McCaughan, Gil Nathanson, John Perepezko, Jeroen Rietveld, Jim Skinner, Don Stone, Ned Tabatabaie, Worth Vaughan, Frank Weinhold, Jim Weisshaar, Frank Worzala, and Hyuk Yu have contributed their expertise to this project; Deans John Bollinger, Phil Certain, Greg Moses, and John Wiley; individuals associated with the UW—Madison General Chemistry program: Diana Duff, Gery Essenmacher, Don Gaines, John Harriman, Fred Juergens, Lynn Hunsberger, John Moore, Janice Parker, Bassam Shakhashiri, Mary Kay Sorenson, and Paul Treichel, and our graduate student teaching assistants; and students and colleagues at Beloit College: Laura Parmentier, Brock Spencer, Rona Penn, Jill Covert, Megan Reich, and Pete Allen, who have helped us develop and test these instructional materials. We thank David Boyd of the University of St. Thomas and Todd Trout of Mercyhurst College for developing solutions to the exercises. We also thank our many colleagues around the country who have field-tested materials in the Companion.

Constructive reviews of the Companion by Wayne Gladfelter, Hans-Conrad zur Loye, Mary Castellian, Andy Bocarsly, Allen Adler, Francis Galasso, Norman Craig, and several anonymous reviewers are deeply appreciated.

Finally, on a personal note as chair of the committee, I would like to acknowledge my family, colleagues, and students, who have been the inspiration for this project. I am grateful to my wife, Susan Trebach, to our children, Joshua and Margot, and to our families for their advice and support. I thank my colleagues on the Ad Hoc Committee and in the Department of Chemistry and the Materials Science Program at UW—Madison for their contributions and support. My student research co-workers during the period of this project, Melissa Baumann, Bob Brainard, Ann Cappellari, Kathy Gisser, Pat James, Keith Kepler, Glen Kowach, Dale Moore, Kris Moore, Don Neu, Joel Olson, Dean Philipp, Ed Winder, and John Zhang, and my general chemistry students are thanked for their help and feedback, which, I believe, has greatly improved the final product.

All of us connected with the project hope that succeeding generations of students and teachers will be the beneficiaries of introductory chemistry courses that more effectively engage, prepare, and inspire. As Robert Browning wrote, "our reach should exceed our grasp. Or what's a heaven for?" *(6)*.

Arthur B. Ellis
Chair, Ad Hoc Committee for Solid-State Instructional Materials
University of Wisconsin—Madison
1101 University Avenue
Madison, Wisconsin 53706
August 1993

References

1. "Report on the National Science Foundation Undergraduate Curriculum Development Workshop in Materials," October 11–13, 1989. Pub. April 1990.
2. The development of this project has been summarized in an article, "Symposium Session on Educational Issues," In *Materials Chemistry: An Emerging Subdiscipline*; Interrante, L. V.; Ellis, A. B.; Casper, L., Eds. Advances in Chemistry Series; ACS, submitted. A sketch of the project has also been published: Ellis, A. B.; Geselbracht, M. J.; Greenblatt, M.; Johnson, B. J.; Lisensky, G. C.; Robinson, W. R.; Whittingham, M. S.; *J. Chem. Educ.* **1992**, 69, 1015.
3. Ellis, A. B. *Proceedings of the 1992 Frontiers in Education Conference*, Nashville, TN, p 638.
4. Droske, J. *J. Chem. Educ.* **1992**, *69*, 1014.
5. See for example, Trefil, J. *Meditations at 10,000 Feet*; Scribners: New York, 1986.
6. From *Andrea Del Sarto*. I thank David Pennington of Baylor University and the staff at the Armstrong Browning Library in Waco, TX, for identifying the source of this quotation.

Laboratory Safety

DISCLAIMER

Safety information is included in each chapter as a precaution to the readers. Although the materials, safety information, and procedures contained in this book are believed to be reliable, they should serve only as a starting point for laboratory practices. They do not purport to specify minimal legal standards or to represent the policy of the American Chemical Society. No warranty, guarantee, or representation is made by the American Chemical Society, the authors, or the editors as to the accuracy or specificity of the information contained herein, and the American Chemical Society, the authors, and the editors assume no responsibility in connection therewith. The added safety information is intended to provide basic guidelines for safe practices. Therefore, it cannot be assumed that necessary warnings or additional information and measures may not be required. Users of this book and the procedures contained herein should consult the primary literature and other sources of safe laboratory practices for more exhaustive information. As a starting point, a partial list of sources follows.

General Information on Chemicals

A Comprehensive Guide to the Hazardous Properties of Chemical Substances by Pradyot Patniak. Van Nostrand Reinhold: New York, 1992. This guide classifies chemicals by functional groups, structures, or general use, and the introductions to these chapters discuss the general hazards. Cross-indexed by name-CAS number and CAS number-name.

Catalog of Teratogenic Agents by Thomas H. Shepard. 6th ed. Johns Hopkins University Press: Baltimore, MD, 1989. Among books on this subject, this is perhaps the most thoughtful source list.

Handbook of Reactive Chemical Hazards by Leslie Bretherick. 4th ed. Butterworths: Stoneham, MA, 1990. This book is a "must" for chemistry researchers or for anyone "experimenting" in the laboratory. It has a very useful format for determining the possible explosive consequences of mixing chemicals.

Fire Protection Guide to Hazardous Materials. 10th ed. National Fire Protection Association, 1991. Laboratory personnel are not the primary audience of NFPA publications, but this document has clear, concise, and easily accessed information on most commercial chemicals and a number of laboratory chemicals.

Safe Storage of Laboratory Chemicals edited by David A. Pipitone. 2nd ed. John Wiley and Sons: New York, 1991. The scope of this book is far broader than the title suggests. Techniques for the proper dispensing of flammable solvents and spill procedures are discussed.

Cryogens

Standard on Fire Protection for Laboratories Using Chemicals (NFPA 45). National Fire Protection Association, 1991. Especially relevant is Chapter 8, Section 8-3, "Cryogenic Fluids".

Handbook of Compressed Gases by the Compressed Gas Association. 3rd Ed. Van Nostrand Reinhold: New York, 1990.

"Cryogenic Fluids in the Laboratory" Data Sheet I-688-Rev. 86. National Safety Council, 1986.

Improving Safety in the Chemical Laboratory: A Practical Guide edited by Jay A. Young. 2nd ed. John Wiley and Sons: New York, 1991. Section 11.11, "Low Temperature and Cryogenic Systems", pp 192–195.

Lasers

American National Standard for the Safe Use of Lasers. ANSI Z136.1—1986. American National Standards Institute: New York, 1986. Provides guidance for the safe use of lasers and laser systems.

Light, Lasers, and Synchrotron Radiation edited by M. Grandolfo et al. Plenum Press: New York, 1990. The chapter titled "Laser Safety Standards: Historical Development and Rationale" puts ANSI Z136.1—1986 in perspective.

"Lasers—The Nonbeam Hazards" by R. James Rockwell, Jr. In *Lasers and Optronics*, August 1989, p 25.

Improving Safety in the Chemical Laboratory: A Practical Guide edited by Jay A. Young. 2nd ed. John Wiley and Sons: New York, 1991. Section 11.3.5.1, "Laser Radiation", pp 177–178.

High Voltage

"Electrical Hazards in the High Energy Laboratory" by Lloyd B. Gordon. In *IEEE Transactions on Education,* Vol. 34, No. 3, August 1991, pp 231–242. This article contains primarily physics laboratory safety; it gives excellent information on electrical hazards. Highly recommended.

Improving Safety in the Chemical Laboratory: A Practical Guide edited by Jay A. Young. 2nd ed. John Wiley and Sons: New York, 1991. Section 11.5, "Electric Currents and Magnetic Fields", pp 181–182.

Vacuum Techniques

The Laboratory Handbook of Materials, Equipment, and Technique by Gary S. Coyne. Prentice Hall: New York, 1992. Chapter 7, "Vacuum Systems", pp 275–408.

Improving Safety in the Chemical Laboratory: A Practical Guide edited by Jay A. Young. 2nd ed. John Wiley & Sons, 1991. Section 11.7, "Vacuum and Dewar Flasks", p 183.

Compiled by Maureen Matkovich, American Chemical Society.

Chapter 1

An Introduction to Materials Science

This chapter provides an overview of materials chemistry and some context for the concepts developed in subsequent chapters. The first part of this chapter articulates the interdisciplinary nature of materials chemistry and illustrates the roles of synthesis and processing in defining the properties of a material.

The second section of the chapter provides some perspective on the impact that materials have on our environment. A material's life cycle, comprising its origin, use, and disposal, is an increasingly important consideration in evaluating its utility. These issues are explored, using a now-recyclable industrial hydrodesulfurization catalyst as an example.

The final part of the chapter is meant to convey some of the excitement and relevance of modern materials chemistry. The Age of Materials is characterized by new analytical methods, including atomic-scale imaging; unprecedented chemical control of interfaces; and the purposeful design of new materials, even of "smart" materials that respond in predictable ways to stimuli. Some "high-tech" materials, those developed or produced with the aid of the latest technology, and advanced devices are already an integral part of our lives. The trajectory of the materials chemistry enterprise suggests that many more soon will be!

What Is Materials Science?

Materials chemistry is a broad, chemically oriented view of solids—how they are prepared and their physical and chemical characteristics and properties. Often, the term "materials science" is used to describe the understanding of solids that emerges from the combined viewpoints of

2725–X/93/0001$06.00/1

chemistry, physics, and engineering, and, for biomaterials, the biological sciences.

Materials science has been represented by the tetrahedron shown in Figure 1.1 *(1)*. Each vertex of the pyramid is equally important in this holistic portrait of materials science: "synthesis and processing" refers to the preparative conditions that determine atomic structure and microstructure (the arrangement of micrometer-sized groups of atoms) of a material; "structure and composition" is the arrangement and identity of atoms derived from the particular synthetic conditions employed to prepare the material; "properties" refers to the characteristics (optical, mechanical, electrical, etc.) of the material; and "performance" represents the conditions under which the material maintains its desirable characteristics. Materials chemistry is the key to unifying this entire picture, a theme that will be presented in this overview chapter and reinforced throughout the text.

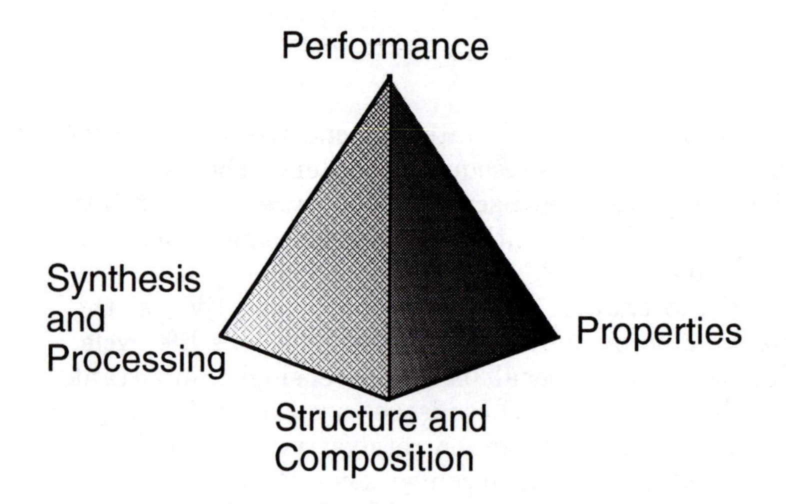

Figure 1.1. The materials tetrahedron.

A particularly striking example of materials chemistry that animates this tetrahedral interdependence is the behavior of polyurethane foam, pictured in Figure 1.2. Polyurethane is prepared as follows:

$$n\ HOCH_2CH_2OH + n\ O{=}C{=}N(CH_2)_mN{=}C{=}O \rightarrow [-OCH_2CH_2O\overset{O}{\overset{\|}{C}}NH(CH_2)_mNH\overset{O}{\overset{\|}{C}}-]_n$$

The desirable features of the foam, for example, good long-term performance as a thermal insulator and as an air filter, are linked to the network of pores in the foam. In turn, the pore network reflects the degree of polymerization (the distribution of n values) and the microstructure of the polymer, both of which are controlled by the synthetic conditions employed.

Commercially available samples of polyurethane have a mechanical property that is typical of most materials commonly encountered: When opposite ends of the foam are pulled apart, placing the foam under tension, it becomes longer in the direction that it is pulled and constricts in the perpendicular directions (Figure 1.2A and 1.2B); a common rubber band provides another graphic example of this effect.[1]

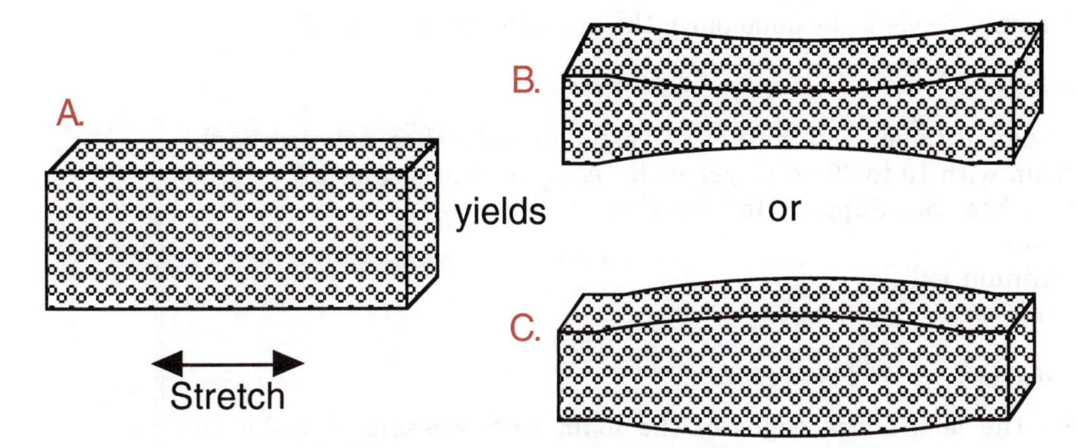

Figure 1.2. A block of polyurethane foam (A) will constrict (B) or bulge (C) in the middle when stretched, depending upon how it has been processed.

The importance of processing in materials chemistry is dramatically illustrated by converting this foam to so-called "re-entrant foam" *(2)*. If the polyurethane foam is compressed uniformly in all directions (isotropically) at a temperature of about 180 °C, below the melting point of the polymer, the applied pressure and heat can collapse the pores in the foam. Such a material will be substantially denser than it was before processing. Moreover, placing the pores under tension in one direction should cause them to expand in all directions, as sketched with the idealized pore shown in Figure 1.3. This expansion of the microstructure should result in a macroscopic expansion of the sample in all directions.

Figure 1.2C shows the realization of this prediction: when the re-entrant foam is stretched, it counterintuitively expands in the perpendicular directions as well! In the context of the materials tetrahedron, although the chemical composition of the solid remains unchanged, the processing step has increased the density of the solid, affected its microstructure, and completely altered the mechanical properties of the polymer.

---

[1] Such materials are described as having a positive Poisson ratio. This ratio, often symbolized by μ (mu), is defined as

$$\mu = \frac{\text{fractional decrease in width}}{\text{fractional increase in length}} = \frac{\Delta w/w}{\Delta l/l}$$

As defined, the Poisson ratio is usually positive, reflecting the fact that the change in the width is negative (the width shrinks) as the length increases. A negative Poisson ratio indicates the opposite effect, a relatively unusual expansion of the width as the material is stretched.

Demonstration 1.1. Re-entrant Foam

Part A. Preparation of Re-entrant Foam
(This only needs to be done once; the samples can be reused.)

Materials

Low-density open-cell (filter) polyester or polyether polyurethane foam with 10 to 20 pores per inch. A convenient size is 1.5 × 1.5 × 6 inches. *See* Supplier Information.
Scissors
Aluminum foil
Drying oven

Procedure

- Trim the lengthwise edges of the foam with scissors in order to make the sample easier to roll.
- Compress a piece of foam lengthwise to about one-third or one-half of its length, and then compress it axially by the same amount while rolling it in a 1 foot × 1 foot piece of aluminum foil. To optimize the effect, the foam should be compressed isotropically (equally in all directions) by about a third to a half.
- Twist the foil ends to hold the foam in its compressed shape. Figure 1.4 shows a sample before and after compression in aluminum foil.
- Heat the foil-wrapped foam in an oven for about an hour at about 175–195 °C. Remove it from the oven and let the sample cool before opening.

Part B. The Effect of Processing

Materials

Thick rubber band that has been cut to be a flat strip
Untreated foam
Re-entrant foam

Procedure

- Place each material, in turn, on an overhead projector. Stretch each by pulling opposite ends of the material apart. The rubber band and the untreated foam will become thinner, but the re-entrant foam will become thicker in cross-section upon stretching.

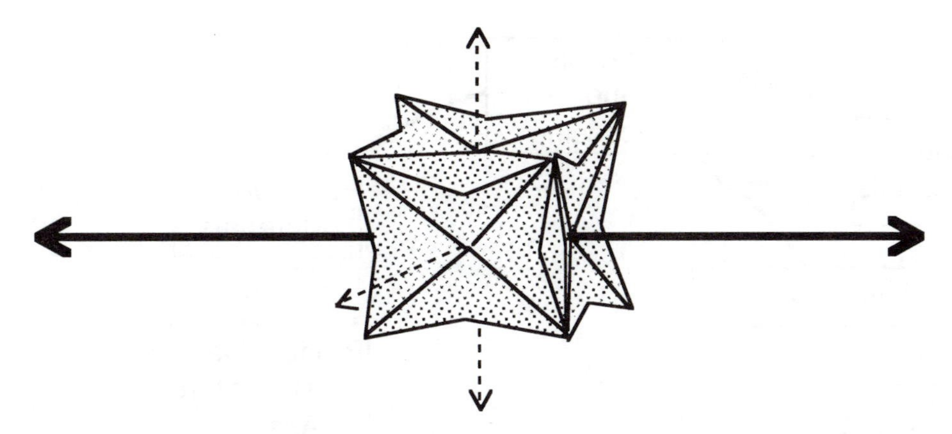

Figure 1.3. An idealized re-entrant pore, viewed as a cube with each of the six faces collapsed inward. After the appropriate processing, a stretching force applied in the horizontal direction (dark arrows) will cause the remaining four faces of the cube to expand (dashed arrows).

Figure 1.4. Untreated foam and the same sample compressed in aluminum foil for heating.

The Life Cycle of a Material

In preparing and using materials, the entire life cycle of a solid needs to be appreciated. Figure 1.5 shows a total materials cycle *(3)*, which begins with extracting raw materials for a targeted solid; continues with the synthesis and processing leading to the tailored material; passes on to the incorporation of the material into a deliverable product for a desired application; and ends with the fate of the waste derived from the material. In all steps of this cycle, energy costs and the effect on the environment need to be addressed. This kind of holistic thinking is rapidly becoming part of the culture of materials chemistry.

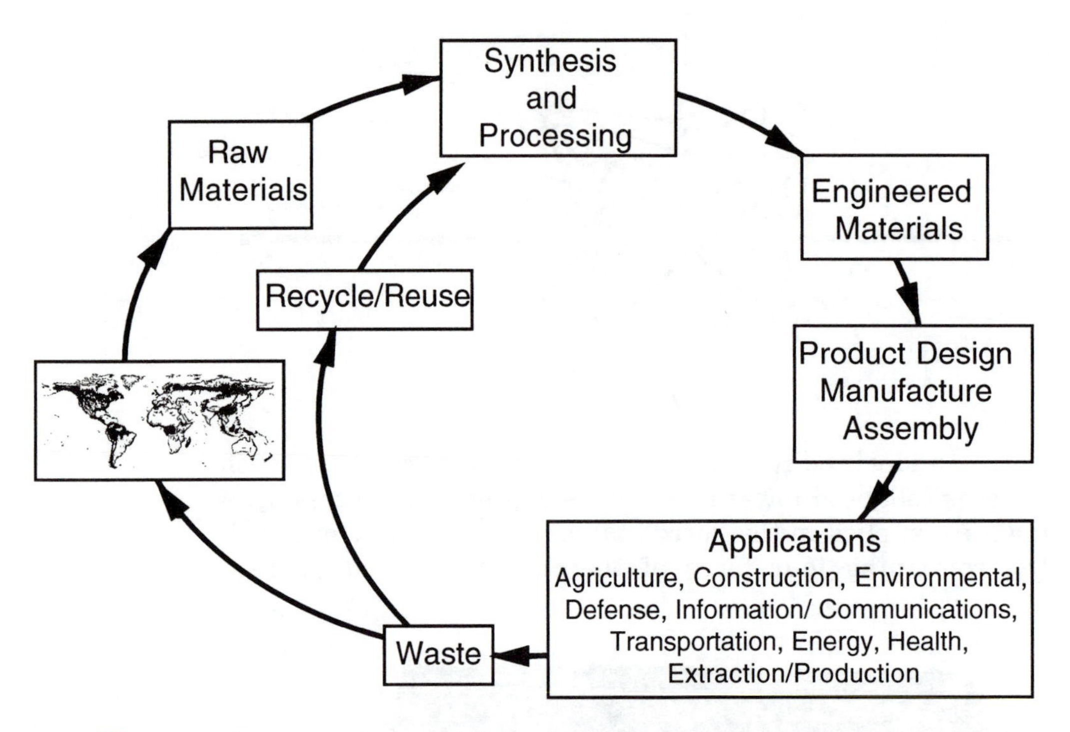

Figure 1.5. A total materials cycle. (Adapted from reference 3).

This cycle is illustrated better by looking at a specific example: the production and use of a catalyst for the hydrodesulfurization (HDS) process, a process for removing sulfur from petroleum feedstocks. If the sulfur is not removed before combustion of the petroleum, it contributes to acid rain. The typical catalyst used in the HDS process is MoS_2 supported on alumina; the activity of this catalyst is promoted by the addition of small amounts of cobalt. There is some dispute as to the exact chemical form of the cobalt, that is, whether it is present as the sulfide, Co_9S_8, or somehow more intimately connected with the MoS_2.

Figure 1.6 illustrates the synthesis and recovery cycles involved with an HDS catalyst. The production of this catalyst begins with the extraction of the components from the natural ores. Molybdenum occurs naturally as MoS_2, the mineral molybdenite. Cobalt sulfides and arsenides ($CoAs_2$, CoAsS, and Co_3S_4) are the compounds most frequently found in cobalt-containing ores, which usually also contain nickel, copper, and lead. The aluminum ore, bauxite, occurs as a mixture of aluminum hydroxide and aluminum oxide hydroxide, the fraction of each depending on the location. The ores are processed and refined to extract the purified elements in some usable form *(4)*.

The synthesis and processing of the working HDS catalyst involves slurrying the high surface area (finely divided) alumina support with solutions of molybdenum and cobalt, followed by firing to yield a mixture of the oxides ($CoO–MoO_3–Al_2O_3$) *(5)*. This oxidic form of the catalyst is then converted to the catalytically active sulfidic form during the HDS process, as it extracts sulfur from the petroleum feedstocks. After some time, the HDS activity of the catalyst drops, as a result of poisoning of the surface by deposits of carbonaceous material and by metals such as vanadium and nickel that are also extracted from the feedstock. The catalyst can be partially regenerated by burning off the carbonaceous material. However, the regeneration step does not remove the deposited metals, and eventually the activity can no longer be restored.

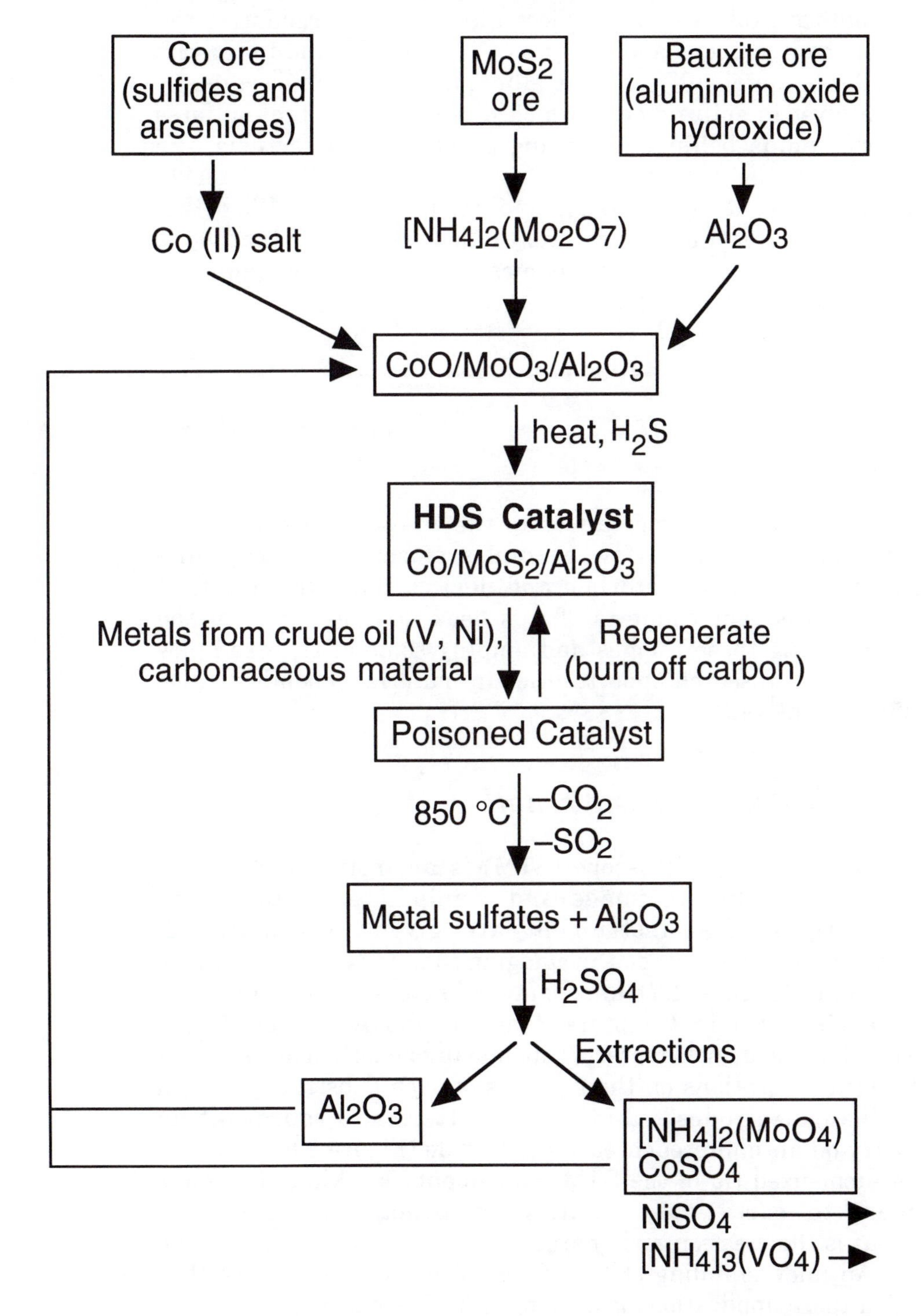

Figure 1.6. The life cycle of the components of a hydrodesulfurization catalyst.

In years past, the spent catalysts were committed to landfills; however, environmental concerns resulted in a ban on this practice in the mid-1980s. Recently, Metrex opened a plant in Heerlen, the Netherlands, that recovers the metals from spent HDS catalysts *(6)*. In the recycling process, the spent catalysts are first subjected to a heat treatment at 850 °C to remove the hydrocarbons and sulfur as CO_2 and SO_2 and to convert the metal sulfides to metal sulfates. The alumina support is recovered by leaching the catalyst in sulfuric acid, which dissolves the metal sulfates and leaves the alumina behind. By using a series of extractions, the individual metals can be separated and recovered as cobalt and nickel sulfates, ammonium vanadate, and ammonium molybdate. With this process, Metrex estimates a metals recovery efficiency greater than 90%. The metals and alumina are sold to companies that can use them, often in the preparation of additional batches of HDS catalysts.

The Age of Materials

Probably the best way to develop a feel for the vitality of the materials chemistry enterprise is with a few examples of current high-profile research activities and with materials and devices that already affect everyday life. Subsequent chapters of this book address many of the concepts underpinning these themes and amplify some of the examples given. Summaries of research opportunities in materials chemistry also were published recently *(3)*.

Analytical Methods: Thinking Small

The scanning tunneling microscope (STM) is an instrument that is revolutionizing the ability to image and manipulate matter. The principles underlying this instrument are given in Chapter 2. Researchers have already succeeded in imaging individual surface atoms on common metals (Figure 1.7) and semiconductors, whose physical structures are discussed in Chapters 3 and 5 and whose electronic structures are discussed in Chapter 7; in moving adsorbed atoms and molecules to desired locations on the surfaces of such substrates; and in fashioning "nanoelectrodes" and "nanobatteries," representing electrochemistry on an unprecedentedly small scale (Figure 2.8).

Use of the atom-sized tip of the STM instrument as a kind of "atomic hole puncher" to carry out atomic-scale lithography is under development, as is the creation and characterization of wires that are one atom thick! An understanding of the effects of physical structure that control many of these applications is developed in Chapter 6.

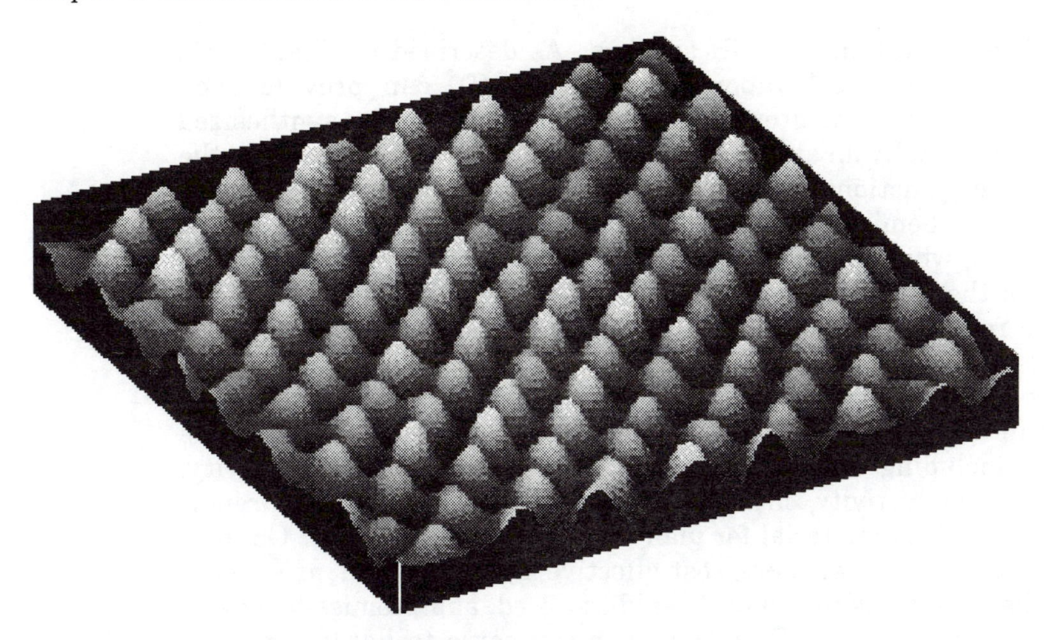

Figure 1.7. An STM-generated image of the surface of a gold crystal. (Original image courtesy of Burleigh Instruments, *see* Supplier Information.)

The pH sensor is another example of an analytical tool amenable to miniaturization. By placing dyes whose fluorescence is pH-dependent on the tips of optical fibers that are as small as 100 nm, the pH inside single biological cells and their substructures can be measured: The dyes can be excited by light-emitting diodes (LEDs) or diode lasers whose emitted light travels down the optical fibers; and the light emitted by the dye can be analyzed to determine the pH value of the medium in which it is immersed. Principles by which these semiconductor-based light sources and optical fibers operate are outlined in Chapters 8 and 10.

Machines represent a third rapidly moving area of analytical miniaturization, one that is heavily dependent on materials chemistry. Micromachines such as micromotors that are the size of a human hair have been prepared using polymer photoresists (for micrometer-scale patterning) and metal electroplating techniques. These machines are being proposed for use as on-line sensors in, for example, human bloodstreams, and as miniature gas chromatographs. Even smaller structures, nanostructures, are on the horizon.

Control of Interfaces

Interfaces, the zones where one material comes in contact with another, play a critical role in many devices based on electronic materials like semiconductors, metals, and superconductors. These solids, in which electrons are the mobile charge carriers, are discussed extensively in Chapters 5, 7, 8, and 9. The state of the art in growing certain solids like semiconductors is illustrated by the LEDs and diode lasers that have been mentioned in connection with pH sensing. As described in Chapter 10, techniques like chemical vapor deposition (CVD) can provide such exquisite control over the growth process that solids can be synthesized by depositing virtually an atomic layer at a time. This method permits the preparation of junctions that are essentially atomically abrupt. By controlling the chemical composition of these solids through the use of solid solutions, which are described in Chapter 3, such properties as the frequency of the light emitted by diodes may be tuned continuously over substantial regions of the visible and near-infrared spectrum.

Coatings and thin films play critical roles in the control of interfaces. An example of a coating with enormous potential is diamond. As discussed in Chapters 2, 5, and 7, diamond has a variety of remarkable properties, including exceptional hardness, high thermal conductivity, low electrical conductivity, chemical inertness, and optical transparency, that make it an ideal material for many industrial applications. Chapter 10 highlights some new, more cost-effective synthetic and processing routes to diamond films that have been discovered, and promise to extend the range of applications. Diamond films can serve to harden ceramic cutting tools, improve the scratch resistance of lenses in glasses, and enhance the quality of tweeters in loud speakers, to cite a few examples.

New Materials

The re-entrant foam alluded to earlier in the chapter illustrates the notion that we are at a stage where we can often design materials to have desired properties. Ferrofluids, discussed in Chapter 2, are another such example: suspensions of Fe_3O_4 magnetite particles that are roughly 100 Å in size elegantly provide magnetic control over the liquid ambient in which they are suspended through the cooperative interactions of electrons in the solid. This design feature is used to advantage in making rotating seals for high-vacuum equipment, for example.

High-temperature superconductors, which burst upon the scene in the late 1980s, have provided the impetus for preparing, characterizing, and modeling many new perovskite structures, which are described in Chapter 5. Important practical aspects of superconductivity—the unusually high temperatures at which the new ceramic materials lose their electrical resistance and expel magnetic fields from their interiors; the kinds of electrical current densities and magnetic field strengths they can withstand before reverting to nonsuperconducting electromagnetic properties; and strategies for fabricating them into useful shapes like thin films and wires for devices—provide a backdrop for this exciting research area and are covered in Chapter 9. Synthetic routes to these materials, compounds like $YBa_2Cu_3O_7$, are discussed in Chapter 10. Their variable stoichiometry is presented in Chapters 3 and 5.

Like high-temperature superconductors, fullerene-derived materials have had an enormous impact on the materials chemistry community in the short time they have been available for study. Isolated from soot that is generated from graphite electrodes, the parent soccer-ball-shaped C_{60}

compound, buckminsterfullerene, and its derivatives are being proposed as lubricants, organic magnets, frequency doubling materials, and substrates for the growth of diamond films. Samples of C_{60} have been reacted with several alkali metals to make relatively high-temperature superconductors of formula M_3C_{60} that are discussed in Chapters 5 and 9; the alkali metals occupy holes formed by octahedral and tetrahedral arrangements of close-packed C_{60} molecules in the solid state.

"Smart" Materials

An alloy of nickel and titanium exhibits several astonishing properties, described in Chapter 9, which identify it as a "smart" material capable of responding to external stimuli. For example, wires of NiTi "memory metal" can be bent at room temperature and, when gently warmed by a hair blower or hot water, will return to their original linear shape. Or the wire can be heated in a candle flame to give the sample a new shape that it will remember, a consequence of moving defects in the solid, which are discussed in Chapter 6. The solid-state phase change that accounts for these seemingly improbable transformations can also be initiated mechanically, an effect exploited in the manufacturing of eyeglass frames from NiTi that remember their shape. The ability to respond to mechanical and thermal stimuli is being used to create new kinds of actuators and sensors.

Chapter 2 presents electrorheological (ER) fluids as another example of "smart" materials. Ingredients as common as corn oil and flour can be used to prepare a suspension whose viscosity changes markedly with a strong applied electric field. Application of the electric field causes dipoles associated with the suspended particles to align and thereby creates a fibrous structure in the fluid that stiffens it. Automobile manufacturers are interested in such fluids as a means of damping vibrations that might otherwise damage the vehicle's engine block.

Optical fibers and diode light sources, discussed in Chapters 8 and 10, can help to identify structural flaws that might be developing in a material. For example, a network of these fibers, embedded in a structural feature like a concrete beam or aircraft wing, might be used to sense a developing fracture as a deviation in the light beam that is passing through the optical fiber network. Early detection of the fracture would permit the damaged component to be replaced before a catastrophic failure occurs.

Some of What's Already Here

High-tech materials and advanced devices already play a prominent role in everyday life. For example, the LEDs mentioned earlier serve as indicator lights for myriad appliances; and diode lasers are a key component of compact disk (CD) players.

Piezoelectric materials are also widely used. These crystals have the property that when an electric field is applied to them, they mechanically deform; and, conversely, when they are mechanically deformed, they generate an electric field. Variations in electric-charge distribution within such solids are discussed in Chapter 2. Examples of piezoelectric materials are quartz and so-called PZT ceramics, a group of perovskites, described in Chapter 5, that includes $PbZrO_3$ and $PbTiO_3$. These materials are used in a variety of devices, including loudspeakers, buzzers, electrical relays, and pressure gauges. Some torches are ignited by a spark that is produced by the electric field that results when a piezoelectric crystal is squeezed.

Our information technologies are heavily materials-dependent. For example, magnetic domains, which are discussed in Chapter 2, are used extensively in storage media such as floppy disks, magnetic tape, and hard drives for computers.

In medicine, a processing technique called thermal spraying has proven to be useful. Thermal spraying is a generic term for a variety of processes that heat, melt, and spray metallic, ceramic, carbide, or plastic particles onto a surface. The particles flatten upon impact and bond to the substrate. In order for tissue to adhere properly to an implant such as a stainless steel hip replacement, a coating must cover the implant. Hydroxyapatite, $Ca_5(PO_4)_3OH$, a mineral that is a component of teeth and bones, is applied to such implants in a plasma spray process and provides a bioceramic that promotes good tissue adhesion.

The detection of harmful radiation has benefited from developments in materials chemistry. Some polymers are being used as inexpensive detectors for alpha particles emitted by radon gas, because of defects created in the polymer structure by the alpha particles: more rapid chemical etching occurs at the defect sites and permits them to be identified and used as a measure of radon concentration, as discussed in Chapter 6 and Experiment 6.

Our air quality has been aided by the catalytic converter, which has been used to control automobile emissions since 1975, and will be applied to other internal combustion systems in the near future. These devices use metals such as platinum, palladium, and rhodium to catalyze the complete conversion of unburned hydrocarbons and to decrease the emission of nitrogen oxides. The catalyzed reactions occur at the surface of the metal particles that are dispersed onto ceramic (metal oxide) substrates that have highly porous honeycomb ceramic monoliths (Figure 1.8).

Thin film technologies have been used to make new kinds of jewelry and sunglasses from metal oxides. Titanium oxides are being used to create brightly colored, highly reflective earrings, for example; and metal oxides are used to make highly reflective coatings for "one-way" sunglasses. The high reflectivity of metals is discussed in Chapter 7.

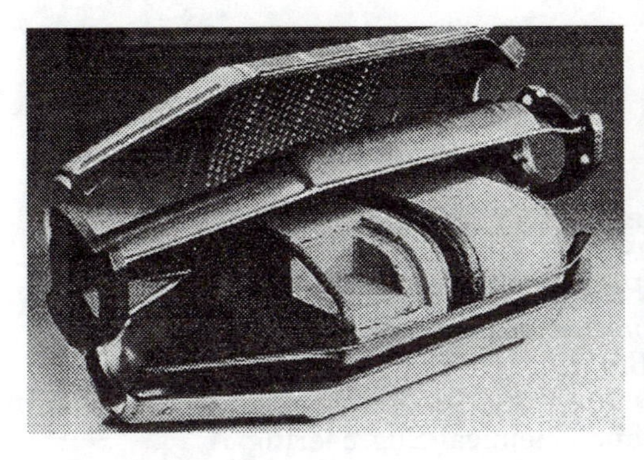

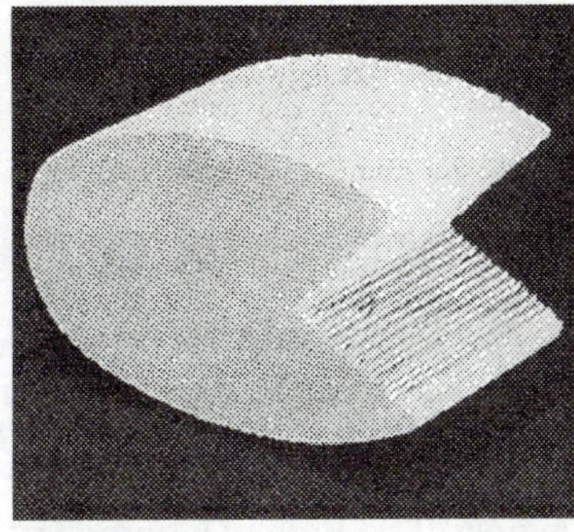

Figure 1.8. Cutaway view of Allied-Signal automotive catalytic converter (left) shows the interior of ceramic honeycomb monolith. The honeycomb (enlarged, right) contains 300 to 400 square channels per square inch, each coated with a porous, high-surface-area layer, such as activated alumina, on which precious metal catalysts are dispersed. Carbon monoxide and hydrocarbons in exhaust gases, which must pass through the channels, are catalytically converted to carbon dioxide and water. (Reproduced from reference 6. Copyright 1992 American Chemical Society.)

Finally, a popular product that is marketed in toy stores is Magic Sand. Control of the surface chemistry of sand particles by coating them with organosilanes causes them not to wet when placed in water. This effect is similar to that observed when fabric is sprayed with a silicone-based water repellent and provides another illustration of the importance of interfaces *(7)*.

In short, the Age of Materials presents extraordinary opportunities and challenges. The intent of the remainder of this volume is to make materials chemistry accessible to teachers and students by connecting solids to fundamental chemical and physical principles.

References

1. *Materials Science and Engineering for the 1990s: Maintaining Competitiveness in the Age of Materials*; National Research Council, National Academy Press: Washington, DC, 1989.
2. Lakes, R. *Science (Washington, DC)* **1987**, 235, 1038–1040.
3. *Advanced Materials and Processing: The Federal Program in Materials Science and Technology*; Committee on Industry and Technology/COMAT, National Institute for Standards and Technology: Gaithersburg, MD.
4. Greenwood, N. N.; Earnshaw, A. *Chemistry of the Elements*; Pergamon Press: New York, 1984.
5. Furimsky, E. *Catal. Rev.-Sci. Eng.* **1980**, 22, 371.
6. O'Sullivan, D. *Chem. Eng. News* October 26, 1992, p 20.
7. Vitz, E. *J. Chem. Educ.* **1990**, *67*, 512.

Additional Reading

- Advanced Materials and Processing: The Federal Program in Materials Science and Technology; *A Report by the FCCSET Committee on Industry and Technology; Office of Science and Technology Policy. To Supplement the President's 1993 Budget. (Available from Committee on Industry and Technology / COMAT; Room B309, Materials Building; Gaithersburg, MD 20899; phone (301) 975–5655.)*
- Critical Technologies: The Role of Chemistry and Chemical Engineering; *A Report by the Board on Chemical Sciences and Technology; National Academy Press: Washington, DC, 1992. (Available from Board on Chemical Sciences and Technology; National Research Council; 2101 Constitution Ave, N.W. ; Washington, DC 20418.)*
- Materials Science and Engineering for the 1990s: Maintaining Competitiveness in the Age of Materials; *National Research Council, National Academy Press: Washington, DC, 1989.*
- *Drexler, K. E.* Nanosystems: Molecular Machinery, Manufacturing, and Computation; *Wiley: New York, 1992.*
- MRS Bulletin, *issued monthly by the Materials Research Society, Pittsburgh, PA.*

Exercises

1. To which corner or corners of the materials tetrahedron (Figure 1.1) can the following be assigned?
   a. The reaction used to prepare polyurethane foam.
   b. The mechanical properties of a sample of polyurethane foam.
   c. The temperature below which a material becomes a superconductor (its critical temperature).
   d. Conversion of C_{60} to K_3C_{60}.

2. Sketch a figure, similar to the one shown in Figure 1.3, that you think might illustrate the behavior of the pores in an untreated polyurethane foam as it is stretched.

3. If re-entrant foam is compressed instead of stretched, what do you predict will happen to the size of its cross-section? Explain your answer in terms of the idealized cubic re-entrant pore, Figure 1.3.

4. How might re-entrant foam be used? What advantages does it offer if it is to be inserted into a hole as a kind of fastener?

5. Investigate the life cycle of a material of interest to you: how the raw materials are obtained; the processing needed to make the material; the use(s) of the material; how the material is disposed of; its recyclability; its environmental impact; and the energy costs associated with the material. You have the option of researching your topic individually or in small groups over a period of several weeks. Optional: Present your results at a poster session in which you share what you have learned.

Chapter 2

Atoms and Electrons

The story line in chemistry is atoms, the building blocks of matter, and their electrons. The sharing and transferring of electrons permits atoms to combine with one another in chemical reactions that lead to new forms of matter. Until recently, we had only indirect evidence for the existence of atoms, obtained by such methods as stoichiometric relationships, spectroscopy, diffraction methods, and thermodynamics. As noted in Chapter 1, however, it is now possible to directly image atoms on the surface of a material. This chapter illustrates the use of solids to provide direct and indirect evidence for atoms.

Direct Evidence for Atoms: The Scanning Tunneling Microscope

Progress in the development of analytical instrumentation has enabled scientists to characterize and manipulate matter with ever-greater spatial resolution. Nowhere is this more evident than in the development of the scanning tunneling microscope (STM). Developed in the mid-1980s, the STM permits direct imaging of atoms. An STM specifically designed for undergraduate laboratory and classroom use has recently been marketed,[1] and a paper describing a student experiment and containing references to a number of applications of scanning tunneling microscopy has been published *(1)*.

To put the STM experiment in context, consider the simple circuit shown in Figure 2.1. When the circuit is complete, the light bulb will light:

---

[1]The Burleigh instructional STM (*see* Supplier Information)

2725–X/93/0015$09.50/1

electrons are pumped through the circuit as the battery's chemical energy is converted to electricity. If the wire is cut, however, and the cut ends are not in contact, the light is extinguished, because there is no longer a complete circuit for the electrons. If the wire ends are brought into contact, the bulb can again glow. This experiment raises an interesting question: How close must the two ends of the cut wire be for electrical current to flow? The question is difficult to address experimentally because of the unimaginably large numbers of Cu atoms on the two surfaces of the wire ends. Each surface is probably quite rough on the atomic scale, and thus the atoms of the two surfaces are collectively at a variety of distances from one another.

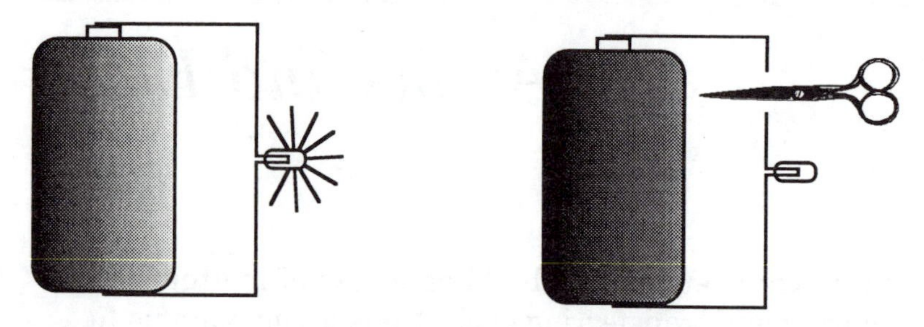

Figure 2.1. A simple electrical circuit containing a battery and a light bulb. When the connecting wire is cut, the circuit is open and the bulb will not light.

Now, however, imagine that one of the wire ends can be sharpened to a tip of one-atom dimension, an "atomic tip", and brought close enough to the surface that current can again flow, as shown in Figure 2.2. Furthermore, imagine that this tip can be moved in atomic-scale increments laterally across the surface. As it moves, it samples the atomic surface topography through variations in current as the tip-to-surface height varies (Figure 2.3). Alternatively, the tip-to-surface height can be continuously adjusted with atomic-scale resolution so that a constant current flows while the tip moves along the surface. By either method, the spatial arrangement of atoms on the surface is determined by the variation in electron density sensed by the probe tip as it traverses the surface. This is, in fact, how the STM operates.

How close does the tip need to be to the surface? Only within a few angstroms (10^{-8} cm) because of a quantum mechanical phenomenon called tunneling. An easy way to look at tunneling is to begin by considering one atom. The electrons surrounding the atomic nucleus are not confined to a hard shell but rather have a smoothly varying distribution, meaning that the edge of an atom is indistinct. If the quantum mechanical equations describing the probability of finding an electron at a given distance r from the nucleus at any time are solved, the probability distribution of the electron, $P(r)$, would look something like

$$P(r) = A\, e^{-r/r_o} \quad (1)$$

where A is a constant and r_o is the Bohr radius. From this equation it can be shown that the electron spends most of its time near the nucleus and less time at larger, but unbounded, distances. This probability distribution falls off rapidly—exponentially—with distance from the nucleus.

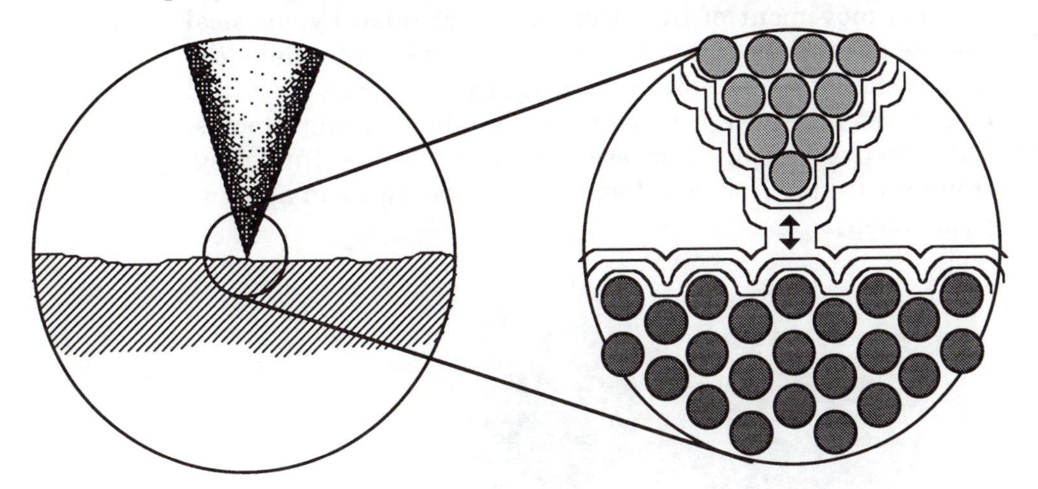

Figure 2.2. A sketch of an atomically sharp tip near a surface.

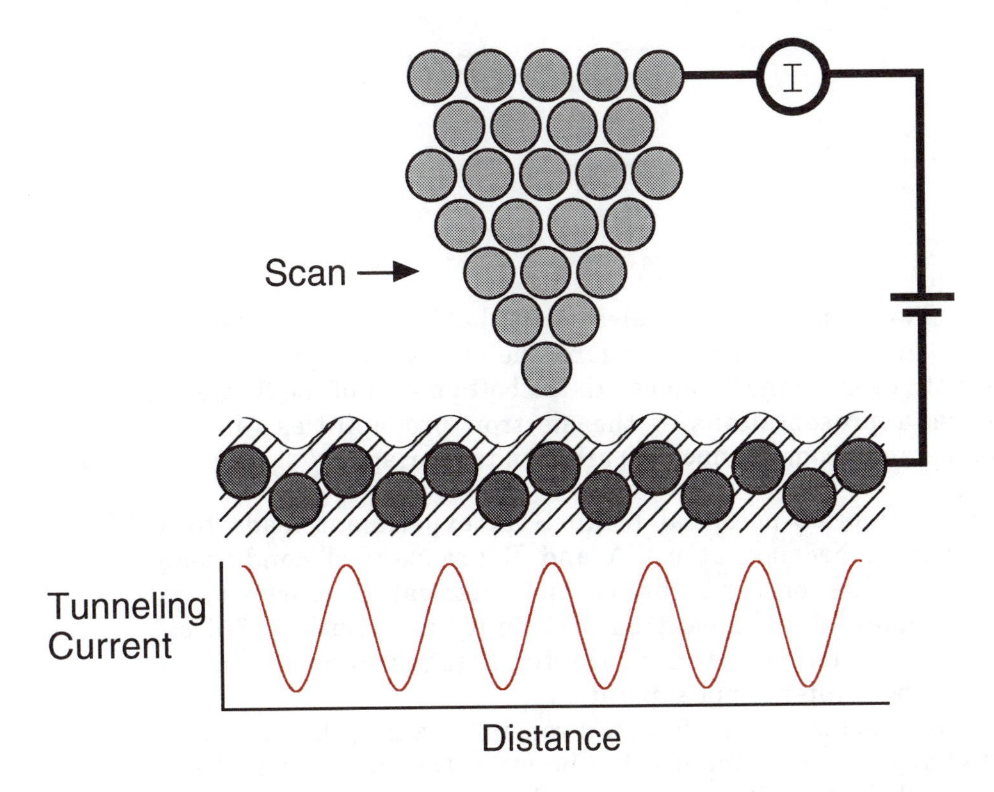

Figure 2.3. A plot of tunneling current as a function of horizontal probe tip position. The absolute vertical position is held constant. When the tip is nearest the surface atoms, the current is highest. The wavy line above the shaded circles represents the contour of the surface.

Now consider two atoms A and B that are within angstroms of one another, with their associated electron probability distributions, *P(r)*, as sketched in Figure 2.4. An electron initially on atom A can move through the region of overlapping electron density to become part of atom B's electron cloud. This movement of the electron is forbidden by classical physics, because the electron does not have sufficient energy to make the transfer between atoms; however, electrons actually behave as described by quantum mechanics and have a non-zero probability for the process. The motion of the electron through a classical barrier is called tunneling. Thus, an electron may tunnel from an atom in the probe tip to an atom on the surface (or vice versa).

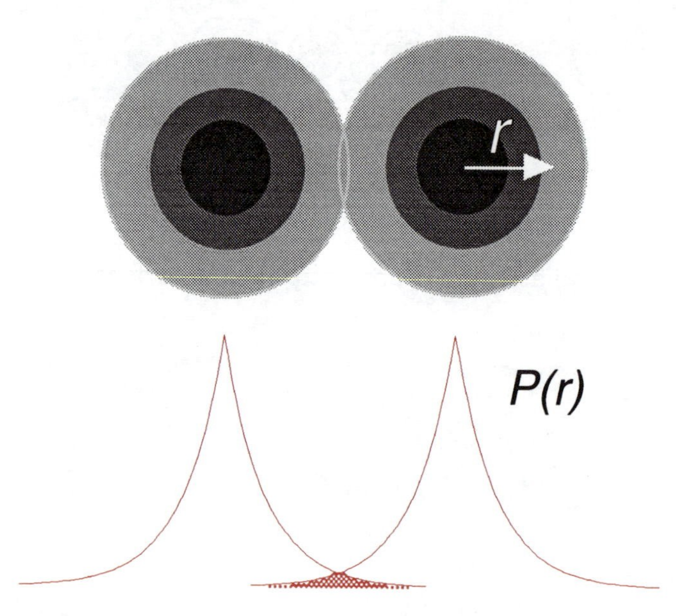

Figure 2.4. Two atoms with their electron probability clouds slightly overlapping. In the top part of the figure, the atoms are represented as spheres with overlapping volumes. In the bottom part of the figure, the graphical representations of the electron probabilities (as a function of distance from the nucleus) are seen to overlap.

The use of an applied potential in an STM experiment leads to a continuous current, because atoms A and B are part of conducting materials like metals or semiconductors (electrical conductivity is discussed in Chapter 7). The electron probability distributions fall off exponentially with distance; therefore, the tunneling current provides a very sensitive probe of interatomic separation.

As the tip moves across the surface at a constant height, the tunneling current will change because the overlap between the tip atom and the surface atoms will change. When the tip is directly over a surface atom, the tunneling current will be higher than when the tip is between two atoms. Higher currents mean that the substrate being investigated has a higher surface height and a shorter tunneling distance.

A problem with studying the surface in this manner is that because the tip is only a few angstroms from the surface, it is easy to crash it into the sample unless the substrate is smooth on the atomic scale. To circumvent this problem, the STM is usually operated in constant-current mode, which makes use of a feedback controller. The feedback controller measures the actual tunneling current and compares it to a desired value that the operator can control. (For an example of a feedback controller from physiology, consider trying to balance a yardstick positioned vertically in the palm of your hand. It is much easier to do when your eyes are open—the eye provides feedback that allows the hand to move to compensate for the shifting of the yardstick.)

The tip is connected to a positioner that is in turn connected to the feedback controller. Voltages from the feedback controller change the length of the positioner so as to pull the tip away from the surface if the current is higher than desired, or to lower the tip toward the surface if the current is lower than desired. In this way, the desired current value is maintained as the tip scans above the surface. The change in the positioner's length is directly proportional to the applied voltage, so that the changes in voltage produced by the feedback controller are directly proportional to the changes in surface height.

A computer is used to control the location of the STM tip, and at each location on the surface, the computer measures the surface height based on the voltage applied to maintain constant tunneling current. This two-dimensional array of numbers, representing heights at different positions in the area surveyed, is often displayed through gray-scale imaging: high locations are represented as white, and low locations as black, giving a picture that is essentially what a photograph of the surface would look like. An image of a surface of silver atoms is shown in Figure 2.5. The silver atoms on this surface are close-packed (Chapter 5). The STM does not really give the physical positions of the atoms, but senses their electrons and bonds. As a result, the current detected depends on both the tip-to-atom distance and the chemical identity of the surface atoms.

How is an STM tip prepared so it terminates in a single atom and how does the positioner work to permit control of the tip's position on the atomic scale? The tips are usually either made of W or Pt. For STM studies in vacuum environments, W is usually preferred because of the relative ease in preparing a single-atom-terminated tip. If the STM is to be operated in air or liquid, the ease with which W oxidizes precludes its use, and Pt or Pt–Ir alloys are preferred, although tips from these materials are more difficult to prepare and generally are not as atomically sharp.

Tips are usually made by an electrochemical etching process, starting with fine wire. With W, the wire is held vertically and partially immersed in an aqueous solution of base. When a 5-V potential is applied to the wire, the W surface is oxidized to a dense, soluble oxide. The tungsten oxide flows down past the tip and prevents further electrolysis except where the wire enters the solution. For Pt and Pt–Ir alloys, an alternating current etching procedure is employed. In either case, localized etching results in a "neck" in the wire. Eventually, the wire will be etched through at this thinnest part, and the bottom part falls off. The tip often terminates in a small cluster of atoms or the desired single atom.

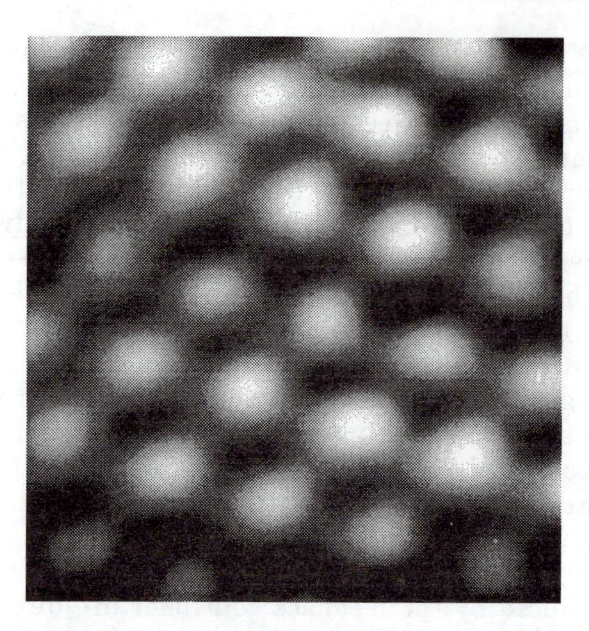

Figure 2.5. The STM image of a close-packed layer of Ag atoms. (Photograph courtesy of Robert Hamers.)

The tip's position is controlled with a piezoelectric scanner. Piezoelectric materials are ceramic materials that will expand or contract when a voltage is applied to them (conversely, mechanical deformation can yield a voltage; *see* Chapter 1 and a later section of this chapter for further discussions of piezoelectricity). The most common way of using piezoelectric materials for tip positioning is to take a single tube of piezoelectric material and to place four metal electrodes on the outside of the tube and a single electrode on the inside (Figure 2.6). Appropriate combinations of voltages applied to these electrodes can make the tube extend or contract along its length and/or bend in any direction. The tip is mounted on the axis of the tube, at its end. When the tube expands or contracts, the surface-to-tip separation decreases or increases, respectively. When the tube is made to bend, the amount by which it bends is small compared to its length, so that the tip moves mainly sideways. A typical piezoelectric scanner requires voltages on the order of 100 V, and the tip can move inside a square scan area with 4-μm (0.004-mm) sides. Because the tunneling current can be detected only when the sample and tip are less than about 10 Å apart and the positioner can only move a few micrometers, levers, gears or other piezoelectric translators are used to bring the tip to within a few micrometers of the sample without crashing the tip into it. In designing an STM instrumental setup, the apparatus must be isolated from room vibrations and temperature fluctuations, which can cause crashing and smear data.

In principle the STM can be applied to a variety of problems spanning the spectrum of science and engineering. In practice, many studies have focused on the properties of "clean" surfaces or surfaces that have been modified in some controlled way. Even at a pressure of 10^{-6} torr (10^{-4} Pa), each atom on a surface will undergo one collision with a gas-phase molecule per second, and a surface will likely react or become contaminated in about a second. Pressures of 10^{-10} torr (10^{-8} Pa) are needed to study surfaces for periods of hours without contamination. STMs that are compatible with such high-vacuum conditions require special materials.

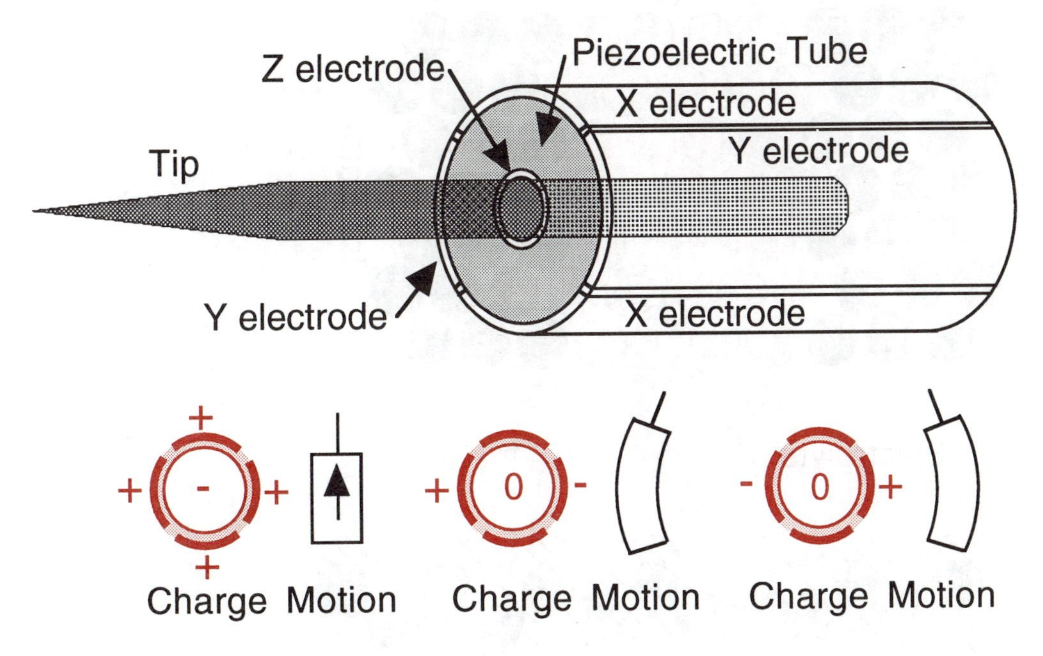

Figure 2.6. A piezoelectric tube. Top: Enlarged view. Bottom: The effect of various applied voltages on the position of the tip. (Adapted with permission from reference 2.)

The STM can also be used to study electrode surfaces immersed in liquid electrolytes. In this case, there can be faradaic currents resulting from solution electrochemical reactions in addition to tunneling currents. To image the electrode surface ideally requires insulating the tip except for the atoms at the tip end (Figure 2.7). In practice, if the total exposed tip is smaller than about 1 µm, typical electrolyte concentrations and applied voltages produce faradaic currents that can be neglected compared to the tunneling current.

The application of STM to biological molecules has been of interest and even proposed as a method for gene sequencing. For any molecule to be observed with the STM, however, it must be attached to a rigid, flat, conductive material. Thus, the molecule–substrate interaction must be strong enough to anchor the molecule so that it can be imaged, but not so strong that the molecule is altered by the anchoring. Large molecules have been imaged either by depositing them onto flat substrates like graphite or gold and then doing the STM experiment in vacuum; or they can be adsorbed in electrolyte solutions and the STM experiment performed in situ. Finding substrates that will hold biological molecules rigidly without tearing them apart has been a problem; some early images claimed to be DNA were later shown to be features in the graphite surfaces. The STM being marketed for educational use can be used to image a long-chain hydrocarbon ($C_{32}H_{66}$) that orients itself on a graphite substrate.

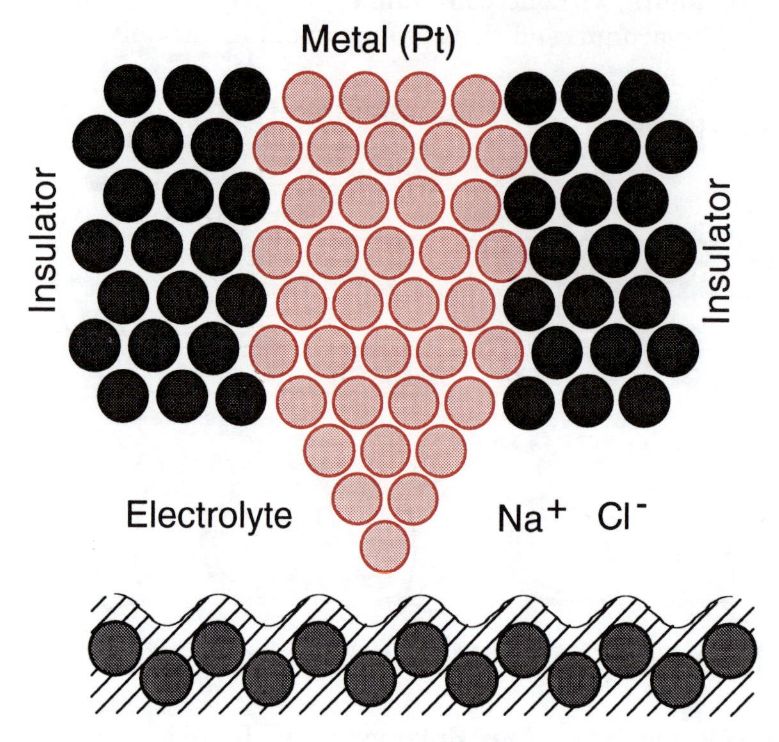

Figure 2.7. An insulated atomic tip can be used to probe electrode surfaces. The wavy line above the shaded circles represents the contour of the surface.

As noted in Chapter 1, the STM can even be used to position atoms. Researchers at IBM were able to manipulate xenon atoms on a metal surface to spell out "IBM." This was done by cooling a single crystal of nickel metal to 4 K under high vacuum and condensing some Xe atoms onto the surface. The Xe atoms were initially at random locations and were detected with the techniques described earlier.

To move an atom, the tip was placed above the atom and lowered to a closer distance than is normally used for scanning. The tip was then moved across the surface at a speed of 4 Å per second, dragging along the xenon atom. Most likely, the main interaction between the tip and the xenon atom is due to van der Waals forces *(3)*.

These same techniques may be used to fabricate atomic-scale devices and new structures. In fact, the STM has been used to prepare a "nanobattery." The battery (Figure 2.8) consists of two copper pillars and two silver pillars that are deposited sequentially on a graphite surface by electrochemical reduction of solutions of copper sulfate and silver fluoride. An STM tip serves as an electrode that is fine enough to deposit these very small pillars. This nanometer-scale galvanic cell (about the size of a common cold virus) is estimated to generate about 0.020 V.

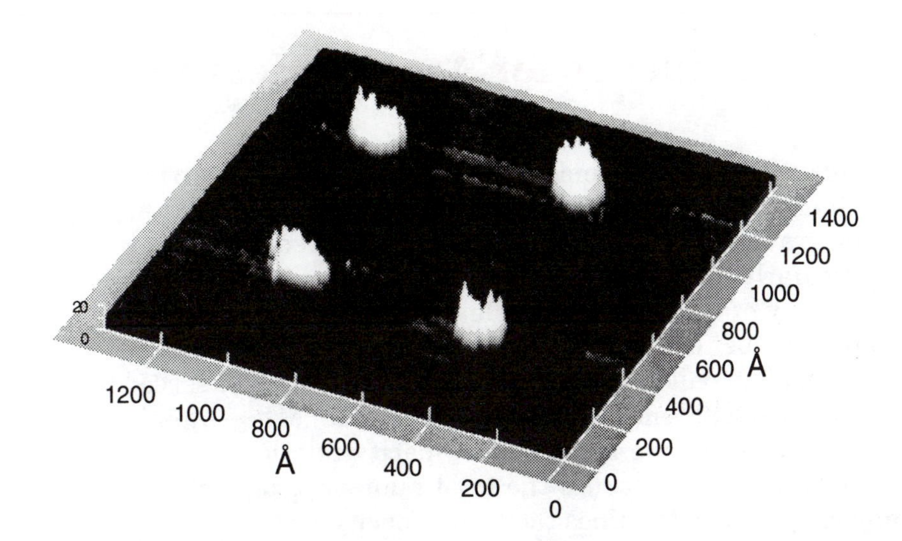

Figure 2.8. STM images of a copper–silver nanometer-scale cell on graphite. The two structures on the right are electrochemically deposited silver, and the two structures on the left are electrochemically deposited copper. In a copper solution the cell is galvanic; the copper pillars decrease in size and copper metal plates onto the silver pillars. (Reproduced from reference 4. Copyright 1992 American Chemical Society.)

A close relative of the STM that is being applied to many of the same experimental systems is the atomic force microscope (AFM). The AFM operates by sensing the force between surface atoms and the atom in a tip rather than by measuring the current. Thus, the AFM can be used to study electrically insulating materials. To measure interatomic force, the tip of the AFM is mounted on the end of a small lever. As the interatomic force varies, the deflection of the lever can be sensed by bouncing a laser beam off the back of the lever and measuring displacements with a pair of photosensors.

The STM can be operated in air. Stable surfaces are provided by a variety of materials, including layered solids like graphite and molybdenum disulfide. The layered nature of these solids and the weak van der Waals forces holding the layers together permits layers to be removed easily to create fresh surfaces (Chapter 5 gives more information on graphite and molybdenum disulfide).

Indirect Evidence for Atoms: Heat Capacities

Given the tremendous diversity in elemental solids and their properties, it is surprising to find that many of them share a common feature: At sufficiently high temperatures (often room temperature is adequate) many of the elements, particularly the metallic elements, have approximately the same molar heat capacity of ~25 J/mol-deg. Heat capacity must be more strongly related to the number of atoms of matter present than to the mass of matter. This finding, often called the law of Dulong and Petit *(5),* served as the basis for an important empirical method of estimating atomic masses in the 1820s. Moreover, the molar heat capacity relationship can be extended to many ionic solids, because the total number of particles in the solid, either atoms or ions, determines the heat capacity.

To understand the experiment and its implications regarding atoms, consider first a collection of gaseous atoms. The average kinetic energy of a mole of a monoatomic gas is 3/2 RT, where R is the gas constant and T is the absolute temperature in kelvins. This fact illustrates the equipartition theorem: At thermal equilibrium, there is on average 1/2 RT of energy for each independent degree of freedom in a chemical system; because the gas atoms can move freely in three mutually perpendicular dimensions (x, y, and z), their average energy is 3/2 RT. A temperature change of a degree thus corresponds to a molar change in heat content or molar heat capacity for the gas of 3/2 R or ~12.5 J/mol-deg. For polyatomic gas molecules, additional contributions to the heat capacity can come from rotational and vibrational motion *(6)*.

Now consider a crystal of metal atoms in which each atom is connected to its neighbors through a network of imaginary springs. For most metallic elements, room temperature provides sufficient thermal energy to cause all the atoms to vibrate as they absorb the ambient thermal energy. The total energy of these vibrating atoms comprises both kinetic and potential energy contributions that are continuously interconverting. According to the equipartition theorem, the six degrees of freedom (three for the three dimensions of motion associated with kinetic energy and three for the three dimensions associated with potential energy) lead to an average thermal energy of $6 \times 1/2\ RT = 3RT$, and thus a molar heat capacity of $3R$, ~25 J/mol-deg.

The molar heat capacity value of 25 J/mol-deg is a high-temperature limit that can be achieved only if all the normal vibrational modes of the solid can be excited. For several of the lighter elements like Be and C (diamond), the vibrational frequencies are relatively high; thus, temperatures substantially above room temperature are needed to activate these modes and reach the limiting heat capacity. (The frequency of a simple harmonic oscillator varies as $1/\sqrt{m}$ where m is the mass, and room temperature does not supply enough ambient thermal energy to activate all of the vibrational modes.)

Figure 2.9 illustrates the temperature dependence of the molar heat capacities for diamond, aluminum, lead, and copper. Aluminum, lead, and copper are typical metallic elements that reach the high-temperature limit by room temperature. As temperature is reduced, atomic vibrations and their contribution to heat capacity are no longer activated. At very low temperatures, the much smaller heat capacity due to freely mobile electrons in metallic solids (Chapter 7) begins to be the dominant contributor to the measured heat capacity.

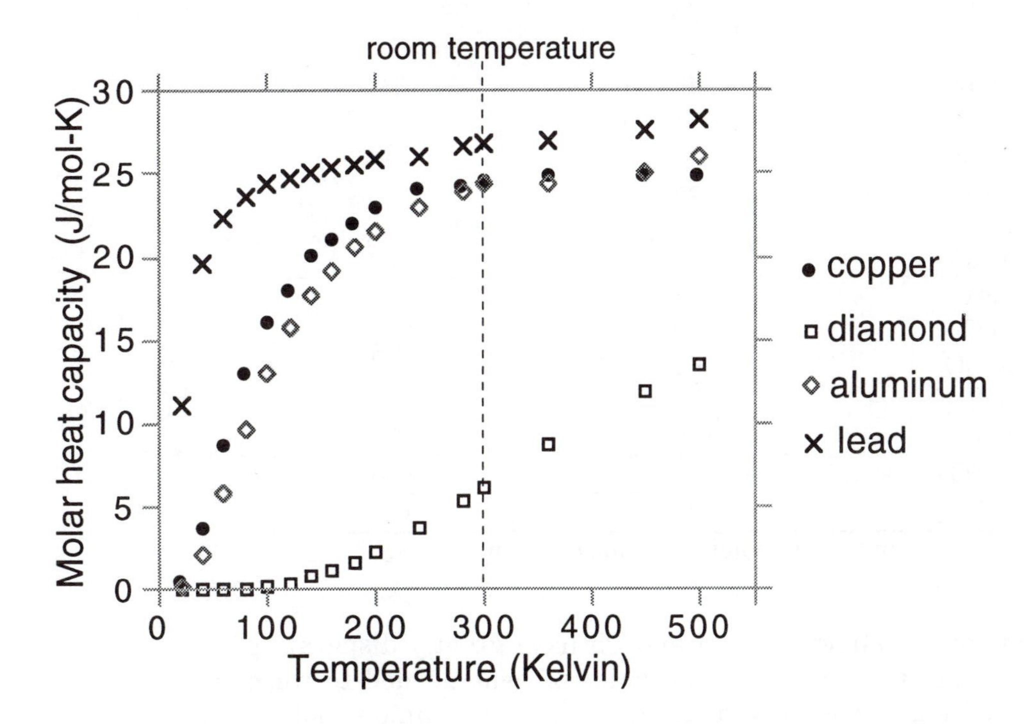

Figure 2.9. Plot of molar heat capacity versus temperature for copper, aluminum, lead, and diamond. (Data taken from references 7 and 8.)

Monatomic solids have heat capacities of $3R$. Based on the preceding discussion, we predict that polyatomic solids like salts would have a high-temperature heat capacity limit of $3R \times p$, where p is the number of atoms in the chemical formula of the solid. Thus, salts like NaF, NaCl, NaBr, NaI, KF, KCl, KBr, KI, MgO, and CaO are all predicted to have a molar heat capacity of $3R \times 2 = 6R$ (~50 J/mol-deg), because there are two atoms per formula unit; likewise, CaF_2 and SiO_2 are predicted to give values of $3R \times 3 = 9R$ (~75 J/mol-deg); and $BaSO_4$ is expected to have a molar heat capacity of $3R \times 6 = 18R$ (150 J/mol-deg). All of these solids have measured molar heat capacities that are, at room temperature, good to better than within ~85% of the predicted values (Table 2.1).

Table 2.1. Molar Heat Capacities of Selected Solids at 298 K

Solid	Experimental Value	Theoretical Value
Al	24	
Sb	25	
Be	16	
B	11	
Ga	26	25
Mg	25	
Pb	26	
Te	26	
NaF	47	
NaCl	55	
NaBr	52	50
NaI	54	
MgO	44	
CaO	53	
CaF_2	66	
SiO_2	73	
Na_2O	69	75
$SrBr_2$	75	
KIO_3	106	125
$BaSO_4$	128	150
K_2SO_4	131	175

NOTE: All values are given in joules per mole-kelvin.
SOURCE: Adapted from reference 9.

Demonstration 2.1 shows that Al and Pb have greatly disparate gram specific heats (J/g-°C, the number of joules of heat needed to raise the temperature of 1 g of the substance by 1 °C) but common molar heat capacities (J/mol-°C, the number of joules of heat needed to raise the temperature of 1 mole by 1 °C).

Demonstration 2.1. Specific Heat and Molar Heat Capacity

Materials

- Aluminum and lead blocks with equal masses (masses in the range of 100 g to 1 kg work well) or equal moles (optional). *See* Supplier Information.
- Double-pan balance (or digital balance)
- Large beaker (must simultaneously contain both metal blocks)
- Hot plate
- Tongs or string
- Thermometer (ideally a digital thermometer, or one that can be read by the entire class)
- Two smaller beakers (to contain Al and Pb blocks separately)
- Water

Two glass rods

Procedure

- Place the metal blocks on a double-pan balance to show that they have equal masses. Alternatively, weigh them separately and compare the masses to show that they are the same.
- Place the metal blocks in a large beaker of boiling water. Make sure that there is enough water to completely cover the blocks. Maintain this temperature for several minutes to ensure complete heating. Measure the temperature of the boiling water.
- Place equal amounts of water at room temperature in the smaller beakers. Choose the beaker and water volumes so that placement of the metal blocks in the beakers will not cause the water to overflow. Measure the temperature of the water.
- Using the tongs (or, alternatively, a string that has been tied to each object), quickly transfer the metals, one to each of the smaller beakers of water.
- Stir the water with the glass rods. Measure the temperature of the water in each beaker, recording the highest temperature reached. The bath containing the aluminum will reach a higher temperature than the bath containing the lead (the same mass of Al has about eight times as many atoms, based on the relative atomic weights of the two elements).

Variations

- Knowing the actual masses of the metals, the volume of water used in the smaller beakers, the temperature of the boiling water bath, and initial temperature and the highest temperature reached by the baths containing the metals, the gram specific heats and molar heat capacities for each metal can be calculated to a reasonable level of accuracy, as described in most introductory chemistry texts. If the mass of water into which the metal is immersed equals the mass of the metal, the calculation of specific heat is simplified, as described in Experiment 1.
- The same demonstration can be performed with blocks of metal that contain the same number of atoms. This will produce the same temperature change in equal amounts of water.

Laboratory. A laboratory experiment that permits a class to pool its data for a variety of metal elements to discover these heat capacity relationships is described in Experiment 1.

Electrons in Solids

Just as the scanning tunneling microscope and the measurement of heat capacity provide evidence for the existence of atoms, the magnetic and electrical properties of solids provide evidence for the existence of electrons. Electrons may exhibit localized behavior, meaning that they are associated with one particular atom, bond, or ion. Electrons can also exhibit delocalized behavior, being shared among several atoms as in benzene or among larger numbers of atoms as in an entire crystal. The electromagnetic properties of a solid reflect not only the bonding character of its electrons, but their spin and charge, as well.

Electrons have two spin states, corresponding to the m_s quantum numbers +1/2 and −1/2, spin up and spin down. A single electron, e.g., spin up, without a corresponding spin down electron, is called an unpaired electron and in both molecules and solids can produce magnetic effects that, as shown later, can be used to augment traditional presentations of electronic configurations and formal oxidation states in introductory chemistry courses. The rich diversity of magnetic properties arising from unpaired electrons can also be demonstrated with paramagnetic and ferromagnetic materials.

Because electrons are mobile electrically charged particles, they are easily displaced in a solid by applied electric fields or by mechanical deformation. This can give rise to interesting electrical effects, the simplest of which is electrical conductivity. Described in this chapter is an electrorheological fluid which, as discussed in Chapter 1, is an example of a "smart" material; and common devices based on piezoelectric solids, which are the crystals used in watches and as strikers for lighters. These materials, whose use depends on induced electric dipoles in the solids, can be used to complement traditional discussions of molecular dipole moments in introductory courses.

Paramagnetism

In the majority of molecules and solids, all of the electrons are paired, and the molecule or solid is said to be diamagnetic. Several types of magnetic behavior are possible, but the largest magnetic effects result from the existence of unpaired electrons. The simplest type of magnetism, paramagnetism, is often demonstrated in introductory chemistry courses by observing the behavior of liquid oxygen in a magnetic field. As a result of the presence of two unpaired electrons in the O_2 molecule, the liquid is attracted to a strong magnet, and if a large horseshoe magnet is used, the liquid oxygen appears to hover, trapped between the poles of the magnet *(10)*.

A family of solid manganese oxides can be used to establish a connection between the formal oxidation state of the manganese atoms in the solids, their electron configurations and number of unpaired electrons, and their relative response to a magnetic field.

Demonstration 2.2. Paramagnetism

Materials

MnO_2
Mn_2O_3
$KMnO_4$
Strong magnet such as a rare earth or cow magnet (*see* Supplier Information)
Empty gelatin pill capsules (available from pharmacies or health food stores)
Overhead projector and screen

Procedure

- Prior to the demonstration, place each solid to be studied in a separate pill capsule. Use about 0.01 mol (simply filling the capsules or using equimolar amounts gives the same qualitative results).
- To do the demonstration, lay the capsules end-to-end in a lengthwise row (with gaps in between them) on the overhead projector.
- Place the magnet on the projector's glass surface and slowly approach one of the capsules with the magnet. Note the effect of the magnet on the capsule (e.g., no attraction, slight attraction when the magnet is very close, or the capsule rolls toward the magnet).
- Repeat the procedure with each capsule. Rank the relative strengths of the interactions. The strength of the interaction will increase as the number of unpaired electrons associated with the manganese atoms increases. This result matches the order of the magnetic moments for these compounds, as determined by more quantitative techniques, Table 2.2.

Variation (requires enclosed analytical balance)

- Place the magnet on a stack of foam cups so that the magnet is near the roof of the balance (*see* Figure 2.10). **Caution! Maintain a sufficient distance between the magnet and the balance electronics. The magnet should not be placed on the balance pan or permanent damage to the balance may result.**
- Close the doors and tare the balance.
- Place the sample on the outside of the balance enclosure and observe the decrease in weight as the magnet is attracted to the sample. More unpaired electrons give larger weight changes.

Table 2.2. Magnetic Moments

Formula	Formal Mn Oxidation State	Number of Unpaired Electrons, n	$\sqrt{n(n+2)}$	Magnetic Moment[a]
Mn_2O_3	+3	4	4.90	4.9
MnO_2	+4	3	3.87	3.8
$KMnO_4$	+7	0	0	0.2

[a]Room-temperature experimental magnetic moment in units of Bohr magnetons, based on quantitative weight measurements with and without a calibrated magnetic field. Most inorganic chemistry laboratory manuals give details. In the simplest case, the predicted magnetic moment is $\sqrt{n(n+2)}$, where n is the number of unpaired electrons.

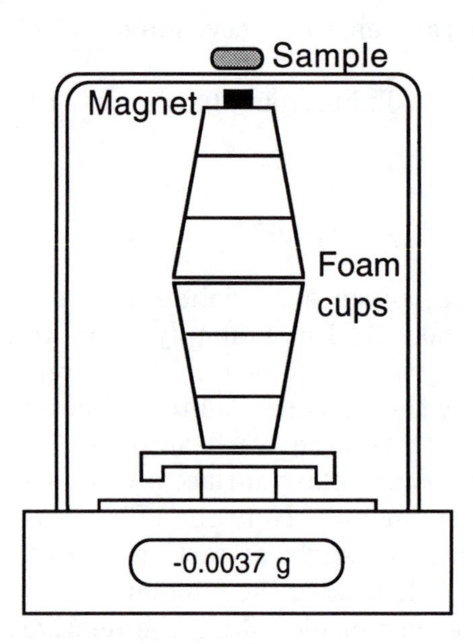

Figure 2.10. Enclosed balance used to observe the change in weight of a magnet attracted to a sample. (Based on reference 11.)

Ferromagnetism

Cooperative effects, in which many unpaired electrons communicate and interact with one another, can lead to more complex magnetic behavior in an extended solid than that observed for molecules in the gas or liquid states. An excellent example of cooperative behavior in solids is ferromagnetism, a property that is technologically important and extensively exploited in the use of permanent magnets, magnetic recording media, and transformers.

The difference between simple paramagnetism and ferromagnetism is shown in Figure 2.11. (Unpaired electrons or spins are depicted in this figure by arrows.) There are unpaired spins in a simple paramagnet (Figure 2.11A), but in the absence of a magnetic field, the spins are randomly oriented relative to one another because of thermal motion and do not interact. The application of a magnetic field to a paramagnet aligns the spins along or in opposition to the magnetic field, and the overall magnetization is small (Figure 2.11B). This property is due to a lack of communication or magnetic ordering between the spins and some thermally induced disordering of the spins. Moreover, the spins disorder when the field is removed.

In a ferromagnet, the unpaired electrons strongly communicate with one another and align (even in the absence of a magnetic field) in large regions known as magnetic domains (Figure 2.11C). The sizes of the domains vary with the material, but are quite large with respect to single atom sizes. In the absence of a magnetic field, the individual domains are ordered randomly relative to one another so that the net magnetization of a macroscopic piece of ferromagnet is small or even zero. The application of a magnetic field (Figure 2.11D) aligns all of the spins in the direction of the magnetic field, leading to a large, overall net magnetization.

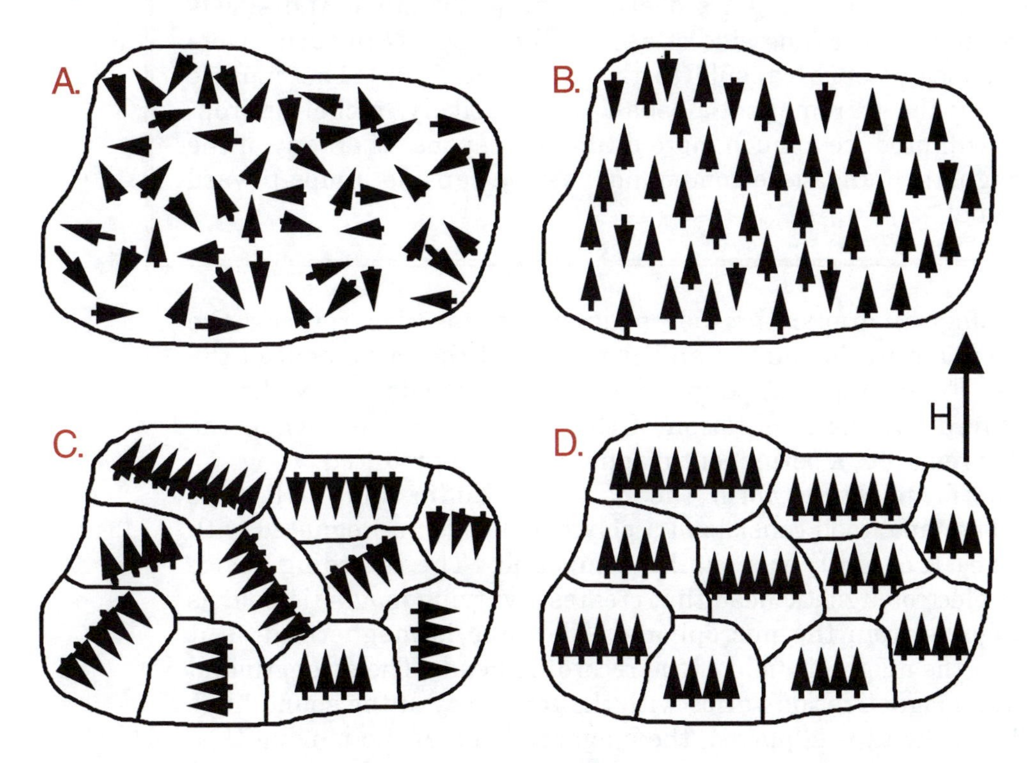

Figure 2.11. A diagram showing the orientation of electrons in (A) a paramagnetic solid in the absence of a magnetic field, H; (B) a paramagnetic solid in the presence of a magnetic field; (C) a ferromagnetic solid in the absence of a magnetic field; and (D) a ferromagnetic solid in the presence of a magnetic field.

Demonstration 2.3. A Comparison of Paramagnetism and Ferromagnetism

Materials

One or more of the gelatin capsules containing paramagnetic manganese compounds described in Demonstration 2.2
Gelatin capsule containing cobalt, nickel, or iron metal powder
Strong magnet such as a rare earth magnet
Overhead projector and screen

Procedure

- Place the gelatin capsules end-to-end in a lengthwise row on the overhead projector. Leave space between them.
- Lay the magnet on the glass overhead projector stage and slowly move it toward the long side of each of the capsules in turn. Note the response of each capsule to the magnetic field. The capsule containing the ferromagnetic cobalt (alternatively, nickel or iron) powder will produce a much more dramatic response than any of the paramagnetic manganese compounds, as the capsule jumps toward the magnet.

The technological applications of magnetic properties rely on the existence of domain walls and the ability to control the movement of the domain walls. For example, in permanent magnets, the domain walls are not easily moved. If the orientation of the spins is retained when the applied field is removed, a permanent magnet has been produced.

Ferromagnetic materials form the basis for audio and VCR tape technology. The tape itself consists of a polymer that is impregnated with crystals of γ-Fe_2O_3 or CrO_2 or a similar compound. The recording device consists of an electromagnetic head that creates a varying magnetic field as it receives signals from the microphone. The tape is magnetized as it passes through the magnetic field of the recording head. The strength and direction of the magnetization varies with the frequency of the sound to be recorded. When the tape is played, the magnetization of the moving tape induces a varying current whose signal is amplified and sent to the speakers.

Ferromagnetic materials also can form the basis for digital recording media such as computer disks. Computers store and manipulate data using binary numbers consisting of the digits 0 and 1. The letters SCN, for example, are typically represented, respectively, by the binary numbers 01010011, 01000011, and 01001110. The power of binary numbers is that their bits can be represented and processed by transistors (off represents 0; on represents 1) or stored using a variety of media that contain regions that can have one of two states. Magnetic media such as computer disks use very small domains to hold binary bits. If a region is magnetized, it represents a 1; if it is not magnetized, it represents a 0.

Recently, magnetic materials based on the garnet structure (see Experiment 14) have begun to replace traditional ferromagnetic materials. In magnetic bubble domain memory storage, the ability to move the domain walls underlies the writing capabilities in the read–write mode. Materials chemists are continuing to search for new storage media that can be written (some bits turned on) and erased (each bit turned off) quickly, that can store huge volumes of information in a small space, and that are stable and retain their information with reliability.

Laboratory. A laboratory experiment involving the preparation and characterization of a series of rare earth iron garnets with interesting magnetic properties is presented in Experiment 14.

The magnetic behavior of a ferromagnet is temperature-dependent because of the interplay of thermal energy and the stability gained by aligning the electrons in the ferromagnetic state. Ferromagnetism is an example of a solid-state phase transition (*see* Chapter 9) with a characteristic transition temperature. Above an ordering or critical temperature, known as the Curie temperature (T_{curie}), thermal energy is sufficient to break the alignment of the spins, and the material exhibits simple paramagnetism. At and below T_{curie}, the aligning forces overcome thermal randomization, the spins are locked in alignment, and the material becomes ferromagnetic. Table 2.3 lists some common ferromagnets and their Curie temperatures *(12)*.

Table 2.3. Some Ferromagnets and Their Curie Temperatures

Material	T_{curie} (K)
Fe	1043
Co	1388
Ni	627
Gd	293
$CrBr_3$	37
EuO	77
EuS	16.5

Demonstration 2.4. The Curie Point of Nickel

Materials

Nickel spheres (6–16-mm diameter; Aldrich 21,577-5)
Common bar magnet or cow magnet. *See* Supplier Information.
Meker burner
Canadian nickel minted prior to 1982 (optional)
Tongs (optional)
Heat-proof pad (optional)

Procedure

- Place one or more nickel spheres in the grating that is at the top of a Meker burner.
- Move the magnet near a sphere. The attraction between them will allow the sphere to be picked up by the magnet.
- Remove the nickel sphere from the magnet and return it to the grating.
- Light the burner with the nickel sphere on top.
- Wait 30–60 seconds and turn off the burner. Move the magnet near a sphere. If the sphere is above the Curie temperature (627 K) of nickel, it will no longer be ferromagnetic and will not be strongly attracted to the magnet. As the sphere cools below the Curie temperature, the ferromagnetism will return, and the nickel sphere can once again be picked up by the magnet. **Warning! The nickel sphere will be hot.**

Variation

Canadian nickels minted prior to 1982 (these contain pure nickel, while nickels dated since 1982 contain 75% Cu and 25% Ni) can be used in place of nickel spheres. If a Canadian nickel is used, hold it in the flame with tongs. Remove it periodically, set it on a heat-proof pad, and note its interaction with the magnet.

Demonstration 2.5. The Curie Point of Iron (modified from reference 13)

Materials

Small magnet
String
Ring stand and clamp
28-gauge steel wire (available at hardware store, may be galvanized)
Two nails

Small block of wood
Variable voltage supply with leads

Procedure

- Pound the nails into the block of wood about 3 inches apart.
- Stretch a section of the wire across the nails and wrap it tightly just below the heads of the nails. Make sure it is taut.
- Fasten the magnet to the string (perhaps with another magnet) and suspend from a clamp on a ring stand. Figure 2.12 shows a picture of the apparatus. Arrange the magnet so that it is pulled toward, but does not touch, the steel wire. If the magnet is touching the wire, the magnet will act as a heat sink and the wire will not reach the Curie temperature at the point of contact.
- Make sure the variable voltage supply is off and unplugged. Attach the leads from the voltage supply to the base of the nails.
- Turn the voltage supply to the zero setting and plug it in. Slowly increase the voltage *while watching the wire. The maximum voltage needed is 5–6 V.* The wire will begin to turn orange, and when the Curie temperature is reached, the magnet will fall away from the wire. Immediately return the voltage to zero and watch the magnet return as the wire cools. Repeat several times. **Warning: The wire will be hot. Too high a voltage or continued current flow will melt the wire. Do not touch the nails or wire while the circuit is energized.**

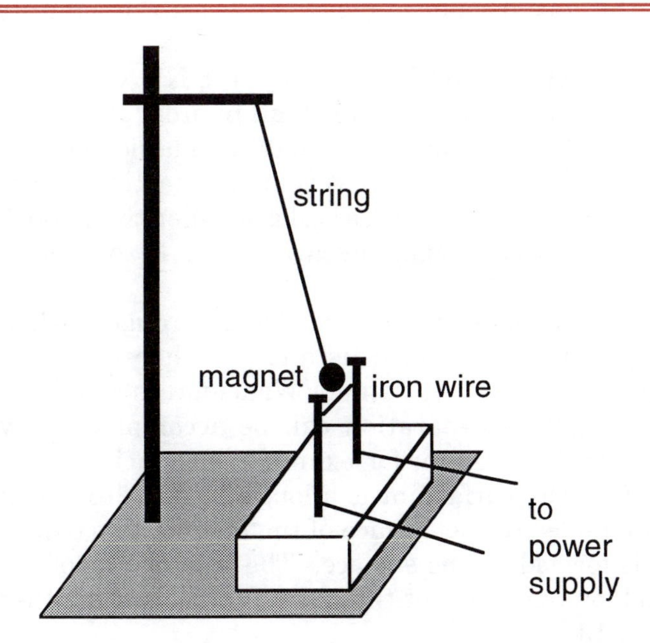

Figure 2.12. The apparatus for the demonstration on the Curie point of iron.

Many other extended solids exhibit interesting magnetic behavior with a characteristic onset temperature. The magnetic ordering can become quite complex, depending on the composition of the solid and the three-dimensional structure in which the atoms are arranged.

Ferrofluids

In recent years, researchers have prepared ferrofluids, which have the fluid properties of a liquid and the magnetic properties of a solid *(14)*. The ferrofluids actually contain tiny particles (~100-Å diameter) of a magnetic solid suspended in a liquid medium.

Ferrofluids were originally discovered in the 1960s at the NASA Research Center, where scientists were investigating different possible methods of controlling liquids in space *(15)*. The benefits of a magnetic fluid were immediately obvious: The location of the fluid could be precisely controlled through the application of a magnetic field, and, by varying the strength of the field, the fluids could be forced to flow. Researchers have prepared ferrofluids containing small particles of ferromagnetic metals, such as cobalt and iron, as well as magnetic compounds, such as manganese zinc ferrite, $Zn_xMn_{1-x}Fe_2O_4$. ($0 \leq x \leq 1$; this is a family of solid solutions, which are described in Chapter 3). But by far, the most work has been conducted on ferrofluids containing small particles of magnetite, Fe_3O_4.[2]

Ferrofluids containing magnetite can be prepared by combining the appropriate amounts of an Fe(II) salt and an Fe(III) salt in basic solution, a combination that causes the mixed valence oxide, Fe_3O_4, to precipitate from solution:

$$2\,FeCl_3 + FeCl_2 + 8\,NH_3 + 4H_2O \rightarrow Fe_3O_4 + 8\,NH_4Cl$$

However, the particles of magnetite must remain small in order to remain suspended in the liquid medium. To keep them small, magnetic and van der Waals interactions must be overcome to prevent the particles from agglomerating. Thermal motion of magnetite particles smaller than 100 Å in diameter is sufficient to prevent agglomeration due to magnetic interactions.

The van der Waals attraction between two particles is strongest when the particles approach each other at close distances. Therefore, one method of preventing agglomeration due to van der Waals forces is to keep the particles well separated. This separation can be accomplished by adding a surfactant to the liquid medium; for example, oleic acid can be used for oil-based ferrofluids. The surfactant is a long-chain hydrocarbon with a polar head that is attracted to the surface of the magnetite particle; thus a surfactant coating is formed on the surface. The long chains of the tails act as a repellent cushion and prevent the close approach of another magnetite particle (Figure 2.13).

---

[2]Magnetite is ferrimagnetic: There are two spin sets in opposing directions that do not cancel and therefore leave a net spin.

A. $2\ FeCl_3 + FeCl_2 + 8\ NH_3 + 4H_2O \rightarrow Fe_3O_4 + 8\ NH_4Cl$

B. Add *cis*-oleic acid, $CH_3(CH_2)_7CH{=}CH(CH_2)_7COOH$, in oil.

C. Remove water:

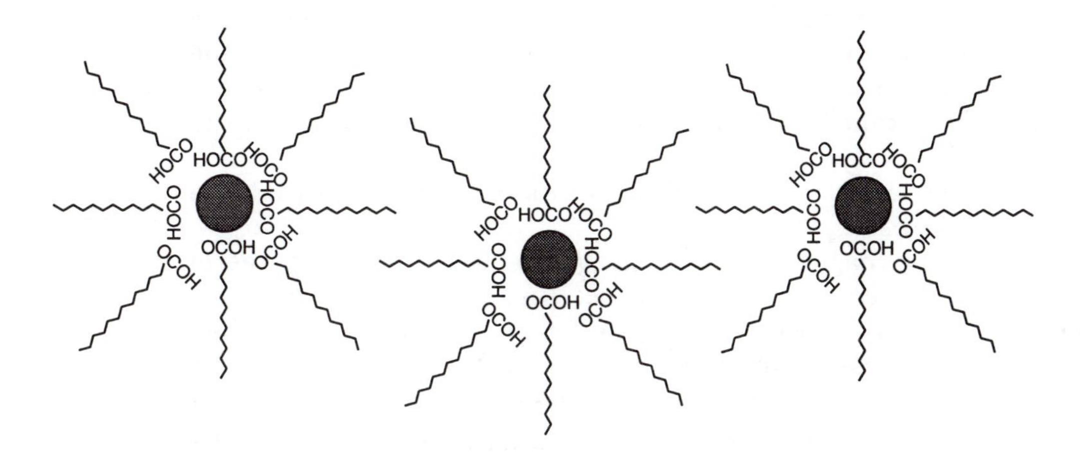

Figure 2.13. The preparation of ferrofluids: (A) the synthetic conditions for production of $Fe_3O_4(s)$; (B) addition of surfactant; and (C) removal of water to give small particles of Fe_3O_4 (shaded circles) stabilized by the interaction of the polar ends of the oleic acid molecules with the magnetite particles, and by the interaction of the nonpolar ends of the oleic acid molecules with the oil serving as the liquid medium.

Demonstration 2.6. Ferrofluids

Materials

A mineral oil-based ferrofluid with a saturation magnetization of 400 gauss obtained from Ferrofluidics[3]

Strong magnet (cow magnet, bar magnet, or rare earth magnet)

Long test tube

Petri dish

Caution! The ferrofluid causes stains and is difficult to remove from skin and fabrics. Keep the fluid off the magnet. It is virtually impossible to remove ferrofluid after direct contact with a strong magnet.

---

[3]Ferrofluidics sells ferrofluids containing magnetite in a variety of liquid media (hydrocarbon, water, mineral oil) and with a variety of magnetic strengths. *See* Supplier Information.

Procedure

- Transfer a small amount of ferrofluid to a long test tube or other long glass tube.
- Using a strong magnet placed outside of the tube, drag the ferrofluid up the side wall of the glass tube. This step demonstrates effectively how the position of the ferrofluid can be controlled with a magnetic field. (Upon removal of the magnet, the ferrofluid tends to coat the glass. Several hours may be required for the ferrofluid to drain sufficiently so that the demonstration may be viewed again in the same tube.)
- Pour a small amount of ferrofluid into a Petri dish, so that the bottom of the dish is covered.
- Bring a strong magnet up *underneath* the Petri dish. Spikes of ferrofluid will rise up from the surface. As the magnet is brought closer to the *bottom* of the Petri dish, first one spike appears, then several spikes appear, until the spikes appear to close-pack (Chapter 5), with any given spike having six nearest neighbors arranged in a regular hexagon. The pattern of spikes appears because the ferrofluid arranges itself along the magnetic field lines of the magnet (as do iron filings). The number of visible spikes reflects the strength of the magnetic field and the surface tension of the medium.
- Place a cow magnet horizontally *underneath* the Petri dish. A pattern of spikes will appear at the two poles of the magnet (*see* Figure 2.14). If no spikes are seen, try using a stronger magnet, as this behavior depends on both the strength of the magnetic field and the magnetic strength of the ferrofluid.
- Place a penny in the ferrofluid in the Petri dish. The penny sinks to the bottom.
- Bring a strong magnet up *underneath* the container. The attraction of the ferrofluid for the magnet forces the penny up and out of the ferrofluid. Repeat this demonstration with water to show that the magnet is not pushing the penny up.

Demonstration 2.7. The Ferrofluidics Bottle Cell

Materials

Bottle cell (Ferrofluidics sells these for demonstration purposes. They consist of a sealed tube that contains a small amount of the dark brown ferrofluid and an aqueous medium. These two liquids occupy the full volume of the tube.)

Strong magnet

Large crystallizing dish, one-third full of water

Overhead projector

Procedure

- Drag the magnet up the side of the bottle cell. The ferrofluid will be attracted to the magnet. (This step basically repeats the first part of the previous demonstration, but in the bottle cell the ferrofluid falls to the bottom of the glass tube immediately upon removal of the magnet so that the demonstration can be shown repeatedly.)
- On an overhead projector, place the bottle cell lengthwise in the crystallizing dish so that it is under water; it may need to be held or taped down.
- Bring the magnet up to the side of the cell. The ferrofluid will be attracted to the magnet and the spiking phenomenon described in the previous demonstration will be visible in profile. (If the bottle cell is not under water, the rounded sides of the bottle cell scatter light and make it difficult to effectively project an image of the ferrofluid.)
- In our experience, a fresh bottle cell works best; the performance deteriorated over several weeks.

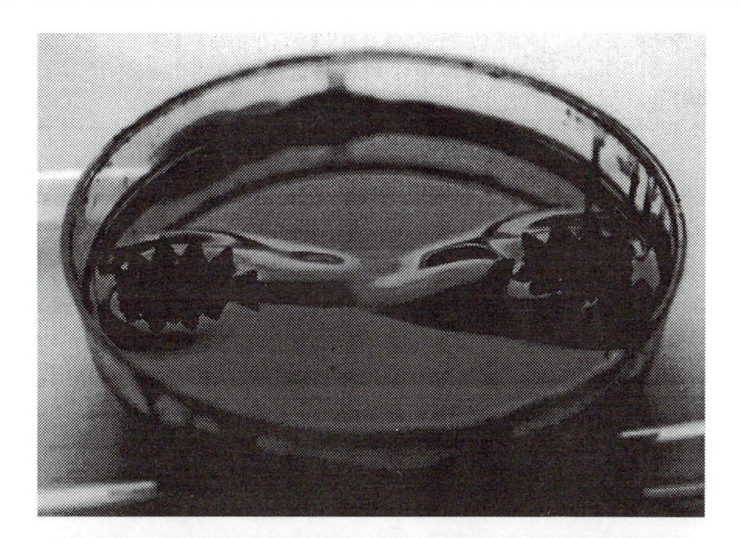

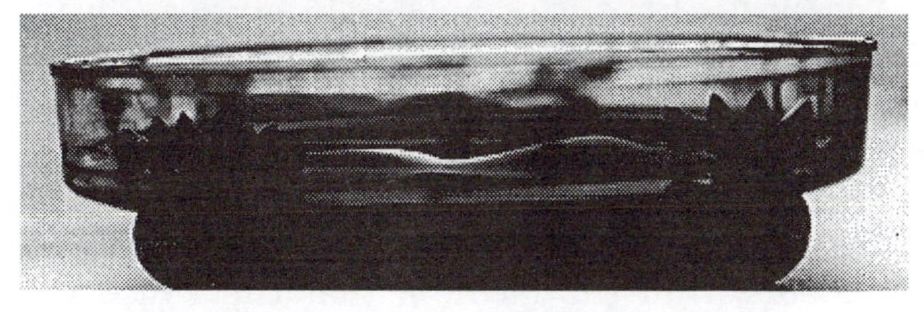

Figure 2.14. Top and side views of the magnetic spiking phenomenon observed when a cow magnet is placed beneath a Petri dish containing a ferrofluid. The ferrofluid aligns with the magnetic field lines of the magnet to produce the spikes.

Although the array of spikes on the surface of the ferrofluid is spectacular, this property is not particularly useful. However, ferrofluids have found a wide variety of applications *(16, 17)*, including use in rotating shaft seals. A ferrofluid can behave as a liquid O-ring where a rotating shaft enters either a low- or high-pressure chamber. The ferrofluid is held in place by permanent magnets and forms a tight seal, eliminating most of the friction produced in a traditional mechanical seal. These rotating shaft seals are found in rotating anode X-ray generators and in vacuum chambers used in the semiconductor industry. Ferrofluid seals are used in high-speed computer disk drives to eliminate harmful dust particles or other impurities that can cause the data-reading heads to crash into the disks.

Another application of ferrofluids is in improving the performance of loudspeakers. In a loudspeaker, electric energy is sent through a coil located in the center of a circular permanent magnet. The magnetic field induced by the electric energy causes the coil to vibrate and thus produces sound and heat. Bathing the electric coil in a ferrofluid, which is held in place by circular permanent magnets, dampens unwanted resonances and also provides a mechanism to dissipate heat from excess energy supplied to the coil. Both of these factors lead to an overall improved sound quality.

Finally, there is much hope for future biomedical applications of ferrofluids. For example, researchers are attempting to design ferrofluids that can carry medications to specific locations in the body through the use of applied magnetic fields. Other ongoing work is investigating the use of ferrofluids as contrast agents for magnetic resonance imaging (MRI).

Electrorheological Fluids

Like ferrofluids, electrorheological (ER) fluids *(18)* comprise suspensions of particles in a liquid medium, but their operation depends on electric rather than magnetic interactions. As noted in Chapter 1, ER fluids are "smart materials," whose viscosity can be tuned by an applied electric field.

ER fluids consist of a colloidal suspension of polarizable particles in a nonpolar solvent. These high-tech materials can be created from low-tech, inexpensive compounds, such as flour suspended in vegetable oil; or alumina or silica suspended in silicone oil.

The mechanism by which ER fluids operate is depicted in Figure 2.15. The presence of a strong electric field formed between two electrodes can induce dipoles in the suspended particles (one end of the particle acquires positive charge, the other end negative charge) that cause them to align with one another in an orientation that is parallel to the applied electric field. The particles stick to each other, much like damp grains of sand, and a fibrous structure is created within the medium that can cause the viscosity of the suspension to be increased by factors approaching 10^5. The dipoles disappear when the electric field is turned off, and the suspension regains its liquid nature. Completing the analogy to sand, this system would be akin to being able to remove the water rapidly from wet sand and thus restore its ability to be poured easily. Figure 2.16 illustrates the change in viscosity accompanying application of a strong electric field to the ER fluid.

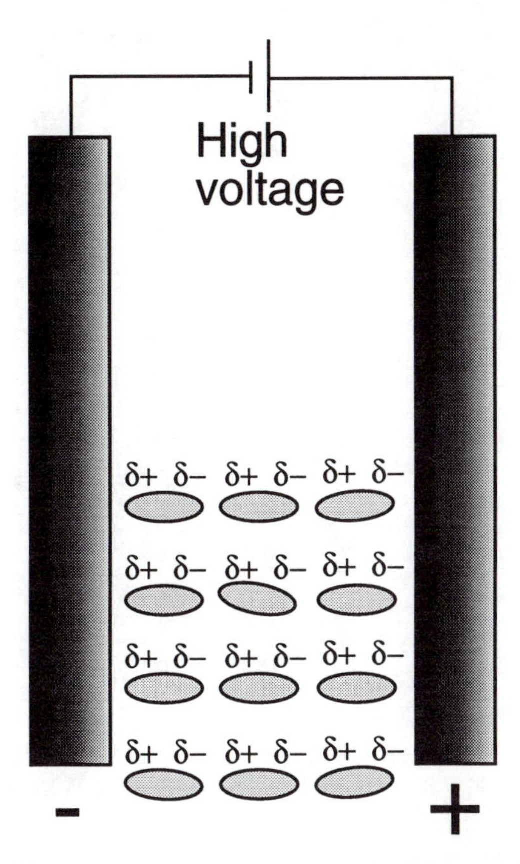

Figure 2.15. When an electrorheological (ER) fluid is placed in a strong electrical field, an induced dipole is created on the particles of the fluid. The particles stick to each other, and the fluid is suspended between the electrodes as long as the field is maintained.

The speed with which ER fluids can respond to applied electric fields is sufficiently fast that they are being considered for use in automobile shock absorbers and clutches. Figure 2.17A illustrates the design of a shock absorber that offers active control over automobile vibrations. Figure 2.17B demonstrates how an automobile clutch might benefit from ER fluids.

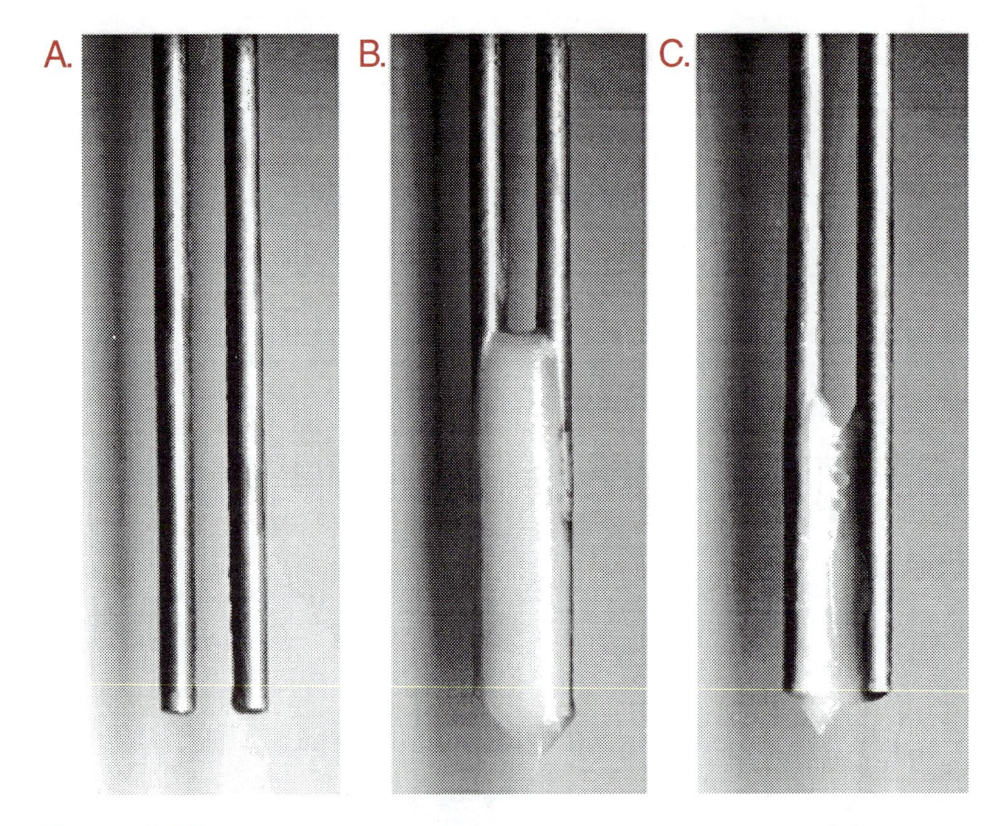

Figure 2.16. (A) The two electrodes with which an electric field can be applied to an ER fluid. No ER fluid is present. (B) When the electrodes are dipped in an ER fluid, flour in corn oil, and a high voltage (~10^4 V/cm) is applied, dipoles are induced in the particles. A fibrous network is formed suspending the material between the electrodes. (C) When the electric field is turned off, the viscosity of the medium decreases and the ER fluid drains off the electrodes.

Piezoelectric Crystals

Even though solids are electrically neutral, they may consist of electrically charged species like the anions and cations that comprise salts (NaCl, for example); or of numerous dipole moments that result from the presence of polar bonds formed between atoms of different electronegativities that make up the solid (quartz or SiO_2, for example). These regions of localized charge concentration notwithstanding, often no net electric dipole moment exists in such materials because the arrangements of the atoms or ions lead to a cancellation of individual dipole moments or charges. A molecular analogy might be to the nonpolar nature of silicon tetrafluoride: despite having individual polar Si–F bonds, the tetrahedral geometry of SiF_4 causes the vector sum of the four individual Si–F dipoles to be zero. Equivalently, because the centers of partial positive and negative charge for the collection of the four individual dipoles coincides, there is no net dipole moment.

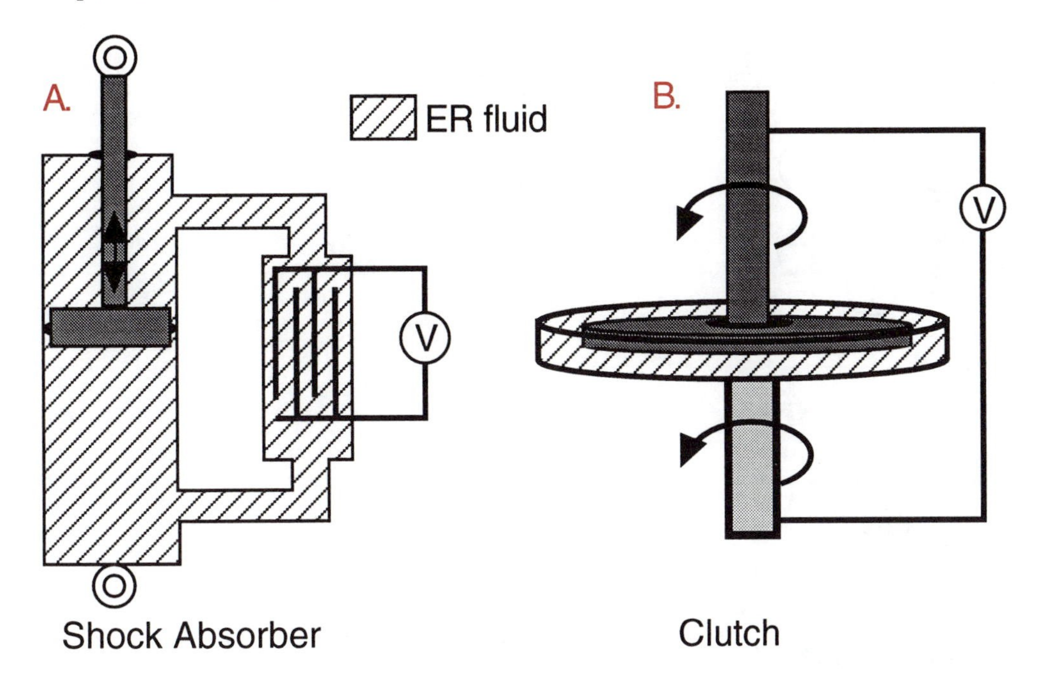

Figure 2.17. (A) A shock absorber damps vibrations by moving a piston (dark shading) through a compressible fluid. In the shock absorber shown, varying the voltage, V, varies the flow of ER fluid through a bypass valve to provide variable cushioning. (B) A clutch mechanically couples two moving parts. In the clutch shown, varying the voltage, V, varies the viscosity of the ER fluid, either letting one plate turn freely or connecting it to the other plate.

Materials like quartz are piezoelectric because they have the ability to develop a net dipole moment if they are mechanically deformed in particular directions relative to the arrangement of atoms in the crystal; and to be mechanically deformable by an electric field having an appropriate direction relative to the atoms in the solid. Both of these effects are reversible.

Figure 2.18 illustrates the mechanically induced creation of a net dipole moment where none is present initially. Shown in the figure are anions and cations from a section of a larger crystalline salt. In their natural state, the geometric centers of the three anions and three cations coincide, and no net dipole moment is expected or observed. If the salt is compressed as shown, however, the ions are displaced in the indicated directions, and the geometric centers of positive and negative charge no longer coincide: a net dipole moment has been generated and will disappear upon release of the pressure. This displacement of electron density corresponds to a voltage that develops across the opposite crystal faces or to a usable electric current if a wire is connected to the opposing crystal faces, because the net movement of negative charge in one direction is reinforced by the net movement of positive charge in the opposite direction.

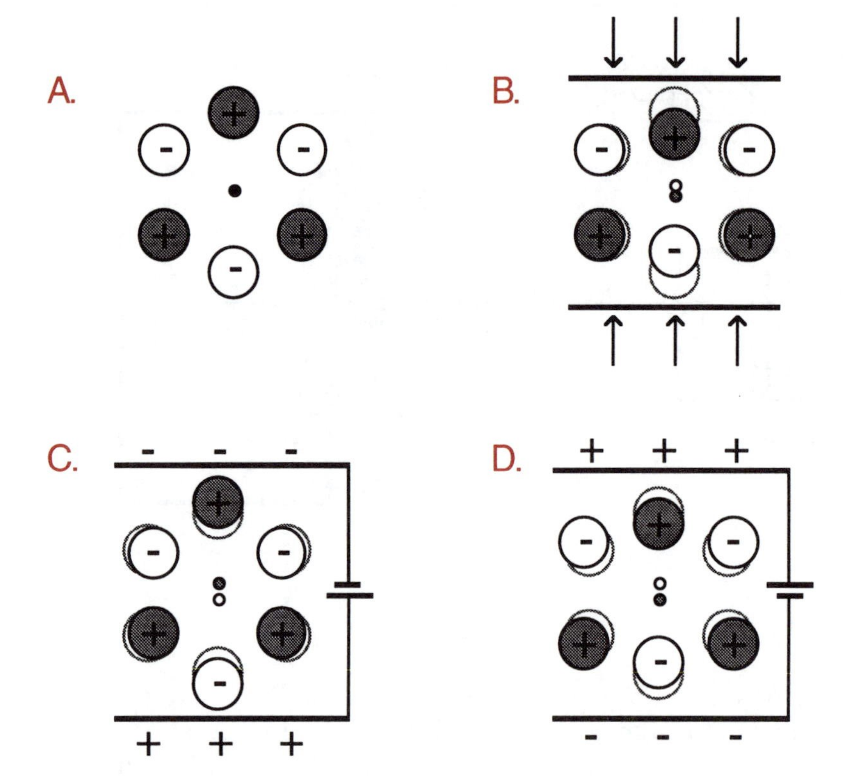

Figure 2.18. A two-dimensional model for a piezoelectric crystal. (A) In its equilibrium shape, the centers of positive and negative charge in the crystal coincide. The black dot in the center of the picture represents these centers of charge. (B) As the crystal is compressed, the ions move as indicated, causing the center of positive charge (small gray dot) to separate from the center of negative charge (small white dot). This separation generates a net dipole moment in the crystal and a movement of charge, that is, a current. As the pressure on the crystal is released, current flows in the opposite direction. (The original ion positions are shown in outline). (C) If a voltage with the indicated polarity is applied, the atoms move as shown, and the result is a net expansion of the crystal. (D) When a voltage of the opposite potential is applied, the atoms move as indicated, and the result is a net compression of the crystal. (Figure based on reference 19.)

Figure 2.18 also shows how an applied voltage can mechanically deform such a crystal. When a voltage is applied across the opposing faces shown, the cations will move toward the negative electrode and the anions toward the positive electrode. With one direction of applied voltage, the electric field compresses the crystal; reversing the direction of the applied voltage causes the ions to move in the opposite direction and thus makes the crystal expand.

Quartz is a piezoelectric material that is commonly used in watches and the strikers of lighters. In watches, an oscillating electric field applied to a quartz crystal makes the quartz crystal oscillate at a natural resonance frequency that is dependent on the size of the crystal. The vibrations of this very specific frequency are counted and used to keep the watch on time.

In strikers, both the creation and loss of the electric dipole correspond to large changes in the electric field. Thus, either rapid squeezing of the piezoelectric crystal or rapid release of the pressure applied to the crystal can cause sparking in the nearby combustible gas and ignite it. The sparking action occurs with the crystal in a commonly available cigarette lighter.

Demonstration 2.8. The Piezoelectric Effect

Materials

A piezoelectric lighter (available from chemical suppliers such as Aldrich or Flinn Scientific) or **empty** Scripto cigarette lighter with no fuel remaining

Procedure

- Squeeze the trigger to cock a spring-loaded hammer.
- When the trigger reaches the end of its cycle, the spring is released and the hammer strikes the piezoelectric crystal.
- A spark that travels several millimeters in air is observed, corresponding to ~10 kV.

References

1. Braun, R. D. *J. Chem. Educ.* **1992**, *69*, A90.
2. Binnig, G.; Smith, D. P. E. *Rev. Sci. Instrum.* **1986**, *57*, 1688.
3. Eigler, D. M.; Schweizer, E. K. *Nature (London)* **1990**, *344*, 524.
4. Wenjie, L.; Virtanen, A.; Penner, R. M. *J. Phys. Chem.* **1992**, *96*, 6529–6532.
5. Atkins, P. W. *Physical Chemistry,* 4th ed.; Freeman: San Francisco, CA, 1990, p 296. See also Conant, J. B. *Harvard Case Histories in Experimental Science;* Harvard University Press: Cambridge, MA, 1957, p 305.
6. Atkins, P. W. *Physical Chemistry,* 4th ed.; Freeman: San Francisco, CA, 1990; Chapter 20.
7. *Specific Heat: Metallic Elements and Alloys;* Thermophysical Properties of Matter Vol. 4, ISI/Plenum: New York, 1970.
8. *Specific Heat: Nonmetallic Solids;* Thermophysical Properties of Matter Vol. 5; ISI/Plenum: New York, 1970.
9. Guinier, A.; Jullien, R. *The Solid State: From Superconductors to Superalloys;* International Union of Crystallography: Oxford, England, 1989, p 34.
10. Shakhashiri, B. Z.; Dirreen, G. E.; Williams, L. G. *J. Chem. Educ.* **1980**, *57*, 373.
11. McHale, J.; Schaeffer, R.; Salomon, R. E. *J. Chem Educ.* **1992**, *69*, 1031.
12. Ashcroft, N. W.; Mermin, N. D. *Solid State Physics;* Saunders College: Philadelphia, PA, 1976; p 697.

13. *Exploratorium Science Snackbook;* Doherty, P.; Rathjen, D., Eds.; Exploratorium Teacher Institute: San Francisco, CA, 1991; pp 34–35.
14. Charles, S. W.; Popplewell, J. In *Ferromagnetic Materials*; Wohlfarth, E. P., Ed.; North Holland Publishing Co.: Dordrecht, Netherlands, 1980; Vol. 2, pp 509–559.
15. Papell, S. S. U.S. Patent 3,215,572, November 2, 1965.
16. Popplewell, J.; Charles, S. *New Sci.* September 25, 1980, 332–334.
17. Rosensweig, R. E. *Sci. Am.* October 1982, 136–145.
18. Halsey, T. C. *Science (Washington, DC)* **1992**, *258*, 761.
19. Holden, A.; Singer, P. *Crystals and Crystal Growing*; Anchor: Garden City, NY, 1960.

Additional Reading

- Binnig, G.; Rohrer, H. *Rev. Mod. Phys.* **1987**, *59*, 615.
- Guinier, A.; Julien, R. *The Solid State: From Superconductors to Superalloys*; Oxford University Press: Oxford, England, 1989.
- Raj, K.; Moskowitz, R.; *J. Magn. and Mag. Mat.* **1990**, *85*, 233–245.
- Holden, A.; Singer, P. *Crystals and Crystal Growing*, Anchor: Garden City, NY, 1960.

Acknowledgments

The section on the scanning tunneling microscope is based on a description supplied by Robert Hamers, University of Wisconsin—Madison, Department of Chemistry, Madison, Wisconsin.

The section on heat capacities was written with assistance from James Skinner, Gil Nathanson, and Hyuk Yu, Department of Chemistry, University of Wisconsin—Madison, Madison, Wisconsin.

The section on magnetic properties was written with assistance from Joel Miller, Department of Chemistry, University of Utah, Salt Lake City, Utah.

The section on ER fluids benefited from comments by Daniel Klingenberg, University of Wisconsin—Madison, Department of Chemical Engineering, Madison, Wisconsin.

We thank Reginald Penner, University of California at Irvine, Department of Chemistry, for supplying a copy of the photograph in Figure 2.8.

We thank Z. S. Teweldemedhin, R. L. Fuller, and Martha Greenblatt, Department of Chemistry, Rutgers University, for suggesting the compounds used in Demonstration 2.2, and for determining their magnetic moments.

Exercises

1. The STM trace of the surface of a niobium crystal is identical in dimension and arrangement of the atoms to that of a tantalum crystal. Why is the density of tantalum equal to 16.6 g/cm^3 while that of niobium is only 8.57 g/cm^3?

2. The radius of a tungsten atom is estimated to be 137 pm. What is the diameter in meters, centimeters, and angstrom units of a perfect STM tip that terminates in a single atom of tungsten?

3. Estimate the molar heat capacity of the following compounds:
   a. RbCl, CaS, MnO, or any other 1:1 salt
   b. MgF_2, $MnCl_2$, TiO_2, Na_2S, or any other 2:1 or 1:2 salt
   c. AlF_3, $FeCl_3$, Li_3P, or any other 1:3 or 3:1 salt
   d. Al_2O_3, Ca_3P_2 or any other 2:3 or 3:2 salt

4. Estimate the gram heat capacity (the amount of heat required to raise the temperature of 1 g of a substance by 1K, also called the specific heat) of any of the substances in Table 2.1.

5. Use a metal from Table 2.1 and its heat capacity for the following problem. A 100-g sample of a metal is heated to 100 °C in boiling water. The sample is then placed in 100.0 g of cool water at 23.4 °C. What will be the final temperature of the water and the metal, assuming no loss of heat to the surroundings?

6. A 70.0-g piece of metal at 80.0 °C is placed in 100.0 g of water at 22.0 °C. The metal and the water come to the same temperature at 26.4 °C. Estimate the approximate atomic weight of the metal, assuming no loss of heat to the surroundings.

7. Sodium bromide dissolves in water but not in ethyl alcohol. Describe an experiment using known masses of aluminum, NaBr, and ethyl alcohol, along with water, that could be used to determine the heat capacity of NaBr.

8. Sketch an STM trace of the front face of any unit cell from Chapter 3.

9. How far does an STM tip traverse in scanning a row of 400 nickel atoms? The radius of a nickel atom is 1.24 Å.

10. Gas pressures of 10^{-10} torr are needed to study metal surfaces for periods of hours without contamination. However, gases with this low a pressure still contain tremendous numbers of molecules. How many molecules are present in 1.0 mL of a gas with a pressure of 1.0×10^{-10} torr at a temperature of 23 °C?

11. Calculate the molar heat capacity of a monatomic gas in Flatland, a country in a hypothetical universe with only two dimensions.

12. If the specific heat of copper metal is 0.38 J/g-°C, estimate the specific heat of silver metal.

13. Given the following measurements for the gram specific heat, calculate and compare the molar heat capacities.

Element	Gram heat capacity, J/g-°C
Bi	0.120
Zn	0.388
Fe	0.460
S	0.787

14. Elemental iron is ferromagnetic, yet an iron nail does not attract iron filings.
    a. Explain.
    b. If, however, a magnet is rubbed over the surface of a nail, the nail will attract iron filings. Why?

15. Although the formula weight of the salt calcium oxide and the atomic weight of elemental iron are about the same, the specific heat of calcium oxide in joules per gram-degree Celsius is twice as large as that of iron. Why?

16. Compare the solids VO and V_2O_5 in their attraction to a magnetic field.

17. Trace the energy path in a piezoelectric lighter, starting from the chemical energy used to move the trigger.

18. Propose a new use for an electrorheological fluid.

Chapter 3

Stoichiometry

Stoichiometry, the study of the quantitative aspects of composition and chemical reactions, is traditionally taught from a nonstructural basis using molecular formulas consisting of easily countable numbers of atoms in small integer relationships. Calculations involve the determination of empirical and molecular formulas, mass–mole relationships, and concentrations, using equations balanced with integer coefficients.

Another aspect of stoichiometry involves the determination of empirical formulas for compounds having extended structures. However, before the stoichiometry of extended structures is addressed, we should ask the following questions: How can the existence of an extended structure be predicted, and how prevalent are extended structures?

Chemists divide the periodic table into two main regions—the metals and the nonmetals—using a staircase-like diagonal line that runs through the table. Elements that are in contact with this line are sometimes called metalloids. Table 3.1 summarizes the structural properties of combinations of these classes of elements. Compounds formed between nonmetals and other nonmetals tend to be relatively small, discrete molecules. In contrast, compounds formed between metals and nonmetals

Table 3.1. Combinations of Classes of Elements

Element Combination	Likely Structure	Examples
Nonmetal and nonmetal	Discrete molecule	CO_2, PCl_3, NO
Metal and metal	Extended (alloys)	CuZn (brass), NiTi
Metal and nonmetal[a]	Extended (salts)	NaCl, ZnS, $CaTiO_3$
Metalloid combinations	Extended or discrete	Si, BN (extended)
		$SiCl_4$, $AsCl_3$ (discrete)

[a]Some combinations of transition metals, in high oxidation states, with nonmetals can produce discrete molecules. For example, OsO_4 and $TiCl_4$ are generally considered to be discrete molecules rather than salts.

2725–X/93/0049$07.75/1

or between metals and other metals usually form solids with extended structures. Pure metals themselves also form solids with extended structures. These extended-structure solids include many metals, minerals, and salts that are commonly encountered.

Determining Stoichiometry from Unit Cells

Many of the solids discussed in this book are properly described as extended solids, because they comprise patterns of atoms that continuously repeat until the boundaries of the solid are reached. This section discusses the common formalism for describing the repeating patterns—the construction of unit cells—and illustrates the connection between unit cells and stoichiometric formulas.

The unit cell of an extended solid is a collection of its atoms, and sometimes fractions of its atoms, that can be systematically translated through space to create the entire solid. The size and shape of a unit cell are described by the lengths of three edges (a, b, and c) and the angles (α, β, and γ) between them. The edges connect points in the solid that have identical environments.

Unit cells are parallelograms in two dimensions and parallelepipeds in three dimensions. The procedure for translating the cell is to move the cell and its contents the length of its edge along the direction in which that edge points. When carried out in all directions defined by the edges of the unit cell (and when repeated from each newly produced cell), this process will leave no gaps between unit cells, and will generate the entire atomic array or crystal.

Figure 3.1 illustrates a two-dimensional example of a unit cell on a square array of identical dots that could represent, for example, a layer of atoms in the metallic element chromium. The square is a unit cell; its edges connect identical points that lie at the centers of squares formed by the dots. When the unit cell is translated parallel to its edges by the length of the edge, in each of the four possible directions, an identical set of four unit cells results, as outlined by the dashed lines. By continuing this process, the entire structure will be produced, as demanded for a unit cell.

In contrast, Figure 3.2 shows a square (formed by the solid lines) that is not a unit cell. Its edges do not connect points with identical environments, and it will not generate the correct array if it is replicated along its edges, since this test unit cell requires the dots to have a closer repeat distance.

The selection of a unit cell is not unique; its corners can be placed anywhere within a crystal. For example, using the same pattern of dots, another kind of square, shown as solid lines in Figure 3.3, would also serve as a unit cell. In this case, the corners are placed at the centers of the dots. Again, translating the square leads to identical unit cells, as shown by dashed lines on the figure.

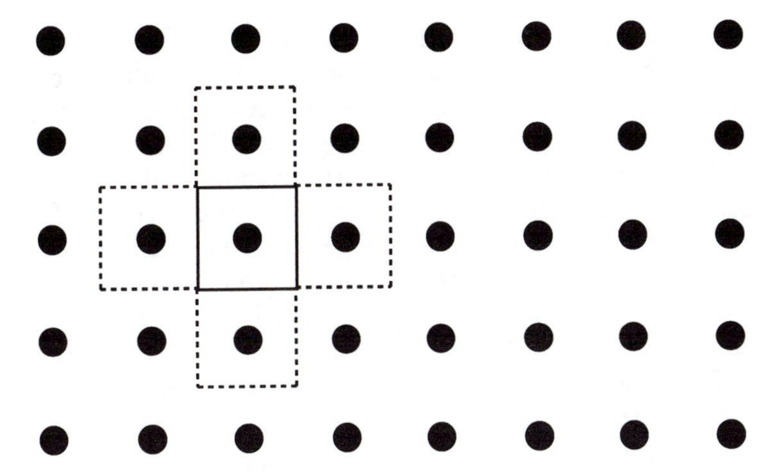

Figure 3.1. An array of dots with a square unit cell superimposed on it. Translation of the unit cell in any direction (represented by the dashed squares) will generate identical cells that, as the process continues through space, will produce the entire structure.

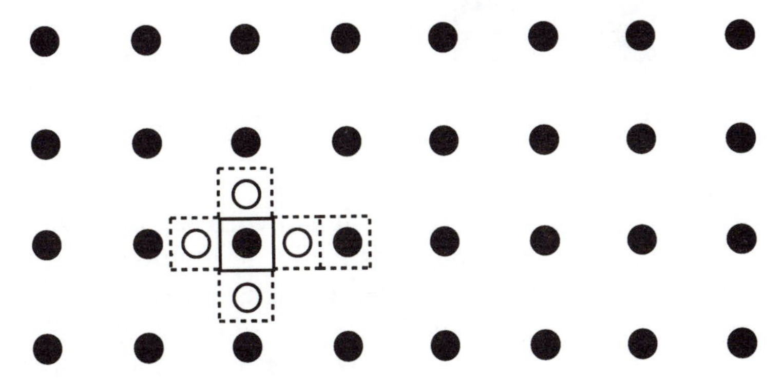

Figure 3.2. If a unit cell is not chosen properly, translation of that cell will not generate the desired array. In this picture translation of the original unit cell (dark outline) produces some dots that do not belong to the array (white circles).

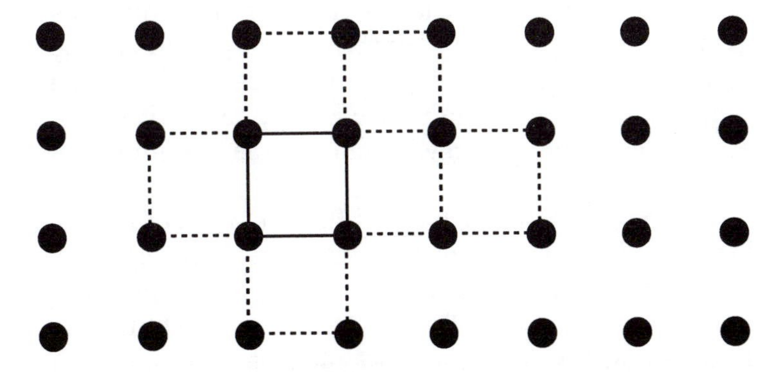

Figure 3.3. Selection of several different unit cells is possible for the same structure. This one differs from Figure 3.1 in that the atoms are at the corners of the unit cell rather than in the middle. However, translation of this unit cell will still generate the complete structure.

Once a unit cell has been determined, the number of atoms in the unit cell can be identified. In Figure 3.1, there is one dot contained entirely within the unit cell and no others associated with the unit cell, leading to a count of one dot (atom) per unit cell. The count should be unaffected by the choice of the unit cell if their areas are the same: In Figure 3.3, no dots are contained entirely within the unit cell, but each of those at the corners of the unit cell is partly within the unit cell. In the expanded view in Figure 3.4, it is apparent that one-quarter of each dot lies within the bounds of the unit cell, and, because there are four such dots (one at each corner), there is $4 \times 1/4 = 1$ dot associated with the unit cell. Another way to think of this condition is that if any dot (atom) is shared by four unit cells, such as the situation here, then it will contribute one-fourth of its total area to each of the cells (one-fourth to any given cell).

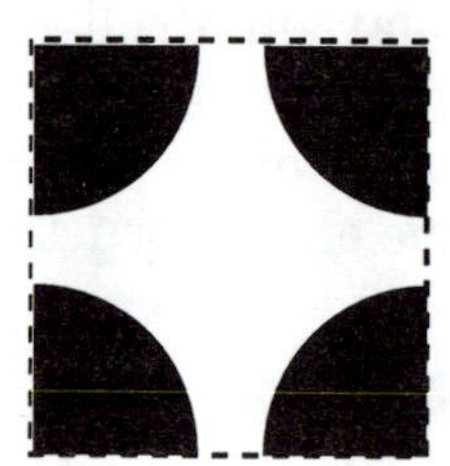

Figure 3.4. Only one-fourth of each corner atom is actually within the unit cell shown in Figure 3.3.

Although it is conventional practice to choose a unit cell with the highest possible symmetry, many different parallelograms could be employed as unit cells. Figure 3.5 shows a unit cell without right angles on the same array of dots. In this case four points lying at the corners of a "squashed" parallelogram are used to define the unit cell. Translation in each of four directions as described above yields identical unit cells (outlined with dashed lines) that can be further replicated to produce the entire structure. In this case, too, a "stoichiometry" of one dot per unit cell is found. Geometric arguments can be used to show that each adjacent pair of dots accounts for a total area of one-half of a dot, and there are two such pairs for a total area of one dot; alternatively, the portions of the four dots defining the parallelogram that are contained within the unit cell can be cut out with scissors and, when pasted together, will be seen to form one complete dot. The sharing formalism can again also be applied: With each dot participating in four unit cells, *on average* it contributes one-fourth of its area to any one of these unit cells.

Moving to an example involving a solid with two different types of atoms, consider a pattern with two different sizes of dots (Figure 3.6). This pattern could represent a layer from the NaCl structure, with the small circles being Na^+ ions and the large circles Cl^- ions. A square with its four corners in the centers of a square of four Na^+ ions may be selected as a unit cell. It contains one Cl^- ion entirely within its bounds and $4 \times 1/4 = 1$ total Na^+ ion, from the collective contributions of the four corner Na^+ ions, for a 1:1 Na^+:Cl^- stoichiometry, corresponding to the empirical formula NaCl.

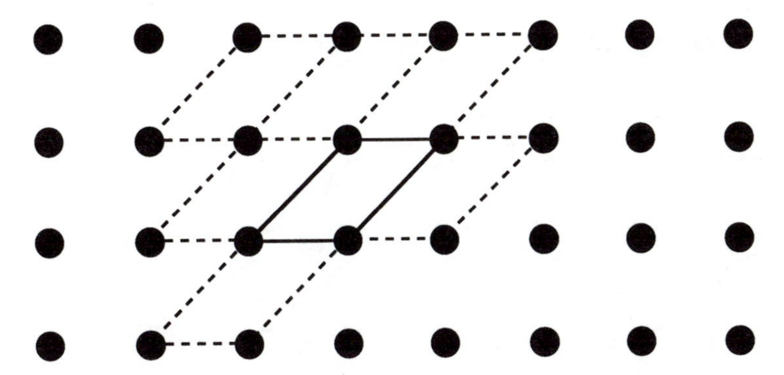

Figure 3.5. Unit cells need not be squares or rectangles; here, a tilted parallelogram serves as a unit cell.

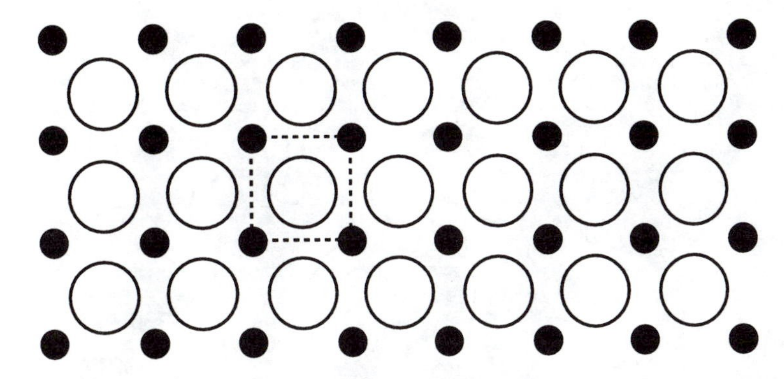

Figure 3.6. A two-dimensional model for a solid such as NaCl, with small circles representing the Na^+ ions and the large circles representing the Cl^- ions. One possible unit cell is indicated by the dashed lines.

A larger unit cell for the same pattern is shown with dashed lines in Figure 3.7. In this case the corners have been placed in the centers of the Cl^- ions, and a square that is larger than the square in Figure 3.6 has been selected as the unit cell. This unit cell contains four Na^+ ions entirely within its bounds and none at the edges or corners. To count the number of Cl^- ions within the unit cell, one is in the center (counts as 1); four are at the corners (counts as $4 \times 1/4 = 1$); and four are on the edges (counts as $4 \times 1/2 = 2$; each of these four ions is half in the unit cell or, alternatively, is shared by two unit cells and thus contributes one-half to this particular unit cell). This gives a total of four Cl^- ions to go with the four Na^+ ions, resulting once again in a 1:1 NaCl stoichiometry.

A variety of two-dimensional patterns, suitable for practice in finding unit cells, is found in wallpapers, tile floors, and Escher prints. The Escher print shown in Figure 3.8 has two possible unit cells outlined upon it.

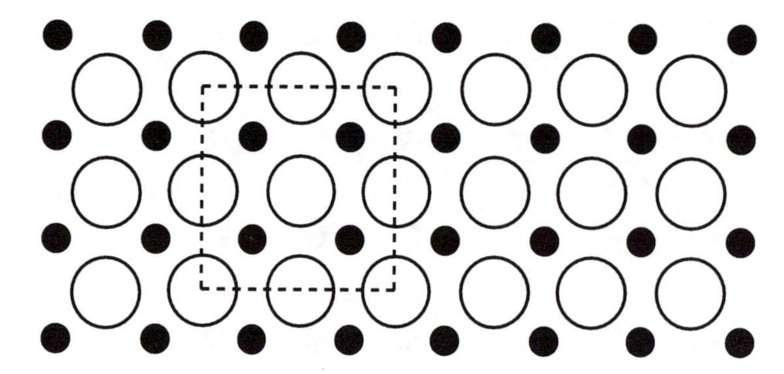

Figure 3.7. A larger unit cell for the structure shown in Figure 3.6.

Figure 3.8. Unit cells may be identified in Escher prints. The top unit cell contains one black and one white bird, and the lower unit cell contains two black and two white birds. The notations at the edges of the figure are Escher's own marks.

The concepts articulated for two dimensions are readily extended to three dimensions[1]. Although all crystalline solids have unit cells that are parallelepipeds in three dimensions, many simple compounds (and some not-so-simple compounds) have a so-called cubic unit cell for which the parallelepiped has three equal, perpendicular edges ($a=b=c$; $\alpha=\beta=\gamma=90°$). One metal (polonium) crystallizes with a simple cubic (sc) structure: the metal atoms sit only at the unit cell corners. Shifting the cube in all three mutually perpendicular directions, by the length of a side of the cube, will generate six identical unit cells, which can be further replicated to create the entire structure. The unit cell of polonium is sketched in Figure 3.9 but is most easily seen with a model kit: From the cut-away view shown, each sphere contributes one-eighth of its total volume to the indicated unit cell. The total sphere volume from the eight spheres defining the cube is $8 \times 1/8 = 1$ sphere. Alternatively, extending the structure makes it apparent that any given sphere is shared by eight unit cells and thus contributes one-eighth of its volume to any particular unit cell.

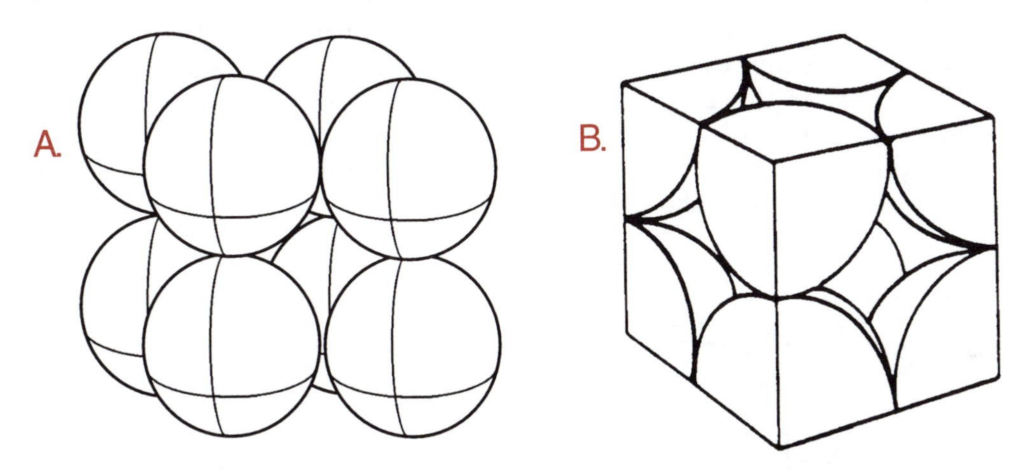

Figure 3.9. A: A simple cubic unit cell showing the eight atoms that are part of the unit cell. B: Only one-eighth of the volume of each of the eight spheres contributing to the unit cell is contained within it.

Demonstration 3.1. Corner-Shared Atoms in a Simple Cubic Unit Cell

Materials

Knife
Orange or apple
Toothpicks (optional)

Procedure

---

[1]A copy of the ICE Crystal Structure Solid-State Model Kit (SSMK) or a similar kit or software is recommended as an aid to visualizing structural features.

- In front of the class, cut the piece of fruit into eighths.
- Arrange four pieces on a flat surface as shown for the bottom layer of a simple cubic unit cell (Figure 3.9B).
- Hold the remaining four pieces in place as the second layer or fasten them in place with toothpicks. This will show that the eight eighths form the corners of a cube.

Variation

A similar procedure may be used to show that one-fourth of the volume of an edge atom is within a given unit cell, and one-half of a face-centered atom is within a given unit cell (see below).

A more common structure for metallic elements is the body-centered cubic (bcc) structure (Figure 3.10), with a cubic unit cell having atoms at the corners of the cube and an identical atom in the center of the cube (the body center). The corners and the center of a bcc unit cell have the same environment. The count in this case will be two atoms per unit cell: now one atom is entirely enclosed by the unit cell by virtue of being in its center, and again the $8 \times 1/8 = 1$ atom results from the contributions of the eight atoms at the corners of the cube.

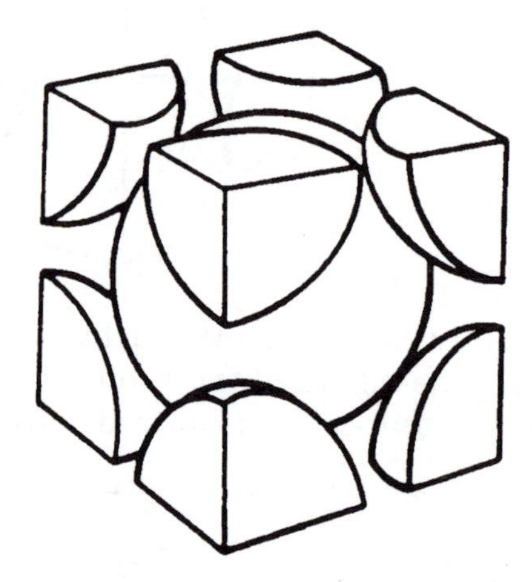

Figure 3.10. A body-centered cubic unit cell showing a full sphere within the unit cell in addition to the eight one-eighth sphere volumes from the corner spheres

Figure 3.11 shows that the body-centered cubic structure is adopted by many elements, including all the alkali metals and a number of transition elements, including V, Cr, Mo, W, and Fe.

Aluminum, copper, silver, gold, and nickel are common examples of the elements that adopt a face-centered cubic (fcc) structure (Figure 3.12). In addition, all of the solid noble gases adopt this same structure. The unit cell of this structure may be described with the help of a model kit as having identical atoms at each corner of the cube as well as in the center of each face (hence the term face-centered, for which the corners and face-centered positions possess identical environments). The count in this case will be four atoms per unit cell: the corner atoms make their usual contribution of a single atom (8 × 1/8 = 1); and each of the remaining six atoms, one lying in the center of each cube face, contributes one-half its volume to the unit cell (each of these atoms is shared by two unit cells, thus contributing one-half to the unit cell of interest), making 6 × 1/2 = 3 atoms, for a total of four atoms.

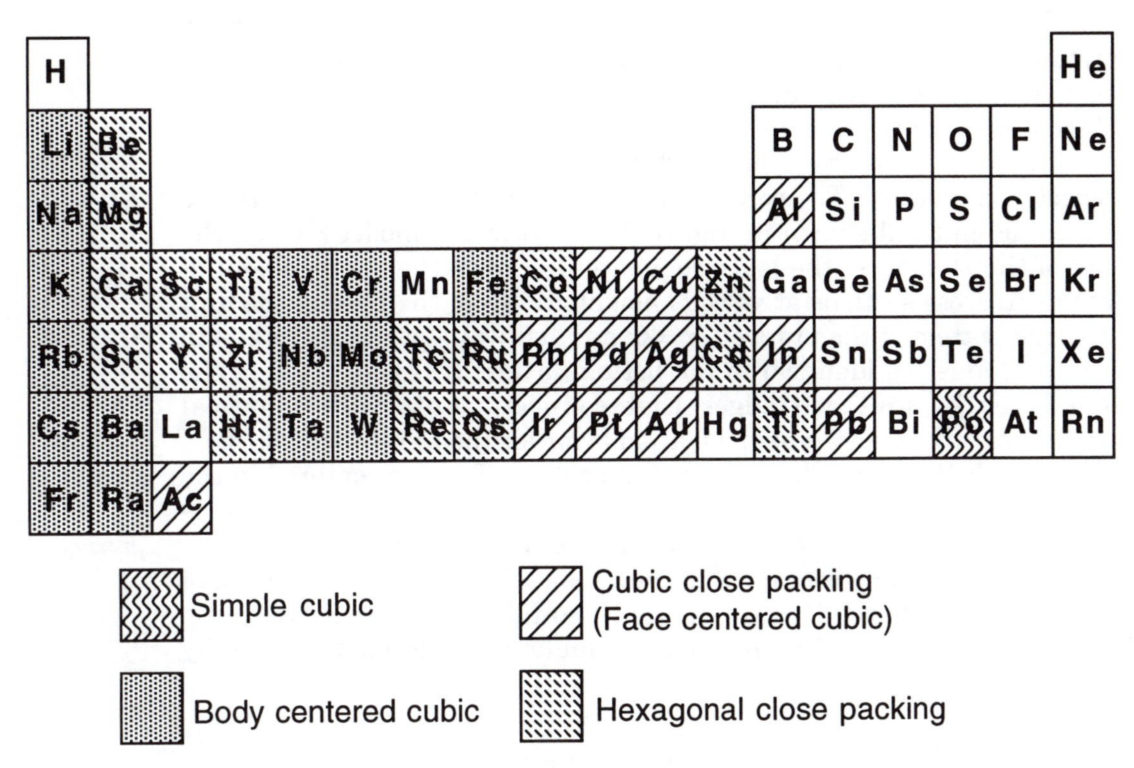

Figure 3.11. Periodic table showing the metallic elements that have one of the four indicated packing arrangements. (Data from reference 1.)

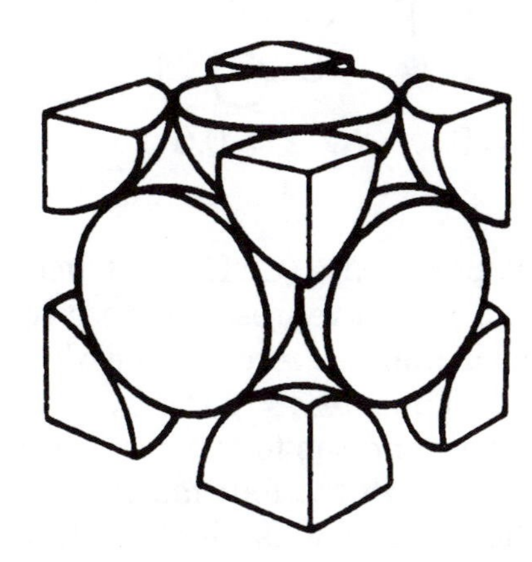

Figure 3.12. A face-centered cubic unit cell, showing the half-sphere volumes on the six faces of the cube in addition to the eight one-eighth sphere volumes from the corner spheres.

Demonstration 3.2. The FCC Structure Videotape.
A videotape showing different viewing perspectives of the FCC structure is available. For information on obtaining a copy, write Professor Nathan S. Lewis, Dept. of Chemistry, California Institute of Technology, Pasadena CA 91125.

The stoichiometry of solid compounds can also be determined by counting the atoms in their unit cells. Calcium titanate, which is found to occur naturally as the mineral perovskite, is a good example. The stoichiometry of perovskite and its structure are the basis for a variety of important solids ranging from high-temperature superconductors to piezoelectric materials. The perovskite structure can be described with the simple cubic unit cell shown in Figure 3.13 and in the model kit. In the picture of the structure, the layer sequence describes the arrangement of the atoms in cross sections at various heights (z) from the base of the unit cell. At the bottom of the unit cell, $z = 0$; and halfway up, $z = 1/2$. The top layer ($z = 1$) must be identical to the bottom layer: It is connected to the bottom layer by unit cell edges that connect points with identical environments, and it serves as the bottom layer for the unit cell above it.

There are Ca^{2+} ions at the corners of the cube, a Ti^{4+} ion centered in the cube, and O^{2-} (oxide ions) in the middle of each face of the cube. This arrangement corresponds to a stoichiometry of one Ca^{2+} ion ($8 \times 1/8 = 1$), one Ti^{4+} ion (completely contained in the cube), and three O^{2-} ions (each of the oxide ions is shared by two unit cells, so that $6 \times 1/2 = 3$ oxide ions belong to this unit cell). The resulting formula of $CaTiO_3$ is thus readily established from the unit cell.

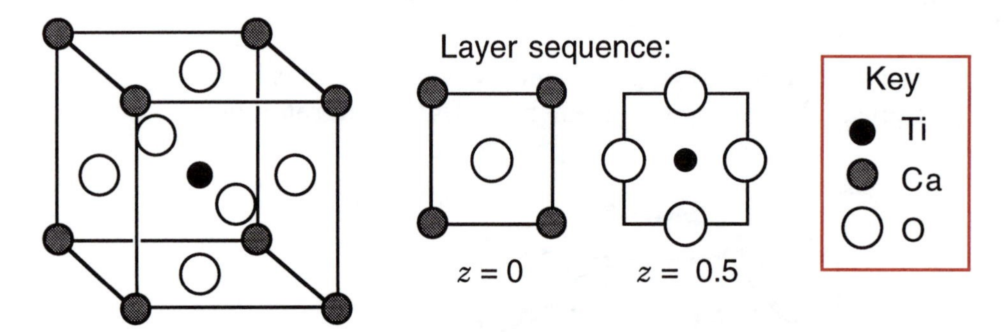

Figure 3.13. One unit cell of the perovskite ($CaTiO_3$) structure. The layer sequence shown next to the unit cell represents cross sections of the unit cell that would be seen if planes were passed through the cubic unit cell parallel to the base. Thus, at $z = 0$ the base of the unit cell contains an oxide ion at its center and the four bottom corner Ca^{2+} ions. At $z = 0.5$, the plane passes through the four face-centered oxide ions and the center Ti^{4+} ion. The $z = 1$ view is identical to the $z = 0$ view, because the top face of the unit cell is the bottom face of the unit cell above it.

Solid Solutions

As already noted, discrete molecules have easily countable numbers of atoms, and small integer relationships exist between the numbers of different kinds of atoms composing the molecule. Extended crystalline solids provide far broader stoichiometric vistas than discrete compounds. An analogy might be that moving from discrete molecules to extended solids is somewhat akin to discovering fractions after having dealt exclusively with positive integers. In addition to the discussion of metallic and ionic solids presented earlier, two classes of extended crystalline solids—substitutional and variable stoichiometry solid solutions—are particularly interesting complements to introductory discussions of stoichiometry. These materials are the basis of commonly encountered solids such as the semiconductors used in light-emitting diodes and diode lasers, metal alloys like brass[2] and bronze, and battery electrodes.

Substitutional Stoichiometry

As will be seen in Chapter 7, the ability to continuously tune physical properties using solid solutions can be thought of as a means of effectively expanding the periodic table. For example, the number of molecular compounds of silicon and hydrogen is presently limited to about a dozen, including SiH_4, Si_2H_6, and Si_3H_8. In contrast, a vast number of extended solids of formula Si_xGe_{1-x} ($0 \le x \le 1$) exists. Both silicon and germanium have the structure of diamond (Chapter 5), as do all of the Si_xGe_{1-x} solid solutions. As x varies in this stoichiometric formula, many of the properties of this family of solids are continuously tunable between the extremes of pure silicon and pure germanium. This tunability is extremely important in the semiconductor industry, in which, for example, the colors of light absorbed and emitted by these solids can be tuned with their composition, that is, with the value of x (*see* Chapter 7).

Solids like Si_xGe_{1-x} often form disordered substitutional solid solutions; that is, germanium atoms randomly replace silicon atoms in elemental silicon (or vice versa). Thus, we can speak only of the probability of finding a substituting germanium atom at sites in the crystal structure where the exchangeable silicon atoms would usually be found: A formula like $Si_{0.283}Ge_{0.717}$ means that at any site where a silicon or germanium atom could be found in the crystal structure, there is a 28.3% chance that the atom is silicon and a 71.7% probability that the atom is germanium; the number of significant figures reflects the quality of the analytical data. The formula weight for stoichiometric purposes would be the weighted average,

[2]As the ratio of zinc to copper changes, the structure of brass changes. Consequently, the name brass actually represents a number of phases that can have different structures. Within a given range of compositions, however, brass is a solid solution. *See* reference 2.

calculated as [(0.283) × (atomic weight of Si)] + [(0.717) × (atomic weight of Ge)] = 60.00 g/mol. An analogy can be drawn between this calculation and the determination of the atomic weight for an element with two isotopes.

Figure 3.14 shows two-dimensional solid solutions of elements A and Z of varying stoichiometries, along with the "pure" A and "pure" Z structures from which the solid solutions are derived.

Demonstration 3.3. Disordered Binary Solid Solutions

Materials

Coins or sets of objects (marbles or identical sheets of paper, for example) that are distinguishable by having one of two colors.
Opaque bag or container, one for each student
Solid-State Model Kit (SSMK) (optional)

Procedure

- Individual students are assigned an atomic site in, for example, an fcc structure for which they will decide the identity of the atom at that site. If a disordered binary solid solution of composition $A_{0.5}Z_{0.5}$ is to be modeled, have each student flip a coin. A toss of "heads" puts an A atom at that student's atomic position, and a "tails" puts a Z atom at that position. To model other stoichiometries, have students place two types of visually distinguishable but otherwise identical objects in an opaque container in the stoichiometric ratio to be modeled: for example, to model $A_{0.25}Z_{0.75}$, three marbles of one color (Z atoms) and one of another color (A atoms) are placed in the container, and one marble is drawn at random by the student. After the drawings, the total number of A and Z atoms selected can also be counted to compare the agreement with the theoretical probability of three Z atoms to one A atom.
- To model a two-dimensional solid, draw a grid on the board with an arbitrary number of rows and columns. As each coin is tossed or marble is drawn by the student assigned to a particular atom, place a symbol (one representing A; one representing Z) in one of the boxes in the grid. A single row could be used to model a one-dimensional structure in the same way.

Variation for SSMK

- As noted in the kit's instructions, a Sharpie permanent marker can be used to color some of the large spheres in the SSMK (the color can later be removed with 95% ethanol). This step permits three-dimensional disordered solid solutions to be built for a variety of structures such as fcc and diamond (other structures amenable to forming solid solutions are described later): The choice of what color sphere to place at a site is again dictated by the results of student coin tosses or draws, as described.

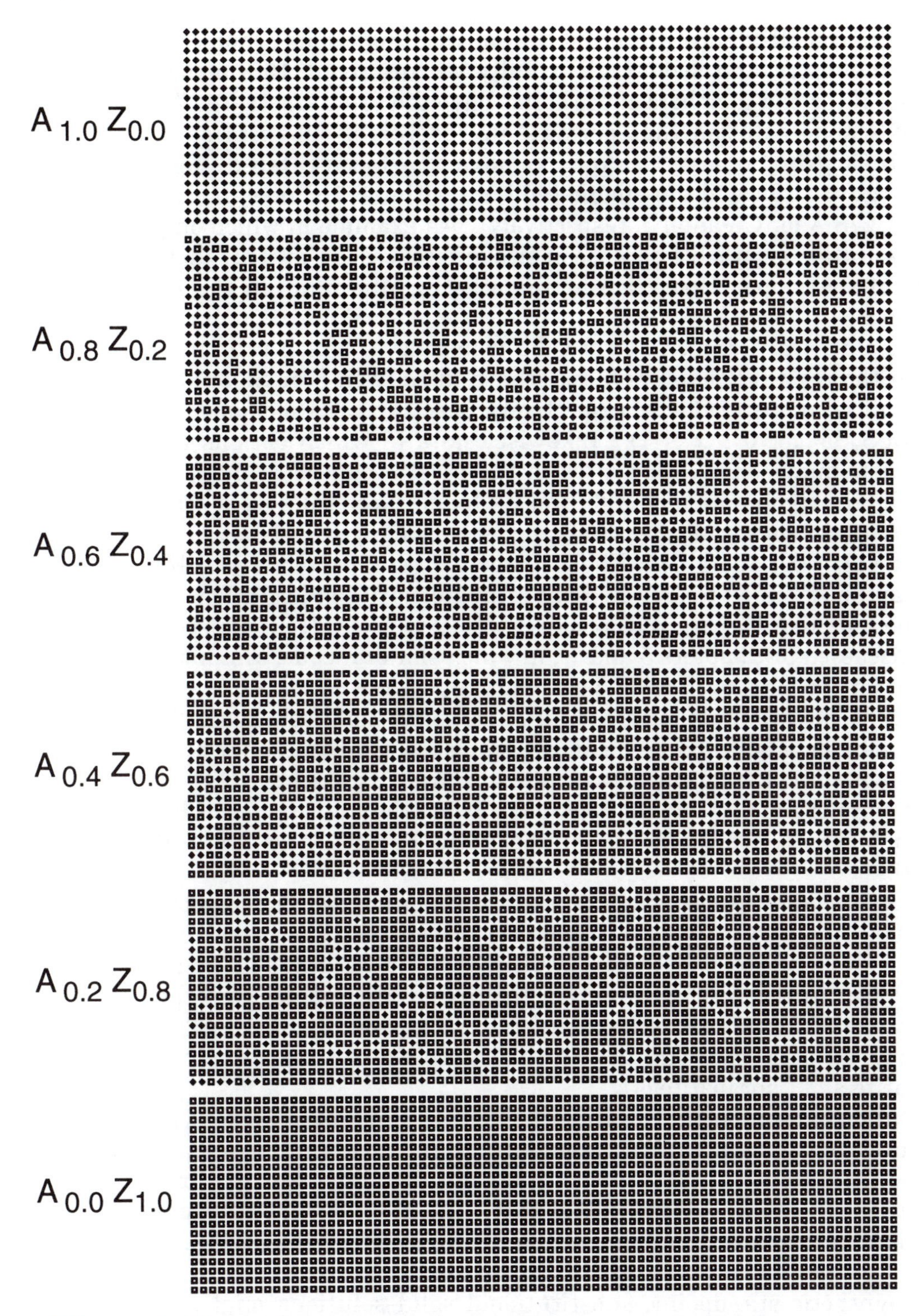

Figure 3.14. Substitutional solid solutions $A_{1-x}Z_x$ with varying stoichiometries (varying values of x). Diamonds represent atom A and squares represent atom Z.

Two other examples of disordered solid solutions come from metal alloys and salts. In general, two solids are most likely to be completely miscible in one another (x taking on any value from 0 to 1, that is, from no substitution to complete substitution) if the substituting atoms or ions are similar in chemistry and size (radii usually within about 15% of one another) and if the end members of the series have the same structure. An example of a solid solution that permits any degree of substitution is cupronickel, Cu_xNi_{1-x}: both Cu and Ni have the face-centered cubic structure (Figure 3.12), and the atomic radii are within 3% of one another (1.246 and 1.278 Å for Cu and Ni, respectively). An example in which only limited substitution is possible because of size differences is that only up to about 20% of the copper atoms can be replaced by aluminum in the fcc structure.

Salts like CdS and CdSe can form solid solutions in which the anion positions in the crystal (Chapters 5 and 7) are filled with S or Se anions with probabilities of x and $(1 - x)$, respectively, based on the stoichiometric formula CdS_xSe_{1-x} ($0 \leq x \leq 1$). Likewise, salts in which cations are randomly distributed onto cation sites in the crystal can be formed, as illustrated by the combination of MgO and FeO, and these salts form solid solutions of stoichiometry $Mg_xFe_{1-x}O$ ($0 \leq x \leq 1$), having the structure of NaCl, rock salt (Chapter 5). These three-element or ternary solid solutions can be extended to four-element or quaternary solid solutions in some compounds like $Ga_xAl_{1-x}P_yAs_{1-y}$ ($0 \leq x \leq 1$; $0 \leq y \leq 1$; zinc blende structure discussed in Chapters 5 and 7), in which simultaneous, unlimited substitution for the Groups 13 and 15 elements is possible.

In geology, some minerals form partial or complete solid solutions. In the mineral olivine, Mg^{2+} and Fe^{2+} can be completely substituted for each other. Consequently, the formula for olivine is written as $Mg_xFe_{2-x}SiO_4$ ($0 \leq x \leq 2$) or more commonly as $(Mg,Fe)_2SiO_4$. Thus, the limiting formulas range from Mg_2SiO_4 (the mineral forsterite) to Fe_2SiO_4 (the mineral fayalite) *(3)*.

The mineral hydroxyapatite, $Ca_5(PO_4)_3OH$, is the principal component of tooth enamel. Addition of fluoride in water or from toothpaste causes the formation of fluorapatite, $Ca_5(PO_4)_3F$, which is more resistant to cavity formation. Substitution of fluoride for hydroxide is tunable over the complete range of formulations $Ca_5(PO_4)_3(OH)_xF_{1-x}$ ($0 \leq x \leq 1$). This substitution is also true in geological samples of the minerals hydroxyapatite and fluorapatite *(4)*.

Solid solutions are important in radiochemistry. Radium is present in uranium ore, but in such small amounts that it cannot be easily precipitated from solution. To overcome this disadvantage, barium ions are added to the solution in addition to sulfate. The radium will substitute for the barium in the barium sulfate precipitate. In this case the barium is acting as the carrier for the radium *(5)*.

From a synthetic standpoint, substitutional solid solutions might be expected to form whenever the exchangeable atoms or ions undergo similar chemical reactions and are both present in the reaction medium. Of course, the relative quantities of exchangeable atoms or ions present in the medium used for synthesis may or may not be preserved in the resulting solid. General routes to preparation of substitutional solid solutions include comelting, coprecipitation, codeposition from gas and solution phases, and codecomposition of volatile precursor molecules (chemical vapor deposition, CVD). (Chapter 10 gives a further description of solid-state synthetic techniques.) If the reactivities of the exchangeable species in the solid solution are similar, parallel reaction paths for these atoms or ions will generally be found in decomposition reactions of the solid solutions. Examples are given in the exercises.

Laboratory. Students can grow solid solutions derived from colorless crystals of alum [$KAl(SO_4)_2 \cdot 12H_2O$] and purple chrome alum [$KCr(SO_4)_2 \cdot 12H_2O$]. As these compounds form a complete solid solution, the color of the [$KCr_xAl_{1-x}(SO_4)_2 \cdot 12H_2O$] crystals can be tuned by varying the composition. The experiment is written so that the students can design their own experiments to explore the effect of reaction conditions on crystal growth (*see* Experiment 3).

Variable Stoichiometry

As a result of the influence of Dalton's Law of Multiple Proportions, most contemporary chemists have viewed materials as having fixed compositions. During the 18th century, however, Berthollet held the opposite view from Dalton: Berthollet proposed that materials could have variable compositions, depending on the conditions of formation. Many such systems have been discovered in modern times, and the terms "berthollide" or "nonstoichiometric" are sometimes used to describe compounds that exhibit variable stoichiometry (*6–8*).

Compounds with variable stoichiometry are solids in which the stoichiometry can vary from a simple integral relationship, typically over a range of composition, but the basic crystal structure does not vary over the same range. The two most common ways in which variable stoichiometry arises are from interstitial atoms, which are atoms inserted into the void spaces in a structure, or from vacancies of atoms in the structure. Variable stoichiometry is most common in compounds of the *d*-, *f*-, and *p*-block metals containing soft (polarizable) anions such as S^{2-} ions. It is also prevalent in many transition metal oxides, where the valence of the transition metal can readily change to maintain charge neutrality. Compounds containing hard anions such as fluorides, chlorides, sulfates, and nitrates rarely have variable stoichiometry. Table 3.2 lists some common materials with variable stoichiometry.

Table 3.2. Materials with Variable Stoichiometry

Compounds	Range of x	
Hydrides		
TiH_x	1.0–2.0	
ZrH_x	1.5–1.6	
HfH_x	1.7–1.8	
NbH_x	0.64–1.0	
GdH_x	fluorite[a] type, 1.8–2.3	hexagonal[a], 2.85–3.0
ErH_x	fluorite type, 1.95–2.31	hexagonal, 2.82–3.0
LuH_x	fluorite type, 1.85–2.23	hexagonal, 1.74–3.0
Oxides		
TiO_x	rock salt[a] type, 0.7–1.25	rutile type[a], 1.9–2.0
VO_x	rock salt type, 0.9–1.20	rutile type, 1.8–2.0
NbO_x	rock salt type, 0.9–1.04	
Fe_xO	rock salt type, 0.88–0.95	
UO_{2+x}	0–0.25	
Sulfides		
ZrS_x	0.9–1.0	
YS_x	0.9–1.0	
Cu_xS	1.77–2.0	

[a]Fluorite, hexagonal, rock salt, and rutile are different structure types. *See* Chapter 5.
SOURCE: Adapted from reference 9.

Transition metal hydrides are classic examples of compounds with variable stoichiometry. These compounds are best thought of as metals with variable amounts of hydrogen inserted in the void spaces of the metal crystal. For example, palladium is a transition metal that forms a metal hydride; a maximum of 0.8 hydrogen atoms per palladium atom can be inserted in the octahedral holes (Chapter 5) of the fcc palladium crystal. In many cases, the formation of metal hydrides is readily reversible. For example, hydrogen may be removed from many metal hydrides by raising the temperature. The reversible formation and decomposition of transition metal hydrides has been an area of active research, as these compounds can be used as hydrogen storage materials.

Similarly, materials of variable composition that contain lithium have been of interest in the development of new solid-state battery materials. As Ti^{4+} ions are reduced to Ti^{3+}, lithium ions can be reversibly inserted into the octahedral holes (Chapter 5) found between two sulfide layers in TiS_2 to yield Li_xTiS_2 ($0 \le x \le 1$). This material is a promising candidate for use in lightweight solid-state lithium batteries.

In many compounds of variable stoichiometry, a change in composition has a dramatic effect on material properties. For example, the sodium tungsten bronzes, Na_xWO_3 ($0 \le x \le 1$), show a wide variety of color over the range of composition (*see* Experiment 8). Even a seemingly immeasurable variation in stoichiometry can cause a material to be highly colored. For example, trapping an electron at a halide vacancy in an alkali halide crystal ($MX_{1-\partial}$, where ∂ is 10^{-3} to 10^{-6} or less) leads to the intense color of the so-called F-center (Chapter 6). Variable stoichiometry can also have a large effect on the electrical properties of a material. Variations in the oxygen stoichiometry in the high-temperature superconductors play a critical role in determining whether a material is superconducting or not. For example, $YBa_2Cu_3O_{6.9}$ is a superconductor with a high T_c (onset temperature of superconductivity), whereas when more of the oxygen atoms are removed, as in $YBa_2Cu_3O_{6.1}$, superconductivity is destroyed.

Finally, if stoichiometrically equivalent numbers of normally occupied metal and nonmetal sites are vacant (e.g., when one Ca^{2+} and two F^{-} sites are empty in CaF_2), the compound will appear to be stoichiometric (the metal–nonmetal ratio will be a simple whole-number ratio) even though it is not a perfect structure. Careful measurements of the density of a single crystal, however, can reveal such missing atoms. A classic example is titanium monoxide, TiO. At a titanium:oxygen ratio of 1:1, as many as 15% of each of the ideal cation and anion sites may be vacant.

A Caveat for Classroom Use

In discussing material from this chapter, it is important to differentiate substances that have tunable compositions from those that do not. Certainly the composition of a molecule like water is not tunable. Thus, when determining empirical formulas for molecules, only integers should appear in the formula. Fractional numbers of atoms arise logically, as shown, for solids that are clearly identified as having tunable stoichiometry.

References

1. Wells, A. F. *Structural Inorganic Chemistry*; Clarendon Press: Oxford, England, 1984, Chapter 29.
2. Greenwood, N. N.; Earnshaw, A. *Chemistry of the Elements*; Pergamon Press: Oxford, England, 1984; pp 1369–1370.
3. Klein, C.; Hurlburt, C., Jr. *Manual of Mineralogy,* 20th ed., John Wiley and Sons: New York, 1977.
4. Deer, W. A.; Howie, R. A.; Zussman, J. *Rock Forming Minerals: Volume 5. Non-Silicates.* John Wiley and Sons: New York, 1962; p 324.
5. Wulfsberg, G. *Principles of Descriptive Inorganic Chemistry;* University Science Books: Mill Valley, CA, 1991; p 122.
6. Feinstein, H. I. *J. Chem. Educ.* **1981**, *58*, 638.
7. Greenwood, N. N. *Ionic Crystals, Lattice Defects, and Nonstoichiometry*, Butterworths: London, 1968.
8. Bevan, D. J. M. "Nonstoichiometric Compounds: An Introductory Essay" in *Comprehensive Inorganic Chemistry*; Bailar, J. C.; Emeléus, H. J.; Nyholm, R.;

Trotman-Dickenson, A. F., Eds.; Pergamon Press: Oxford, England, 1973; Vol. 4, pp 453–540.
9. Shriver, D. F.; Atkins, P. W.; Langford, C. H. *Inorganic Chemistry;* Freeman: New York, 1990; p 577.

Additional Reading

- Galasso, F. S. *Structure and Properties of Inorganic Solids*; Pergamon Press: Oxford, England, 1970.
- Smart, L.; Moore, E. *Solid State Chemistry: An Introduction*; Chapman and Hall: London, 1992.
- West, A. R. *Solid State Chemistry and Its Applications*; Wiley: New York, 1984.

Exercises

1. The unit cell pictures in this problem can be used for questions regarding unit cell occupancy or stoichiometry. For example, verify the stoichiometry for each of the structures. Information about some of the materials is also included.

   a. *Cesium chloride*. (Other compounds that crystallize with the CsCl structure: CsCl, CsBr, CsI, TlCl, TlBr, TlI, and many intermetallic compounds such as AgCd, AgZn, AlFe, AuZn, CoFe, HgMg, and NiTi.).

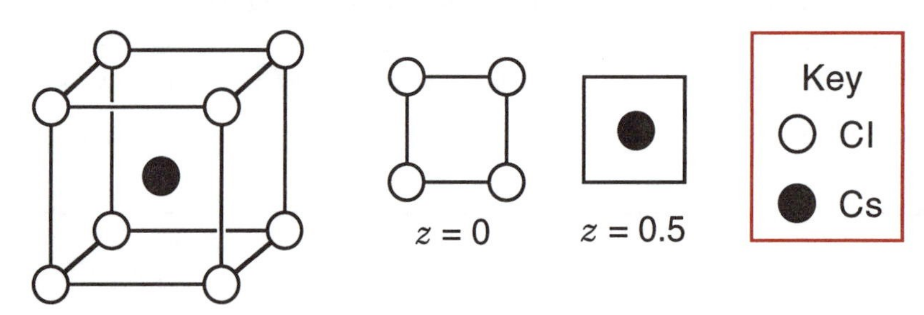

   b. *Cuprite* (Cu_2O). This compound is a p-type semiconductor and was used in the construction of the first photocells. Ag_2O and Pb_2O have the same structure.

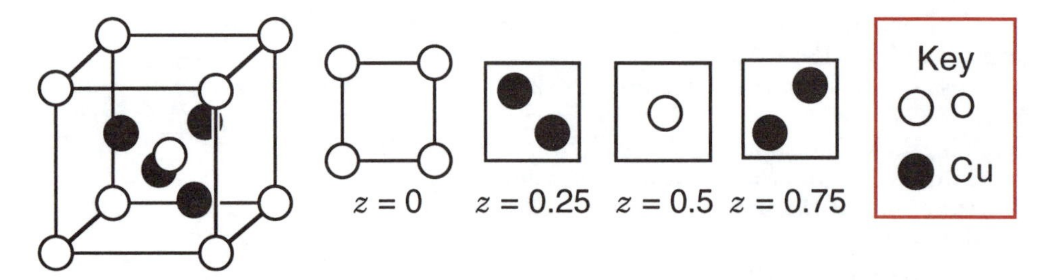

c. *Cu_2AlMn*

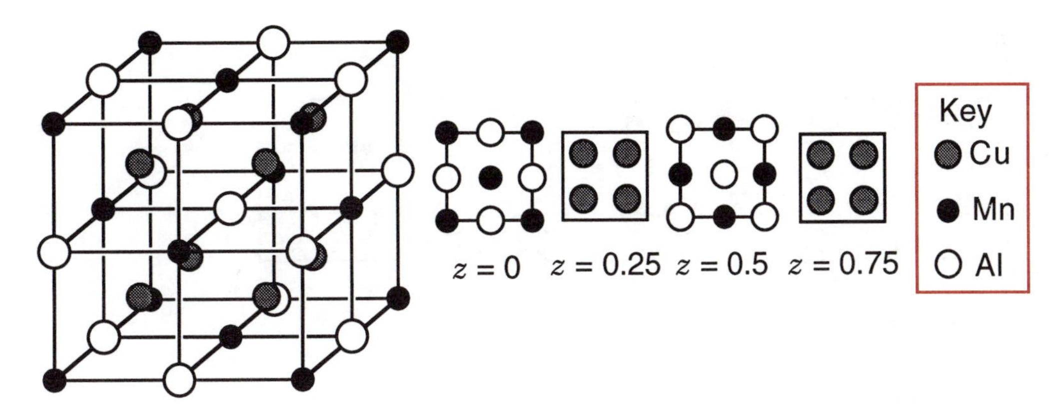

d. *Rutile* (TiO_2). This compound is the main pigment in white paint. Other compounds with this structure include the difluorides of Co, Fe, Mg, Mn, Ni, Pd, and Zn; and GeO_2, IrO_2, RuO_2, and SnO_2.

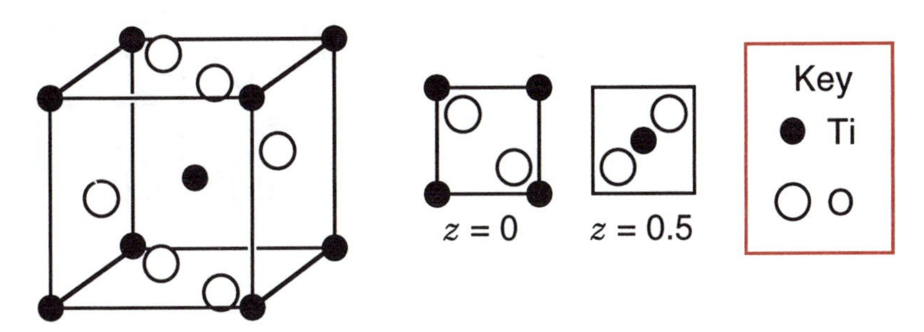

e. *Cu_3Au*

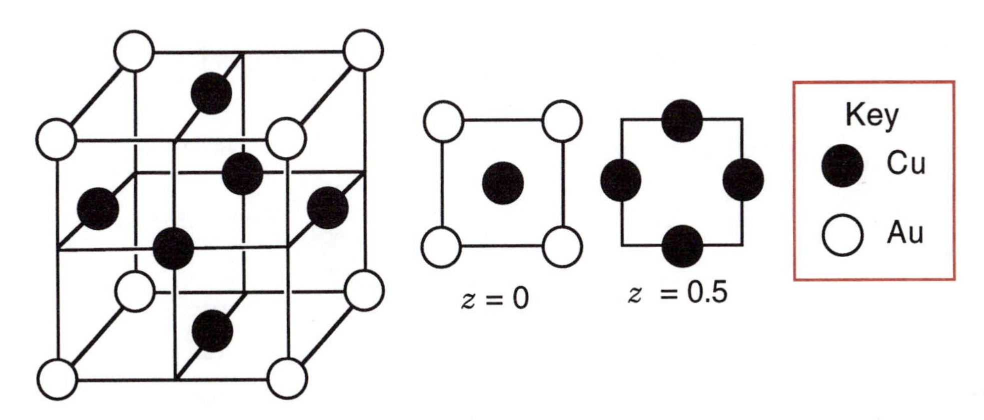

f. *Cadmium iodide* (CdI_2).

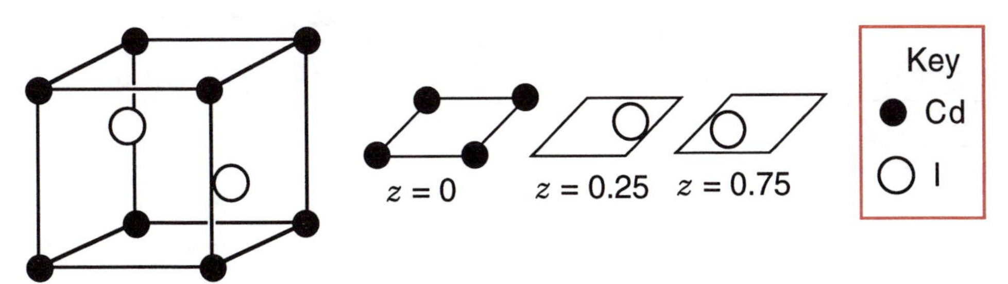

g. *ReO*$_3$. Other compounds with this structure include MoF_3 and NbF_3.

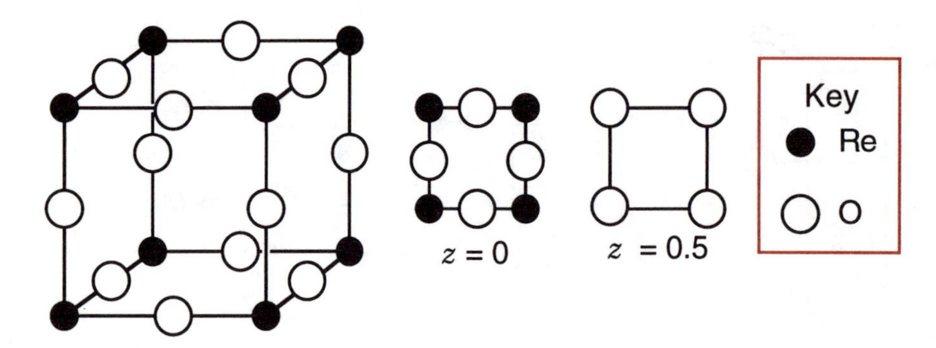

h. *ZnO* is a semiconductor. It is used in the production of rubber to shorten vulcanization times, as a paint pigment, and as an agent to improve the chemical durability of glasses. This structure is also adopted by AlN and BeO.

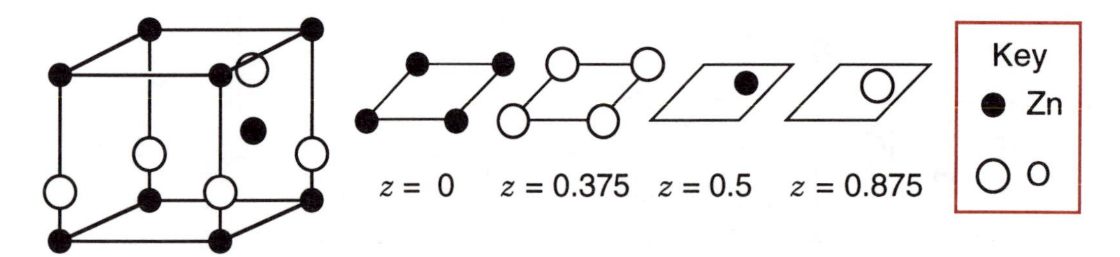

i. *Nickel arsenide* (NiAs).

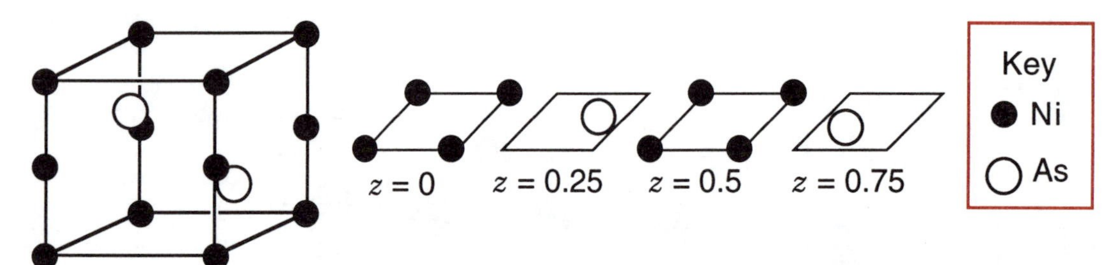

j. *MgAgAs*

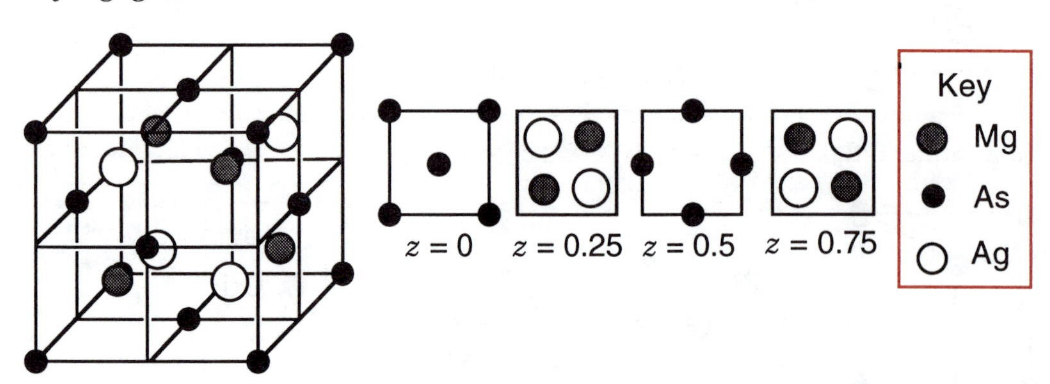

k. *Zinc blende* (ZnS). This mineral is the main source of the element zinc. Many of the important semiconductors have this structure (namely GaP, GaAs, GaSb, InP, and InSb). These semiconductors have also been used as infrared detectors, photoresistors, photocells, and transistors. Fluorescent lamps make use of the highly luminescent ZnS and CdS that are doped with Cu^{2+} and Mn^{2+}, for example. Other compounds that crystallize with the ZnS structure are ZnS, ZnSe, BeS, BeSe, BeTe, HgS, HgSe, HgTe AlP, GaP, InP, SiC, AgI, CuBr, and CuCl.

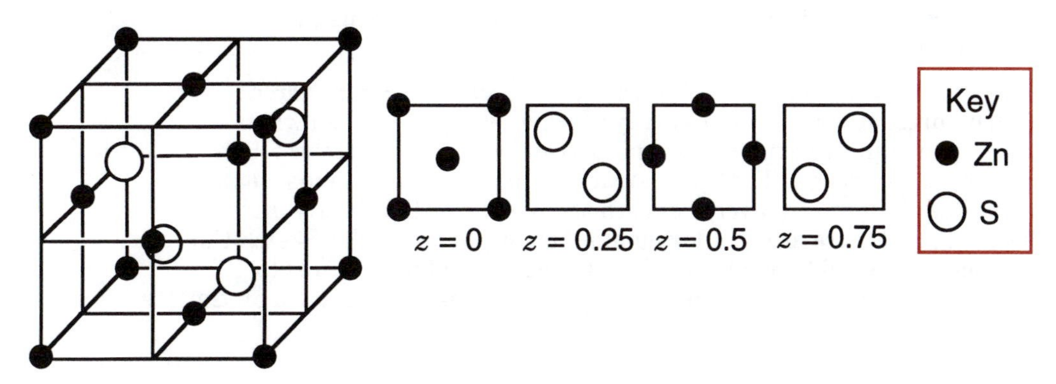

l. *Calcium carbide* (CaC_2). This compound has an extremely high melting point (2300 °C) and reacts with water to liberate acetylene. BaC_2, MgC_2, SrC_2, and UC_2 adopt this structure, as do the peroxides BaO_2, CaO_2, and SrO_2.

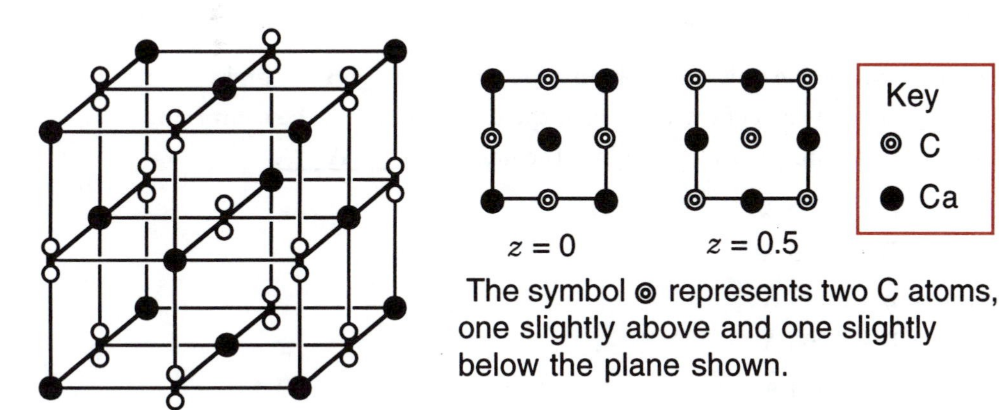

The symbol ◎ represents two C atoms, one slightly above and one slightly below the plane shown.

2. What is the formula of the compound that crystallizes in a cubic unit cell with the following (a–e)?

   What is the percent composition of the compound that crystallizes in a cubic unit cell with (a–e)?

   What mass of ions is required to prepare 5.00 g of the compound that crystallizes in a cubic unit cell with (a–e)?

   a. a rubidium ion on each of the corners of the unit cell and a bromide ion in the center of the unit cell? (Other compounds that crystallize with the CsCl structure: CsCl, CsBr, CsI, TlCl, TlBr, TlI, and many intermetallic compounds such as AgCd, AgZn, AuZn, and HgMg.)

   b. nickel ions at each of the corners and the middle of each face and oxide ions in the middle of the edges and the center of the unit cell? (Other compounds that crystallize with the NaCl structure: AgCl, LiF, LiCl, LiBr, LiI, NaF, NaBr, NaI, KCl, KBr, KI, NaH, KH, RbH, MgO, CaO, SrO, CoO, NiO, CaS, CaSe, CaTe, MnS, MnSe, SnSe, SnTe, TiC, VC, ScN, TiN, and VN.)

c. sulfide ions at each of the corners and the middle of each face and four cadmium ions within the unit cell? (Other compounds that crystallize with the zinc blende structure: ZnS, ZnSe, BeS, BeSe, BeTe, CdS, CdSe, HgS, HgSe, HgTe AlP, GaP, InP, SiC, AgI, CuBr, and CuCl.)

d. a titanium ion at each of the corners of the unit cell, oxide ions in the middle of each edge, and a strontium ion in the center of the cell? (Other compounds that crystallize with the idealized perovskite structure: $BaTiO_3$, $CsCaF_3$, $CsCdBr_3$, and $CaVO_3$.

e. a rhenium ion at each of the corners of the unit cell and oxide ions in the middle of each edge?

3. For all of the simple cubic (sc), body-centered cubic (bcc), and face-centered cubic (fcc) structures, at least a square array of spheres would be seen when looking down upon a unit cell cube face, as illustrated here using circles for the atoms in the three structures. For the sc structure, no additional atoms would be added, because all atoms in layers beneath the first layer lie directly under the atoms shown, and there are no additional atoms in the unit cell. Sketch the additional atoms that would have to be added to the bcc and fcc pictures to make them accurate representations of what you would see looking down on the structure from above.

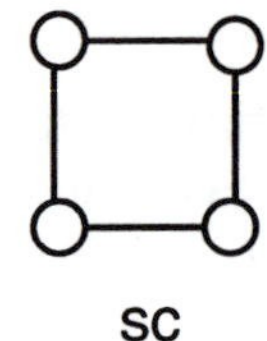

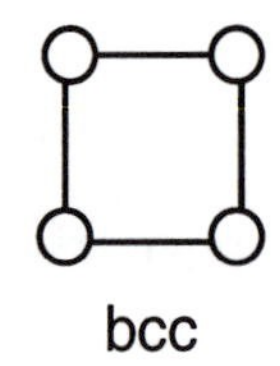

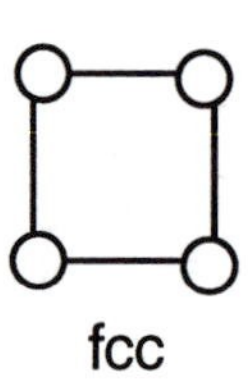

4. ___A___ crystallizes in a cubic unit cell with an edge length of ___B___ and with a(n) ___A___ atom on each of the corners of the unit cell and in the center of the unit cell. Using this information calculate the density of ___A___.

A = barium	B = 5.025 Å
A = lithium	B = 3.509 Å
A = molybdenum	B = 3.147 Å
A = tantalum	B = 3.306 Å
A = vanadium	B = 3.024 Å
A = tungsten	B = 3.165 Å

5. Aluminum crystallizes in a cubic unit cell with an edge length of 4.050 Å and with an aluminum atom on each of the corners of the unit cell and in the center of each of the faces. Using this information calculate the density of aluminum.

6. Given the unit cell length of a cubic compound, its density, and the contents of the unit cell and their atomic weights, determine Avogadro's number. For example, determine the value of Avogadro's number from the atomic weights of calcium and fluorine and the following information. The unit cell length of CaF_2 is 5.46295 Å. The density of CaF_2 is 3.1805 g/cm^3. CaF_2 crystallizes with calcium ions at each of the corners and the middle of each edge of the unit cell and with eight fluoride ions within the unit cell.

7. In this chapter it was stated that more than one unit cell can be chosen for a structure. If they represent the same structure, however, the unit cells must have the same types of atoms (with the same stoichiometric relationships between them) and the same coordination geometry around each type of atom. The following are two possible representations of CsCl. Do they represent the same CsCl structure?

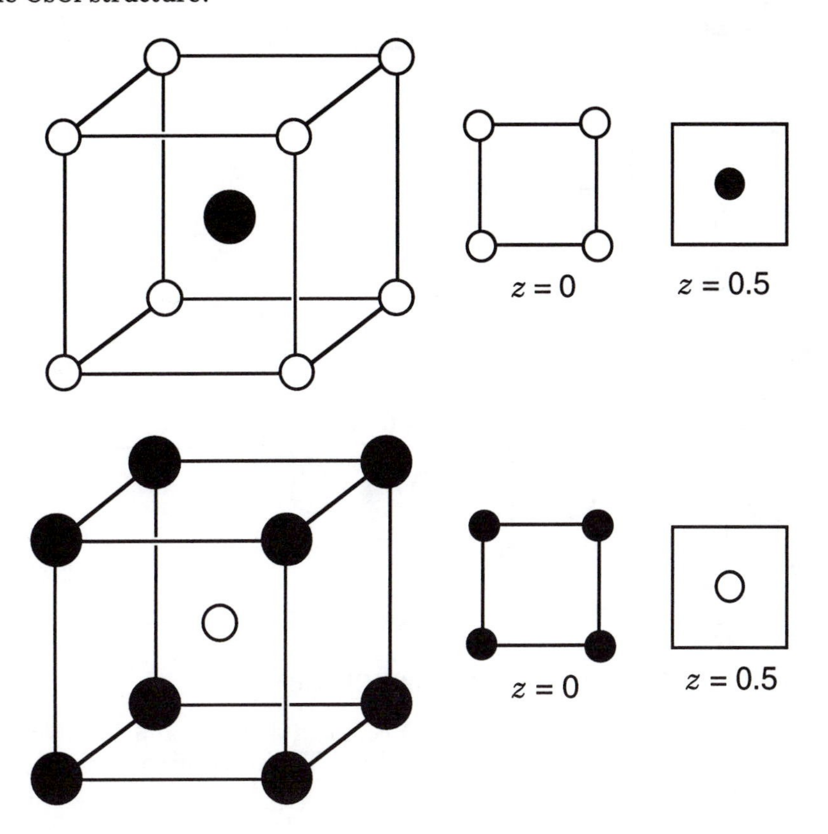

8. What is the maximum and minimum percent of Ga in the Ga(P, As) solid-solution system?

Other Solid Solutions
K (Cl, Br)
Biotite [$K(Mg,Fe^{II})_3(OH,F)_2(Si_3AlO_{10})$]
Ga(P, As)
Sphalerite is tunable from ZnS to $Zn_{0.68}Fe_{0.32}S$
$Y_{3-x}Gd_xFe_5O_{12}$
bcc: V_xMo_{1-x}, W_xTa_{1-x}, W_xNb_{1-x}
hcp: Zn_xRe_{1-x}, Zn_xRu_{1-x}, Ru_xRe_{1-x}, Ru_xOs_{1-x}, Os_xRe_{1-x}
(hcp is hexagonal close-packed, see Chapter 5.)
fcc: Rh_xPt_{1-x}, Pd_xAu_{1-x}, Pd_xAg_{1-x}

9. The following are two representations that have been used for the perovskite structure. By the criteria given in problem 7, do they represent identical structures?

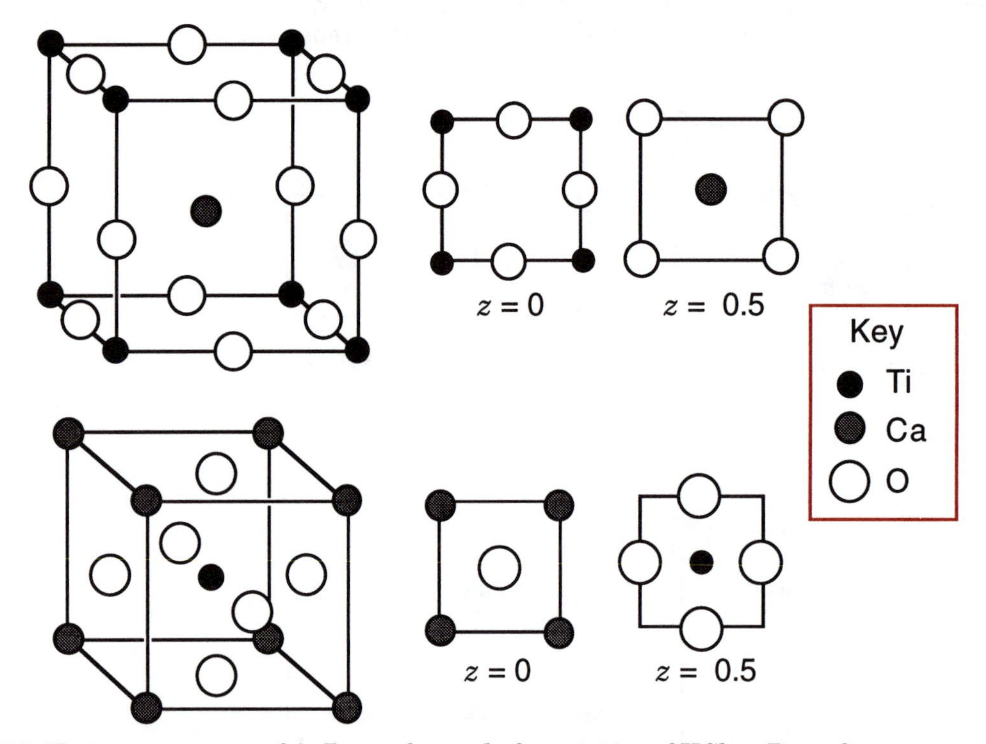

10. How many grams of AgBr can be made from 1.00 g of $KCl_{0.30}Br_{0.70}$?

11. The iron oxides can exhibit variable stoichiometry. A compound of iron and oxygen forms a cubic unit cell of edge length 4.29×10^{-8} cm. The crystalline compound has a density of 5.66 g/cm^3. Calculate:
   a. the mass of one unit cell
   b. the average number of ions of iron and oxygen in a unit cell if the atomic ratio of iron to oxygen is 0.932:1.000
   c. The percentage of occupied sites of each type of ion if FeO crystallizes in the NaCl structure. (Hint: NaCl contains four ions of each type in the unit cell.)

12. Titanium oxides are known to exhibit variable stoichiometry. A titanium oxide (TiO_x) is analyzed and found to be 70.90% Ti by mass. What is the value of x in the formula for the compound?

13. What mass of ___A___ must be processed to isolate ___B___ of ___C___?

A = $Cd_{0.010}Zn_{0.99}S$	B = 1.000 metric ton (1000 kg)	C = Cd
A = $UO_{2.10}$	B = 1.000 metric ton (1000 kg)	C = U
A = $Cu_{1.79}O$	B = 1.00 g	C = Cu

14. Calculate weight percentages of the elements in the following solid solutions:

   % Cd, % S, % Se in $CdS_{0.29}Se_{0.71}$

   % Na, % W, % O in $Na_{0.24}WO_3$

15. Determine the values of x and y in $Ga_{0.46}Al_xP_{0.37}As_y$.

16. What mass of sodium fluoride, the active ingredient in some fluoride toothpastes, is required to convert 1.00 g of $Ca_5(PO_4)_3OH$ to $Ca_5(PO_4)_3(OH)_xF_{1-x}$? (Choose a value for x between 0 and 1.)

$$Ca_5(PO_4)_3OH + (1-x)NaF \rightarrow Ca_5(PO_4)_3(OH)_xF_{1-x} + (1-x)NaOH$$

17. What is the total mass of oxide formed when 0.5356 g of Si_xGe_{1-x} is converted into a mixture of SiO_2 and GeO_2? (Choose a value for x between 0 and 1.)

18. Some of the calcium dihydrogen phosphate used in fertilizers is prepared from the reaction of $Ca_5(PO_4)_3(OH)_xF_{1-x}$ with sulfuric acid. The HF produced in the reaction is recovered and used in the production of other compounds.

$$2\ Ca_5(PO_4)_3(OH)_xF_{1-x} + 7\ H_2SO_4 \rightarrow$$
$$3\ Ca(H_2PO_4)_2 + 7\ CaSO_4 + 2x\ H_2O + (2-2x)HF$$

What mass of HF is produced by the reaction of 1.000 metric tons of $Ca_5(PO_4)_3(OH)_xF_{1-x}$? (Choose a value for x between 0 and 1.)

19. What mass of cadmium chloride is formed when 0.5356 g of CdS_xSe_{1-x} is converted into $CdCl_2$? (Choose a value for x between 0 and 1.)

$$CdS_xSe_{1-x} + 2\ HCl \rightarrow CdCl_2 + x\ H_2S(g) + (1-x)H_2Se$$

20. What volume of dry H_2S at STP can be formed when 0.5356 g of CdS_xSe_{1-x} is converted into $CdCl_2$? (Choose a value for x between 0 and 1.)

$$CdS_xSe_{1-x} + 2\ HCl \rightarrow CdCl_2 + x\ H_2S(g) + (1-x)H_2Se$$

21. Which contains more hydrogen, by mass, in one cubic centimeter, lithium hydride or liquid hydrogen? Liquid hydrogen has a density of 0.070 g/cm^3. Lithium hydride is a solid that crystallizes with a cubic unit cell (a = 4.085 A) with lithium ions at each of the corners and the middle of each face and hydride ions in the middle of the edges and the center of the unit cell.

22. When the chlorides of rubidium and potassium are mixed and heated to about 800 °C, the following reaction will occur:

$$KCl + RbCl \rightarrow 2K_{0.5}Rb_{0.5}Cl$$

How many grams of each starting material should be mixed in order to prepare 10.0 g of the final product?

23. Classify the following substances as discrete molecules, alloys, or salts.
   a. TiO_2
   b. C_5H_{12}
   c. NaK
   d. ClF_3
   e. $SrBr_2$

24. Which of the following combinations of atoms do you expect to yield a discrete molecule and which should produce an extended solid?
   a. Ba and O
   b. N and O
   c. Cu and Ni
   d. Na and F

25. The table that follows lists the crystal structure and atomic radius of several metallic elements. Select pairs of atoms that are likeliest to form a complete family of substitutional solid solutions using the indicated structures and justify your answer.

Element	Structure	Atomic radius, Å
Cu	fcc	1.28
Mg	hcp	1.60
Mo	bcc	1.36
Ni	fcc	1.25
V	bcc	1.31

26. In the pattern, sketch and label unit cells that
a. contain one P and one Q in a single unit cell
b. contain two Ps and two Qs in a single unit cell.

P Q P Q P Q P Q P Q P
Q P Q P Q P Q P Q P Q
P Q P Q P Q P Q P Q P
Q P Q P Q P Q P Q P Q

27. The following is a simple cubic (sc) unit cell. In the sc structure, a given sphere, labeled X, has six nearest neighbors, three of which are shown and labeled A. The next nearest neighbor spheres to X are across the diagonal of a face of the cube (a face diagonal). One such sphere is shown and labeled B.
a. How many such spheres are there at this same distance from sphere X?
b. The next nearest neighbor to sphere X is labeled C and lies across the so-called body diagonal of the cube. How many such spheres are there at this same distance from sphere X?

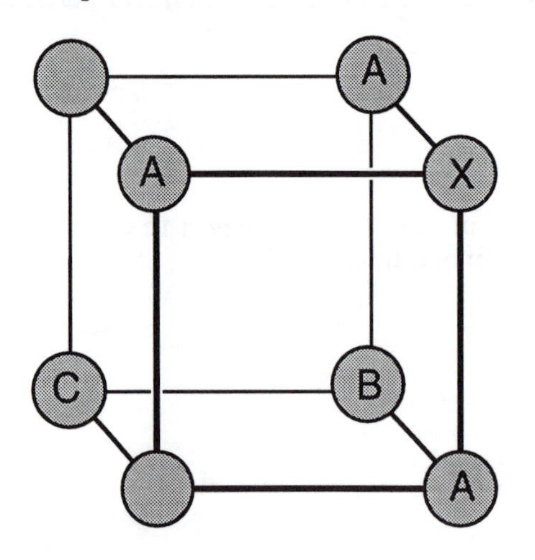

28. In the body-centered cubic structure, the sphere in the center of the cube has eight nearest neighbors, which lie at the corners of the cube. The next nearest neighbor spheres to the sphere in the center of the cube are at the centers of adjacent cubic unit cells. How many such spheres are there?

29. Which of the following combinations of atoms is most likely to form a salt (an ionic compound)?
a. Ba, Ne
b. Cs, Cl
c. S, N
d. Ti, Hf

30. Because of the similar size and chemistry of As and P and the fact that GaAs and GaP have the same kind of crystal structure, it is possible to form solid solutions of tunable stoichiometry with the composition of $GaAs_{1-x}P_x$, where x takes on any value between 0 and 1, that is, $0 \le x \le 1$. What is the formula weight in grams per mole for such a solid when $x = 0.25$?

31. The formula weight in grams per mole for the substitutional solid solution $CdS_{0.10}Se_{0.90}$ is which of the following?
a. 144
b. 149
c. 187
d. 191

32. Is a solid solution like $Cu_{0.5}Ni_{0.5}$ distinguishable from a 1:1 equimolar mixture of Cu and Ni by elemental analysis? How might a magnet distinguish them?

Chapter 4

Determination of Structure Using Diffraction Data

Underpinning our understanding of the chemical and physical properties of solids is our knowledge of the structures they adopt. In this chapter, we will discuss how diffraction data is used to determine structures of crystalline materials. The same diffraction experiments that provide our window on the structures of extended solids also permit the elucidation of molecular structures, ranging from the simple (triatomic molecules like carbon dioxide) to the complex (biopolymers like proteins).

Diffraction

Atomic dimensions are on the order of angstroms (10^{-8} cm), and therefore unraveling the atomic positions of a solid requires a physical technique that operates on a similar spatial scale. The scanning tunneling microscope (STM) described in Chapters 1 and 2 provides direct images on this spatial scale but is limited to probing certain kinds of surfaces. Historically, X-ray diffraction has been the technique that has provided most of our information on the molecular-level structures of bulk solids; electron and neutron diffraction are also important techniques that obey similar physical laws *(1, 2)*. These diffraction techniques do not provide the direct atomic images accessible with the STM, but rather patterns from which atomic positions must be calculated using the physical principles of diffraction.

The role of X-rays in diffraction experiments is based on the electromagnetic properties of this form of radiation. Electromagnetic radiation is associated with electric and magnetic fields whose time-

2725–X/93/0077$06.00/1

varying magnitudes describe sine waves that are oriented perpendicular to each other and to the propagation direction. Electromagnetic radiation can thus be regarded as a wave associated with a wavelength λ and a frequency ν that obey the relationship $c = \lambda\nu$, where c is the speed of light (~3×10^{10} cm/s in a vacuum). The wavelengths of X-rays are on the order of angstroms, or about the same length as interatomic spacings.

Maxwell von Laue recognized at the beginning of this century that X-rays would be scattered by atoms in a crystalline solid if there were a similarity in spatial scales *(1, 2)*. As illustrated in the top portion of Figure 4.1, this property leads to a pattern of scattered X-rays (the diffraction pattern) that can be mathematically related to the structural arrangement of atoms causing the scattering.

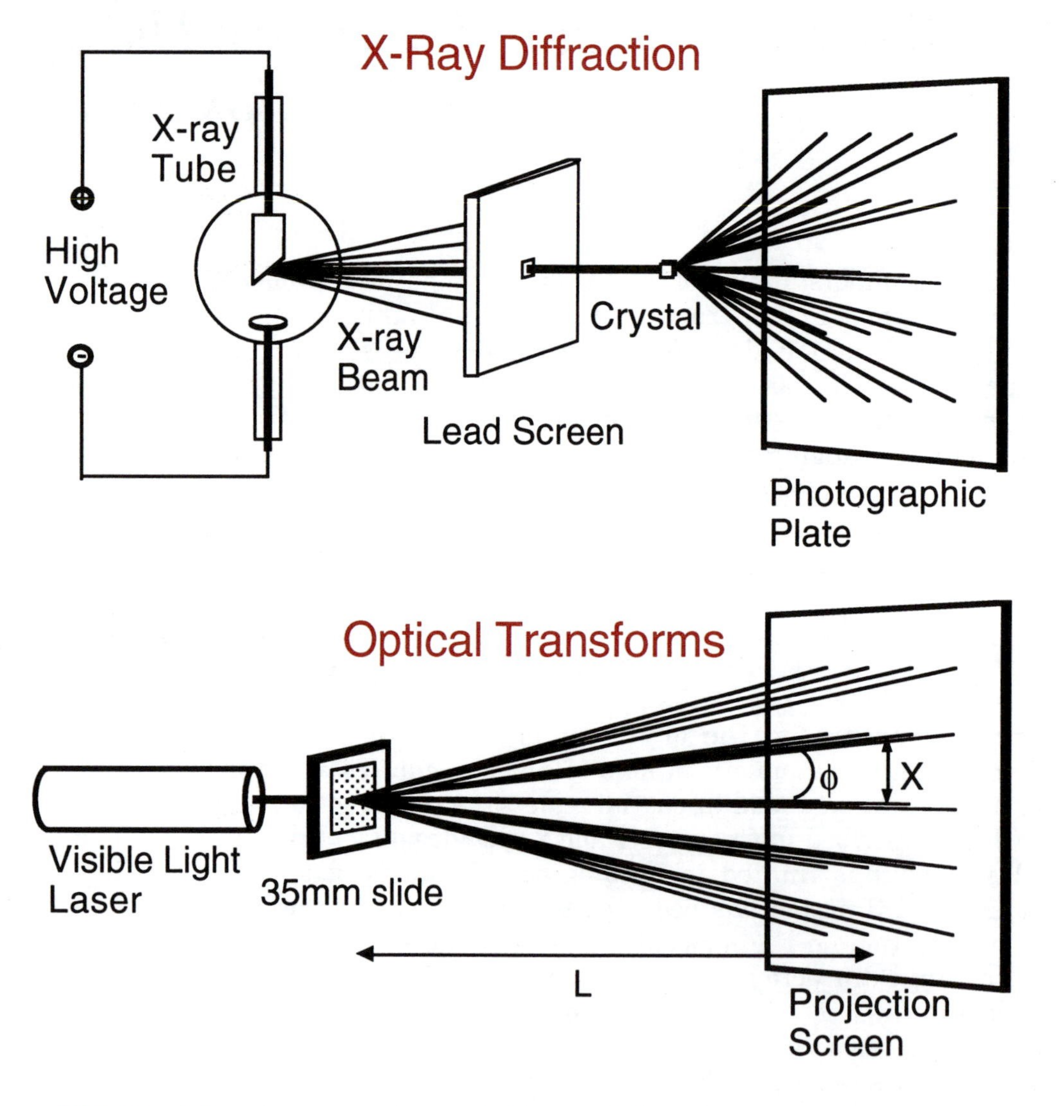

Figure 4.1. A comparison of X-ray diffraction and the optical transform.

Some time later, Sir Lawrence Bragg recognized that the experiment could be simulated inexpensively and safely by an increase in scale. He suggested the use of visible light with wavelengths thousands of times larger than those of X-rays; and arrays of dots or pinholes mimicking atomic arrangements on a scale magnified by thousands from interatomic spacings. The experiment has benefited from the technological advances of the past half-century: laser printers and lasers have superseded Bragg's heroic use of pinhole-derived patterns and multicomponent optical assemblies to produce coherent light. This kind of experiment, which yields diffraction patterns similar to X-ray diffraction patterns, is illustrated in the bottom portion of Figure 4.1, and is called an optical transform experiment. It is easily brought into the classroom or laboratory, as will be described.

The diffraction experiments shown schematically in Figure 4.1 reflect the fact that when electromagnetic radiation from several sources overlaps in space simultaneously, the individual waves add (the principle of linear superposition). The limiting cases are sketched in Figure 4.2 for two waves that are identical in amplitude, wavelength, and frequency. If the waves are in phase, reaching maximum amplitude at the same time, they reinforce one another, a condition known as constructive interference. Conversely, if the waves are completely out of phase (separated by half a wavelength, $\lambda/2$), with one at maximum amplitude while the other is at minimum amplitude, they "annihilate" one another or sum to zero, a condition known as destructive interference.

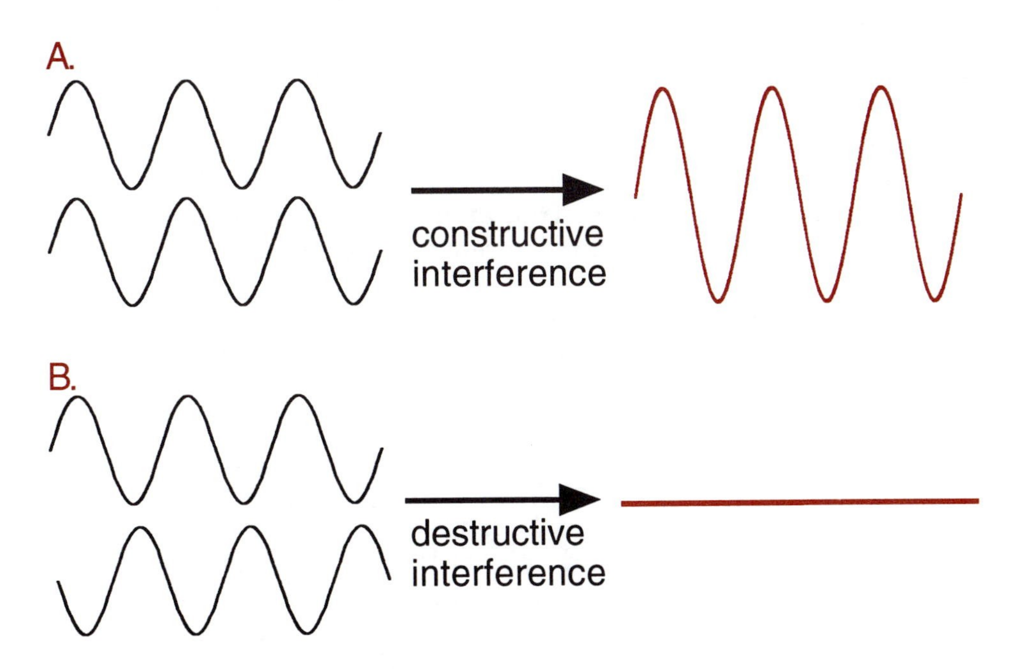

Figure 4.2. Electromagnetic radiation overlapping in space.

Exposing crystalline solids to X-rays yields diffraction patterns because the atoms in the crystal scatter the incoming radiation, and interference occurs among the many resulting waves. For most directions of observation, destructive interference will occur, but in specific directions constructive interference will be found. Electron and neutron beams also have a wavelike nature (de Broglie wavelength; $\lambda = h/mv$, where h is Planck's constant and m and v are the mass and velocity of the particle, respectively), and thus these beams lead to similar diffraction effects.

One condition for constructive interference, the Bragg diffraction condition, is illustrated in Figure 4.3 and is often described in terms of reflection from parallel planes of atoms because the angle of incidence equals the angle of diffraction. As the figure shows, if the atomic planes (represented by the rows of dots) are separated by a distance d, then X-rays reflected at a given angle θ relative to the plane will arrive at a detector in phase, if the additional distance traveled by the lower light ray relative to the upper light ray is an integral number of wavelengths, $n\lambda$ ($n = 1, 2, 3 ...$; $n = 1$ is called first-order diffraction, $n = 2$ is second-order diffraction, etc.). From the trigonometry indicated in Figure 4.3, this extra path length traversed by the lower light ray is $2(d \sin \theta)$, and leads to Bragg's law, $2(d \sin \theta) = n \lambda$.

Diffraction patterns thus constitute evidence for the periodically repeating arrangement of atoms in crystals. Their overall symmetry corresponds to the symmetry of the atomic packing, and the use of a single wavelength (monochromatic light) of X-ray radiation directed at the solid permits the simplest determination of interatomic distances. From the Bragg equation, it can be seen that the angle of diffraction θ is related to the interplanar spacing, d. This interplanar spacing may equal the distance between two parallel faces of a unit cell, which is the length of one edge of that unit cell. Consequently, by changing the orientation of the crystal in the X-ray beam, the lengths of all three of the edges and angles of the unit cell parallelepiped (*see* Chapter 3) can be determined. Furthermore, the intensity of the diffracted beams depends on the arrangement and atomic number of the atoms in the unit cell. Thus, the intensities of diffracted spots calculated for trial atomic positions can be compared with the experimental diffraction intensities to obtain the positions of the atoms themselves.

Bragg's version of the diffraction experiment, the optical transform experiment based on visible light, is an example of a more general phenomenon called Fraunhofer diffraction. As shown in Figure 4.3, if the light transmitted through an array of scattering centers is viewed at what is effectively infinite distance, the condition for constructive interference is $d \sin \phi = n \lambda$, where the spacing (d) between atoms and scattering angle ϕ are defined in the figure; the scattered rays in the figure are in phase if the lower ray travels an additional distance ($d \sin \phi$) that is an integral number of wavelengths λ. Mathematically, the equations for Fraunhofer and Bragg diffraction have a similar functional dependence on d, λ, and the scattering angle (in Bragg diffraction, the angle between the incident and diffracted beams is 2θ).

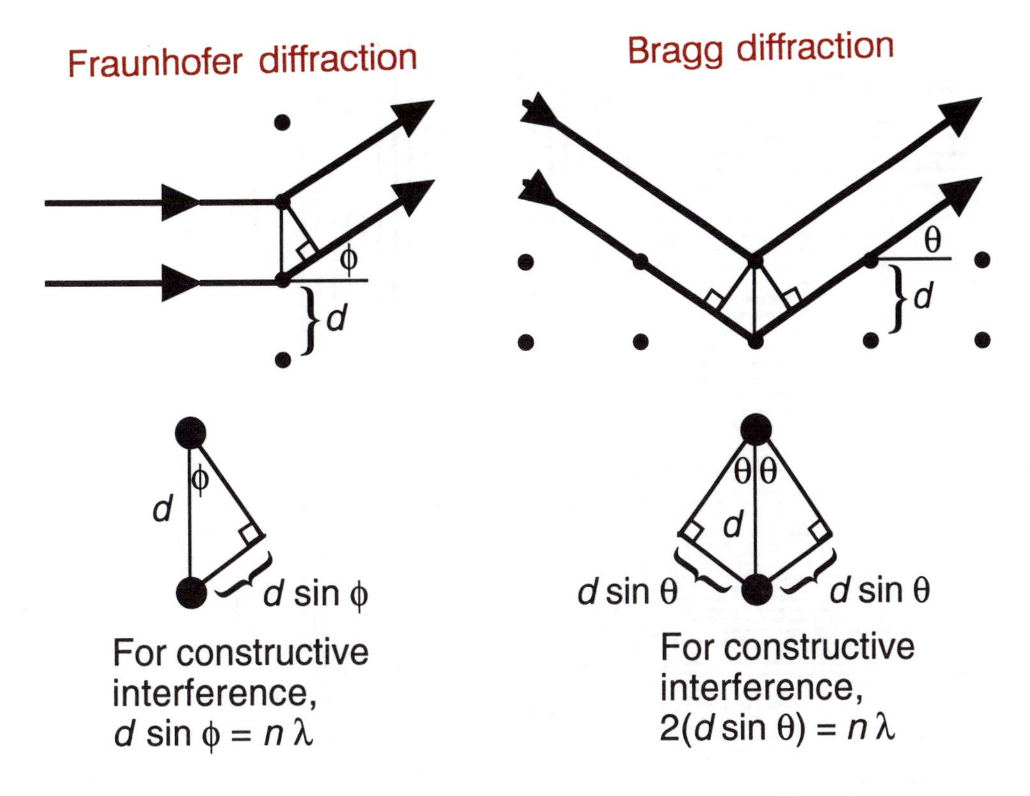

Figure 4.3. A comparison of Fraunhofer and Bragg diffraction.

Optical Transforms

The Institute for Chemical Education (ICE) makes available a set of inexpensive optical transform slides suitable for illustrating diffraction experiments, as part of an Optical Transform Kit (*see* Supplier Information for ordering information). These slides contain patterns that have been created with a laser printer (in registry with one another to permit direct comparisons of different patterns) and then photographically reduced. One such slide, a Discovery slide, contains the arrays shown in Figure 4.4. This introductory slide contains eight panels consisting of vertical lines and horizontal lines with two different spacings, and four dot patterns with the dots arranged in squares of increasing size. Figure 4.5 presents the resulting diffraction patterns obtained from each panel.

Patterns a and c of Figure 4.4 contain horizontal lines at larger and smaller separations, respectively. Each of these patterns yields a diffraction pattern of vertical spots, and by moving between these patterns (shifting the pattern relative to the light source), the viewer will note that a larger spacing between the rows on the slide results in a smaller spacing between the vertical spots in the diffraction pattern and vice versa (Figure 4.5, panels a and c).

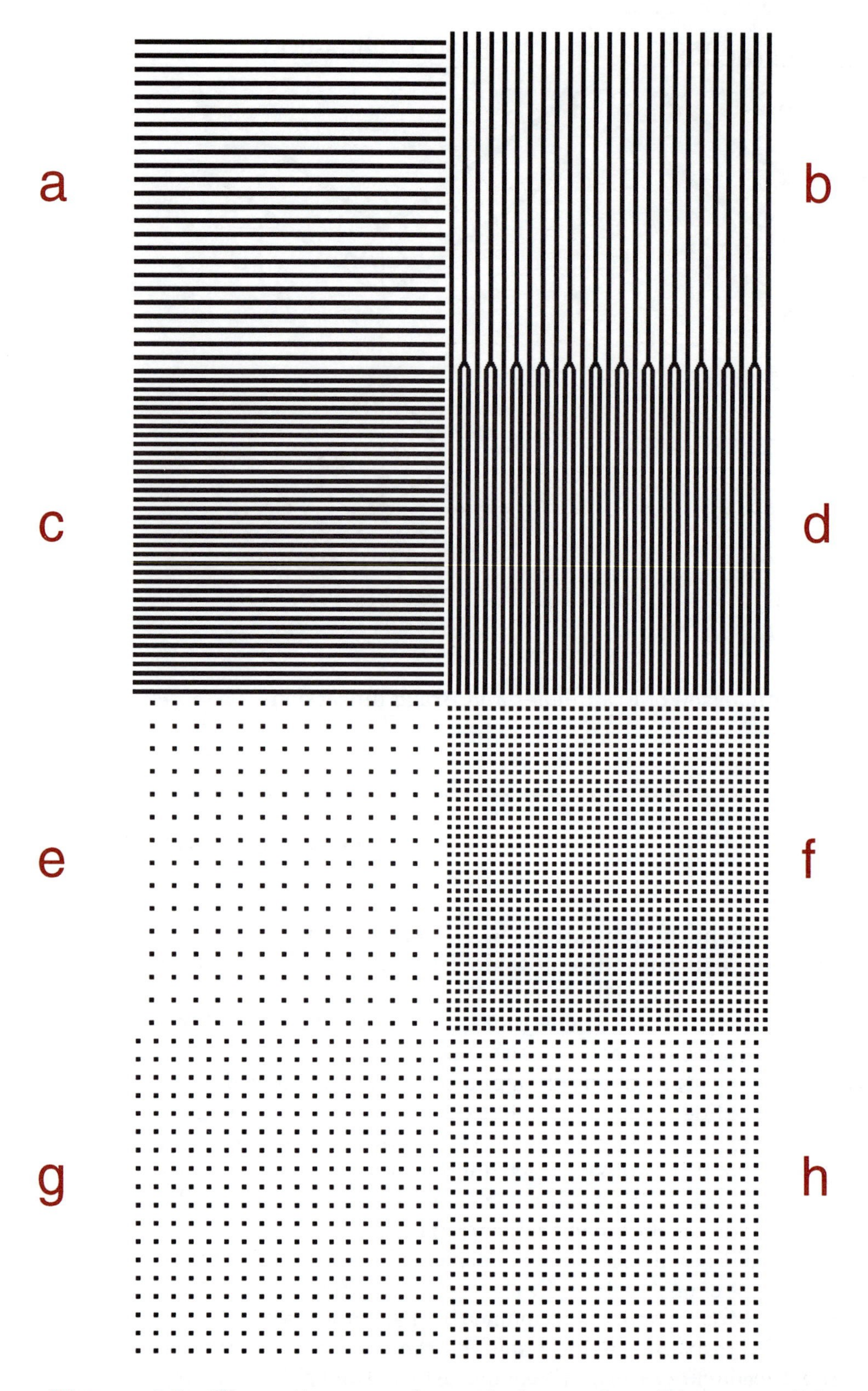

Figure 4.4. The patterns on the optical transform Discovery slide, available from ICE, oriented so that the ICE logo is on the right side as you look through the slide.

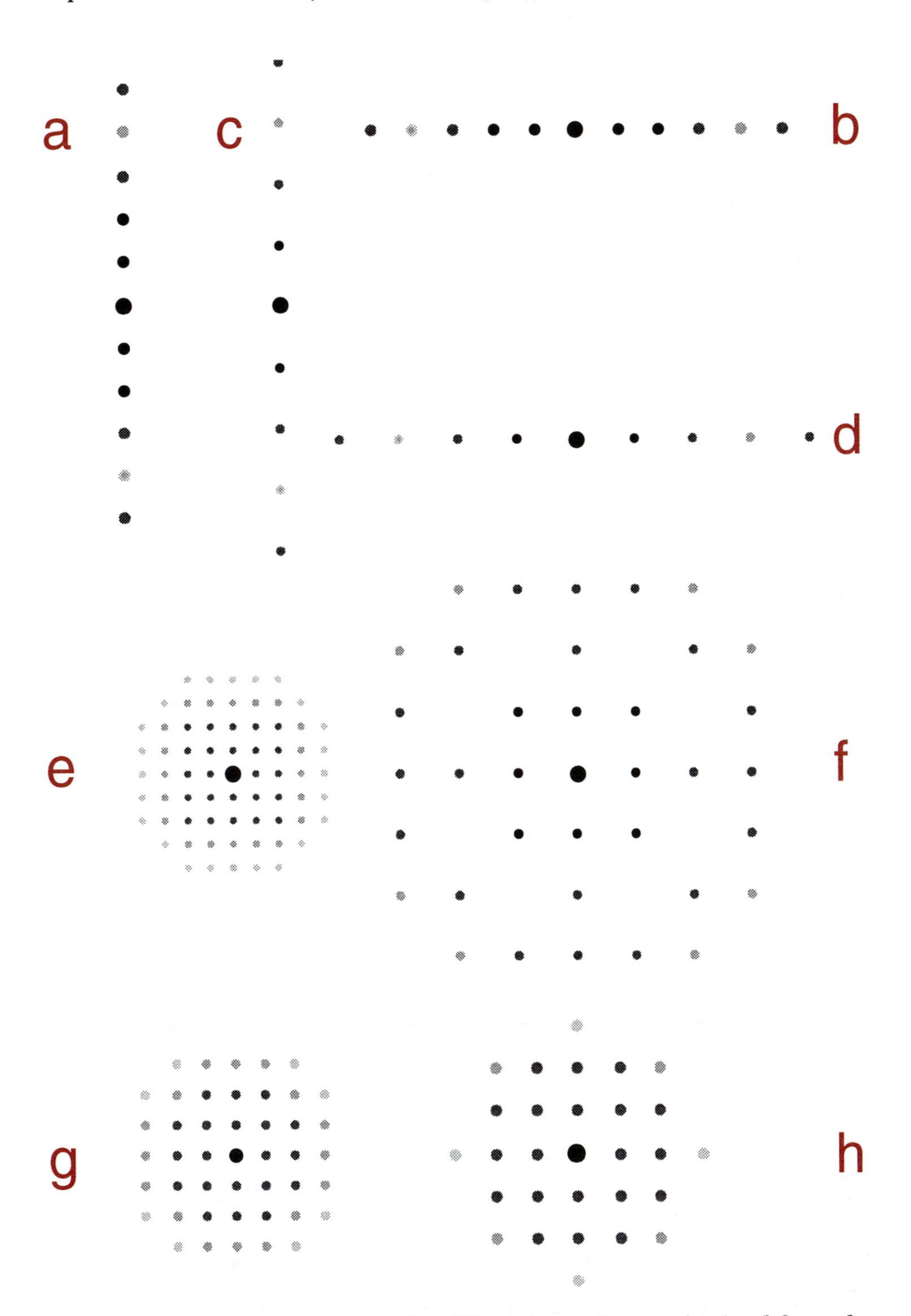

Figure 4.5. A simulation of the diffraction patterns obtained from the Discovery slide, with the indicated letter corresponding to the array (Figure 4.4) from which the diffraction pattern is produced. (The number and intensity of spots observed will vary with the intensity of light used and the darkness of the room in which the patterns are observed.)

Similarly, patterns b and d of Figure 4.4 comprise vertical lines at large and small separations, respectively, leading to a diffraction pattern of horizontal spots whose spacing varies inversely with the separation of the vertical lines on the mask (Figure 4.5, panels b and d). To lend chemical relevance to these observations, the rows and columns on the masks could be regarded as oriented polymer chains.[1]

Patterns e–h of Figure 4.4 can be viewed in rapid, counterclockwise succession, beginning with panel e (e, g, h, f), to demonstrate that as a square array of dots becomes smaller, a square diffraction pattern is also seen, whose size moves in the opposite direction, becoming largest for pattern f. Chemically, these patterns can be related to a plane of atoms in the cubic face of a simple cubic or body-centered cubic structure (Chapter 3).

The general observation illustrated by the Discovery slide is that a line of diffraction spots is seen in a direction perpendicular to and inversely spaced relative to the original rows or columns of patterns on the slide. The size of the overall diffraction pattern increases with the wavelength of light used to generate it. These findings can be amplified using other patterned 35-mm slides, as will be described later.

---

[1]The optical transform pattern produced is perpendicular to the lines in the scattering pattern. This result can be explained in the following way: Consider a pattern of horizontal lines such as those in Figure 4.4a. In the vertical direction, the repeat distance d is the separation between any two lines. Using monochromatic light, constructive interference will result when the product of the sine of the diffraction angle ϕ times d corresponds to a multiple (n) of the wavelength, according to the diffraction equations in Figure 4.3. Because there is a finite repeat distance in the vertical direction, there will be constructive and destructive interference in this direction.

Again considering pattern 4.4a, the repeat distance is infinite in the horizontal direction. This condition requires that $\sin \phi$ and ϕ must be zero. Consequently, n must equal zero, meaning that there will be no constructive interference in the horizontal direction.

In summary, horizontal lines will result in interference in the vertical direction, and vertical lines will result in a horizontal interference pattern. Furthermore, as d becomes larger, $\sin \phi$ (and consequently ϕ) must become smaller. Thus, the spacings in the transform pattern will be smaller as the repeat distance gets larger. These relationships are closely related to the notion of the reciprocal lattice in crystallography.

Demonstration 4.1. The Optical Transform Slide

Materials

Discovery slide or other optical transform slide*

Laser: A battery-powered, 5 mW, 670 nm pocket diode laser, sold as a pointer, is an effective light source and available from a number of suppliers.* For a large, darkened lecture hall, either a 10 mW, 633 nm He–Ne laser (red light) or a 0.2 mW, 544 nm He–Ne laser (green light, to which the eye is more sensitive) is effective. Both lasers are available from Melles Griot.*

Overhead transparency of the pattern on the slide (such as Figure 4.4)

Overhead projector and screen

Procedure

- In a darkened room, shine the laser light through the slide so that the diffraction pattern is displayed on a screen at a distance of several meters from the source. The laser should be clamped or taped in a fixed position. **CAUTION: Take care to ensure that the viewers will not inadvertently look directly at the beam where it emerges from the laser. Clamp or tape the laser in a fixed position so that the beam will not accidentally shift. Aiming a laser at a person or looking directly at a source of laser light can be harmful to the eye!**
- Each slide contains eight patterns. Move the slide so that it intercepts the beam in different places. This procedure will allow the patterns to be displayed in order according to preference. Simultaneous display of the overhead transparency containing the enlarged image pattern will allow the pattern of diffraction spots to be discussed in light of the array of dots that generated them. Some comparisons that can be made:
- Shine the laser through patterns a and c of Figure 4.4. This will show the effect of varying the spacings of horizontal lines on the diffraction pattern.
- Shine the laser through patterns b and d of Figure 4.4. This will show the effect of varying the spacings of vertical lines on the diffraction pattern.
- Shine the laser sequentially through patterns e, g, h, and f of Figure 4.4. This will show the effect of changing the size of a square unit cell of dots on the diffraction pattern.

**See* Supplier Information.

Demonstration 4.2. Individualized Use of Optical Transform Slides

Materials

Optical transform slides, one for each person in the class (ICE sells the optical transform slides in bulk. *See* Supplier Information.)
Laser (*see* Demonstration 4.1)
Point source of white light (a small flashlight bulb or miniMaglite—a small flashlight that is available at camping or hardware stores—with the lens unscrewed)
Magnifying glass or hand-held lens (a 30× Micronta Microscope from Radio Shack works well) (optional)

Procedure

- Make a 35 mm optical transform slide available to each class member. In addition to having the instructor provide an enlarged view of the slide's contents, a small magnifying glass, hand-held lens, or microscope can be used to verify the patterns on the slide.
- Have the students visually orient their slides by noting the relative transparency of different regions (patterns c and d of Figure 4.4, with the densest packing of lines, will appear darkest, for example).
- Project a laser beam on the screen (without passing it through a slide). **CAUTION: Take care to ensure that the viewers will not inadvertently look directly at the beam where it emerges from the laser. Clamp or tape the laser in a fixed position so that the beam will not accidentally shift. Aiming a laser at a person or looking directly at a source of laser light can be harmful to the eye!**
- Have the students view the laser spot through their own 35-mm slides at a distance of at least a few meters from the screen. The diffraction pattern is observed and can be changed at the viewers' convenience as they view the beam through different portions of the slide. Thus, the same diffraction pattern appears whether the laser beam passes through the slide and off a projection screen to reach the eye (*see* previous demonstration), or whether the laser beam bounces off the projection screen and then through the slide to reach the eye.
- Hold a point source of white light such as the miniMaglite in front of the projection screen. Ask students to view the light through their slides. Instead of seeing monochromatic diffraction spots, as observed when the laser is used, each diffraction spot is dispersed into the colors of the visible spectrum to give a dramatically colorful effect. This latter viewing technique demonstrates that the slide patterns are indeed diffraction gratings.

To quantitatively verify the Fraunhofer relationship directly, any of the patterns in Figure 4.4 may be used. As depicted in Figure 4.1, the size or spread of the diffraction pattern is reflected in the value of tan ϕ, which can be calculated by trigonometry from the spot spacing in the diffraction pattern X and the distance between the pattern and the slide L, that is, tan $\phi = X/L$; the value of ϕ is sufficiently small in this experiment that tan ϕ can be well approximated as ϕ (in radians) so that $\phi = X/L$. Knowing λ, the diffraction equation can then be used to calculate d and to compare it with the known values.[2] Or, assuming that the array spacing d is known, λ can be calculated. Because sin ϕ is also approximately equal to ϕ (measured in radians) for the small angles observed in this experiment, ϕ should be roughly proportional to λ.

The relationship can be demonstrated qualitatively by using the white light source described: each diffracted spot contains the full visible spectrum, with red at the greatest distance from the center (the largest diffracted angle ϕ) and violet at the shortest distance from the center (the smallest diffracted angle ϕ). If several different lasers are available (like red and green He–Ne lasers), the quantitative form of the Fraunhofer equation can be experimentally verified at each wavelength.

In contrast to the direct relationship between sin ϕ and λ, d and sin ϕ should be inversely related from the Fraunhofer equation. Indeed, the comparisons of panel a with c; of panel b with d; and of panels e through h all demonstrate that as the spacing between lines or dots on the masks diminishes, the size of the diffraction pattern increases.

Laboratory. A laboratory experiment that explores X-ray diffraction using the optical transform slides is described in Experiment 4. A second experiment uses X-ray data to determine the interlayer spacing in a layered solid (Experiment 5).

In addition to the optical transform experiments already described, several other slides are available from ICE. Figure 4.6 displays eight additional arrays of laser-printer-written, photographically-reduced dots, and Figure 4.7 presents the corresponding diffraction patterns. Although these arrays are two-dimensional, they mimic what would be observed for diffraction from particular three-dimensional structures that are viewed in projection perpendicular to a face that is a parallelogram. For example, Figures 4.6b and 4.6d are the arrays of atoms in the sc structure and the projection of a fcc structure; the centered array of Figure 4.6a mimics the projection of a bcc structure and a projection of the diamond structure

[2]The patterns are drawn pixel-by-pixel and printed on a laser printer. The actual size of the pattern may vary slightly from one slide to another because each 35-mm slide is shot individually. The size of the pixel (x) is about 0.030 mm on the slide. Thus, in Figure 4.4, the repeat distance in patterns c, d, and f is $2x$; in a, b, and h, $3x$; in g, $4x$; and in e, $5x$.

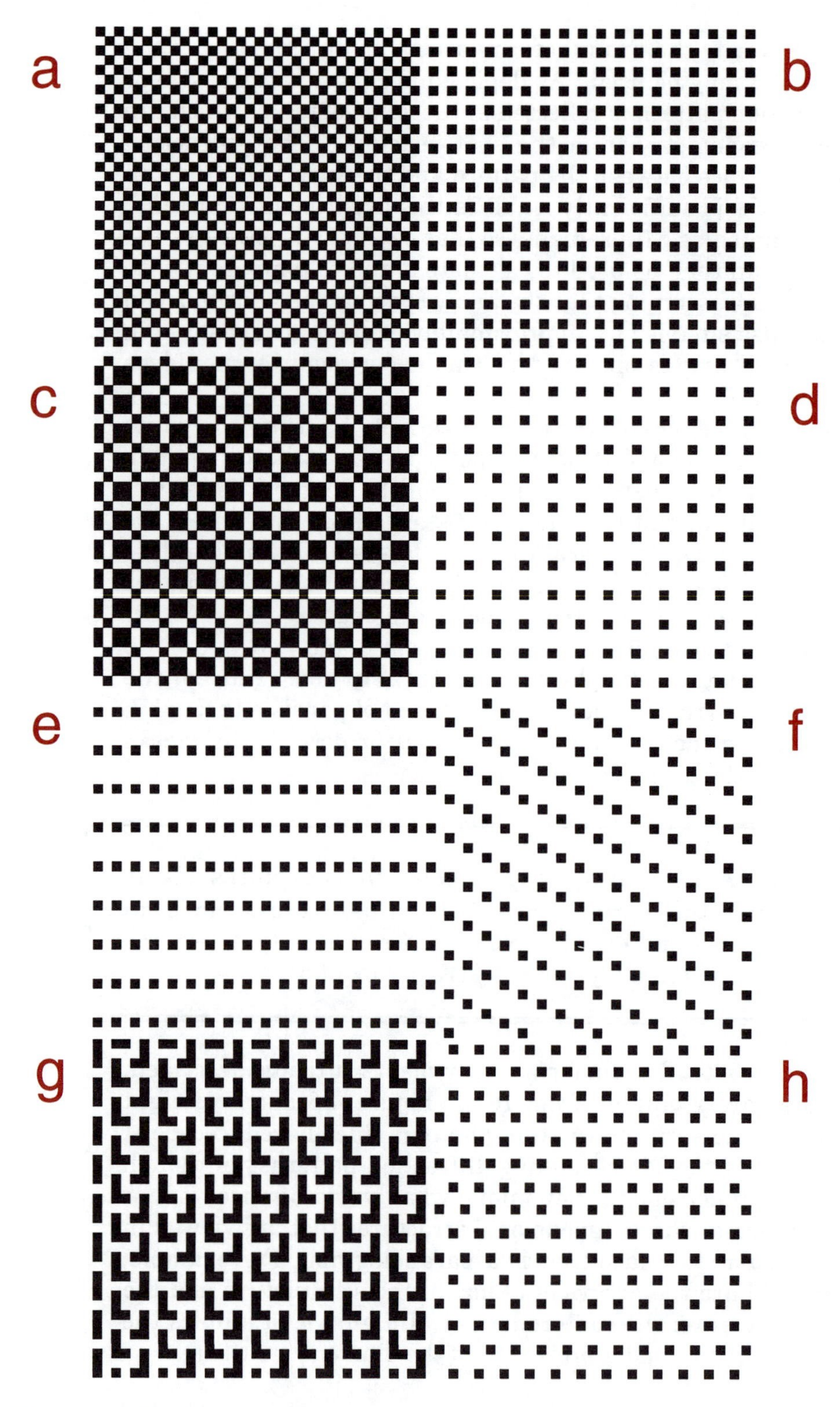

Figure 4.6. The patterns on the optical transform Unit Cell slide, available from ICE, (oriented so that the ICE logo is on the right side as you look through the slide).

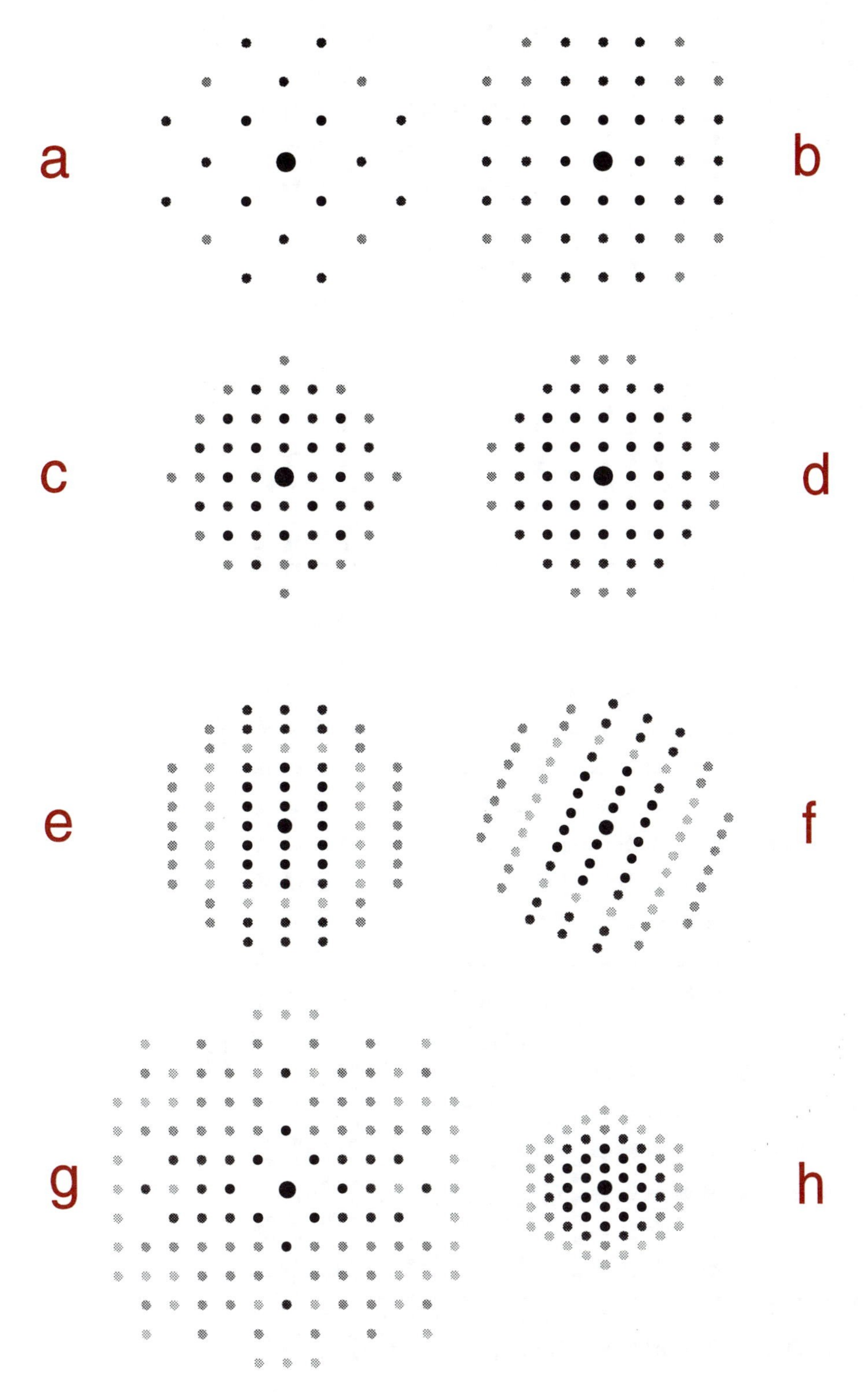

Figure 4.7. A simulation of the diffraction patterns from Figure 4.6. The letters correspond to the array from which the diffraction pattern is derived.

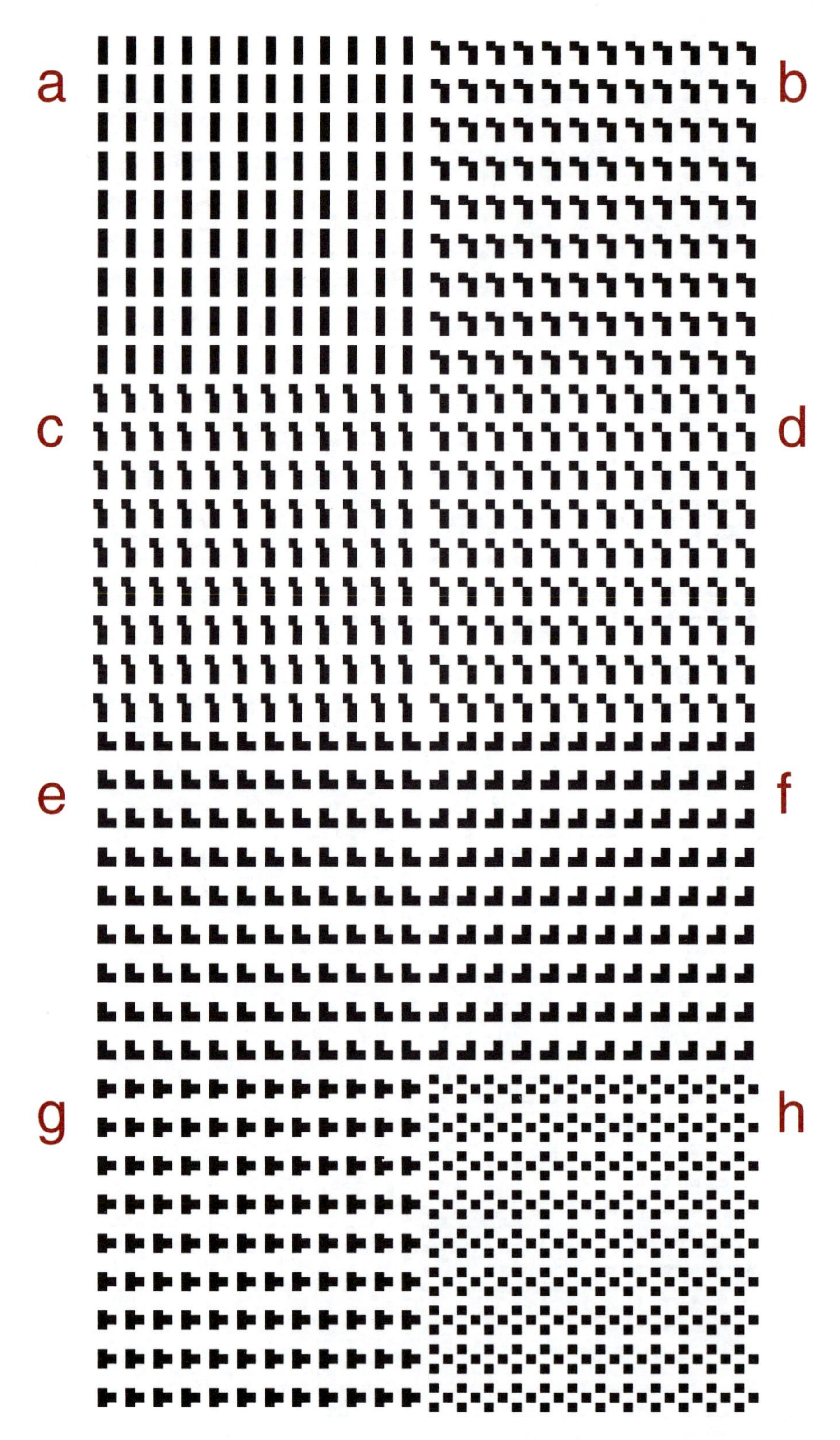

Figure 4.8. The patterns the optical transform VSEPR slide, available from ICE, (oriented so that the ICE logo is on the right side as you look through the slide).

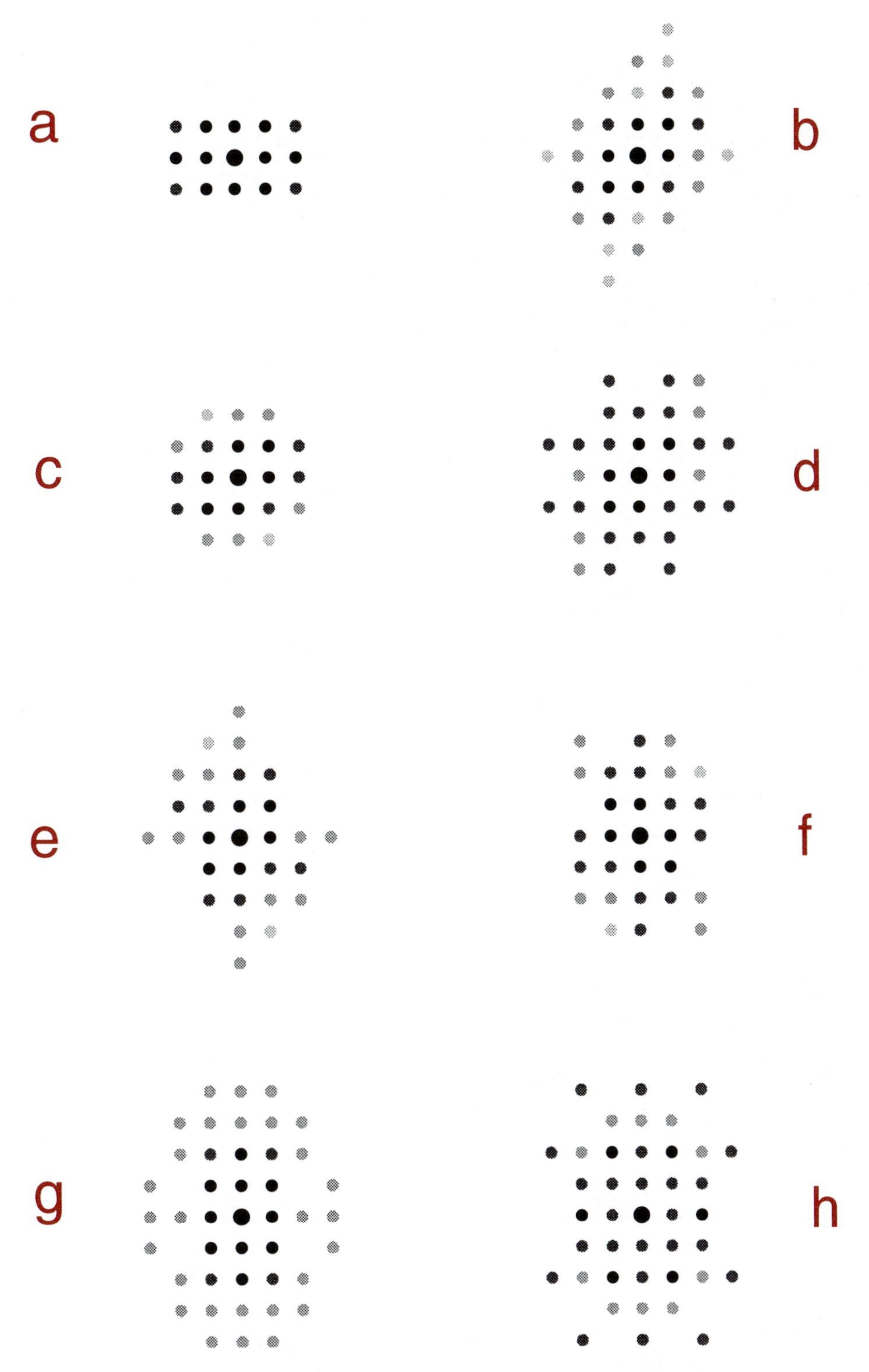

Figure 4.9. A simulation of the diffraction patterns from Figure 4.8. The indicated letter corresponds to the array from which the diffraction pattern is produced.

(Chapter 5); the array of Figure 4.6c simulates the projection of the cubic zinc blende structure (Chapter 5), having two different kinds of atoms, each surrounded by four atoms of the other type in a tetrahedral arrangement (this is also the array that would be seen for a single layer of the NaCl rock salt structure).

Comparison of the diffraction patterns from Figures 4.6a and 4.6b shows the effects of placing an identical atom in the center of the unit cell (every other diffraction spot is eliminated, Figures 4.7a and 4.7b), while comparison of Figures 4.6c and 4.6d shows the effects of placing a different atom in the center of the unit cell (every other diffraction spot is changed in intensity, Figures 4.7c and 4.7d). Additional examples include the rectangular structures of Figure 4.6e and 4.6g which are derived from so-called orthorhombic unit cell structures, wherein all angles are 90°, but the sides are of unequal length; the array of Figure 4.6f mimics a so-called monoclinic structure, where a non-90° angle is in the plane of the paper. Note the change in orientation between the unit cell and its diffraction pattern, where the long direction becomes the short direction and the short direction becomes the long direction. Figure 4.6h is a closest-packing structure, the basis for layers of hexagonal close-packed (hcp) and cubic close-packed (ccp) structures (Chapter 5). Finally, diffraction arising from the packing of L-shaped molecules related by glide symmetry (the symmetry of footprints on a beach) is illustrated by Figure 4.7g.

Also available from ICE is a slide for illustrating how molecular geometries are determined, which is meant to complement discussions of the valence-shell electron-pair repulsion (VSEPR) rules used to predict molecular structures. This slide and its corresponding diffraction patterns, Figures 4.8 and 4.9, simulate molecular solids comprised exclusively of triatomic molecules. The molecular shapes correspond to bond angles ranging from 180° (array a) down through about 60° (array h). The unit cell for all of these arrays is identical, a rectangle of 3×4 picture elements (pixels).

The diffraction patterns that result from these arrays all have the same geometry and reflect the size of the unit cell they have in common. However, the relative intensities of the different diffracted spots vary substantially and produce strikingly different patterns. The point can be made that the overall symmetry of the diffraction pattern reveals the symmetry of the molecular packing, and the relative intensities of the spots can be analyzed to determine the molecule's structure.

References

1. Laue, M. *Sitzungsber. Kais. Akad. Wiss., München* **1912**, *1912*, 263.
2. Schwartz, L. H.; Cohen, J. B. *Diffraction from Materials*, 2nd ed.; Springer-Verlag; New York, 1987.
3. Lisensky, G. C.; Kelly, T. K.; Neu, D. R.; Ellis, A. B. *J. Chem. Educ.* **1991**, *68*, 91–96.

Additional Reading

- Harburn, G.; Taylor, C. A.; Welberry, T. R. *Atlas of Optical Transforms*; Cornell University Press: Ithaca, NY, 1975.
- Taylor, C. A.; Lipson, H. *Optical Transforms*; Cornell University Press: Ithaca, NY, 1964.
- Welberry, T. R.; Thomas, J. M. *Chem. Br.* **1989**, 383–388.

Acknowledgments

This chapter is based in part on reference 3 and on discussions with Richard Matyi and Thomas Kelly, University of Wisconsin—Madison, Department of Materials Science and Engineering.

Exercises

1. What is the frequency of an X-ray with a wavelength of 1.54 Å, the wavelength produced by an X-ray tube with a copper target? (The speed of light is 2.998×10^8 m/s). What is the frequency of an X-ray with a wavelength of 0.7107 Å, the wavelength produced by an X-ray tube with a molybdenum target?

2. What is the angle of the second-order diffraction ($n = 2$) of 1.54-Å X-rays, the wavelength produced by an X-ray tube with a copper target, from each of the following sets of planes? (a is the length of the cubic unit cell)
   a. The unit cell faces of CsCl, $a = 4.123$ Å
   b. The unit cell faces of NaCl, $a = 5.6406$ Å
   c. The unit cell faces of Po, $a = 3.35$ Å
   d. The unit cell faces of Na, $a = 4.291$ Å
   e. The unit cell faces of Ni, $a = 3.524$ Å

   (Alternate questions could use any of the data in Appendix 5.6, or could use 0.7107-Å X-rays, the wavelength produced by an X-ray tube with a molybdenum target.)

3. What is the angle of first-order diffraction ($n = 1$) of 1.54-Å X-rays from the sets of closest-packed planes in each of the following metals?
   a. Be (layer spacing $d = 1.79$ Å)
   b. Cd ($d = 2.809$ Å)
   c. Mg ($d = 2.60$ Å)
   d. Ni ($d = 2.16$ Å)
   e. Zn ($d = 2.473$Å)

4. First-order diffraction (n = 1) is observed from the closest-packed planes of the following metals. If 1.54-Å X-rays were used in the diffraction study, determine the spacing between the closest-packed planes from the given diffraction angles.
   a. Be (θ = 25.48°)
   b. Cd (θ =15.91°)
   c. Mg (θ = 17.23°)
   d. Ni (θ = 20.88°)
   e. Zn (θ = 18.14°)
5. What is the frequency of a 670 nm laser?
6. Sketch a unit cell of any of the patterns in Figures 4.6 and 4.8.
7. Sketch a qualitative representation of the diffraction pattern produced by the following two-dimensional arrays.

a. 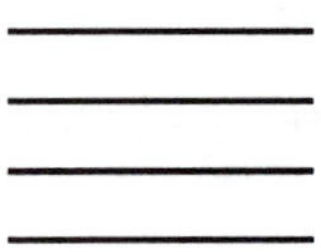 b. 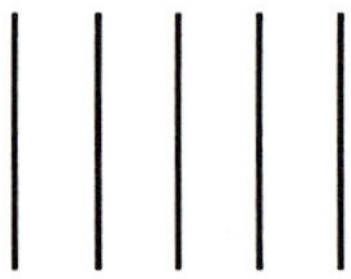 c.

8. Sketch a qualitative representation that shows the difference in the diffraction pattern produced by the following pairs of two-dimensional arrays.

a.

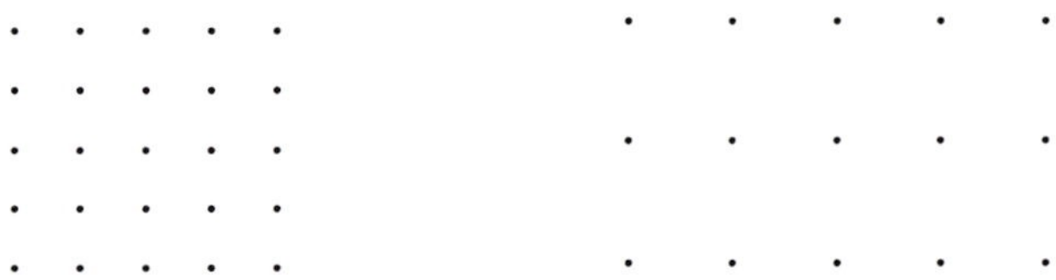

b.

c.

d.

9. The two arrays of dots, **a** and **b** are identical except that in array **b** there is an extra dot in the middle of each rectangle formed by four neighboring dots in array **a**.

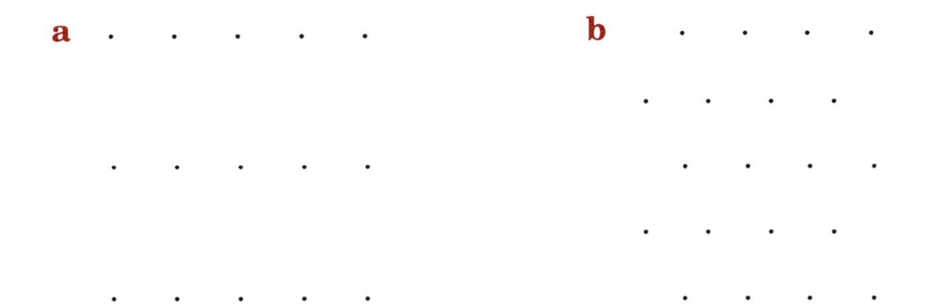

The diffraction patterns are identical except that every other diffraction spot is missing from one of the patterns, as follows:

a. Which of the two diffraction patterns above corresponds to array **a** and which to array **b**?
b. Explain why having twice as many dots (array **b** relative to array **a**) in the array has the effect you describe in part a on the diffraction pattern.
c. Interpret the change in diffraction patterns in going from array **a** to array **b** in terms of the unit cells of the two arrays.

10. Infrared light and ultraviolet light can be used to generate a diffraction pattern from an array like that in Exercise 7a. If your eyes were sensitive to infrared (IR) light with a wavelength of 8000 Å and to ultraviolet (UV) light with a wavelength of 3000 Å, what difference would you see in the patterns generated with these two types of light?

11. Why must X-rays be used in crystal-structure determinations rather than visible light?

12. Why does X-ray diffraction give more information about the three-dimensional structure of a crystalline solid than does scanning tunneling microscopy?

13. A student is using an optical transform slide and is attempting to determine the spacing between images on the slide. If she is using a red He–Ne laser (λ = 633 nm) and is standing 6.0 meters from the projection screen, the distance from the large center spot to the nearest diffraction spot is 3.8 cm. What is the spacing between images on the slide?

14. Use the Bragg equation to explain the observation that as the spacing between atoms decreases, the spacings in the resulting diffraction pattern increase.

15. The X-ray diffraction patterns from a crystal oriented to view the diffraction from a face of the cubic unit cell are shown.

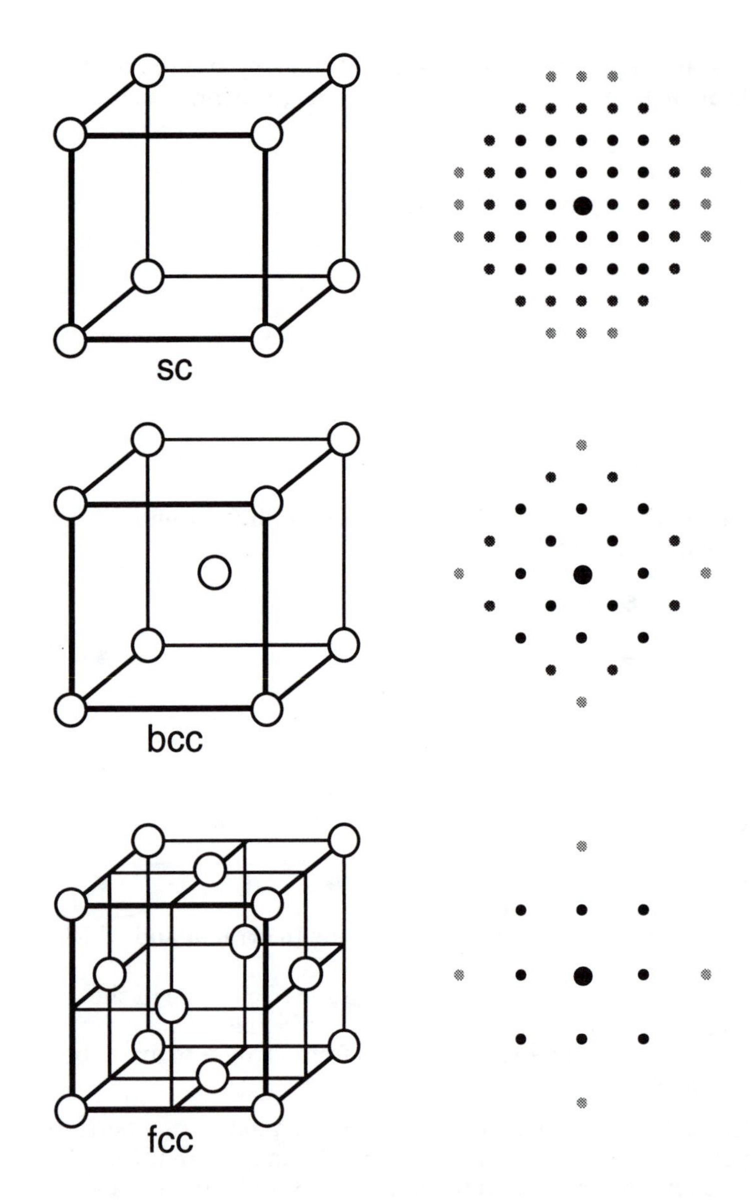

a. What happens to the diffraction spot pattern in moving from simple cubic to body-centered cubic or face-centered cubic?
b. Predict the diffraction patterns for the same orientation of CsCl and Cu_3Au.

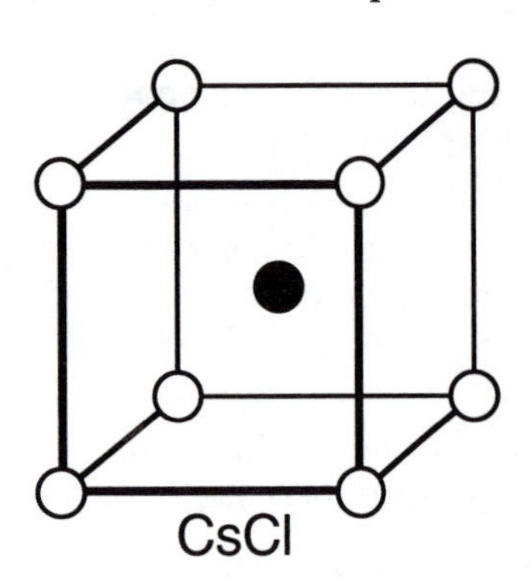

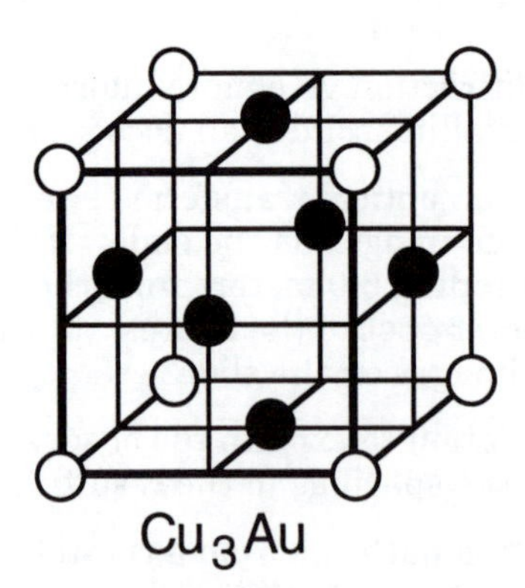

Chapter 5

Common Crystalline Structures

The X-ray diffraction experiment has been used to establish the structures of many different crystalline solids. In this chapter we describe some common crystalline extended structures, encompassing, in turn, materials that contain metallic, ionic, and covalent bonding. Also highlighted are the geometries found in extended solids, some of which are also present in discrete molecular species.

Because of the three-dimensional nature of these structures, the use of the ICE Solid-State Model Kit (SSMK, *see* Appendix 5.1), or a kit with comparable capabilities[1], is strongly recommended. *Structures whose assembly directions are included in the SSMK are indicated in this chapter by* ♦.

In presenting this material to a large classroom audience, a closed-circuit television or a video camera could be used. Styrofoam spheres[2] up to 1 foot in diameter permit construction of large models suitable for use in a large lecture hall. Data files containing the coordinates for the structures in the SSMK are available for the MacMolecule program (*see* Appendix 5.2).

Metals

Metals have played a central role in technology since before recorded history. For most of this time, the mechanical properties of metals have been the underlying reason for their usefulness. For example, gold, which

[1]Ordering information for other model kits and a computer program suitable for displaying solid-state structures is given in Supplier Information.
[2] Sold by Molecular Models (Supplier Information gives complete details).

2725–X/93/0097$14.55/1

is most notable for its beautiful luster and resistance to corrosion, would nevertheless be of little value were it not for the fact that this metal can be easily worked into thin sheets and intricate shapes. The modern widespread use of metals as electrical conductors would not be practical if metals could not be drawn into wires. The ductility of metals (their ability to be formed into new shapes) derives from the nondirectionality of the metallic bond. In this section we explore the basic structures of metals that result in their remarkable mechanical properties. Microstructural defects in metals, which provide the mechanistic basis for these properties, are discussed in Chapter 6.

Metal atoms can be regarded as hard spheres for purposes of modeling them as they pack together to form a solid (recognizing that their size is actually "fuzzy," as noted in the STM description of Chapter 2). Most of the elemental metal crystal structures can be modeled by simple sphere-packing arrangements♦.

Metallic bonding is, for the most part, nondirectional. For many metals, therefore, the metal atoms arrange themselves to achieve the most efficient packing and the densest structure. Such metals will either be face-centered cubic (e.g., Al, Cu, and Au) or hexagonal close-packed (e.g., Mg, Zn, and Ti). Covalent bonds, such as those in diamond, lead to a relatively open structure dictated by the angles of the bonds. Thus, the types of bonding in materials are reflected in crystal structure.

Two-Dimensional Packing

The most efficient way to arrange spheres (circles in two dimensions) in a plane is hexagonal packing. In this arrangement, six outer circles can be arranged so that they touch a central circle. The outer circles touch so as to form a regular hexagon about the central one. Alternatively, less efficient square packing may be used to arrange the circles. Close-packing and square-packing arrays are shown in Figure 5.1 with a unit cell superimposed on each.

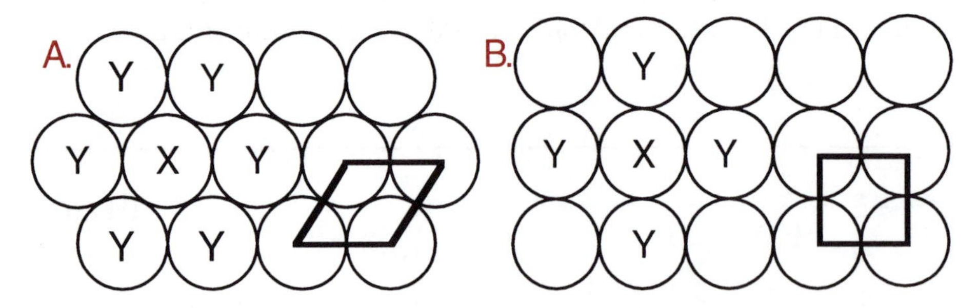

Figure 5.1. Packing arrangements in two dimensions. For an arbitrary atom marked with an X, the nearest neighbors are indicated with a Y. A unit cell is superimposed on each array. A: Close packing. B: Square packing.

The unit cell can be used to determine the "packing efficiency" of each structure. Packing efficiency in two dimensions can be related to the density of atoms (units of atoms per square centimeter):

$$\frac{\text{area of circles in the unit cell}}{\text{area of unit cell}} \times 100\% = \%\text{ packing efficiency} \quad (1)$$

Close packing produces, as the name implies, a denser structure. The unit cell shown for the close-packed hexagonal array (Figure 5.1A) comprises a total of one circle (Chapter 3 presents a discussion of unit-cell occupancy in structures in which the angles are not 90°) and has a packing efficiency of ~91%. For square packing, the packing efficiency is about 79%.

Another important characteristic of packing is the coordination number, or number of nearest-neighbor circles to any of the packed circles. In the close-packed structure, the coordination number is six; any arbitrarily chosen circle within the close-packed structure (one has been marked with an X) is surrounded by six nearest neighbors (marked with a Y). The coordination number is four in square packing because there are four nearest neighbors.

The hexagon-based close packing shown in Figure 5.1 is ubiquitous in nature. In addition to the hexagonal packing of magnetic field lines dramatically illustrated by ferrofluids in Chapter 2, a bubble raft (a coalesced group of hundreds of bubbles) provides a simple, compelling illustration of nature's predilection for this packing arrangement. The bubble raft, like the optical transform experiment, is commonly credited to Sir Lawrence Bragg who developed it almost 50 years ago (*1*), though modern dishwashing detergent has supplanted Bragg's detailed formula for the medium supporting bubble formation in the raft. A diagram of the bubble raft setup is shown in Figure 5.2.

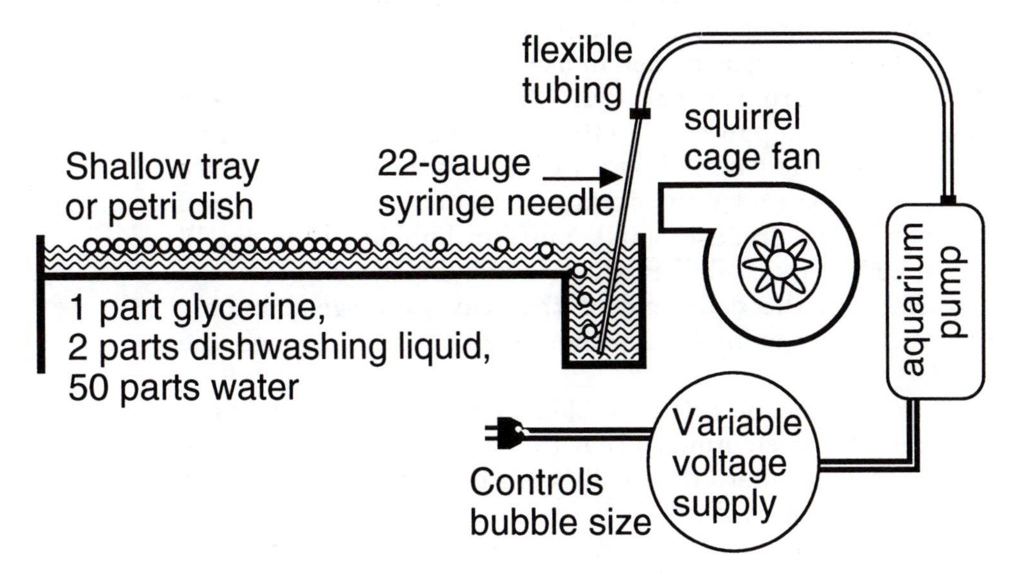

Figure 5.2. A diagram of a bubble raft.

The bubble raft provides an excellent two-dimensional model of a metal on the atomic level; the attraction of small bubbles for one another mimics the attractions between metal atoms in a crystal, bubbles of uniform size correspond to atoms, and the nondirectional forces that make the bubbles stick together correspond to the metallic bonding. The hexagonal arrangement of bubbles is the closest possible packing of bubbles and is caused by surface tension in the bubble raft.

Demonstration 5.1. Demonstrations of Close Packing: Bubble Rafts, BB Trays, and Bubble Wrap

Materials

Bubble raft apparatus—*see* Figure 5.2 (Its construction is shown in Appendix 5.3)
Bubble raft tray
22-gauge syringe needle
Aquarium pump (Hartz, for example)
Variable-voltage transformer
Flexible tubing
Small electric fan (optional)
Bubble solution made of 1 part glycerine, 2 parts thick commercial dishwashing detergent (Joy or Dawn work well), and 50 parts of water *(2)*
Overhead projector and screen

Procedure

- Fill the shallow tray to a depth of about 0.5 cm with bubble solution. For classroom demonstrations place the tray on an overhead projector. The tray design also helps protect the projection plate B (Appendix 5.3) from being scratched in use and storage.
- Connect a 22-gauge syringe needle (with the pointed tip ground off to prevent accidental skin puncture) to the hose (Tygon tubing) from an aquarium pump that is powered by a variable-voltage supply. Place the end of the needle under the surface of the bubble solution and adjust the voltage to produce small, uniform bubbles. (For a 4-W Hartz pump, 20 V works well.) Production of bubbles well below the solution surface in the deep end of the tray gives good bubble uniformity, but a better projection image is obtained through the shallow end.
- To aid in formation of the raft of bubbles, aim a small electric fan over the surface of the solution so that the bubbles are blown to the other end of the tray where they will begin to form the raft. Push aside unwanted bubbles with a Plexiglas slider (shown in Appendix 5.3) or scoop them out. To produce bubbles of uniform size, it is important to keep the needle near the bottom of the tray.

- Once the bubble raft has formed, turn off the aquarium pump and fan. The hexagonal arrangement of bubbles that occurs throughout much of the raft can be observed. Figure 5.3 shows a sketch of a typical raft. Defects in the regular hexagonal pattern are also generally observable and will be discussed in more detail in Chapter 6.

Variations

There are several variants of the bubble raft. A Petri dish works adequately, but the tray designed by Bragg, fabricated from Plexiglas, enhances the quality of the demonstration. Sealing metal BB pellets of uniform size in a plastic tray permits the BBs to be repeatedly shaken and will produce close packing that can be shown by laying the tray on an overhead projector. A detailed description of the construction of this tray is available.[3] Also available for projection on the overhead projector is the bubble wrap commonly used for packing; the bubbles in this plastic wrapping material form a close-packed array. The hexagonal packing arrangement can equivalently be modeled by placing seven identical coins or marbles on an overhead projector.

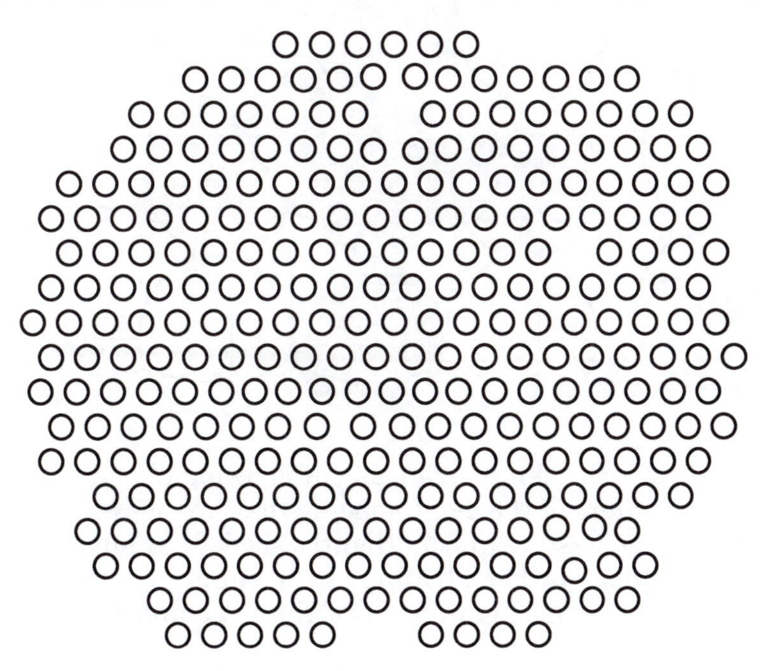

Figure 5.3. A typical bubble raft obtained using the setup shown in Figure 5.2.

---

[3]Assembly instructions for the BB board model are available from the Institute for Chemical Education.

Packing in Three Dimensions

If the spheres composing a square-packing layer (Figure 5.1B) are placed directly on top of one another, the simple cubic crystal structure♦ results (Figure 3.9). This structure is rare (Figure 3.11 shows the packing arrangements of the elements) and is characterized by low atomic density and coordination number. The density reflects the packing efficiency, defined in three dimensions as

$$\frac{\text{volume of spheres in the unit cell}}{\text{volume of unit cell}} \times 100\% = \%\text{ packing efficiency} \qquad (2)$$

The packing efficiency of the simple cubic structure is only ~52%. The coordination number in this structure is six, with the nearest neighbors of any sphere defining the six corners of an octahedron, as partially shown in Figure 5.4.

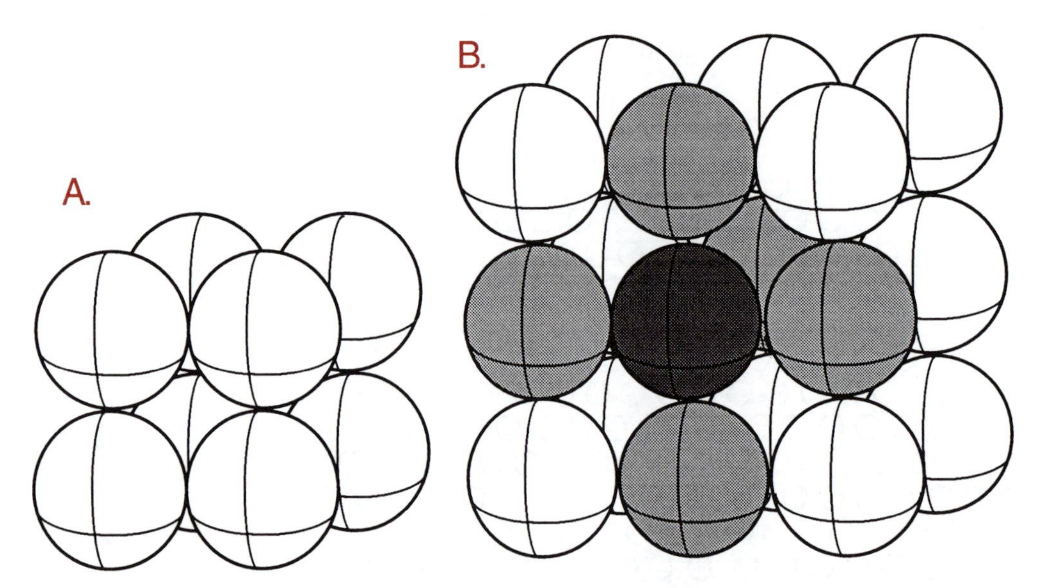

Figure 5.4. The simple cubic-packing arrangement showing (A) one unit cell, defined by the centers of the eight spheres that are shown; and (B) the octahedron of nearest neighbors (lightly shaded) around an arbitrarily selected atom (darker shading) in a simple cubic-packing array. (The sixth neighbor is in front of the page and has been omitted for clarity).

A more common packing arrangement results if the spheres in a square-packing layer are uniformly moved apart so they no longer touch, and the second layer is offset to nest in the spaces between the spheres in the first layer, as pictured in Figure 5.5A. The third layer and each succeeding odd-numbered layer is then in registry with the first layer; and the fourth and succeeding even-numbered layers are in registry with the second, producing a more densely packed body-centered cubic structure♦ (Figure 3.10). In the bcc cubic unit cell the spheres touch only along body diagonals of the cube, as shown in the figure, and all layers are identical, but shifted relative to one another. The bcc packing efficiency is ~68%. Inspection of the structure reveals that the coordination number in the body-centered cubic structure is eight, with the nearest neighbors surrounding any given atom sitting at the corners of a cube.

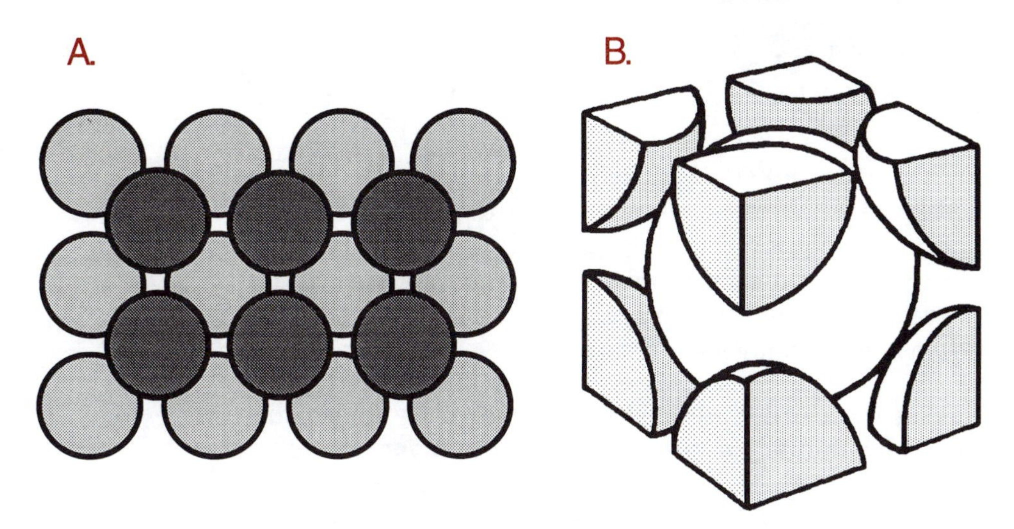

Figure 5.5. The body-centered cubic-packing arrangement. A: The shaded atoms are not touching and form a square-packed layer. The dark atoms form a second layer that is identical in arrangement to the first but shifted so that the atoms sit in the depressions formed by the first layer. The third, fifth, and subsequent odd-numbered layers are in registry with the first, while the fourth, sixth, and subsequent even-numbered layers are in registry with the second. B: A unit cell showing a central atom and its eight nearest neighbors. The latter are centered on the corners of a cube, and one-eighth of each of their volumes lies within the unit cell.

The two most dense three-dimensional packing arrangements of atoms for the metallic elements are hexagonal close packed (hcp) and cubic close packed (ccp). When stacking two-dimensional close-packed layers like that shown in Figure 5.1A, an arbitrarily chosen initial layer can be denoted A. The next close-packed layer, layer B, is placed such that the spheres rest in the hollows of the first layer. If the third layer (and every subsequent odd-numbered layer) is stacked directly over the first layer (layer A), and the fourth (and every subsequent even-numbered layer) is stacked directly over the second layer (layer B), then an ABAB... stacking pattern is formed, defining hexagonal close-packing (hcp) .

The coordination number in the hcp structure♦ (*see* Figure 5.6) is 12—a central sphere is touched by six spheres in its close-packed plane; by three in the close-packed plane above it; and by three more in the close-packed plane below it. The hcp structure has a packing efficiency of ~74%.

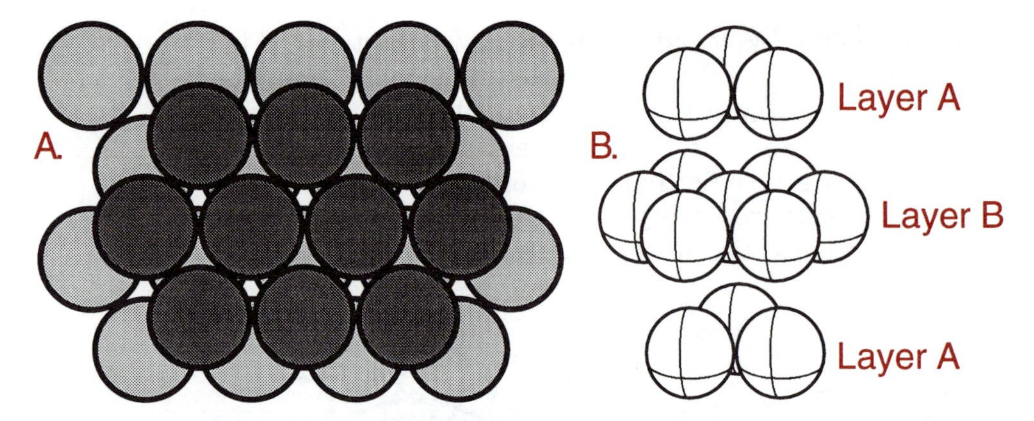

Figure 5.6. The hexagonal close-packed arrangement. A: A second hexagonal close-packed layer is resting in the depressions between atoms in the first close-packed layer. All odd-numbered layers are identical to the first, and all even-numbered layers are identical to the second. B: A second view of the hcp arrangement showing the ABAB packing of the close-packed layers.

Instead of placing the third close-packed layer directly above the first layer, layer A, it can be uniquely positioned to produce layer C, as sketched in Figure 5.7. The structure is then repeated with the fourth layer over the first, the fifth layer over the second, and the sixth over the third, leading to an ABCABC... structure that is called cubic close-packing (ccp)♦. The distinction between hcp and ccp is subtle (the same kind of close-packed layers are stacked but in different sequences), and several metals adopt both structures under different conditions.

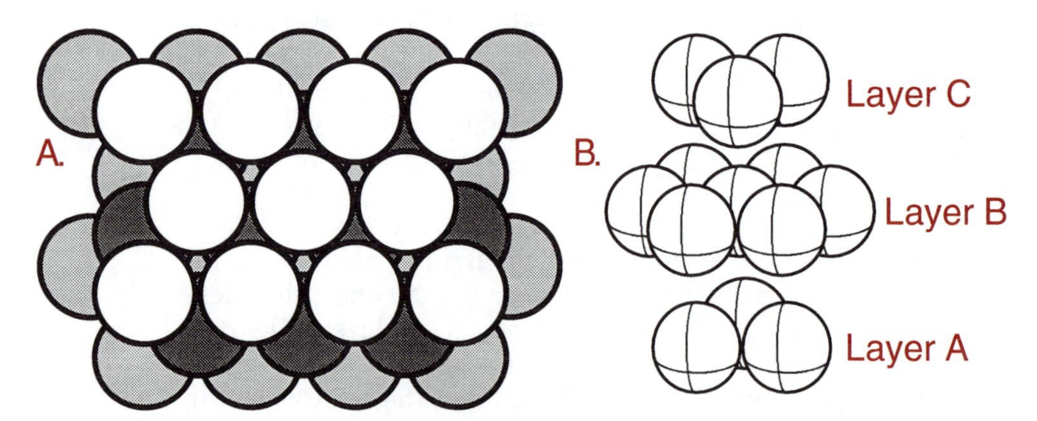

Figure 5.7. The cubic close-packed arrangement. A: The first two layers are identical to those in the hcp arrangement. The third close-packed layer rests in the depressions created by the second layer but in such a way that it does not match the first layer (*see* Figure 5.8). The fourth layer matches the first, the fifth matches the second, the sixth matches the third, and so on. B: A second view of the ccp arrangement that shows the ABCABC... arrangement of the layers.

Demonstration 5.2. Interstices in Close-Packed Structures

Materials

Overhead projector and screen
Thirteen identical coins
Magic markers
Figure 5.8 reproduced on a transparency (optional)
Bubble wrap and Ping-Pong balls (optional)

Procedure

- Arrange seven or more identical coins to form a close-packed layer on a blank transparency placed on the stage of an overhead projector.
- Label the interstices with B and C using a magic marker as shown in Figure 5.8 , or simply give the interstices different colors.
- Cover the interstices labeled B by a second layer of three coins. The interstices labeled C will still be visible. Place a third layer of coins over the original layer to form an hcp structure, in which case the interstices labeled C will still be visible. Alternatively, place the third layer over the interstices labeled C to form a ccp structure, in which case no interstices will be visible.

Variation

Bubble-wrap packing material with 1-inch-diameter bubbles can be placed on the overhead projector and Ping-Pong balls can be used to cover one set or another of the interstices.

The ABCABC packing arrangement is equivalent to the face-centered cubic structure (*see* Figure 3.12). This is readily seen with the SSMK: Two colors (or labels of any sort) are used to distinguish two types of principal spheres to show that a face-centered cube, standing on its body diagonal (the cube is upright on a corner), is camouflaged by the ccp structure, as shown in Figure 5.9A. In the fcc perspective♦ spheres touch along the face diagonals of the cube but not along the edges of the cube. Although the fcc and the ccp structure are equivalent, the close-packed planes in the fcc structure are those that are perpendicular to body diagonals of the cube. Because of the symmetry of the cubic unit cell, there are four such planes, sketched in Figure 5.9B.

The ccp and hcp structures differ only in the stacking sequence of close-packed planes; therefore, the packing efficiency (~74%) and coordination number (12) are identical.

Table 5.1 summarizes packing efficiencies and coordination numbers for the common structures described.

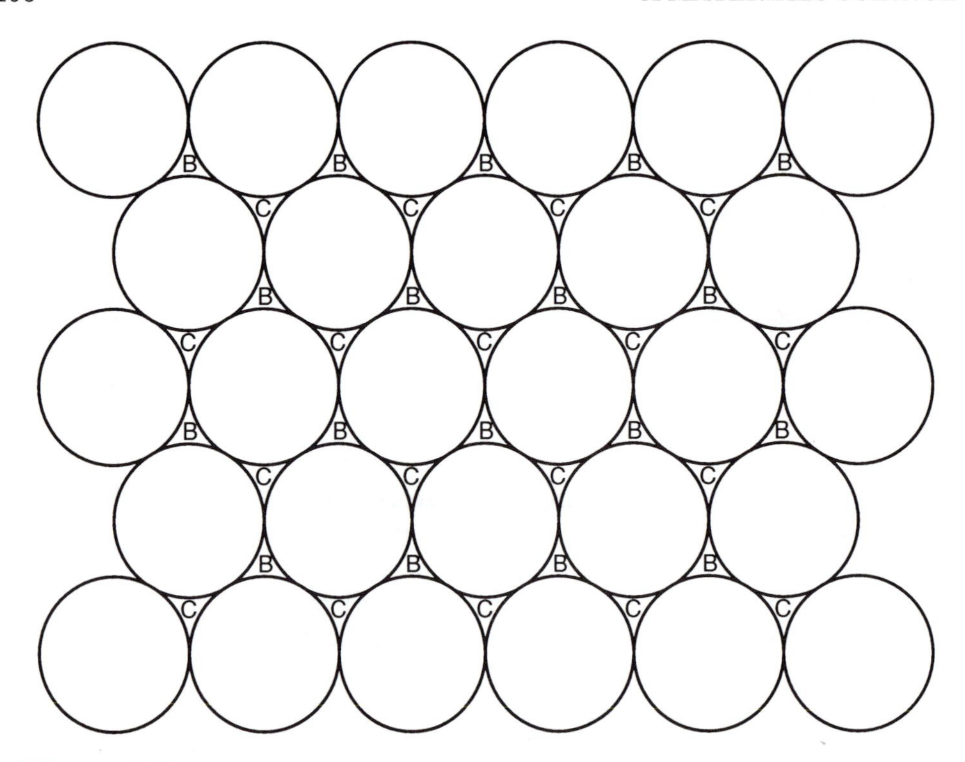

Figure 5.8. A close-packed array of spheres. Around any sphere there are six depressions or interstices labeled B and C that serve as possible resting sites for spheres in the next layer. However, only one set of interstices (set B or set C) can be filled with the next close-packed layer. If the spheres that make up the second layer are placed in the B positions, the C positions will still be visible (or vice versa). There are two ways of placing the spheres to form the third layer. If the spheres are placed so as to match the first layer (leaving the C interstices visible), ABAB... packing (hexagonal close packing) results. If the third layer is placed so as to occupy the C positions, ABCABC... packing (cubic close-packing) results.

Table 5.1. Packing in Metals

Type of Packing	Packing Efficiency[a]	Coordination Number
Simple cubic (sc)♦	52%	6
Body-centered cubic (bcc)♦	68%	8
Hexagonal close-packed (hcp)♦	74%	12
Cubic close-packed (ccp)♦[b]	74%	12

[a]Measurement of the volumes of the cubic unit cells and the volumes of the spheres used to build the structures, for example with the SSMK, will allow an experimental determination of the packing efficiency that is in reasonably good agreement with these values.
[b]Identical to face-centered cubic (fcc).

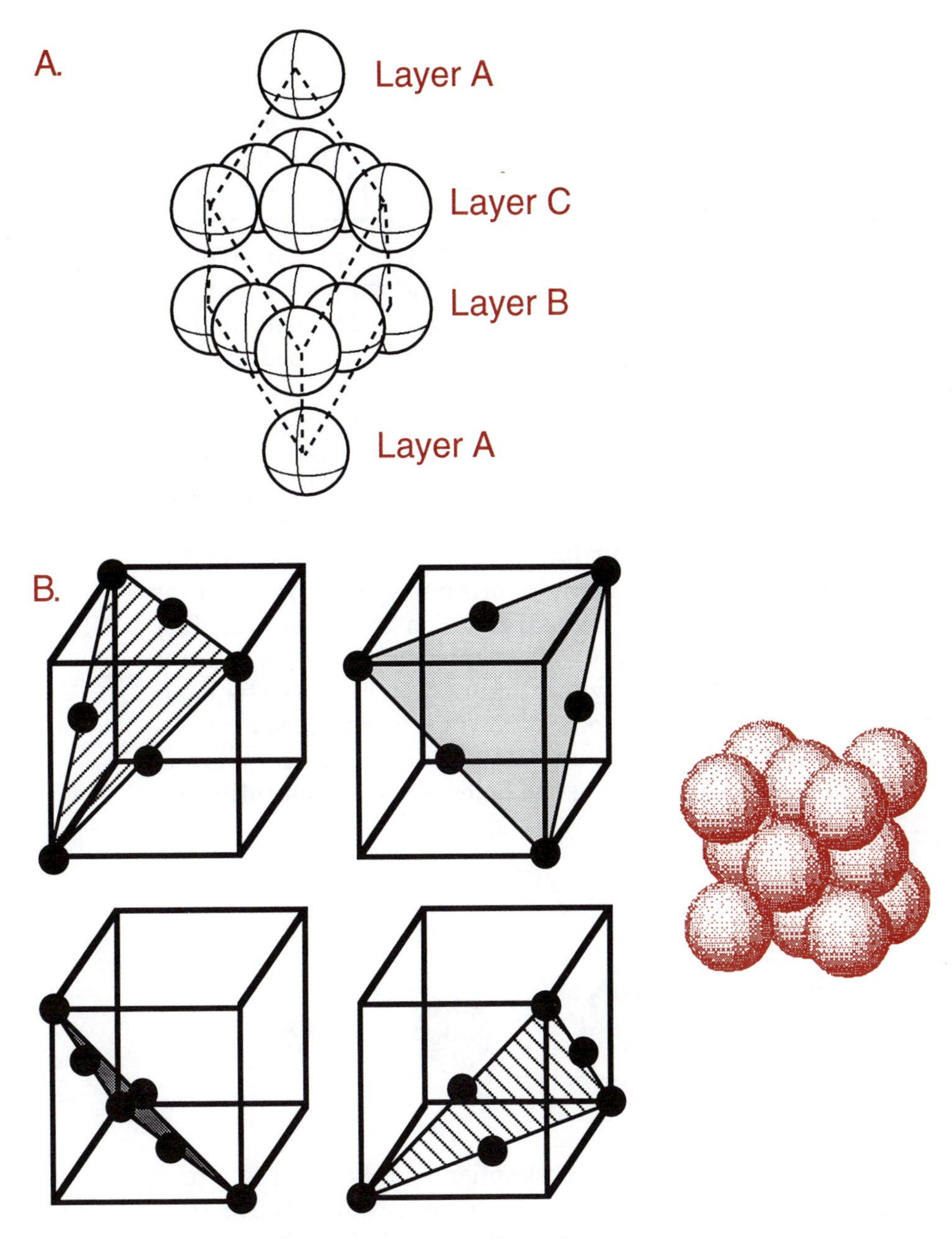

Figure 5.9. A: The cubic close-packed arrangement shown as a face-centered cube (dashed lines). Appropriate selection of one atom in the first and fourth layers and six atoms in each of the second and third layers generates a face-centered cubic unit cell. B: The close-packed layers are perpendicular to the body diagonals of the cubic unit cell.

Ionic Solids

The packing schemes described for metals serve as the basis for describing the structures of several important classes of ionic compounds. A general feature of many of the salts whose structures will be described is a framework comprising ions of one type, often the relatively large anions. These ions might be packed in a simple cubic (sc), hexagonal close-packed (hcp), or cubic close-packed (ccp) arrangement, perhaps "expanded" to move the like-charged ions farther apart from one another. The counterions needed for charge balance, often the smaller cations, are situated in holes or interstitial sites between the framework ions.

The size and shape of these holes reflect the geometry of the framework ions that form them. Holes derived from cubic, octahedral, and tetrahedral arrangements of packed anions figure prominently in describing extended structures. Using the hard-sphere model with the spheres in contact, the size of each of these holes can be calculated relative to the size of the spheres forming it. If the radius of the hole-forming spheres is taken as 1.000, the cubic, octahedral, and tetrahedral holes will accommodate spheres with radii of 0.732, 0.414, and 0.225, respectively. These relative sizes, so-called "radius ratios," are shown in Figure 5.10 and are the basis for the four sizes of spheres used in the SSMK. Furthermore, the SSMK can be used to build cubic, octahedral, and tetrahedral arrangements of spheres to investigate the relative sizes of the holes. The radius ratios are summarized in Table 5.2.

Figure 5.10. The radius ratio. The largest sphere is taken as the size of the close-packed anions and is arbitrarily given a radius of 1.000. The following three spheres can be accommodated in holes formed by arranging the large spheres in the shape of a cube (hole of radius 0.732), an octahedron (hole of radius 0.414), and a tetrahedron (a hole of radius 0.225).

Table 5.2. Radius Ratio Rules

Radius Ratio	Coordination Number	Type of Hole for Cation
0.225–0.414	4	tetrahedral
0.414–0.732	6	octahedral
0.732–1.000	8	cubic

The radius ratio serves as a rough guide to predicting what kind of hole the counterions will occupy.[4] For example, if the cation-to-anion radius ratio is slightly greater than 0.414, the cation should fit in an octahedral hole, representing a cation coordination number of six. As this ratio rises, there is more space around the cations. When the ratio reaches 0.732, there is enough space for eight anions around the cation, and cubic holes are often observed. If the ratio shrinks to below 0.414, the anions are too crowded relative to the smaller cation in an octahedral geometry; their repulsions can be reduced by a smaller coordination number and hole, favoring tetrahedral holes with the cation coordination number of four until a ratio of 0.225 is reached. [If the ratio becomes smaller than 0.225, even the low coordination number of four becomes unfavorable; a molecular analogy is that the thermal instability of tetraiodomethane, CI_4 , has been ascribed to the small ratio of the carbon-to-iodine radii *(6)*.]

Tetrahedral and octahedral geometries are commonly discussed in connection with molecular geometry and bonding in, for example, presentations of valence-shell electron-pair repulsion (VSEPR) rules and hybrid orbitals. Seeing these shapes in the different context of being "holes" in a framework provides a complementary spatial perspective, while underscoring the prevalence of these shapes in nature. Moreover, although the following structures are illustrated with salts, these same packing arrangements are encountered in metal alloys and molecule-derived (molecular) solids.

[4]The radius ratio is a simple and easily understood approximation for explaining the structures that are adopted by ionic compounds, but in many cases it makes the wrong predictions. Nathan (*3*) described 227 compounds and noted that the correct structural prediction is made for 67% of the compounds when the radius-ratio rule is applied. One contributing factor to the erroneous predictions is that the ions are not hard spheres, as assumed in the radius-ratio model, but have sizes that vary with the coordination number. (For example, sodium has an ionic radius that varies from a value of 1.13 Å in structures in which it is four-coordinate to 1.16 Å when six-coordinate, to 1.32 Å when eight-coordinate.) A second factor that causes deviation is that all bonds have some degree of covalency. As the amount of covalency increases, a hard-sphere (ionic) model will be less likely to describe the system accurately (*4*). Other schemes can be used to build structure-sorting diagrams that provide a basis for more reliable predictions, but with a resulting increase in complexity (*5*).

Demonstration 5.3. Models of Cubes, Octahedra, and Tetrahedra

Materials

Transparent dice (cubes)
Octahedra from inorganic model kits or some jacks (Many sets of jacks have one axis that is shorter than the other two.)
Tetrahedra from organic model kits
Octahedra and tetrahedra can be purchased separately in bulk quantities from Darling Models (see Supplier Information).
Paper models from Appendix 5.4
Mineral samples such as pyrite cubes (FeS_2) and fluorite octahedra (CaF_2), available from rock shops (optional)

Procedure

- Distribute the three shapes to each student or to groups of students so that they can examine the shapes while you discuss them.

Demonstration 5.4. Octahedral, Cubic, and Tetrahedral Crystals

Materials

Alum, $KAl(SO_4)_2{\cdot}12H_2O$
Sodium chlorate, $NaClO_3$
Borax, $Na_2B_4O_5(OH)_4{\cdot}8H_2O$
Water
Beakers, watch glass
Hot plate
Thread
Superglue (optional)

Procedure (based on recipes found in reference 7).

- To grow octahedral alum crystals, prepare a supersaturated solution by dissolving 20 g of $KAl(SO_4)_2{\bullet}12H_2O$ in 100 mL of H_2O. Heating will be necessary. When all the solid has dissolved, set the beaker aside in an area of constant temperature and cover with a watch glass. Crystals should appear in a few days, as they grow by slow evaporation.
- To grow cubic $NaClO_3$ crystals, use the same procedure, but dissolve 113.4 g of $NaClO_3$ in 100 mL of H_2O.
- To grow tetrahedral crystals, dissolve 113.4 g of $NaClO_3$ in 100 mL of H_2O. Add 6 g of borax to the solution. The borax is not incorporated into the $NaClO_3$ crystal but does affect the relative growth rate of the crystal faces. *See* Appendix 5.5.

- To grow larger crystals, you can "seed" a supersaturated solution. To prepare the seed, take a small crystal and attach it to a thread with a knot or a small drop of superglue. Prepare a supersaturated solution in which to place the seed. Once crystals begin to come out of the solution, suspend the seed crystal in the center of the solution. Be aware that the seed will dissolve if it is planted before crystals begin to appear. Cover the mixture and allow it to remain untouched and in constant temperature for a few days. Check the growth periodically.

Cubic Holes

Constructing a cubic unit cell with its corners at the centers of the anions in a sc packing arrangement of anions and placing a cation in its center yields a 1:1 cation-to-anion stoichiometry. Inspection of Figures 5.11 and 5.12 or comparison of the two CsCl structures in the SSMK♦ shows that this structure can also be described as comprising an anion in the center of a cube with a cation at each of the corners. From either perspective there is one cubic hole per framework sphere. This structure, which is often called the CsCl structure, can thus be regarded as two

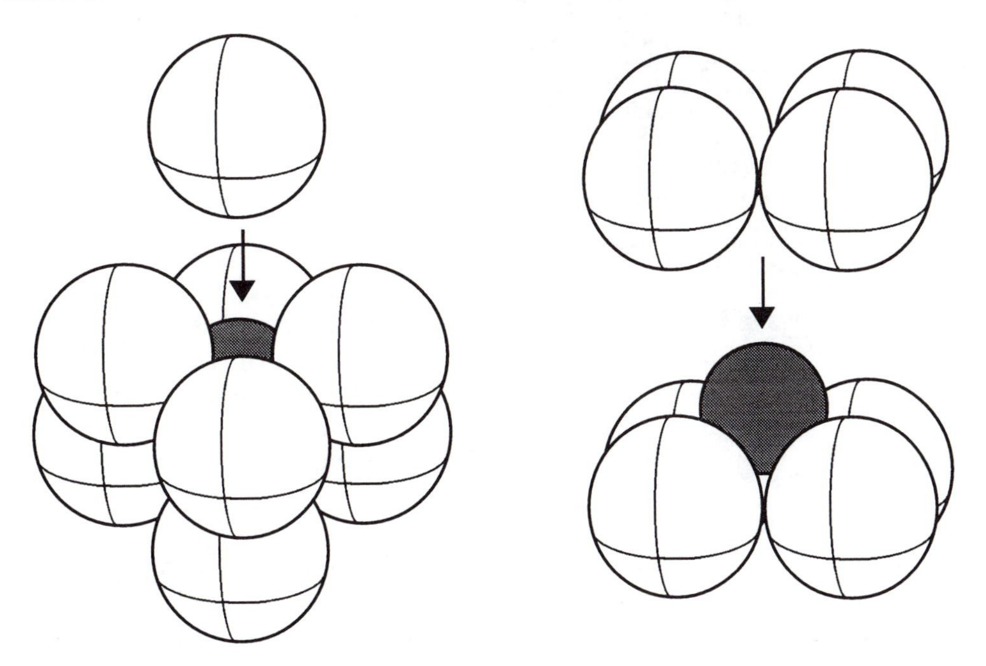

Figure 5.11. Two views of an atom (shaded sphere) in a cubic hole.

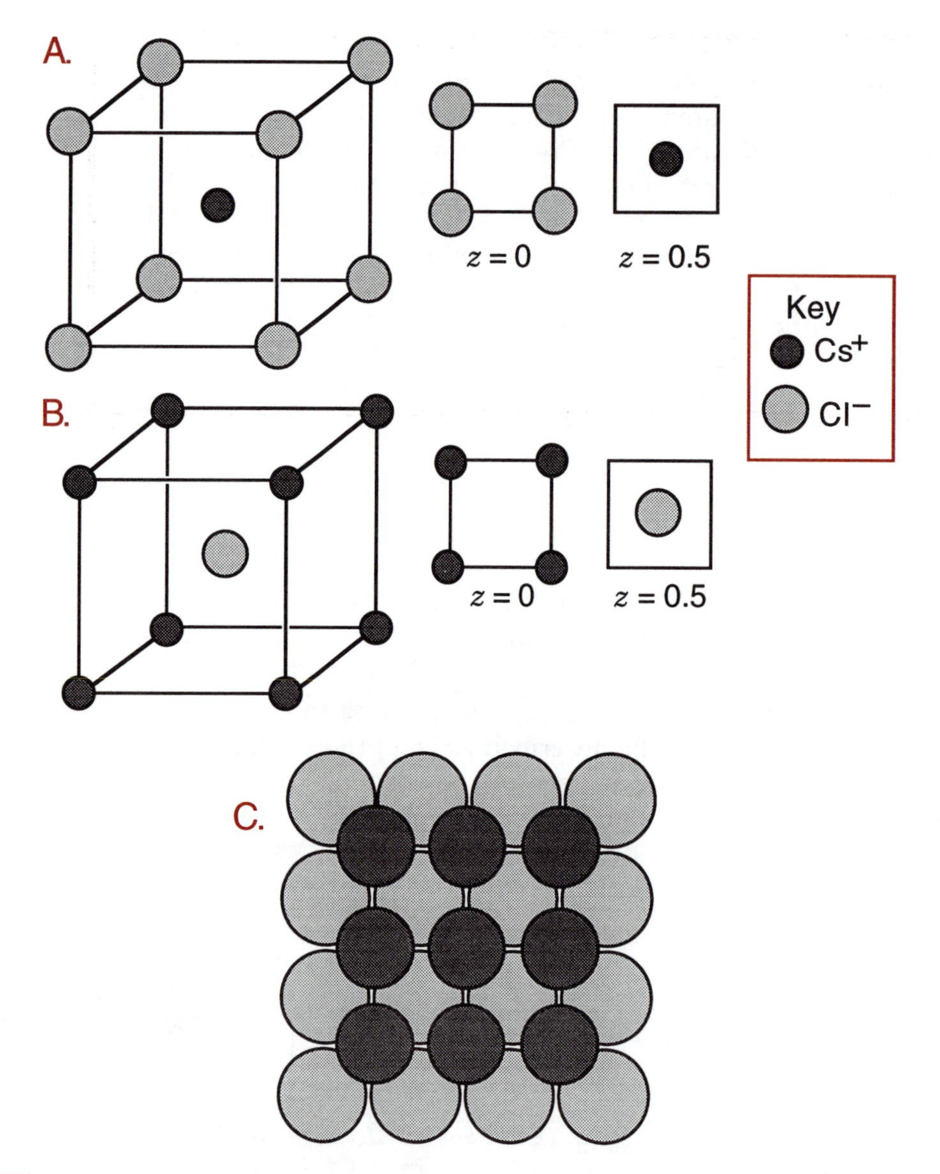

Figure 5.12. Three views of the CsCl structure. A: The cesium ion is in a cubic hole generated by chloride ions. B: Upon shifting origins the unit cell may also be viewed as having cesium ions at the corners of a cube and a chloride ion in the middle. C: A view of the alternating layers of square-packed cesium and chloride ions.

interpenetrating simple cubic frameworks.[5] This aspect of the structure is easier to see if a larger portion of the CsCl lattice is constructed♦. Cations occupy each cubic hole formed by anions; and anions occupy each cubic hole formed by cations. The coordination number of each anion and cation is eight. Both Cs^+ and Cl^- are about the same size, having radii of 1.69 and 1.81 Å, respectively. This same structure is adopted by the high-

---

[5]The CsCl structure is not body-centered cubic. A body-centered cubic structure has an identical environment at the corners and in the center of the unit cell. These environments are different in the CsCl structure.

temperature form of an alloy of NiTi, often called "memory metal" (Chapter 9). Other compounds that crystallize with the CsCl structure are listed in Appendix 5.6.

Octahedral Holes

The prototypical example of a structure featuring octahedral holes is the NaCl structure♦, the structure adopted by a large number of salts having 1:1 cation-to-anion ratios. In NaCl, where the ionic radii are 1.16 and 1.67 Å for Na^+ and Cl^-, respectively, the larger chloride ions may be viewed as sitting in ccp or fcc positions, forming octahedral holes, as shown in Figure 5.13. The structure emphasizing the fcc arrangement of Cl^- ions demonstrates that in the unit cell there is one octahedral hole for each close-packed Cl^- ion, which is filled by the smaller Na^+ ion. A total of four Cl^- ions are in the fcc positions of the unit cell (eight corner atoms times

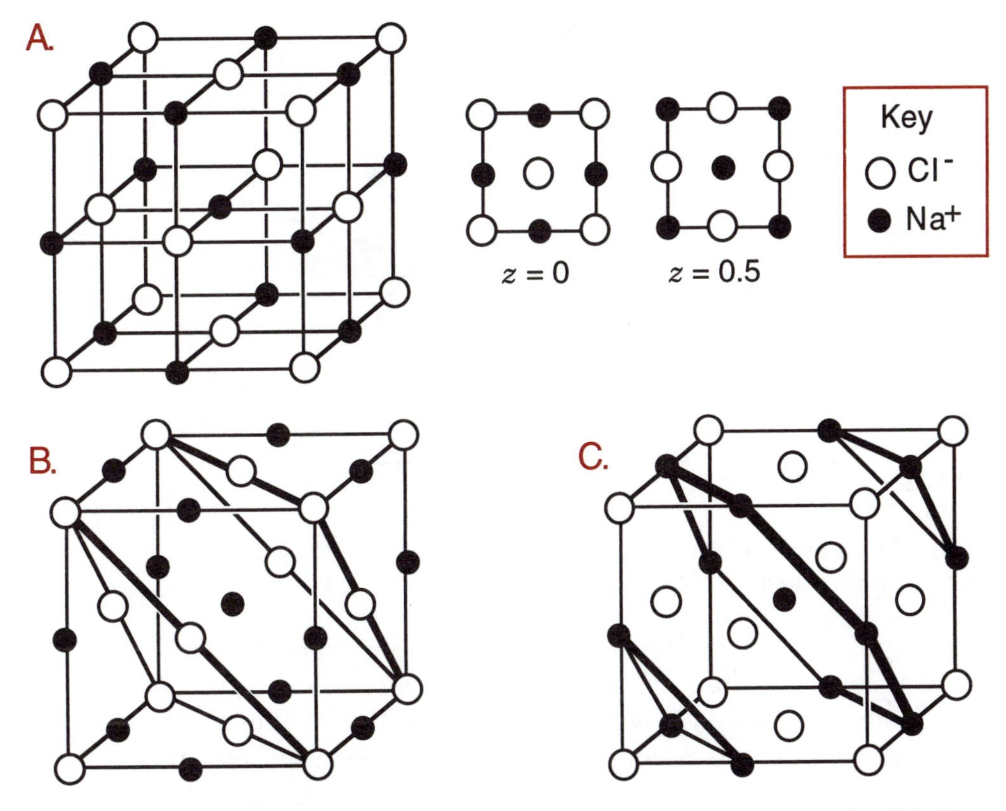

Figure 5.13. The NaCl structure. A: One unit cell of the NaCl structure in which the chloride ions may be viewed as forming a face-centered cubic arrangement with the sodium ions in the octahedral holes. B: A view of the unit cell in which the expanded close-packed planes of chloride ions are outlined. C: A view of the unit cell in which the expanded close packed planes of sodium ions are outlined. In B and C, these planes are perpendicular to a body diagonal of the cube (*see also* Figure 5.9).

1/8, plus six face-centered atoms times 1/2); one Na^+ ion is entirely within the unit cell; and 12 atoms are on the edges of faces of the unit cell (12 × 1/4 = 3), for a total of four. The unit cell thus corresponds to the formula Na_4Cl_4 or to the empirical formula NaCl. Although they are not close-packed, the Na^+ ions are packed in an expanded fcc geometry, with Cl^- ions filling all of the larger octahedral holes formed by the expanded arrangement of cations. This gives interpenetrating fcc arrangements of anions and cations, and each ion has a coordination number of six. Other compounds with the NaCl structure are listed in Appendix 5.6.

The octahedral hole is shown in isolation♦ in Figure 5.14. A view worth emphasizing is the creation of the octahedral hole by two triangular sets of spheres that arise from adjacent close-packing planes (the octahedron is resting on one of its faces that defines an equilateral triangle). The more conventional view of the octahedron resting on one of its vertices♦ is shown in Figure 5.14B.

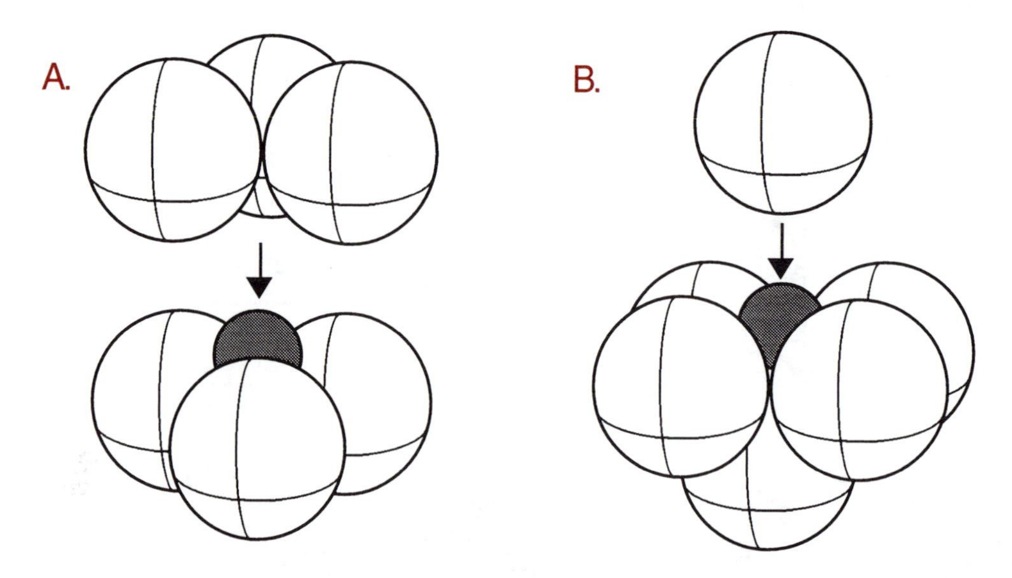

Figure 5.14. Two views of an octahedral hole. A: An octahedron resting on a triangular face. B: An octahedron resting on a vertex.

Fractional filling of octahedral holes also occurs. For example, with $CdCl_2$, the chloride ions are close packed, and only every other layer of octahedral holes (one-half of all such holes) is filled with the smaller Cd^{2+} ions♦, leading to the 1:2 cation-to-anion stoichiometry (*see* Figure 5.15). In this case, the coordination number of Cd^{2+} would still be six, but the coordination number of Cl^- would be taken as three, because each anion is surrounded only by three of the cations.

Since the packing of adjacent layers is identical in hcp and ccp structures, octahedral holes must arise in hcp packing, as well. The structures of NiAs♦ (Figure 5.16) and CdI_2♦ (Figure 5.17) illustrate complete filling and half filling of octahedral holes that result from hcp arrangements of As and I, respectively. Other solids that have these structures are listed in Appendix 5.6.

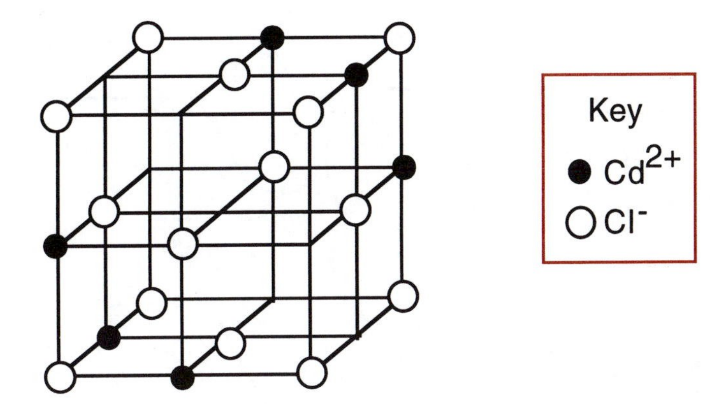

Figure 5.15. The $CdCl_2$ structure. This structure is generated by removing alternate planes of cations from the NaCl structure. The picture shown here does not represent a complete unit cell.

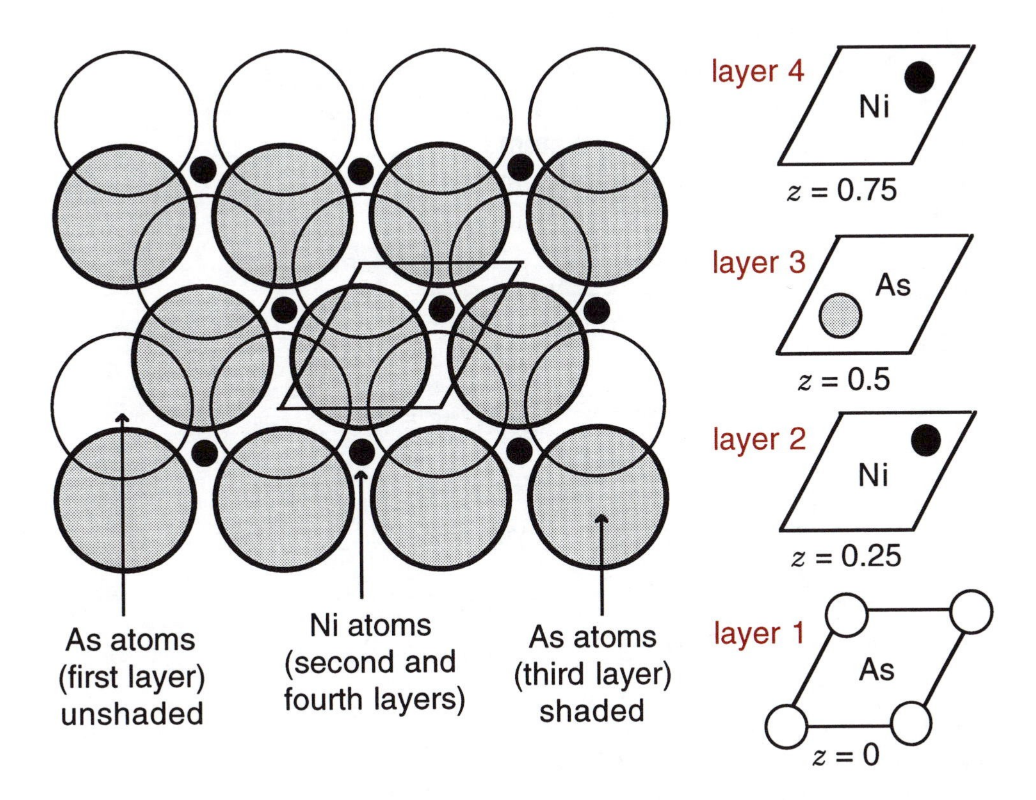

Figure 5.16. The NiAs structure. The arsenic atoms form a hexagonal close-packed arrangement (layers 1, 3, and layer 1 repeated are the ABA layers) while the nickel atoms occupy the octahedral holes. (The nickel atoms in layers 2 and 4 are in registry directly above each other and thus superimposed in the left view above.) The NiAs structure may be viewed as the hcp analog of the NaCl structure. There are two Ni atoms and two As atoms in the unit cell.

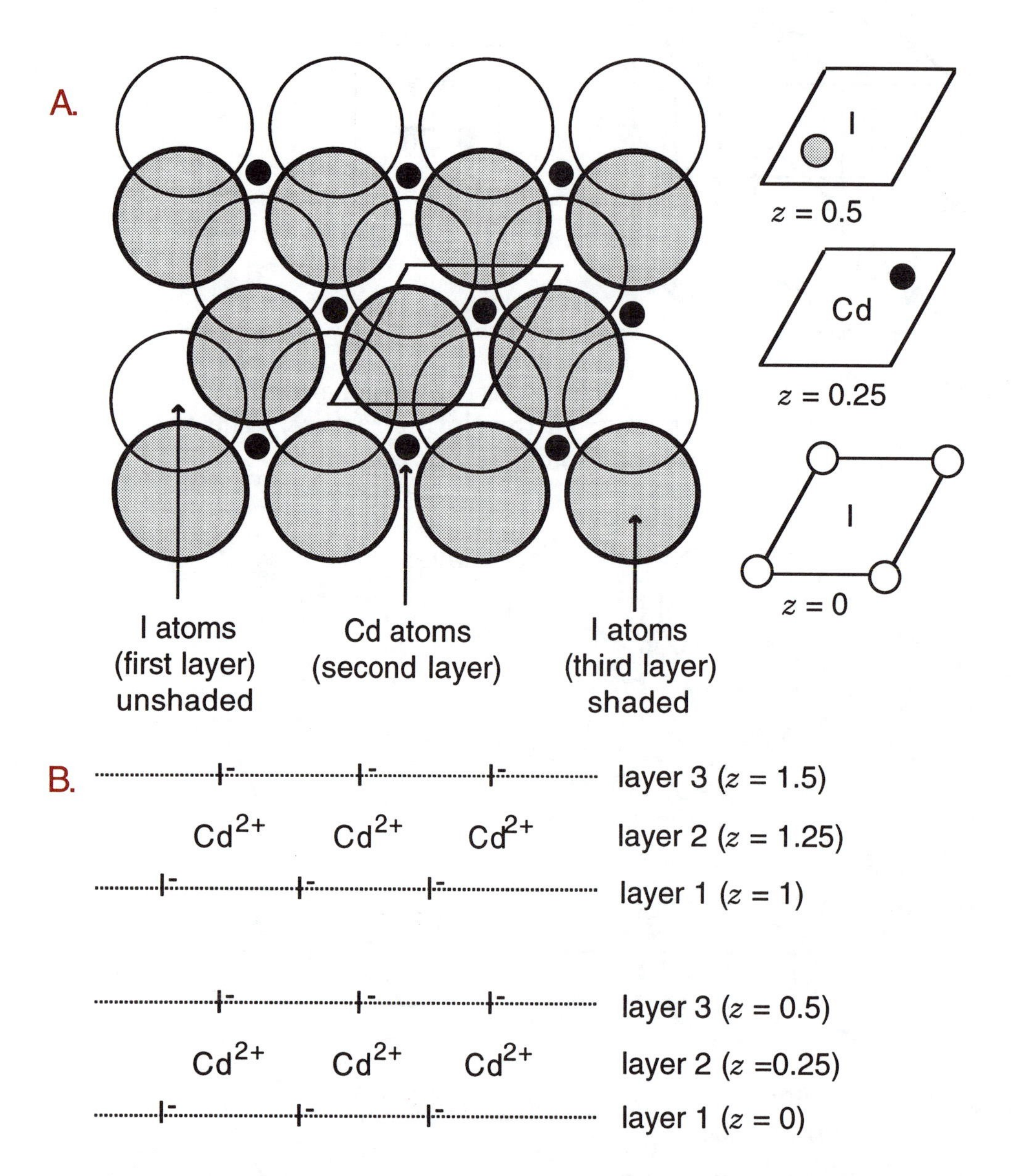

Figure 5.17. The CdI_2 structure. A: The iodine atoms form a hexagonal close-packed arrangement with the cadmium atoms in one-half of the octahedral holes (alternating planes of cations have been removed from the NiAs structure). B: A side view of the CdI_2 structure (this view is at 90° to the view shown in A) and the missing plane of Cd atoms (center). The CdI_2 structure may be viewed as the hcp analog of the $CdCl_2$ structure.

Tetrahedral Holes

As described earlier, tetrahedral holes resulting from close-packed spheres are relatively small. Figure 5.18 illustrates the formation of a tetrahedral hole both by capping an equilateral triangle of three close-packed spheres with a fourth, a view that emphasizes the formation of such holes between close-packed layers of spheres, and by placing two pairs of spheres perpendicular to one another.♦

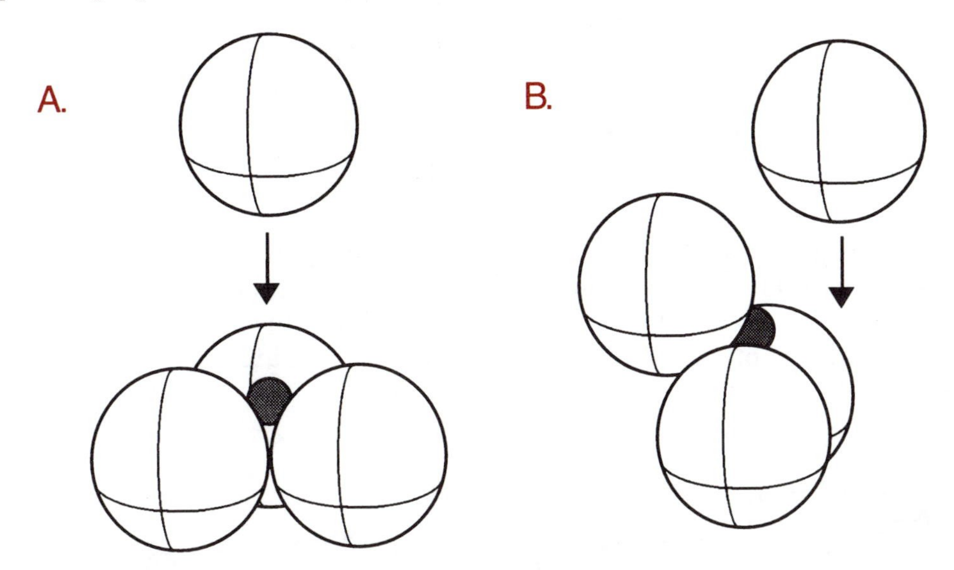

Figure 5.18. Two views of the tetrahedral hole. A: A tetrahedron generated by capping an equilateral triangle of spheres with a fourth sphere. B: A tetrahedron generated by two pairs of spheres with perpendicular orientations.

The antifluorite structure♦ (Figure 5.19) features dianions in an fcc (ccp) arrangement and monocations in all of the tetrahedral holes. This structure is adopted by Li_2O, for which the ionic radii are 0.73 and 1.28 Å for Li^+ and O^{2-}, respectively.[6] Inspection of the unit cell reveals that the four O^{2-} ions (eight corner atoms times 1/8, plus six face-shared atoms times 1/2) composing the fcc unit cell yield eight tetrahedral holes, all within the confines of the cell. There are thus two tetrahedral holes for each close-packed sphere, and in Li_2O they are all filled.

In fluorite, CaF_2♦, the positions of the cations and anions are reversed, as shown in Figure 5.20A. The Ca^{2+} ions are seen to lie in an "expanded"

[6]The radius ratio predicts octahedral coordination for Li^+, but there are not enough octahedral holes (one per oxygen) to match the required stoichiometry. Consequently, Li^+ fills all of the tetrahedral holes instead. This example reinforces the notion that the radius ratio is only a rough guide to predicting structure.

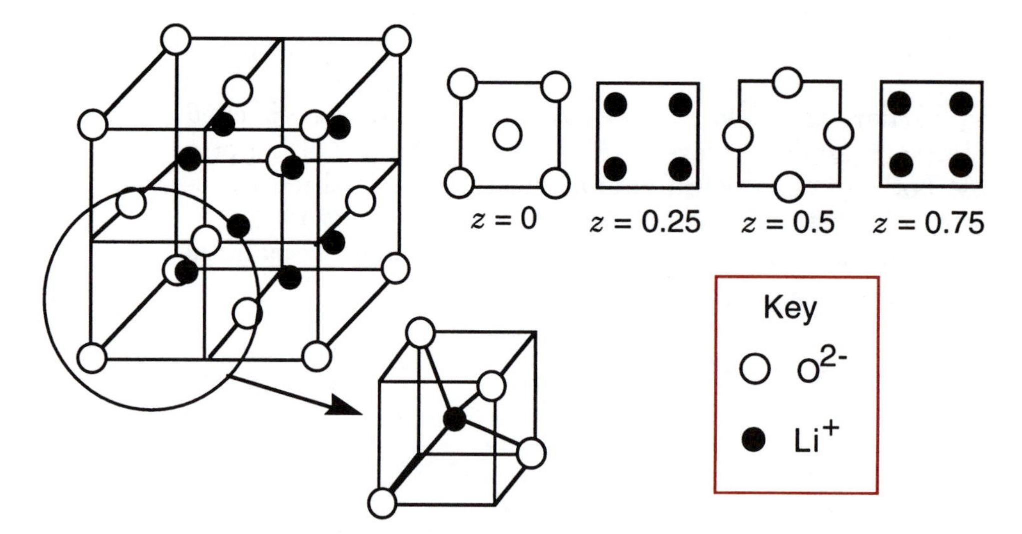

Figure 5.19. The Li_2O (antifluorite) structure. The oxygen atoms form an fcc array, and the lithium atoms occupy the eight tetrahedral holes in the unit cell. Expansion of the circled region shows the tetrahedral coordination of oxygen atoms about the lithium atom in one of the octants of the unit cell.

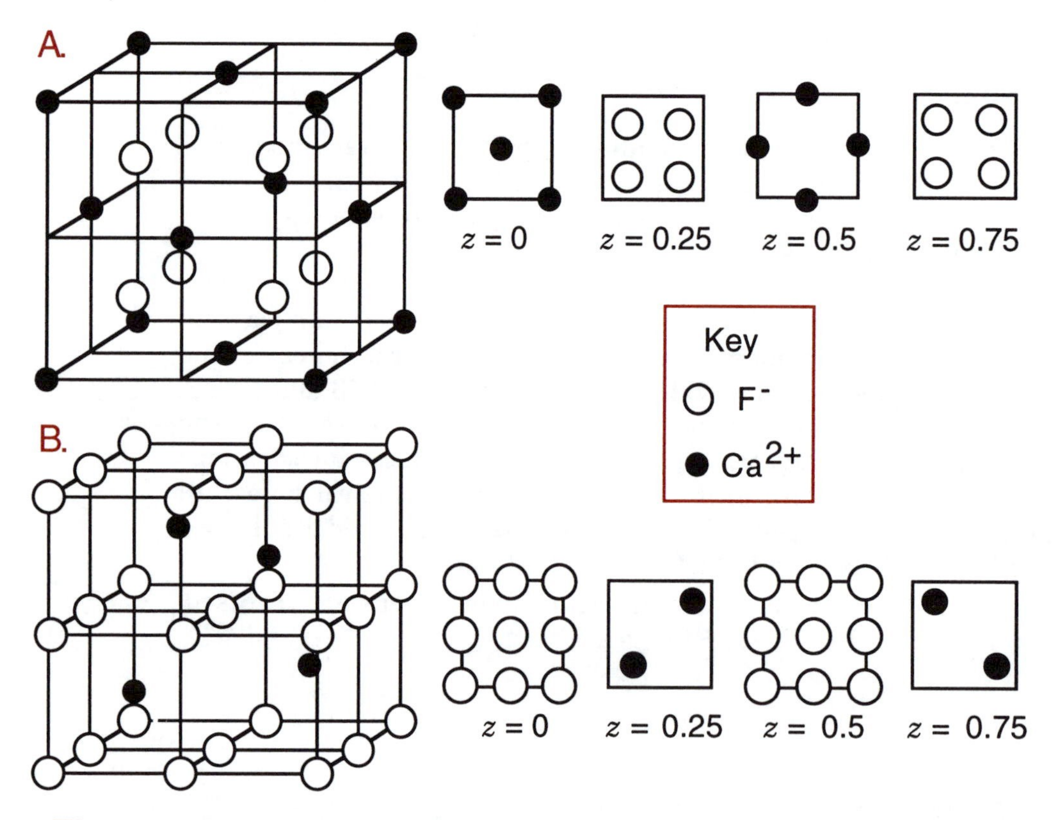

Figure 5.20. A: The CaF_2 (fluorite) structure. B: An alternate view of the CaF_2 structure in which the fluoride ions form a simple cubic arrangement with the calcium ions in half of the cubic holes.

face-centered cubic arrangement;[7] the intervening anions cause their internuclear separation to be a substantial 3.9 Å. This separation distance generates expanded tetrahedral holes for the F^- ions; again, all of the tetrahedral holes are filled, consistent with the 1:2 ion ratio in the salt.[8] Other compounds that have the fluorite or antifluorite structures are listed in Appendix 5.6.

The fluorite structure is also exhibited by a strikingly beautiful purple metallic alloy, $AuAl_2$, often dubbed the "purple plague." In the electronics industry, when Au and Al were sometimes used in close proximity on circuit boards, a purple color often appeared between strips of the two metals. (When the circuit board was used, it would become warm and cause the metals to diffuse and form $AuAl_2$. This alloy is conductive, so the circuit would short.) X-ray diffraction experiments and elemental analysis established the alloy to be $AuAl_2$, with the larger Au atoms (1.34 Å radius) in fcc positions and the smaller Al atoms (1.25 Å radius) in expanded tetrahedral holes.

Appendix 5.7 gives a procedure for preparing a sample of $AuAl_2$. The reactivity of this intermetallic compound provides an excellent illustration of the amphoteric character of aluminum: both acids and bases dissolve the aluminum from the $AuAl_2$, and thus permit the gold to be recovered essentially quantitatively.

Fractional filling of the small tetrahedral holes is illustrated by the important zinc blende structure, adopted by many common semiconductors (Chapter 7). In zinc blende (sphalerite)♦, a mineral form of ZnS, the larger sulfide ions (radius 1.70 Å) are found in a face-centered cubic arrangement similar to the oxide ions in antifluorite. In this case the small Zn^{2+} ions (radius 0.74 Å) are located in exactly half of the tetrahedral holes. The unit cell is shown in Figure 5.21. The Zn^{2+} ions themselves define an expanded tetrahedral hole for the sulfide ions, giving each ion a coordination number of four. The interpenetrating nature of the cations and anions allows each to produce a tetrahedral geometry. Other compounds that have the zinc blende structure are listed in Appendix 5.6.

The mineral wurtzite also has the formula ZnS♦. It differs from zinc blende in that the sulfide ions are now in an hcp rather than a ccp (fcc) arrangement (*see* Figure 5.22). But again, each ion is surrounded by four counterions in a tetrahedral geometry. Other compounds that have the wurtzite structure are listed in Appendix 5.6.

---

[7]Even though CaF_2 has a cubic unit cell, it is often found as octahedral crystals (*see* Demonstration 5.4). External shapes of crystals reflect the relative growth rates of their atomic planes. The octahedral faces of CaF_2 can be related to its cubic unit cell. *See* Appendix 5.5.

[8]CaF_2 may also be viewed as a simple cubic lattice of fluoride ions with the calcium ions in half of the cubic holes. (*See* Figure 5.20B.) The coordination number of the Ca^{2+} ions is 8 and that of the F^- ions is 4 (with tetrahedral coordination). The related structure, antifluorite, may be treated in a similar way: The SSMK shows different views that reveal large anions in fcc positions and small cations in expanded sc positions. Again, half of the cubic holes, now formed by the small cations, are filled by anions, yielding the same overall stoichiometry.

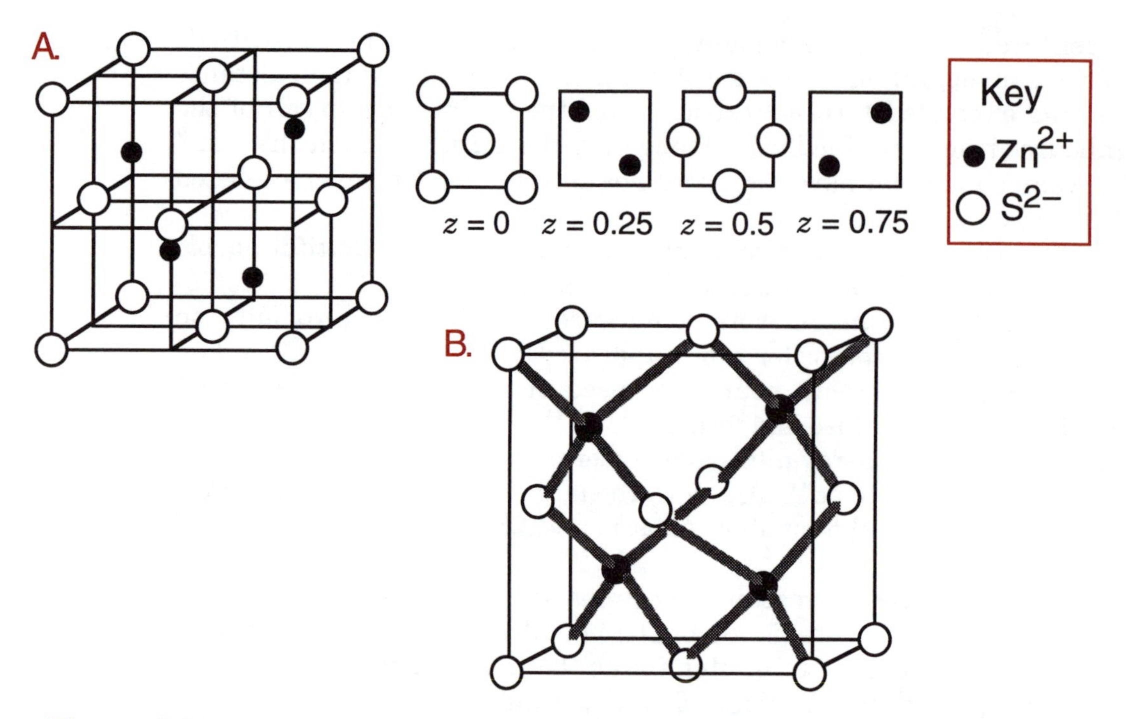

Figure 5.21. The zinc blende (sphalerite) structure. A: A view of the sphalerite structure showing that the sulfur atoms form a face-centered cubic arrangement with the zinc atoms in half of the tetrahedral holes. B: A view of the structure that emphasizes the tetrahedral coordination of the Zn atoms.

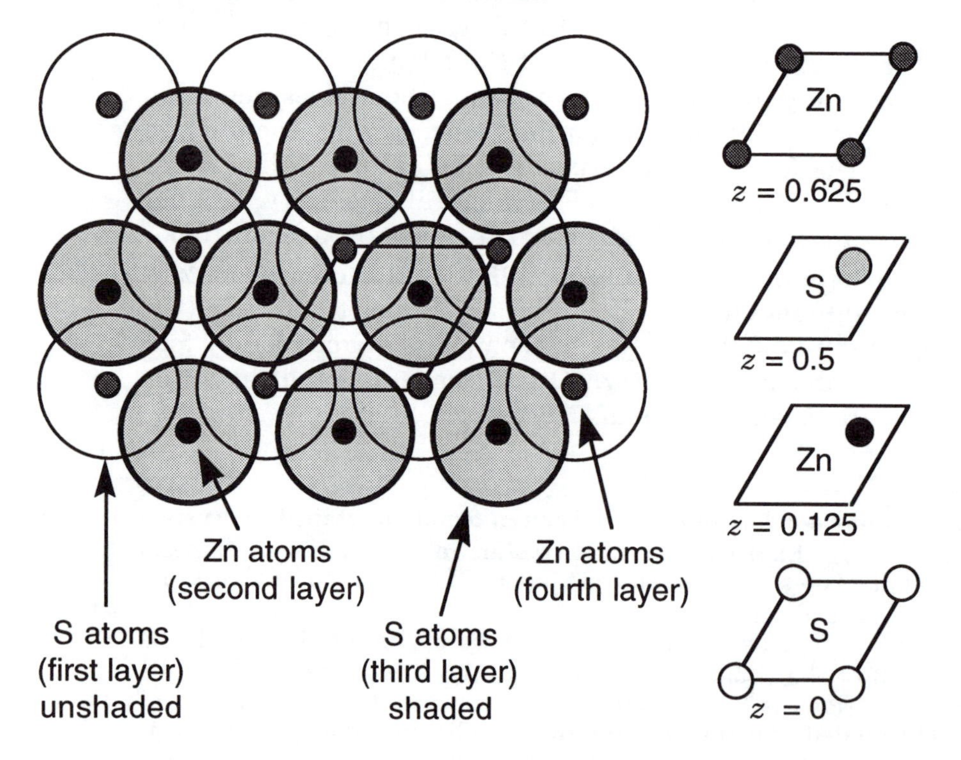

Figure 5.22. The wurtzite structure.

Sphalerite and wurtzite are polymorphic forms of zinc sulfide. In general, polymorphs are compounds (or in some cases, elements) that have the same ionic formulas or repeat units but exist in two or more crystalline phases, differing in atomic arrangement.[9]

Summary of Ionic Structures

In a close-packed structure, there are two tetrahedral holes and one octahedral hole per close-packed atom. The location of these holes is shown in Figure 5.23. Many structures are based on complete or partial occupancy of the tetrahedral or octahedral holes in close-packed arrays of atoms. This viewpoint provides a unifying approach to teaching the many structures that may initially seem unrelated. A summary of structures that can be described in this way is given in Table 5.3; structures that are based on the filling of cubic holes appear in Table 5.4.

Table 5.3. Structures Based on Close Packing of Anions, X

	Type of Close-Packing for X		Coordination Number	
Positions of Cations	ccp	hcp	M	X
All octahedral holes	NaCl♦	NiAs♦	6	6
Half of octahedral holes	$CdCl_2$[a]♦	CdI_2[a]♦	6	3
One-third octahedral holes	$CrCl_3$	BiI_3	6	2
All tetrahedral holes	Li_2O[b]♦	–	4	8
	CaF_2[c]♦	–	8	4
Half of tetrahedral holes	ZnS (zinc blende)♦	ZnS (wurtzite)♦	4	4

[a]Alternating planes of cations are removed (see Figure 5.15). In the case of $CdCl_2$, this corresponds to removal of layers of cadmium ions perpendicular to the body diagonal of the cubic unit cell.
[b]This is the antifluorite structure (anions ccp).
[c]This is the fluorite structure (cations ccp).
SOURCE: References 9 and 10.

Table 5.4. Structures Based on Simple Cubic Packing of Anions, X

		Coordination Number	
Position of Cations	Example	M	X
All cubic holes filled	CsCl♦	8	8
Half of cubic holes filled	CaF_2[a]♦	8	4
Half of cubic holes filled	Li_2O[b]♦	4	8

[a]This is an alternate way of viewing the fluorite structure (anions sc).
[b]This is an alternate way of viewing the antifluorite structure (cations sc).
SOURCE: References 9 and 10.

---

[9]Contrast polymorphs with allotropes, which are different forms of the same element. In this case the number of atoms in the formula units may be also different (e.g., O_2 and O_3). Reference 8 gives a further discussion of the distinction between polymorphs and allotropes.

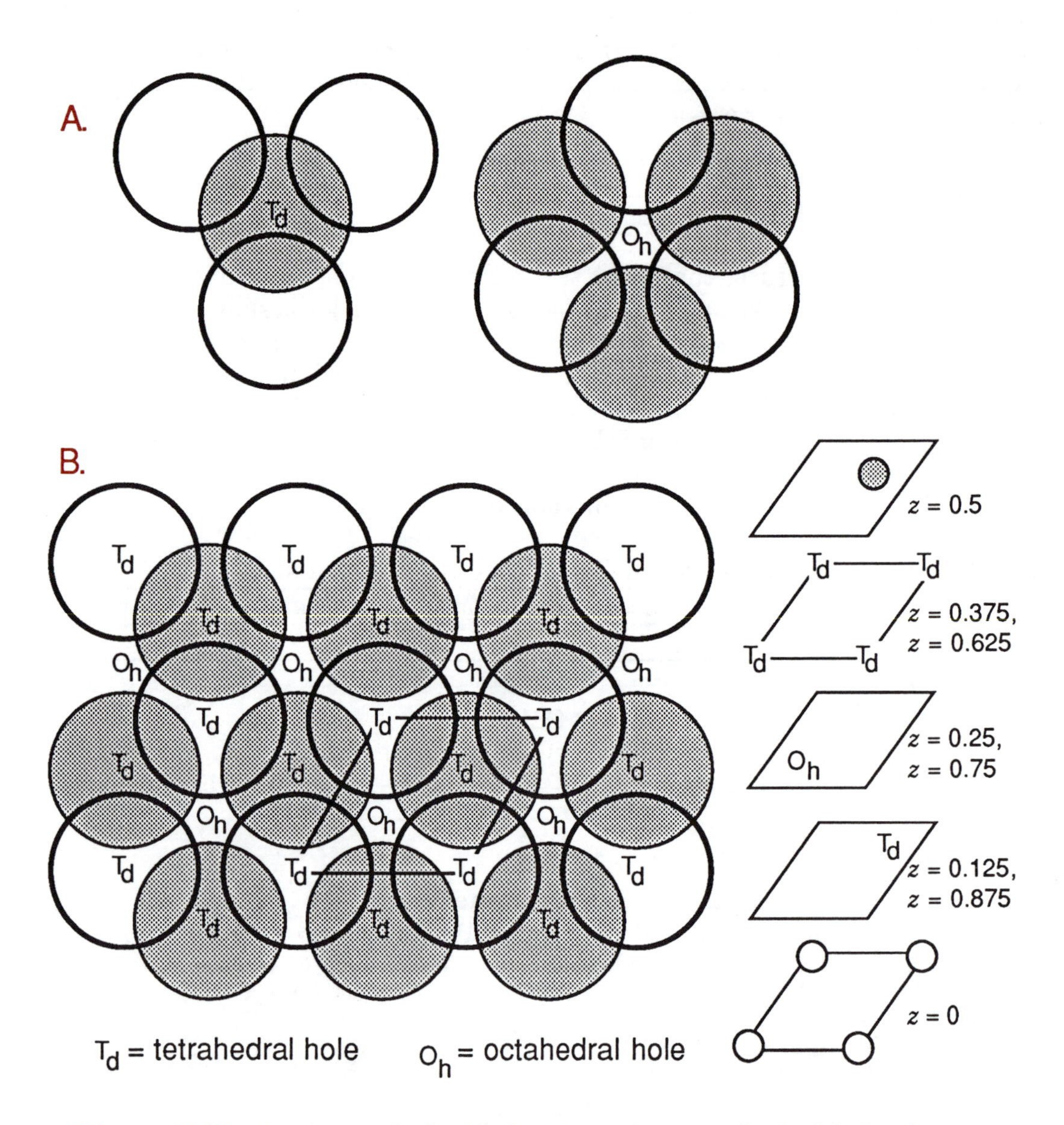

Figure 5.23. A: A tetrahedral hole, T_d, and an octahedral hole, O_h, shown in isolation (*see also* Figures 5.14 and 5.18). B: The first two close-packed layers of atoms in a hcp arrangement of atoms showing the location of octahedral and tetrahedral holes. A parallelogram that defines the base of a unit cell is superimposed. The *z*-layer diagrams for the unit cell show that there are two tetrahedral holes and one octahedral hole per close-packed atom. (Just as atoms on an edge of a unit cell count as contributing one-quarter of their volume to the unit cell, the tetrahedral holes on unit cell edges may be viewed as being one-quarter within a particular unit cell.) There are also two tetrahedral holes and one octahedral hole per atom in a fcc unit cell (*see* Figures 5.13 and 5.19).

Laboratory. Experiment 2 is a laboratory experiment that uses the SSMK to explore the structures of metals and ionic compounds. Topics such as packing efficiency, coordination number, and hole size are also covered.

Covalent Solids

In passing to covalent extended solids, directional bonding replaces the largely nondirectional bonding characteristic of metals and salts. The prototypical example of a covalently bonded solid is the diamond allotrope of carbon. The diamond structure is also exhibited by the elements silicon, germanium, and α-tin. Inspection of the cubic unit cell of the diamond structure♦ (Figure 5.24) reveals it to be a very open structure relative to those of metals and salts. The packing efficiency is, in fact, only about 37%, or about half that of the common metal structures. Each carbon atom is tetrahedrally bonded to four other carbon atoms. Eight atoms compose the unit cell: four are contained entirely within it. The other four are represented by the eight atoms at the corners of the cube and the six in the centers of the cube faces. A comparison with the zinc blende structure (Figure 5.21) demonstrates that the two structures are similar, the difference being that two kinds of atoms are used in zinc blende, with each atom tetrahedrally coordinated exclusively to the other kind of atom.

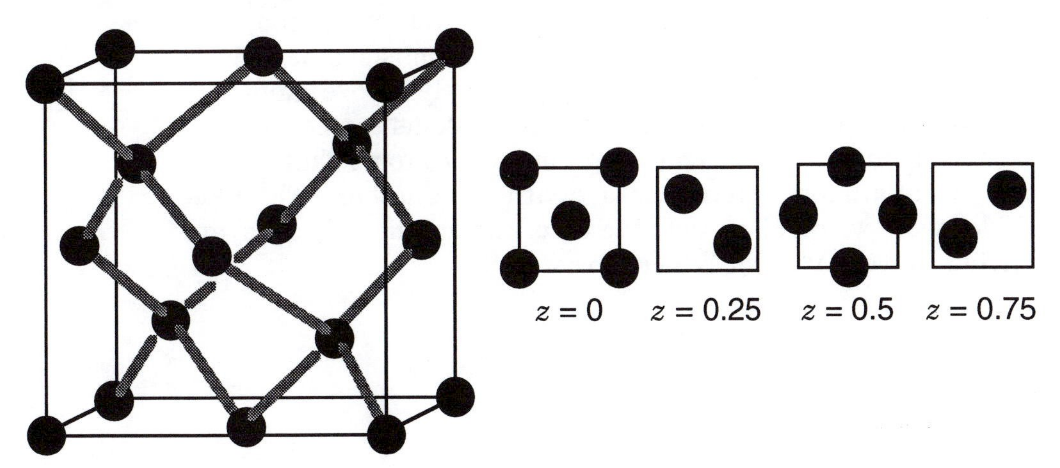

Figure 5.24. The diamond structure.

Another important allotrope of carbon is graphite. Unlike the structures described thus far, graphite's strong covalent bonds lie entirely within planes, making this a two-dimensional layered solid (*see* Figure 5.25). Construction of a model of the structure♦ shows that the spacing between layers (3.35 Å) is substantially larger than the interatomic distance within layers of 1.42 Å. The coordination number of the carbon atoms is only three. Layers are held together primarily by weak van der Waals forces, permitting them to slide relative to one another. Graphite is a lubricant because of this property.

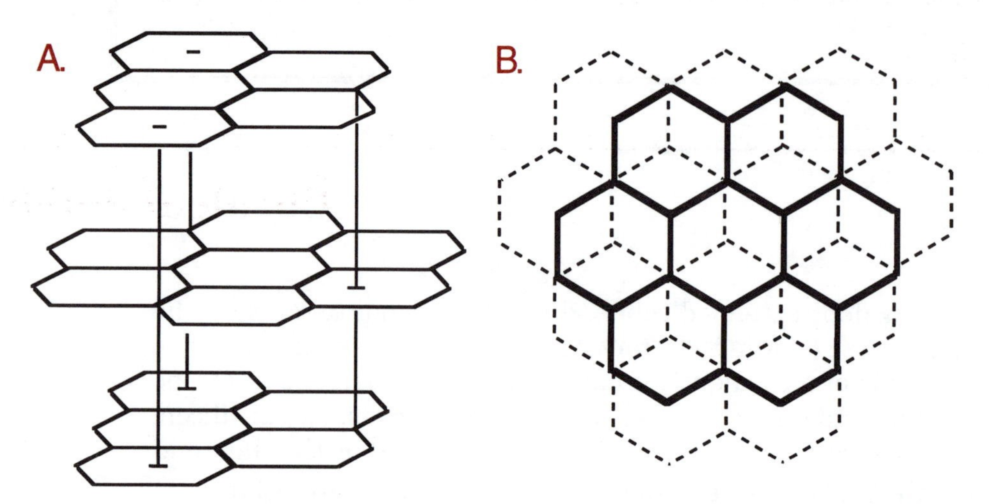

Figure 5.25. Two views of graphite. A: Side view of graphite showing the planar arrays of fused six-membered rings. The second layer is not aligned with the first; however, alternate layers are in registry. B: Top view of graphite showing how the second layer (dashed lines) is oriented with respect to the first layer (solid lines). Half of the atoms have another atom directly below them, and the other half do not.

Another layered solid with a larger component of ionic bonding than graphite is the mineral molybdenite (molybdenum disulfide)♦. As shown in Figure 5.26, the coordination number of the molybdenum is six, defining a trigonal prismatic (one equilateral triangle directly above another) geometry; and that of sulfur is three, because sulfur atoms cap equilateral triangles of molybdenum atoms. The layers of sulfur atoms that do not sandwich molybdenum atoms are in contact with one another. The like-charged sulfide layers would be expected to repel one another; these layers are held together by weak van der Waals forces.

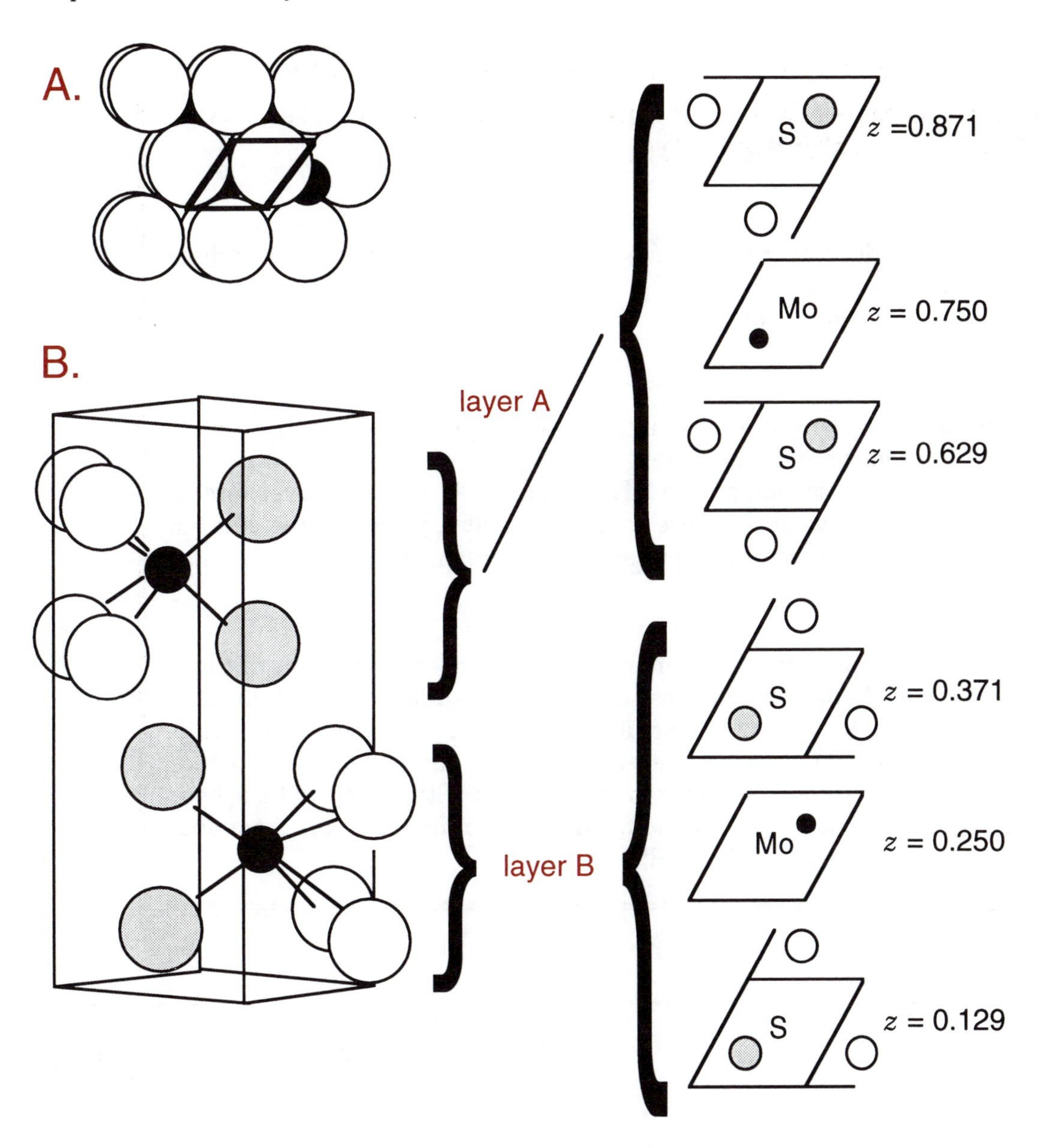

Figure 5.26. The molybdenum sulfide (MoS_2) structure. A: Top view. The structure may be viewed as consisting of pairs of eclipsed hexagonal arrays of sulfur atoms with molybdenum atoms in the trigonal prismatic holes. A unit cell is superimposed. B: Side view. The molybdenum atoms (black) and the shaded sulfur atoms are within the unit cell. The molybdenum atom in layer A is directly above the sulfur atoms in layer B, and the sulfur atoms in layer A are directly above the molybdenum atom in layer B. Layer A is attracted to layer B by weak van der Waals forces. The sulfur atoms in layer A are sitting in the depressions created by the hexagonal array of sulfur atoms in the adjacent layer. Consequently, alternate layers are in registry.

Demonstration 5.5. Removal of Graphite and Molybdenum Sulfide Layers

Materials

MoS_2 (Wards Natural Science, *see* Supplier Information) and/or graphite lumps
Transparent cellophane tape

Procedure

- Lightly press a piece of tape against a sample of MoS_2 or graphite and then remove it. Some layers of the solid will stick to the tape.
- As noted in the section on the STM (Chapter 2), the interlayer forces are sufficiently weak in these materials that a piece of tape can be used to lift off layers from them and reveal fresh surfaces.

Variations

- Students can perform this demonstration themselves if pieces of tape (or a tape dispenser) and samples of the solids are distributed throughout the class. Alternatively, pieces of tape with small amounts of the layered solid already attached can be lightly fastened to a rod or a yardstick for ease in distribution. Students can be asked to lightly fold the tape onto itself, which will transfer layers from one half of the tape to the other half.

Molecular Solids

The molecules in molecular crystals are often held together by weak van der Waals forces. Some of the same packing arrangements discussed earlier are used here, as well. For example, molecules of buckminsterfullerene, C_{60}, pack in the fcc arrangement in the solid state. This packing can be illustrated by simply building the fcc structure♦ with each sphere representing a C_{60} molecule. In K_3C_{60}, all of the octahedral and tetrahedral holes in the fcc arrangement of buckminsterfullerene anions are filled with potassium cations.

More Connections to Molecular Shapes: Extended Shared Polyhedra

The importance of the octahedron and tetrahedron in extended structures is underscored by the multitude of solids that can be constructed by sharing corners, edges (and occasionally faces) of these shapes. For example, isostructural solids like ReO_3 and idealized WO_3♦ can be described as consisting of MO_6 octahedra with the metal atom at the center and shared oxygen atoms at the corners, as sketched in Figure 5.27. The octahedra are combined to create an extended three-dimensional structure by corner-sharing each of the oxygen atoms between two octahedra. The cubic unit cell corresponds to a chemical formula of MO_3, because only one quarter of each oxygen atom belongs to any given unit cell. Coordination numbers for the metal and oxygen atoms are six and two, respectively.

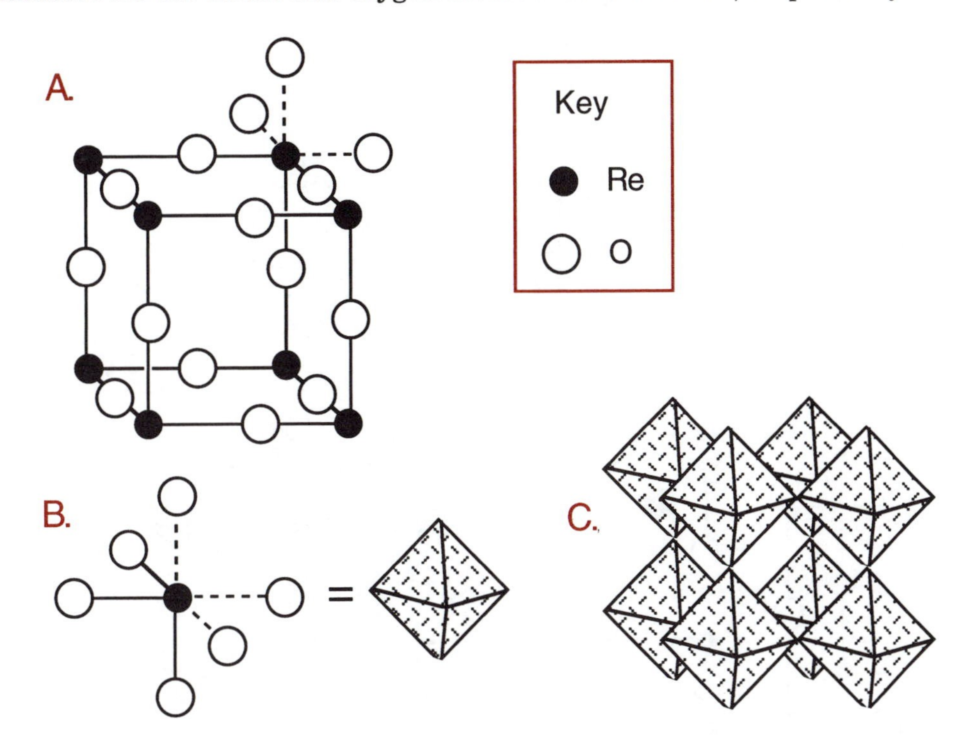

Figure 5.27. Representation of structures by octahedra. A: The ReO_3 unit cell; the dashed lines represent completion of one octahedron of oxygen atoms about a given Re atom. B: Replacement of the ReO_6 coordination sphere with an octahedron. C: The ReO_3 unit cell represented as a series of corner-shared octahedra.

Similarly, the perovskite structure♦ features corner-shared TiO_6 octahedra, with Ca^{2+} ions at the corners of the cubic unit cells in which the octahedra are inscribed, leading to the formula $CaTiO_3$ (*see* Figure 5.28A and 5.28B). A shift in origin that places the larger Ca^{2+} cations at the center of the cubic unit cell leads to another common representation of the structure (Figure 5.28C and 5.28D). Coordination numbers of the calcium, titanium, and oxygen atoms are 12, 6, and 2 (or 6, including contacts of oxygen to both metals), respectively. Among the important examples of related perovskite structures are high-temperature superconductors (Chapters 1 and 9). The sequence shown in Figures 5.29 and 5.30 illustrates how the idealized perovskite structure is altered to produce the superconducting $YBa_2Cu_3O_7$ phase.

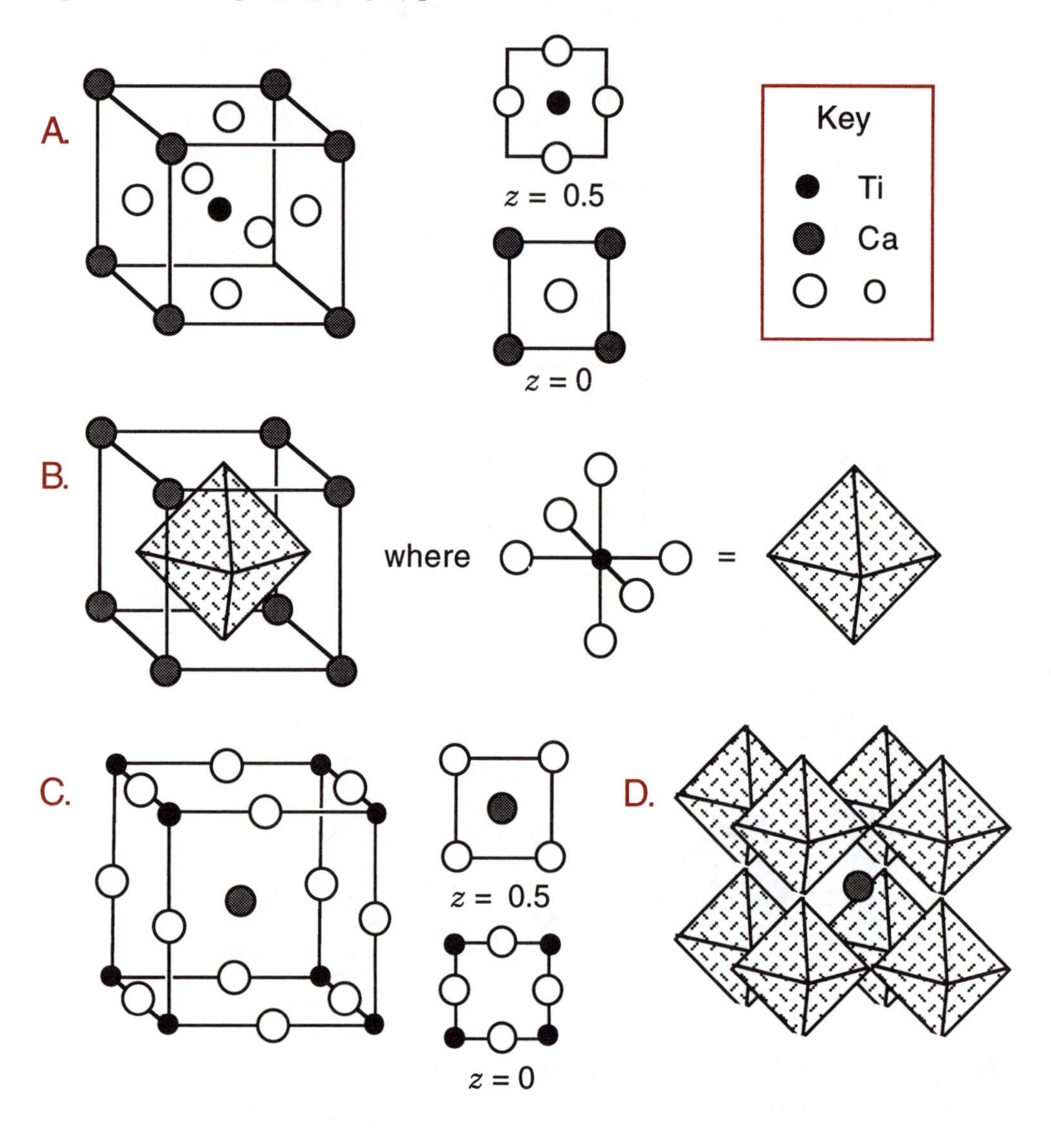

Figure 5.28. The perovskite structure. A: A unit cell with titanium at the center. B: The same unit cell showing the octahedral arrangement of oxygen atoms around a titanium atom. C: Shift in the unit cell to place calcium in the center. D: The unit cell shown in C represented as a set of corner-shared TiO_6 octahedra.

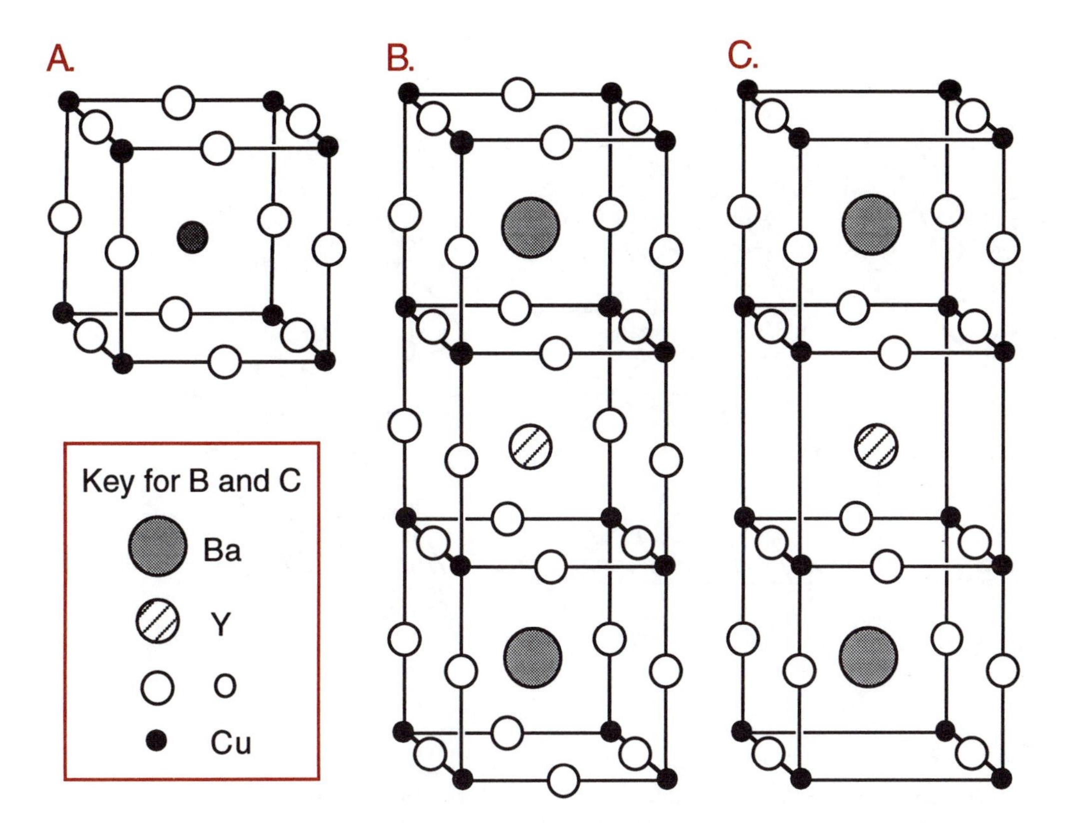

Figure 5.29. Construction of the $YBa_2Cu_3O_7$ superconductor by stepwise alteration of the perovskite structure. A: The perovskite structure as depicted in Figure 5.28C. B: Stacking of three perovskite unit cells followed by replacement of the calcium atoms by barium atoms in the first and third unit cells and by yttrium atoms in the second. In addition, copper is now in the titanium positions. This yields an idealized stoichiometry of $YBa_2Cu_3O_9$. C: Removal of eight edge-shared oxygen atoms (for a total of two oxygen atoms removed) to generate the idealized structure of $YBa_2Cu_3O_7$.

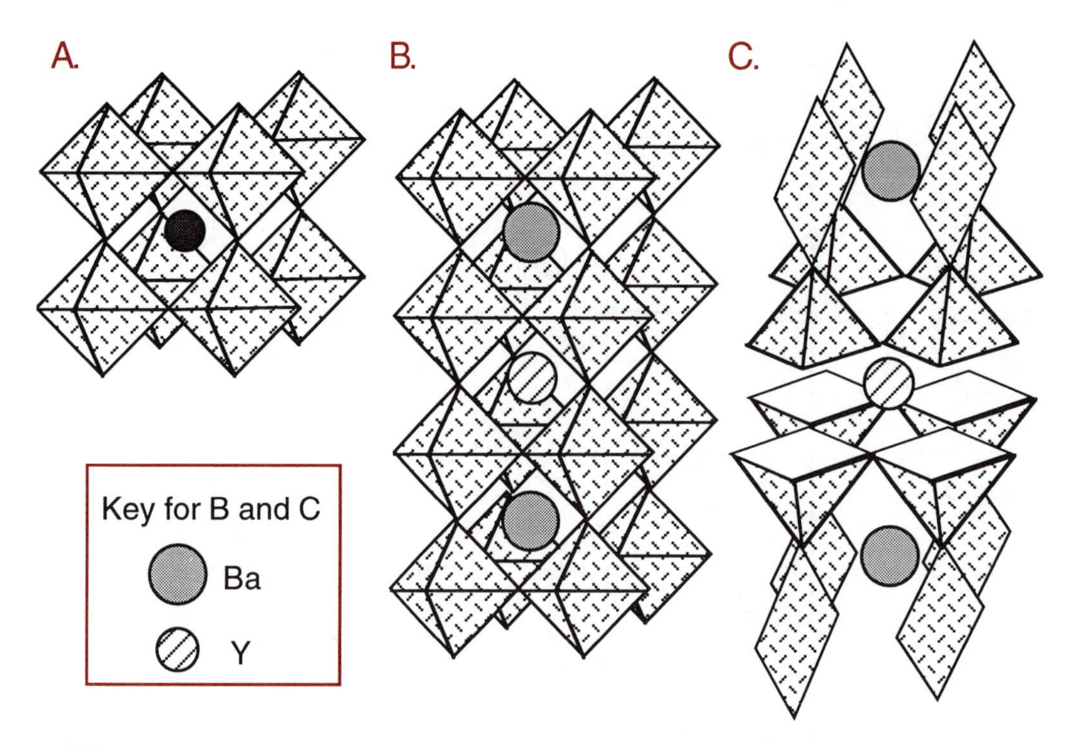

Figure 5.30. Representation of the construction of the $YBa_2Cu_3O_7$ superconductor as a series of fused octahedra. The steps are identical to those described in Figure 5.29. Removal of oxygen atoms in C converts the octahedra to square pyramids and squares.

Like octahedra, tetrahedra can be corner-shared. The most common structures based on linked tetrahedra are undoubtedly those of the silicate minerals *(11)*. The basic building block is the SiO_4^{4-}, or silicate, ion (*see* Figure 5.31A). Though there is not a wide variety of examples, several common minerals contain the silicate anion. These include zircon, $ZrSiO_4$; the garnets, $[(M^{2+})_3(M^{3+})_2(SiO_4)_3]$ where $M^{2+} = Ca^{2+}$, Mg^{2+}, or Fe^{2+}, and $M^{3+} = Al^{3+}$, Cr^{3+}, or Fe^{3+}; and olivine, M_2SiO_4 where $M = Mg^{2+}$ or Fe^{2+}. Both garnet and olivine form extensive series of solid solutions (*see* Chapter 3 and Experiment 14), as a result of substitution at the cation sites.

Successive linking of oxygen atoms at one, two, three, or four vertices of the tetrahedron generates a wide variety of discrete ions, rings, chains, sheets, and three-dimensional structures. If two tetrahedra are linked through the sharing of an oxygen at a common vertex, the result is the disilicate ion, $Si_2O_7^{6-}$ (*see* Figure 5.31B). This ion is not common in nature, although the mineral thortveitite ($Sc_2Si_2O_7$) is one example.

If the silicate ion shares oxygen atoms at two vertices of the tetrahedron, rings or chains may be constructed. Cyclic silicates are known with three, four, six, and eight linked tetrahedra. The most well-known of these is the mineral beryl, which has the formula $[Be_3Al_2(Si_6O_{18})]$ and contains a cyclic array of six tetrahedra (*see* Figure 5.31C). Many examples of linear silicates are known, and minerals with this structural feature are known as pyroxenes. The formula for the silicate repeat unit is $(SiO_3^{2-})_n$ for a pyroxene mineral. Examples include diopside ($CaMgSi_2O_6$) and spodumene ($LiAlSi_2O_6$), the most commercially significant lithium-containing ore.

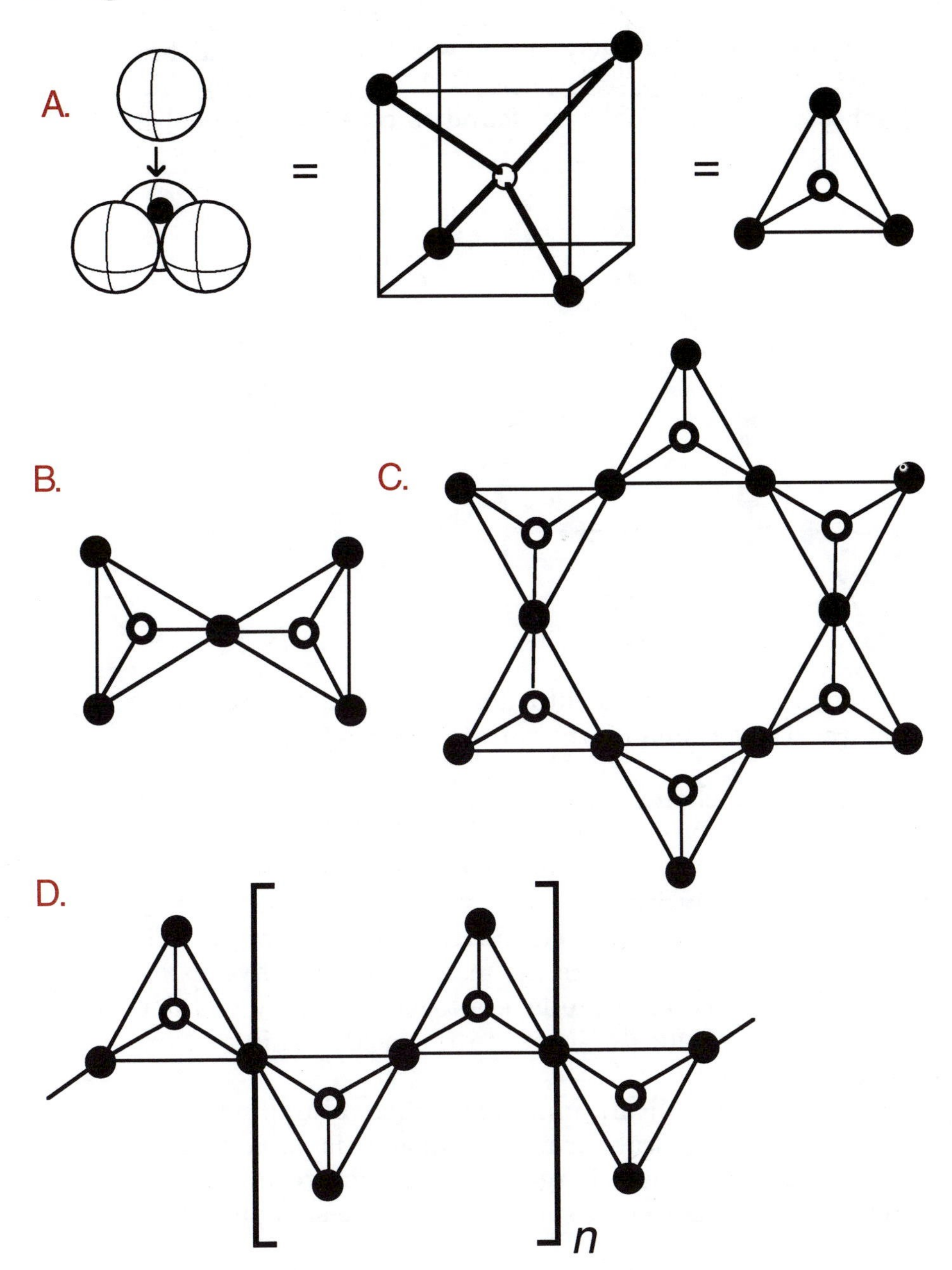

Figure 5.31. Silicate ions. A: Viewing the SiO_4^{4-} tetrahedron down one of the Si–O bonds generates the projection in which the oxygen atoms appear as black spheres at the vertices of a triangle and the silicon atom (white sphere) is below the fourth oxygen. B: The disilicate ion, $Si_2O_7^{6-}$. C: The cyclic $Si_6O_{18}^{12-}$ ion. D: A pyroxene shown without its cations. The silicate portion of the structure has the formula $(Si_2O_6^{4-})_n$.

If the linear chains are linked so as to form double chains with the repeat unit formula $(Si_4O_{11}^{6-})_n$, the series of minerals known as the amphiboles is generated. The silicate tetrahedra share two or three vertices of the tetrahedron (*see* Figure 5.32). A typical example is crocidolite, $Na_2(Fe^{2+})_3(Fe^{3+})_2(OH)_2(Si_4O_{11})_2$. This asbestos mineral has long been valued for its heat and fire resistance. It cleaves in the long thin fibers typical of asbestos, which have been found to constitute a health hazard.

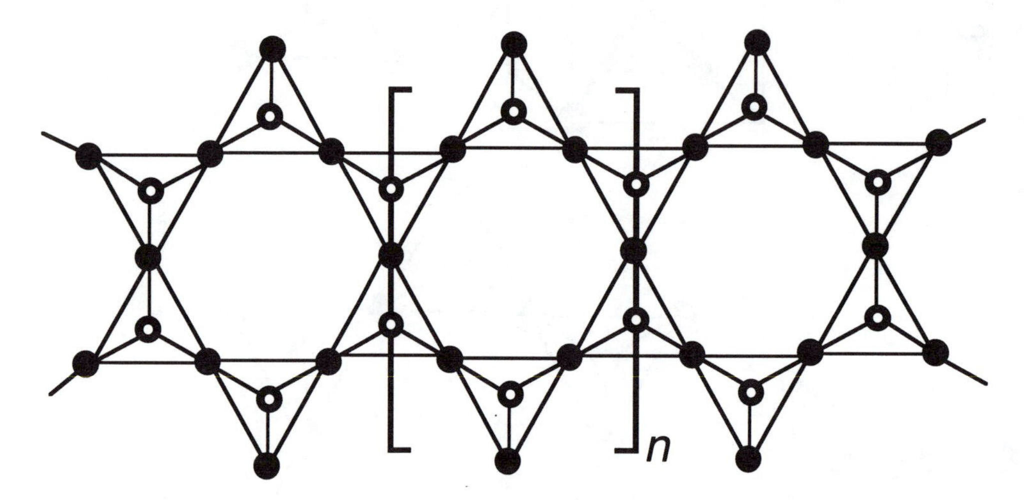

Figure 5.32. Fusing two chains into double chains forms the amphibole structure. The formula for the silicate chain is $(Si_4O_{11}^{6-})_n$.

If the side-to-side linking of chains is continued, a sheet-like structure is produced. The repeat unit for the silicate network is $(Si_2O_5^{2-})_n$ (*see* Figure 5.33). The minerals talc and mica have just such a structure (*see* Figure 5.34A). Its formula is $Mg_3(OH)_2(Si_4O_{10})$, and two parallel sheets actually have the unshared oxygen atoms of the tetrahedra pointing toward each other. In the middle of this sandwich are the magnesium and hydroxide ions, which serve to bind the two silicate layers together. Only van der Waals forces hold one pair of silicate sheets to the next. This feature accounts for the softness of talc.

An important feature of the silicate minerals is that aluminum may substitute for silicon in the tetrahedra. Because this is a formal replacement of Si^{4+} by Al^{3+}, the net charge on the structure must also change. This case is called isomorphous substitution if the structure does not change.

In the mica structure, one of every four silicon atoms is replaced by aluminum, and the formula for the aluminosilicate sheets becomes $AlSi_3O_{10}^{5-}$. These sheets can form a double layer similar to the talc structure. In the mica known as biotite, Mg^{2+} or Fe^{2+}, together with hydroxides, act to hold the sheets together, generating the formula $[(Fe,Mg)_3(OH)_2(AlSi_3O_{10})]^-$. The parentheses around Fe and Mg indicate that they form a solid solution (*see* Chapter 3). A mineral sample may have three iron atoms per formula unit or three magnesium atoms per formula unit (or any combination of the two that adds up to three, including fractional values) and still be classified as biotite.

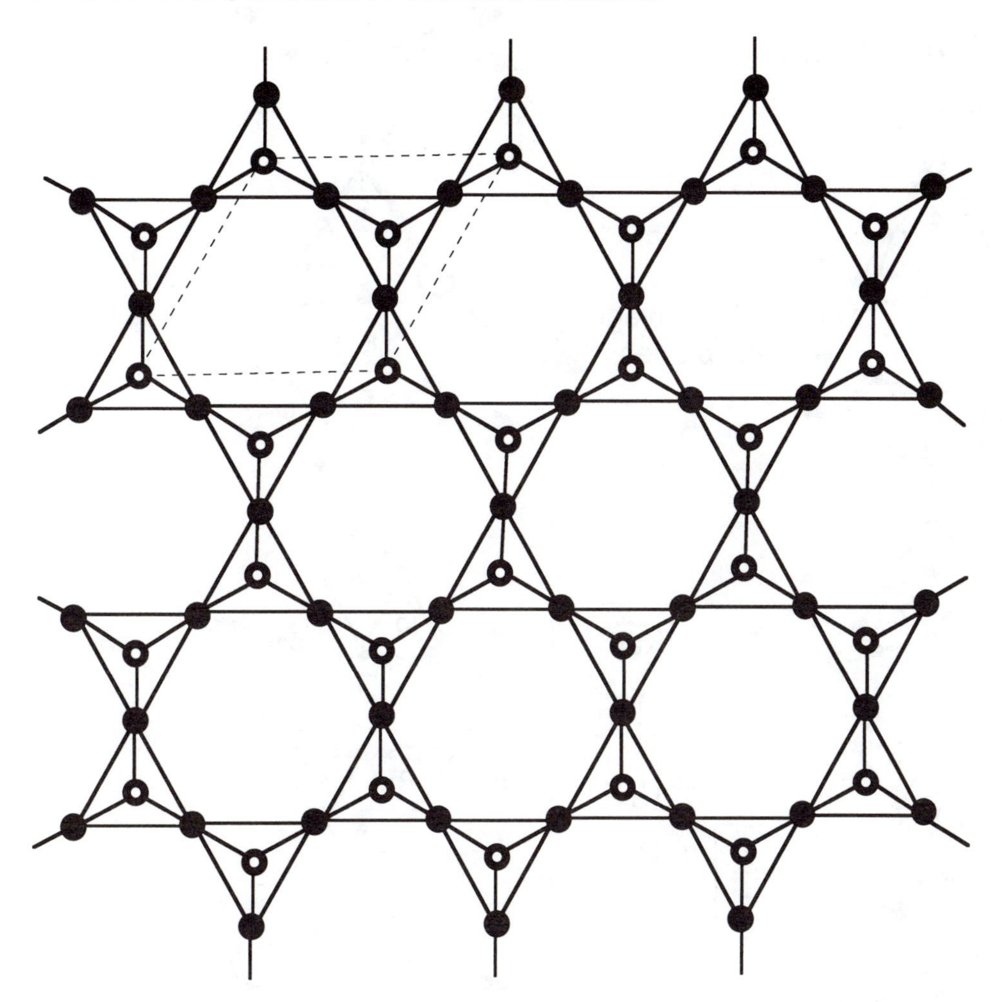

Figure 5.33. Fusing of the double chains forms two-dimensional sheets. The silicate portion of the structure has the formula $(Si_2O_5^{2-})_n$. The dashed lines represent a repeat unit for the sheets.

The exchange of Al^{3+} for Si^{4+} has created an excess of one unit of negative charge per substitution. Electroneutrality requires the addition of a monocation such as K^+, which is located outside the double layers. The presence of the potassium ions helps to hold the double layers to each other and serves to make mica harder than talc (*see* Figure 5.34). However, cleavage still occurs between the double layers. The mineral muscovite $[KAl_2(OH)_2(AlSi_3O_{10})]$ has a structure that is similar to biotite. The clay minerals such as fuller's earth, china clay (kaolin), and vermiculite (a common packing material) also have layered structures.

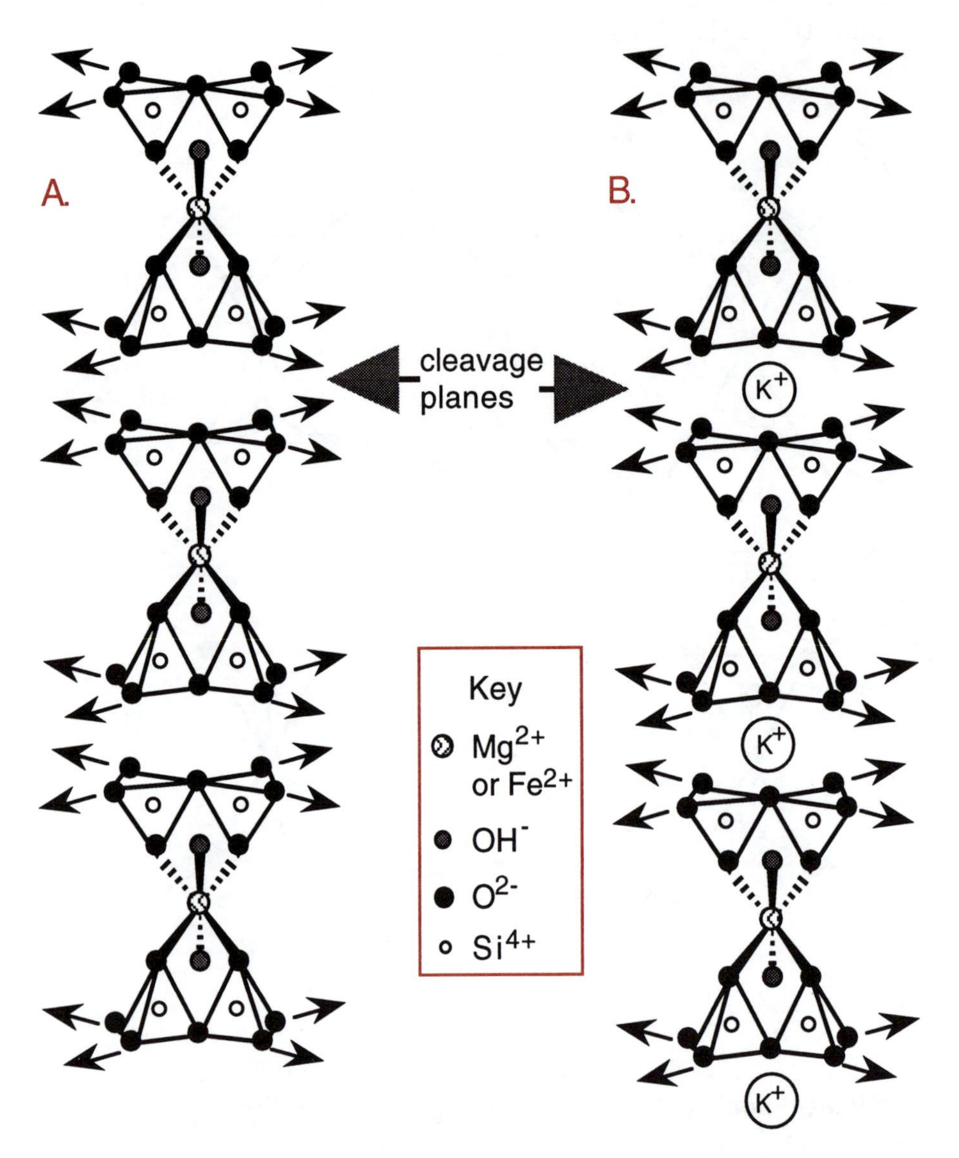

Figure 5.34. A cross-sectional view of the structures of talc and mica. The tetrahedra are portions of sheets as shown in Figure 5.33. The silicate sheets are all parallel, occur in double layer pairs with the unlinked oxygen atoms oriented toward each other, and continue in the direction of the arrows. A: The talc structure. The negatively charged sheets are held together by mutual attraction to a layer of magnesium ions in between. Weak van der Waals forces hold each pair of sheets to the next. B: The biotite mica structure. Replacement of one of every four silicon atoms by aluminum atoms in the sheets creates the need for additional positive charge in order to maintain electrical neutrality. These cations (potassium in biotite) are located outside of the double layer that contains the magnesium ions. Thus they serve to hold one layer to the next. In both cases cleavage involves separation of one double layer from the next.

Laboratory. The use of X-ray diffraction to measure the repeat distance in the stacking direction for layered materials, including these layered silicates, is described in Experiment 5.

Continued linking of silicate or aluminosilicate structures can result in the formation of complex frameworks that contain pores or channels that are large enough for water, small molecules, or cations to fit inside. Such materials are called zeolites (*see* Figures 5.35 and 5.36). They have been used to remove water from organic solvents (molecular sieves), to exchange sodium ions for calcium or magnesium ions in water softeners, and to prepare synthetic gasoline from such feedstocks as methanol. In the latter case, a zeolite catalyst called ZSM–5 was developed by Mobil and is believed to involve sites of high Lewis acidity that activate the methanol toward carbon–carbon bond formation. This catalyst is also shape-selective, permitting only molecules of certain sizes to migrate through the channels.

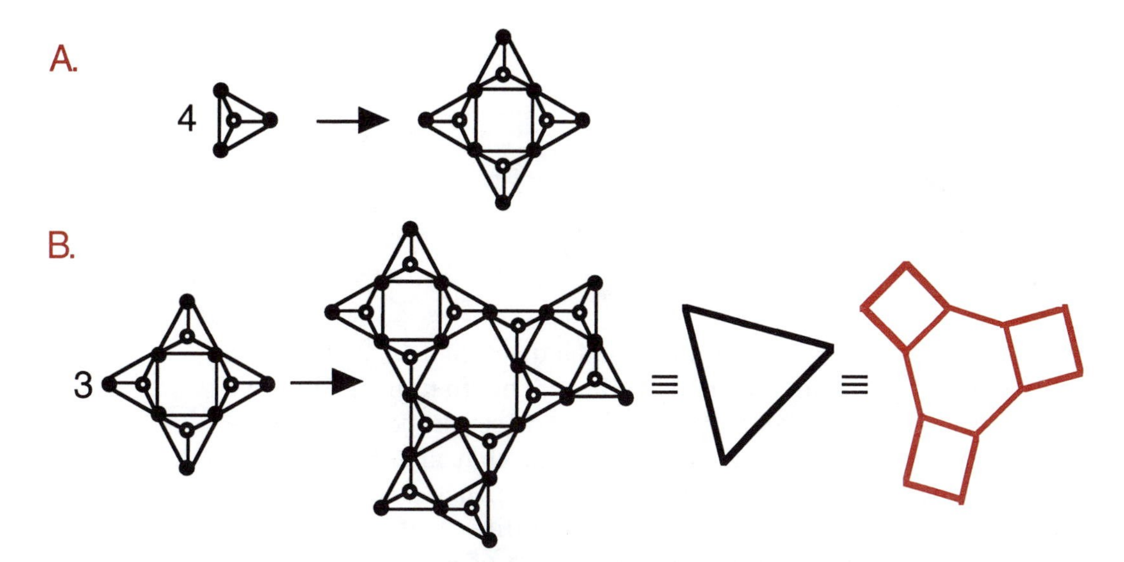

Figure 5.35A–B. Construction of zeolites by linking $SiO_4{}^{4-}$ tetrahedra. A: Linking of four tetrahedra produces a square of tetrahedra. B: Linking of three squares generated in A produces a structure in which an interior hexagon of silicon atoms is formed. Like the triangle shown, this structure is analogous to one face of an octahedron. This shape may also be represented by a hexagon with squares fused to alternate edges; this representation shows only the positions of the silicon atoms, which are located at the vertices of the hexagons and squares. *(Continued on next page.)*

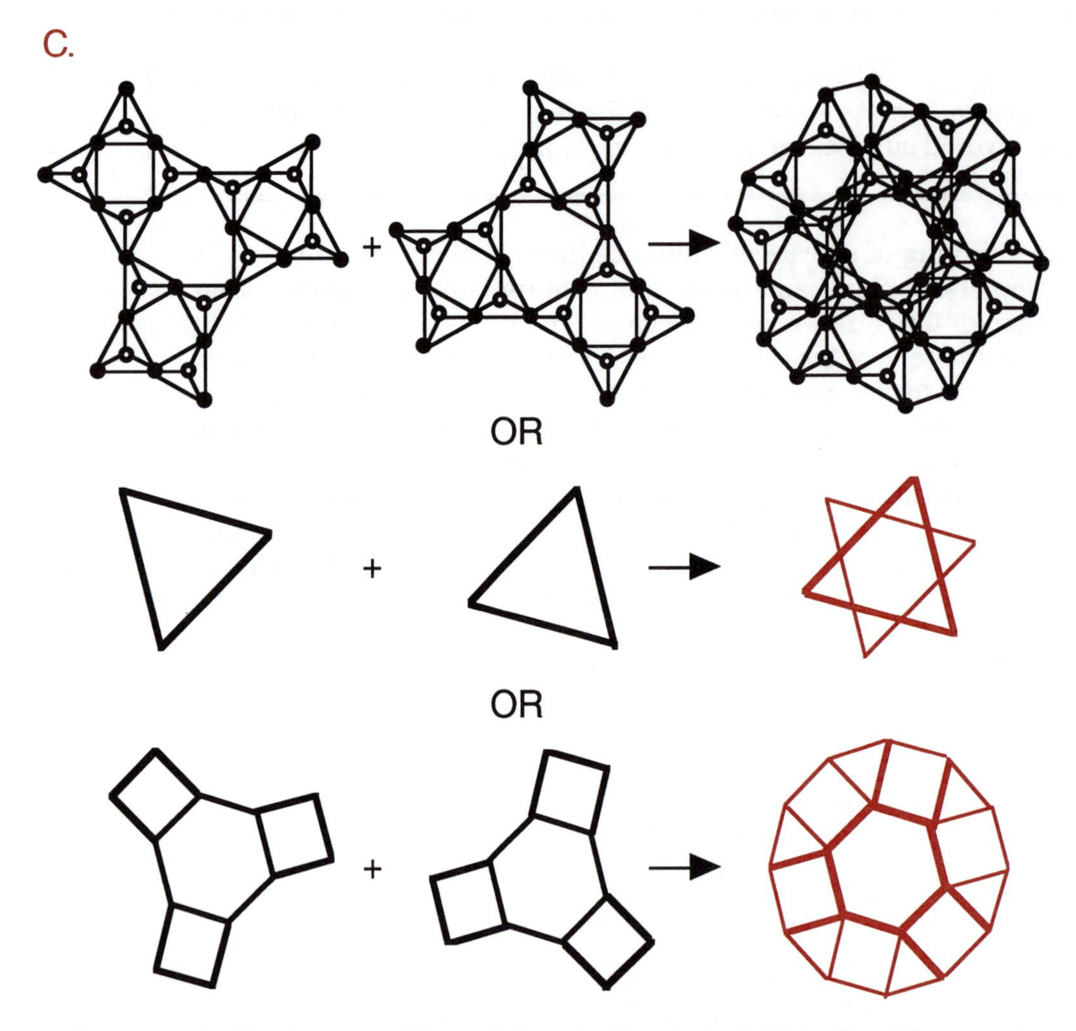

Figure 5.35C. Superposition of the shapes described in B with one oriented at 60° to it and above it, followed by fusion of the two frameworks at the appropriate points, generates shapes that have octahedral symmetry. In the silicate, the shape that is generated has a square of silicon atoms at each vertex of an octahedron and a hexagon on each face. This shape is a sodalite unit of a zeolite structure and is analogous to the superposition of one triangle with another oriented at 60° to it and above it, which generates an octahedron. Generation of the resulting silicate framework may also be viewed by showing only the positions of the silicon atoms.

The linkage of tetrahedra at all four corners generates SiO_2, silica, which exists in at least three crystalline forms—quartz, cristobalite♦ and tridymite♦. The sketches in Figures 5.37 and 5.38 show how the SiO_4^{4-} tetrahedra are linked through their corners to produce the structure of quartz and cristobalite, respectively.

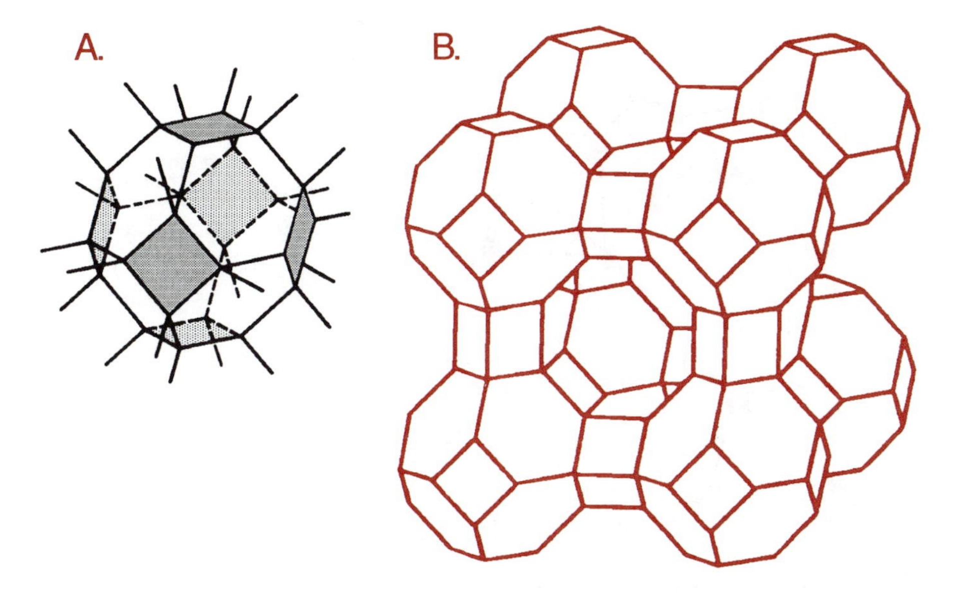

Figure 5.36. A: A sodalite unit. The lines coming out from the shape represent oxygen atoms that are available for additional linking. B: Linking of eight sodalite units generates a zeolite. This structure, similar to a simple cubic lattice, extends in three dimensions. A series of open pores runs through the structure.

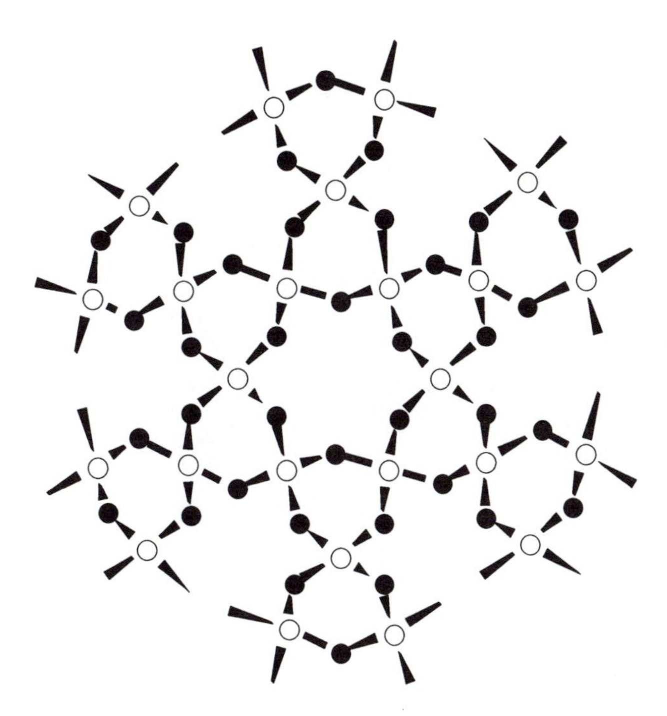

Figure 5.37. The SiO_2 (β-quartz) structure, viewed down the helical axis. The open circles represent silicon atoms and the closed circles represent oxygen atoms.

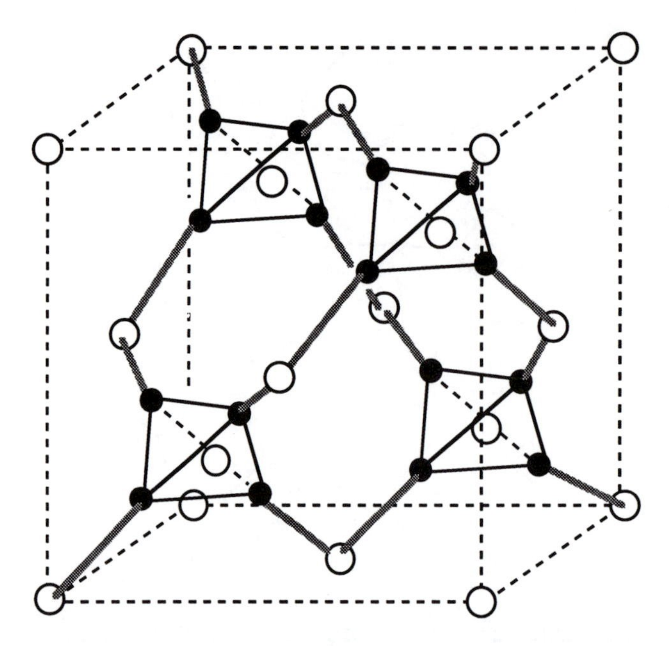

Figure 5.38. The SiO_2 (cristobalite) structure. The atoms in the corners and faces of the cube are silicon (open circles). Some of the surrounding tetrahedra of oxygen atoms (filled circles) have been omitted for clarity.

Appendix 5.1. The ICE Solid-State Model Kit

The SSMK makes use of a base with holes, a template to cover a subset of those holes, and radius-ratio size spheres that slide down rods inserted in holes in the base.

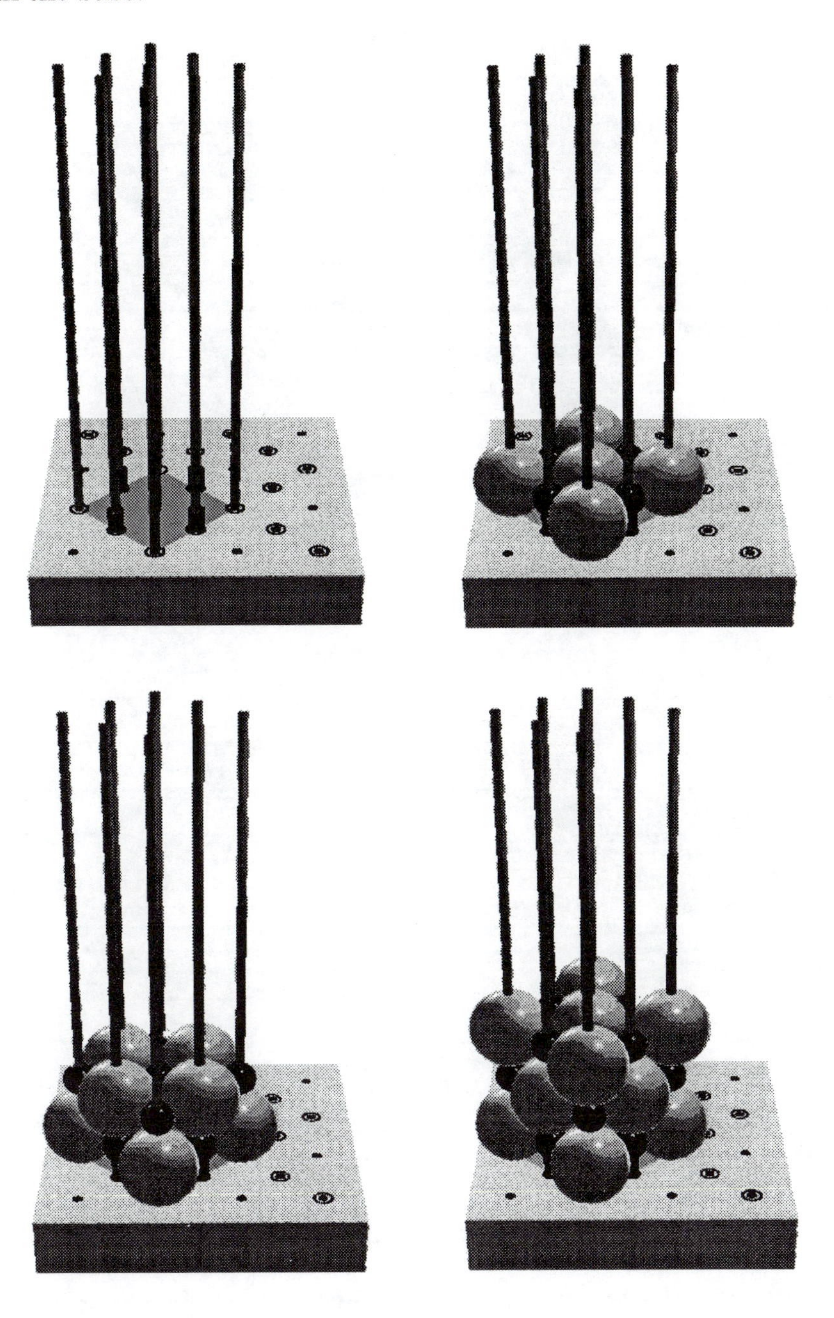

Appendix 5.2. Solid-State Structures and MacMolecule

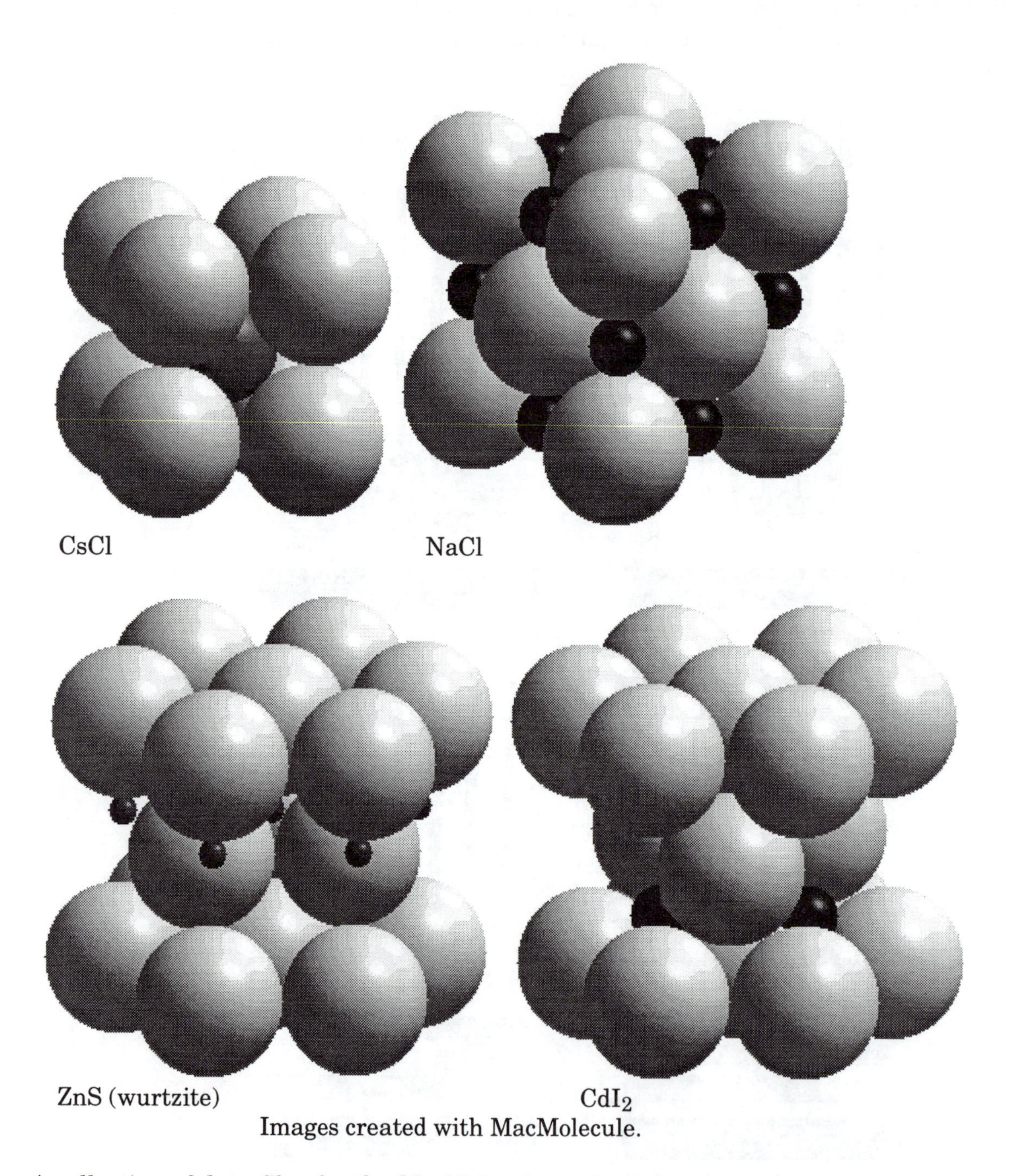

Images created with MacMolecule.

A collection of data files for the MacMolecule program has been prepared by Ludwig Mayer, Department of Chemistry, San Jose State University, San Jose, CA 95192–0101. MacMolecule (by Eugene Myers and Carlos Blanco, Department of Computer Science, and Richard B. Hallick and

Jerome Jahnke, Departments of Biochemistry and Molecular and Cellular Biology, University of Arizona, Tucson, AZ 85721) provides the capability to display, rotate, and examine molecular images in real time. Sample black-and-white images are shown here, but the program runs in color on Macintosh computers. MacMolecule can be obtained by anonymous FTP at joplin.biosci.arizona.edu. User-contributed image files will also be made available at this address in the /pub/MacMolecule directory, and bundled with future releases.

Appendix 5.3. Bubble Raft Construction

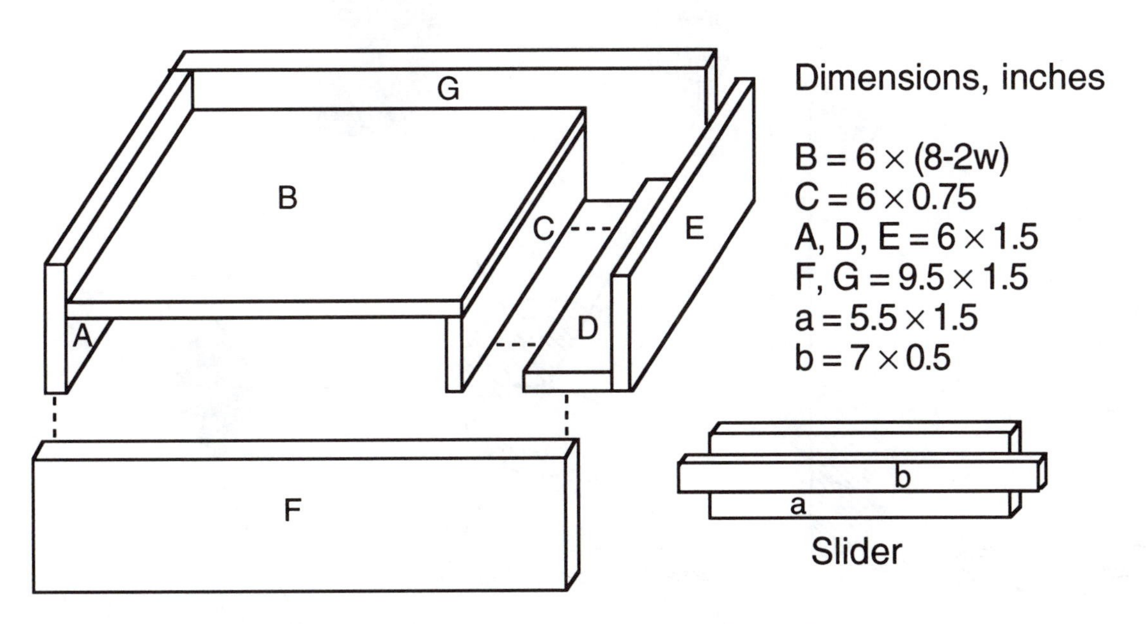

In piece B, *w* is the width of Plexiglas used; 1/8-inch Plexiglas is sufficient for the construction of the bubble raft.

Appendix 5.4. Tetrahedral and Octahedral Models

Cut out model. Use a pen and ruler to score along lines. Fold away from lines to assemble.

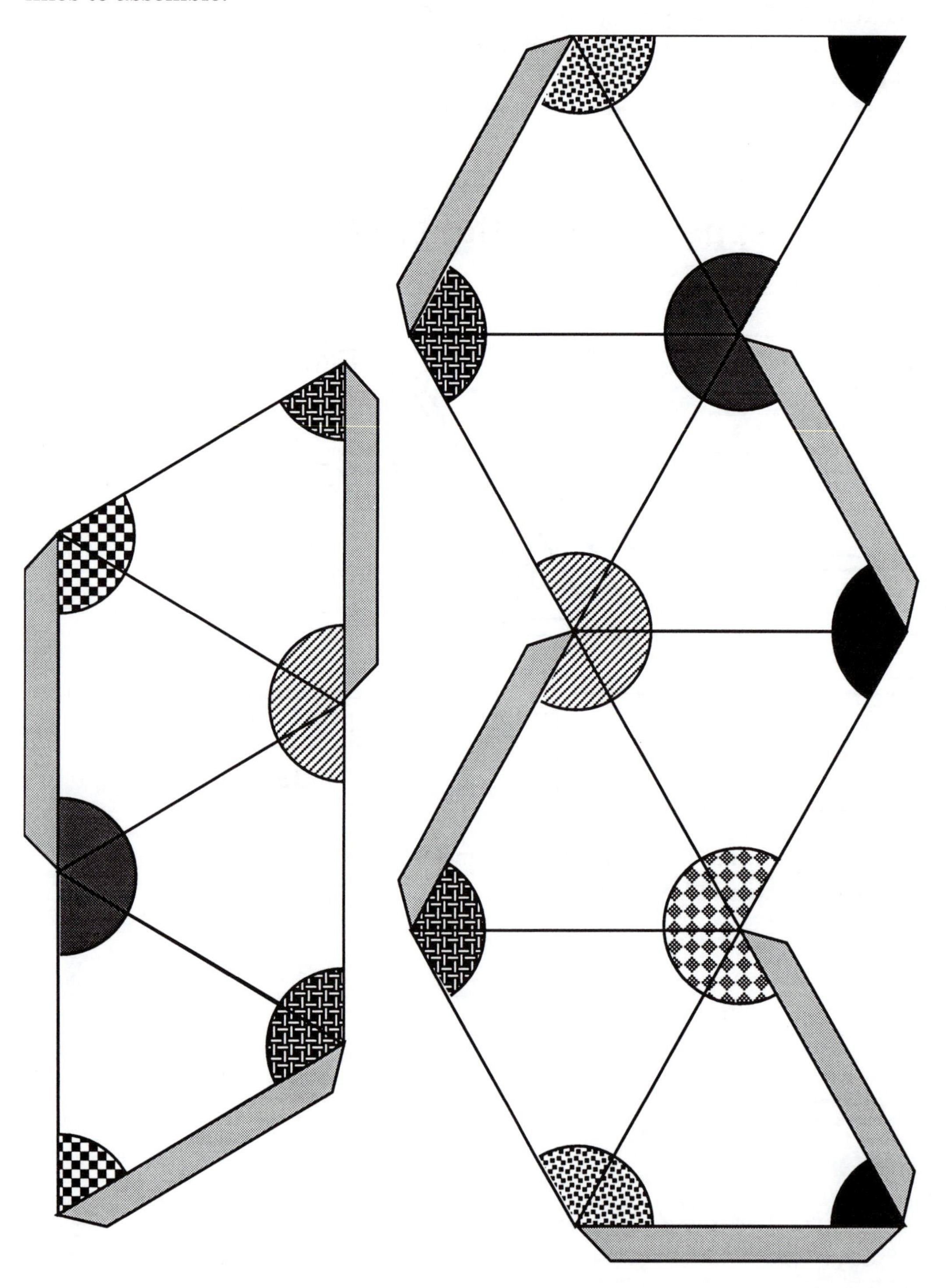

Appendix 5.5. Crystal Habits

The habit or shape of a crystal is determined by the rate at which different faces grow; the fastest growing faces are eliminated. When cubic faces (shaded) grow the fastest, they can produce an octahedral or a tetrahedral habit. (Adapted from reference 7.)

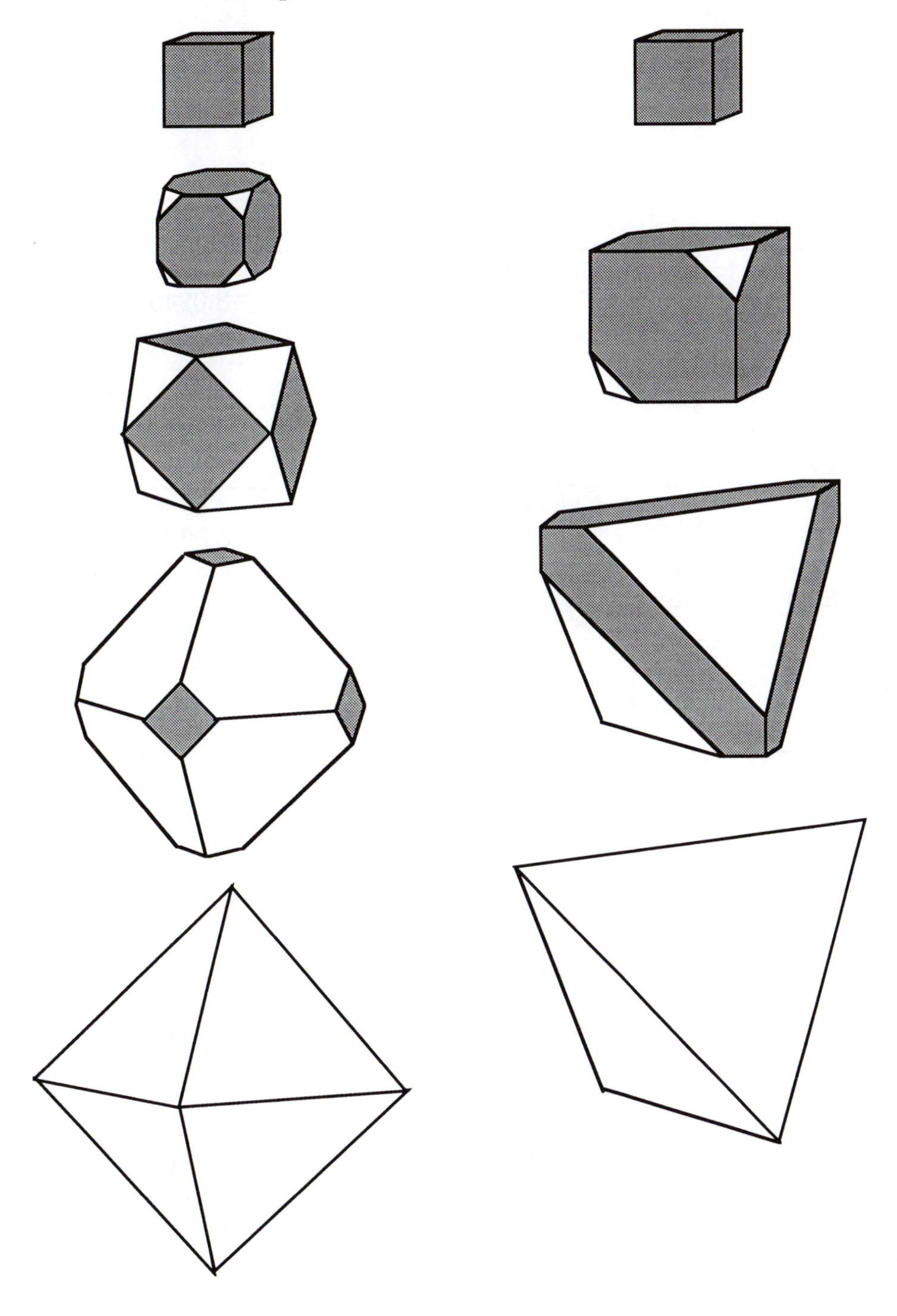

Appendix 5.6. Elements and Compounds That Exhibit the Common Structural Types

Elements

Diamond Structure♦

C	3.56683
Si	5.43072
Ge	5.65754
α-Sn	6.4912

Body-Centered Cubic♦

Ba	5.025	Li	3.5093	Ta	3.3058
Cr	2.8839	Mo	3.1473	V	3.0240
Cs	6.045 (5 K)	Na	4.2906	W	3.16469
α-Fe	2.8665	Nb	3.3004		
K	5.225 (5 K)	Rb	5.585 (5 K)		

Cubic Close-Packed♦

Ac	5.311	In	3.244	Pd	3.8898
Ag	4.0862	Ir	3.8394	Pt	3.9231
Al	4.04958	Kr	5.721 (58 K)	Rh	3.8031
Ar	5.256 (4.2 K)	Ne	4.429 (4.2 K)	Xe	6.197 (58 K)
Au	4.07825	Ni	3.52387		
Cu	3.61496	Pb	4.9505		

Hexagonal Close-Packed♦

Be, β-Ca, Cd, α-Co, Hf, Mg, Os, Re, Ru, Sc, β-Sr, Tc, Ti, Tl, Y, Zn, α-Zr

Compounds

CsCl Structure♦

AgCd	3.33	CsBr	4.286	MgTl	3.628
AgMg	3.28	CsCl	4.123	NiAl	2.881
AgZn	3.156	CsI	4.5667	NiTi	3.01 (austenite)
AuMg	3.259	CuPd	2.988	SrTl	4.024
AuZn	3.19	CuZn	2.945	TlBi	3.98
BeCo	2.606	LiAg	3.168	TlBr	3.97
BeCu	2.698	LiHg	3.287	TlCl	3.8340
BePd	2.813	LiTl	3.424	TlI	4.198
CaTl	3.847	MgHg	3.44	TlSb	3.84
CoAl	2.862	MgSr	3.900		

NaCl Structure♦

AgBr	5.7745	MgO	4.2112	SnAs	5.681
AgCl	5.547	MgS	5.2023	SnSb	6.130
AgF	4.92	MgSe	5.451	SnSe	6.020
BaO	5.523	MnO	4.4448	SnTe	6.313
BaS	6.3875	MnS	5.2236	SrO	5.1602
BaSe	6.600	MnSe	5.448	SrS	6.0198
BaTe	6.986	NaBr	5.97324	SrSe	6.23
CaO	4.8105	NaCl	5.62779	SrTe	6.47
CaS	5.6903	NaF	4.620	TaC	4.4540
CaSe	5.91	NaH	4.880	TaO	4.422
CaTe	6.345	NaI	6.4728	TiC	4.3186
CdO	4.6953	NbC	4.4691	TiN	4.235
CoO	4.2667	NbO	4.2097	TiO	4.1766
CrN	4.410	NiO	4.1684	VC	4.182
CsF	6.008	PbS	5.9362	VN	4.128
CsH	6.376	PbSe	6.1243	YAs	5.786
KBr	6.6000	PbTe	6.454	YN	4.877
KCl	6.29294	PdH	4.02	YTe	6.095
KF	5.347	RbBr	6.854	ZrB	4.65
KH	5.700	RbCl	6.5810	ZrC	4.6828
KI	7.06555	RbF	5.64	ZrN	4.567
LiBr	5.5013	RbH	6.037	ZrO	4.62
LiCl	5.12954	RbI	7.342	ZrP	5.27
LiF	4.0173	ScAs	5.487	ZrS	5.250
LiH	4.085	ScN	4.44		
LiI	6.000	ScSb	5.859		

Fluorite and Anti-Fluorite Structure♦

$AuAl_2$	6.00	K_2S	7.391	$PtGa_2$	5.911
$AuGa_2$	6.063	K_2Se	7.676	$PtIn_2$	6.353
$AuIn_2$	6.502	K_2Te	8.152	$PtSn_2$	6.425
$AuSb_2$	6.656	Li_2O	4.619	RaF_2	6.368
$BaCl_2$	7.34	Li_2S	5.708	Rb_2O	6.74
BaF_2	6.2001	Li_2Se	6.005	Rb_2S	7.65
Be_2B	4.670	Li_2Te	6.504	Rb_2P	5.505
Be_2C	4.33	Na_2O	5.55	ScH_2	4.78315
CaF_2	5.46295	Na_2S	6.526	$SiMg_2$	6.39
CdF_2	5.3880	Na_2Se	6.809	SmH_2	5.376
$CoSi_2$	5.356	Na_2Te	7.314	$SnMg_2$	6.765
$GeMg_2$	6.378	NbH_2	4.563	$SrCl_2$	6.9767
HfO_2	5.115	$NiSi_2$	5.395	SrF_2	5.7996
HgF_2	5.54	β-PbF_2	5.92732	YH_2	5.199
$IrSn_2$	6.338	$PbMg_2$	6.836	ZrO_2	5.07
Ir_2P	5.535	α-PoO_2	5.687		
K_2O	6.436	$PtAl_2$	5.910		

Zinc Blende Structure♦

AgI	6.473	CdS	5.832	HgTe	6.4623
AlAs	5.6622	CdSe	6.05	InAs	6.05838
AlP	5.451	CdTe	6.477	InP	5.86875
AlSb	6.1347	CuBr	5.6905	InSb	6.47877
BAs	4.777	CuCl	5.4057	MnS	5.011
BN	3.615	CuF	4.255	MnSe	5.82
BP	4.538	CuI	6.0427	SiC	4.348
BePo	5.838	GaAs	5.65315	ZnPo	6.309
BeS	4.865	GaP	4.4505	ZnS	5.4093
BeSe	5.139	GaSb	6.0954	ZnSe	5.6676
BeTe	5.626	HgS	5.8517	ZnTe	6.101
CdPo	6.665	HgSe	6.084		

ReO_3 Structure♦

MoF_3	3.8985
NbF_3	3.903
ReO_3	3.734
TaF_3	3.9012

Perovskite Structure♦

$AgZnF_3$	3.98	$KCoF_3$	4.069	$SmFeO_3$	3.845
$BaFeO_3$	4.012	$KFeF_3$	4.122	$SmVO_3$	3.89
$BaMoO_3$	4.0404	$KMgF_3$	3.973	$SrFeO_3$	3.869
$BaPbO_3$	4.273	$KMnF_3$	4.190	$SrHfO_3$	4.069
$BaSnO_3$	4.1168	$KTaO_3$	3.9885	$SrMoO_3$	3.9751
$BaTiO_3$	4.0118 (201 C)	$LiBaF_3$	3.996	$SrSnO_3$	4.0334
$BaZrO_3$	4.1929	$LiBaH_3$	4.023	$SrTiO_3$	3.9051
$CaSnO_3$	3.92	$NaAlO_3$	3.73	$SrZrO_3$	4.101
$CaTiO_3$	3.84	$NaWO_3$	3.8622	$TaSnO_3$	3.880
$CaVO_3$	3.76	$RbCaF_3$	4.452	$TlCoF_3$	4.138
$CaZrO_3$	4.020	$RbCoF_3$	4.062	$TlIO_3$	4.510
$CsCaF_3$	4.522	$RbMnF_3$	4.250	$YCrO_3$	3.768
$CsCdF_3$	5.20	$SmAlO_3$	3.734	$YFeO_3$	3.785
$CsPbBr_3$	5.874	$SmCoO_3$	3.75		
$KCdF_3$	4.293	$SmCrO_3$	3.812		

Wurtzite Structure♦

AgI, AlN, BeO, CdS, CdSe, CuBr, CuCl, CuH, CuI, GaN, InN, MgTe, MnS, MnSe, MnTe, NbN, SiC, TaN, ZnO, ZnS, ZnSe, ZnTe

NiAs Structure♦

AuSn, CoS, CoSb, CoSe, CoTe, CrSb, CrSe, CrTe, CuSn, FeS, FeSb, FeSe, FeTe, IrSb, IrTe, MnAs, MnBi, MnSb, MnTe, NiAs, NiSb, NiSe, NiTe, PdSb, PdTe, PtB, PtSb, PtSn, RhBi, RhTe, ScTe, TiS, TiSe, TiTe, VP, VS, VSe, ZrTe

CdI_2 Structure♦

Ag_2F, CaI_2, CdI_2, $CoBr_2$, CoI_2, $CoTe_2$, $FeBr_2$, FeI_2, GeI_2, HfS_2, $HfSe_2$, $IrTe_2$, $MgBr_2$, MgI_2, $MnBr_2$, MnI_2, $NiTe_2$, PdI_2, $PdTe_2$, PtS_2, $PtSe_2$, $PtTe_2$, $SiTe_2$, SnS_2, $SnSe_2$, α-TaS_2, $TiBr_2$, $TiCl_2$, TiI_2, $TiSe_2$, TiI_2, $TiSe_2$, $TiTe_2$, VBr_2, VCl_2, VI_2, W_2C, ZrS_2, $ZrSe_2$, $ZrTe_2$

NOTE: For cubic unit cells, the edge length is given in angstroms.
SOURCE: Data were obtained from references 11–14.

Appendix 5.7. Synthesis and Reactivity of $AuAl_2$

Synthesis

Aluminum spheres cut to expose fresh surfaces (0.100 g, 0.00371 mol) and powdered gold (0.365 g, 0.00185 mol) are placed in a ~6-mm i.d. quartz tube that has been shaped into a V. One end of the tube is connected to Tygon tubing, through which nitrogen is passed to purge oxygen from the system. While using a very slow nitrogen purge (six bubbles per minute), the metals are heated with a Meker burner in a fume hood. **Caution! Once the aluminum melts, the highly exothermic reaction will proceed quickly** to produce a purple, metallic solid.

Reactivity

$AuAl_2$ will react with 3M HCl to dissolve the aluminum and generate hydrogen gas, leaving behind gold powder. It will also react with 50% NaOH to yield $Al(OH)_4^-{}_{(aq)}$ and gold powder.

References

1. Bragg, L.; Nye, J. F. *Proc. Roy. Soc. London* **1947**, *190,* 474–481.
2. The soap solution is based on recipes found in the following sources: *Scienceworks*; Ontario Science Centre; Addison-Wesley: Reading, MA, 1984; p 84; Katz, D. A. *Chemistry in the Toy Store*, 5th ed.; Doherty, P.; Rathjen, D., Eds.; Community College of Philadelphia: Philadelphia, PA, 1990; *Exploratorium Science Snackbook*; Exploratorium Teacher Institute: San Francisco, CA, 1991; p 16.
3. Nathan, L. C. *J. Chem. Educ.* **1985**, *62*, 215.
4. Smart, L.; Moore, E. *Solid State Chemistry*; Chapman and Hall: London, 1992; p 39.
5. Burdett, J. K.; Price, S. L.; Price, G. D. *Solid State Comm.* **1981**, *40,* p 923.
6. Massey, A. G. *Main Group Chemistry*; Ellis Harwood Ltd. Chichester, England, 1990; p 228.
7. Holden, A.; Singer, P. *Crystals and Crystal Growing*; Anchor: Garden City, NY, 1960.
8. Sharma, B. D. *J. Chem. Educ.* **1987**, *64*, 404–407.
9. Jolly, W. L. *Modern Inorganic Chemistry*; McGraw-Hill: New York, 1984; p 275.
10. Gehman, W. *J. Chem. Educ.* **1963**, *40*, 54–60.
11. Wells, A. F. *Structural Inorganic Chemistry*, 5th ed.; Oxford University Press: Oxford, England, 1984.
12. Wyckoff, R. W. G. *Crystal Structures*, 2nd ed.; Interscience Publishers: New York, 1963, Vol. 1.
13. *CRC Handbook of Chemistry and Physics*, 68th ed.; Weast, R. C., Ed.; CRC Press: Boca Raton, FL, 1987.
14. Galasso, F. S. *Structure and Properties of Inorganic Solids*; Pergamon Press: Oxford, 1970.

Additional Reading

Bubble Raft

- Bragg, L.; Nye, J. F. *Proc. Roy. Soc. London* **1947**, *190*, 474; reprinted in Feynman, R. P.; Leighton, R. B.; Sands, M. *The Feynman Lectures on Physics*; Addison Wesley: Reading, MA, 1964; Vol. II, Chapter 30, p 10.

Structure

- Galasso, F. S. *Structure and Properties of Inorganic Solids*; Pergamon Press: Oxford, England, 1970.
- Smart, L.; Moore, E. *Solid State Chemistry: An Introduction*; Chapman and Hall: London, 1992.
- Wells, A. F. *Structural Inorganic Chemistry*, 5th ed.; Oxford University Press: Oxford, England, 1984.
- West, A. R. *Solid State Chemistry and Its Applications*; John Wiley and Sons: New York, 1984.

Acknowledgments

We thank Jill Banfield, University of Wisconsin—Madison, Department of Geology and Geophysics, for helpful comments on this chapter.

Exercises

1. Determine the packing efficiency of a simple cubic cell that contains one metal atom with a diameter of 1.00 Å as follows:
   a. Calculate the volume of the atoms that are contained in the cell. For a sphere, $V = (4/3)\pi r^3$.
   b. Determine the length of the cell edge in terms of the diameter of the atom contained in the unit cell. Hint: consider where the atoms touch each other.
   c. Determine the volume of the cubic unit cell.
   d. Determine the packing efficiency for the simple cubic structure.
2. Determine the packing efficiency of a body-centered cubic cell that contains metal atoms with a diameter of 1.00 Å as follows:
   a. Calculate the volume of the atoms that are contained in the cell. For a sphere, $V = (4/3)\pi r^3$.
   b. Determine the length of the cell edge in terms of the diameter of the atom contained in the unit cell. Hint: notice that the atoms touch each other along the body diagonal of the unit cell.
   c. Determine the volume of the cubic unit cell.
   d. Determine the packing efficiency for the structure.
3. Determine the packing efficiency of a face-centered cubic cell that contains metal atoms with a diameter of 1.00 Å using the steps described in Exercise 2. Hint: notice that the atoms touch each other along the face diagonal of the unit cell.
4. Which of the following is <u>not</u> characterized by a tetrahedral structure?
   a. CH_4
   b. Zn^{2+} ions in a cubic close-packed structure of S^{2-} ions
   c. a C atom in the diamond crystal structure
   d. Na^+ ions in a cubic close-packed structure of Cl^- ions
5. How does the CsCl structure change as one goes from the unit cell shown in Figure 5.12A to that in Figure 5.12B?
6. Estimate the radius of a hydride ion in LiH using the data in Appendix 5.6. Assume that the hydride ions touch in the structure.
7. Determine the density of any of the compounds or elements in Appendix 5.6 from the formula, structure type, and unit-cell dimension.
8. What is the formula of a compound that crystallizes with zinc atoms occupying one-half of the tetrahedral holes in a cubic close-packed array of sulfur atoms? (Zn and S can be replaced by any of the pairs listed for the Zinc Blende structure in Appendix 5.6.)

9. What is the formula of a compound that crystallizes with lithium ions occupying all of the tetrahedral holes in a cubic close-packed array of sulfide ions? (Li and S can be replaced by any of the pairs listed for the antifluorite structure in Appendix 5.6.)

10. What is the formula of a compound that crystallizes with silver ions occupying all of the octahedral holes in a cubic close-packed array of chloride ions? (Ag and Cl can be replaced by the ions of any of the pairs listed for the NaCl structure in Appendix 5.6.)

11. What is the formula of a compound that crystallizes with thallium ions occupying all of the cubic holes in a simple cubic array of iodide ions? (Tl and I can be replaced by the ions of any of the pairs listed for the CsCl structure in Appendix 5.6.)

12. What is the formula of a compound that crystallizes with barium ions occupying one-half of the cubic holes in a simple cubic array of fluoride ions? (Ba and F can be replaced by the ions of any of the pairs listed for the fluorite structure in Appendix 5.6.)

13. Consider cubic crystals of CsCl, NaCl, ZnS, CaF_2, and $CaTiO_3$.
    a. Sketch the STM trace of the face of the crystal, assuming that the STM tip travels three unit cells in each of the two directions.
    b. Assume that crystals of each compound in part a are cleaved so the atoms at $z = 0.5$ are exposed. Sketch the STM trace of each, assuming that the STM tip travels three unit cells in each of the two directions.
    c. Assume that crystals of CaF_2 and ZnS are cleaved so the atoms at $z = 0.25$ are exposed. Sketch the STM trace of each, assuming that the STM tip travels three unit cells in each of the two directions.

14. Calculate the radius of an atom of any of the metals whose unit-cell dimensions are given in Appendix 5.6.

15. The atomic radii of carbon, silicon, and germanium are approximately 0.77 Å, 1.17 Å, and 1.22 Å, respectively. *Calculate* the covalent radius (r) of carbon, silicon, or germanium by referring to the X-ray diffraction data in Appendix 5.6 and the structure shown in Figure 5.24 (Hint: Each of the four atoms contained within the body of the unit cell touches three face-centered atoms and one corner atom for tetrahedral coordination. Try subdividing the unit cell into eight smaller cubes; the body diagonal of each small cube will have a length of 4r.)

16. Determine the ionic radius of the metal ion in any of the oxides in Appendix 5.6 with the NaCl structure.

17. Assume that the radius of the bromide ion is unknown and determine it from a plot of unit cell volume versus anion volume for KF, KCl, KBr, and KI; or NaF, NaCl, NaBr, and NaI. The radii of the F, Cl, and I anions are 1.19, 1.67, and 2.06 Å, respectively. (This exercise can be repeated for any series of isostructural compounds with one missing ionic radius. For example, the radius of Co may be determined from plots involving oxides with the NaCl structure found in Appendix 5.6 and the ionic radius of 1.26 Å for O^{2-}.)

18. Which of these metals contain undistorted cubic holes? octahedral holes? tetrahedral holes? Al, Ba, Co, Fe, Mo, Pb, Po (*see* Figure 3.11).

19. An anisotropic structure is one that has bonding or structural units along one of its unit cell directions that differ from the bonding or structural units along one or both of the other unit cell directions. From the examples in this chapter give an example of an ionic compound that is anisotropic and one that is not; of a covalent solid that is anisotropic and one that is not.

20. Aluminum is composed of larger and lighter atoms than silicon. Why is the density of aluminum larger than that of silicon?

21. The geometry of a close-packed plane of atoms in a hcp structure is identical to that of a close-packed plane of atoms in a ccp structure. Why do the structures differ?

22. Graphite and diamond contain carbon atoms with coordination numbers of 3 and 4, respectively. However, the radius ratio rule predicts that the coordination number should be 8 or 12 for both substances. Why does carbon in diamond or graphite form differ from the prediction?

23. Several factors can cause a structure to differ from that predicted from the radius ratio rule. Using the radius ratio rule, what coordination number is predicted for silicon in elemental silicon? Why does silicon not exhibit this coordination number?

24. In the figure caption for Figure 5.15 (the $CdCl_2$ structure) the statement is made that the picture shown does not represent a complete unit cell. Show that the picture is not consistent with the $CdCl_2$ stoichiometry.

25. Use the packing diagrams shown in the figures for this chapter to determine the contents of the unit cell and verify the formulas for
    a. NiAs
    b. NaCl
    c. CdI_2
    d. Li_2O
    e. CaF_2

26. What sort of cleavage properties would you expect for a single crystal of CdI_2?

27. Use Figure 5.23 to show that there are two tetrahedral holes and one octahedral hole per atom in a hcp unit cell.

28. Use Figures 5.13 and 5.19 to show that there are two tetrahedral holes and one octahedral hole per close-packed sphere in a fcc crystal.

29. K_xC_{60} has potassium atoms that completely fill the octahedral and tetrahedral holes in a fcc arrangement of C_{60} anions. What is the value of x in K_xC_{60}?

30. Generate a z-layer sequence for ReO_3 (Figure 5.27A).

31. Confirm the formula of the superconductor shown in Figure 5.29C.

32. Confirm the formula of the silicate repeat units in Figures 5.31, 5.32, and 5.33, using the atom sharing formalisms developed for unit cells.

33. The amphibole structure has the silicate repeat unit formula $(Si_4O_{11}{}^{6-})_n$. If there is isomorphous substitution in which aluminum atoms replace (a) 25%, or (b) 50% of the silicon atoms, what will the charge be on the repeat unit?

34. The octahedral ferrocyanide ion, $Fe(CN)_6{}^{4-}$, forms an insoluble salt with Fe^{3+}. This salt is intensely colored and called Prussian blue. In the crystal structure there are linear Fe^{2+}–C–N–Fe^{3+} arrays of atoms.
    a. Crystals of Prussian blue have the NaCl-like structure, with Fe^{3+} in the Na^+ site and $Fe(CN)_6{}^{4-}$ in the Cl^- site. If Prussian blue had the perfect NaCl structure, how many Fe^{3+} ions would surround the $Fe(CN)_6{}^{4-}$ ions?
    b. What is the formula for Prussian blue? In other words, what are x and y in $[Fe^{3+}]_x[Fe(CN)_6{}^{4-}]_y$?
    c. Given your answer in b, why is it impossible for Prussian blue to have a perfect NaCl structure?

35. The text discussed tetrahedral and octahedral holes in close-packed structures. There are also trigonal (three-coordinate) holes with a single layer of a close-packed structure. Draw one close-packed layer and indicate the location of a trigonal planar hole.

36. In α-aluminum oxide, there is a hcp array of oxide ions with two-thirds of the octahedral holes filled with aluminum ions. What is the formula of α–aluminum oxide?

37. A mercury iodide salt can be described as a ccp array of iodide ions with one-fourth of the tetrahedral holes occupied by mercury ions. What is the formula for this salt?

38. A tin iodide salt consists of a ccp array of iodide ions with Sn ions in one-eighth of the tetrahedral holes. What is its formula?

39. Aluminum selenide can be described as selenide ions in a hcp array with aluminum ions in one-third of the tetrahedral holes. What is its formula?

40. An early attempt to develop a set of ionic radii was made by Lande in 1920. He assumed that Li^+ would be the smallest ion found in crystals and that in compounds with the largest halide ions, the lithium ion would be too small to keep the halide ions from touching. Thus, knowledge of the crystal structure and unit-cell dimensions provided a method for an estimation of the lithium and iodide ionic radii.
   a. Draw a picture of the $z = 0$ layer sequence for LiI. (*See* Appendix 5.6 for the crystal structure adopted by LiI and subsequent compounds required for this problem.)
   b. Use the picture you have generated to calculate the ionic radius of iodide, using the unit cell dimensions given in Appendix 5.6 and assuming that the iodide ions are just touching.
   c. Use the calculated ionic radius of iodide and the picture you drew in part a to estimate the radius of Li^+.
   d. Using these two ionic radii, calculate the ionic radii for the alkali metals and the halide ions.

41. The structure of MoS_2 is illustrated in Figure 5.26. In which directions will planes of atoms in MoS_2 slip most easily?

42. The structure of graphite is illustrated in Figure 5.25. In which direction will planes of atoms in graphite slip most easily?

43. Discuss the relationship between the atom positions in the following structures (some of these are shown in the Chapter 3 exercises):
   a. NaCl and CaC_2
   b. MgAgAs and ZnS (Zinc Blende)
   c. CdI_2 and NiAs

44. How many nearest neighbors does each atom have in a bcc structure? in a fcc structure?

45. Graphite can be viewed as a polymerized version of benzene. There is a polymorph of boron nitride that can be viewed as a polymerized version of borazine. Hexagonal boron nitride and graphite are both layered solids that are good lubricants, but boron nitride is white and has a higher oxidation temperature (850 °C).

Benzene   Borazine

a. Draw a layer of hexagonal boron nitride by analogy to graphite.
b. The structure of graphite and hexagonal boron nitride differ since the boron nitride layers stack directly over each other (with alternating B N B N... stacking) and are eclipsed rather than staggered as in graphite. Make a sketch analogous to Figure 5.25B that shows the structure of hexagonal boron nitride.

46. Which of the layer sequences below corresponds to face-centered cubic, hexagonal close packing, and cubic close packing unit cells? How many total atoms are there within each of these unit cells?

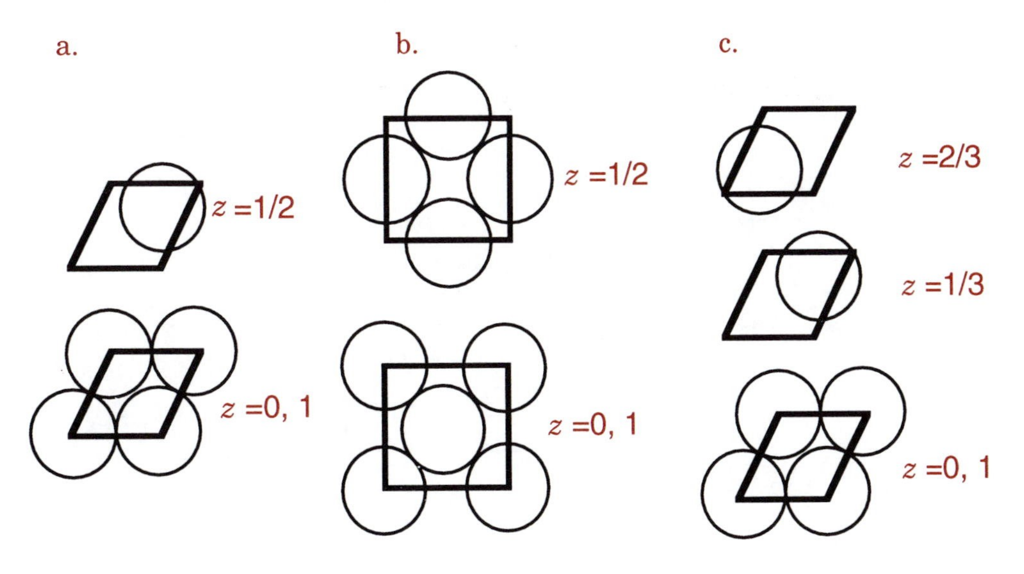

Chapter 6

Defects in Solids

Extended crystalline solids are not perfect repeating structures. All real materials possess defects or imperfections in the regular repeat pattern of the crystal. Defects may be atoms or ions that are missing from the structure, out of position, or replaced by impurity atoms or ions, for example. Defects can exert tremendous influence over the physical and chemical properties of a material. There is, in fact, an old saying among solid-state scientists: "Solids are like people. It is their defects that make them interesting."

In this chapter we present examples of the effects of defects on the behavior and properties of solids. These examples are particularly easy to demonstrate, principally those involving mechanical properties of metals and optical properties of salts. In other sections, defects that lead to the remarkable electronic properties of semiconductors (Chapter 8) and superconductors (Chapter 9) are discussed.

Why should a chemist be interested in defects? As has already been mentioned in Chapter 1, materials science is an interdisciplinary field, and its challenges are increasingly being met by teams whose members have backgrounds that span the scientific and engineering disciplines. One of the primary skills that a chemist brings to such a group is an understanding of synthesis. As scientists seek to control the arrangement of matter on a smaller and smaller scale (as evidenced by the direction of developments in microelectronic circuitry in the past half century), the synthetic control on the scale of individual atoms and molecules that chemists bring to the table will play a crucial role in the creation of new materials and devices.

At present a great deal of research is directed toward what are called nanostructured materials *(1)*. These are particles, layers, or filaments with sizes on the order of 1–100 nm (1 nm = 10^{-9} m). The reduction in size to nanostructured materials has resulted in dramatic changes in properties. In many cases this change results because a significant portion of the solid

2725–X/93/0155$09.00/1

now consists of interparticle and intraparticle interfaces, which are normally considered to be types of defects in crystals. Thus, the solutions to many problems in materials science will involve an understanding of synthesis (arrangement of atoms), the control of defects, and the relationship between structure and properties. This chapter will provide a foundation for such an understanding.

Defects in Metals

Metals can be processed by a number of means (e.g., alloying, cold working, heat treating) to give them high strength or other desirable properties. Such treatments typically produce defects within the metals. These defects range in dimensions from a few nanometers to tens or hundreds of micrometers (1 μm $=10^{-6}$ m), and thus are referred to as microstructures. Microstructures play a critical role in determining the mechanical properties of a material.

A useful context for an initial discussion of defects is provided by comparing the macroscopic mechanical properties of metals with those of salts and extended covalent solids. Application of a mechanical stress (such as stretching) to a metal crystal causes the bonds between planes of atoms to stretch and distort. If the displacement of each atom is small (less than halfway to the next equivalent location in the crystal), the strain is elastic, and the planes relax back to their original positions when the stress is removed. The crystal will deform under the influence of the stress and then return to its original shape when the stress is removed (rather like the behavior of a rubber band). If the stress is large enough, the planes of atoms slide (slip) with respect to each other, giving the atoms new neighbors, and the crystal becomes permanently deformed. This deformation that results from the sliding of planes of atoms is known as plastic deformation; it generally does not reverse itself when the stress is removed.

The ductility of metals, particularly those with face-centered cubic structures, markedly differentiates them from ionic and covalent solids. When a mechanical stress is applied to an ionic solid in certain directions, like-charged ions are forced into proximity with one another, as illustrated in Figure 6.1, and the repulsive electrostatic forces lead to fracturing. In covalent solids, mechanical stress breaks highly directional covalent bonds, again causing fracturing.

Two rows of atoms in a metal sliding past one another pass through a series of energetically stable positions, as shown in Figure 6.2. The plane along which sliding occurs is called a slip plane. Sliding is easiest when one close-packed plane (Chapter 5) slides over another, because the distortion required for the atoms to move from one stable position to another is smallest between close-packed planes (*see* Figure 6.3). Fcc [also called cubic close-packed (ccp), *see* Chapter 5] metals have four sets of close-packed planes in a unit cell (*see* Figure 5.9). As a consequence, fcc metals such as Cu, Al, Au, Pb, Ni, and Pt are relatively easily deformed.

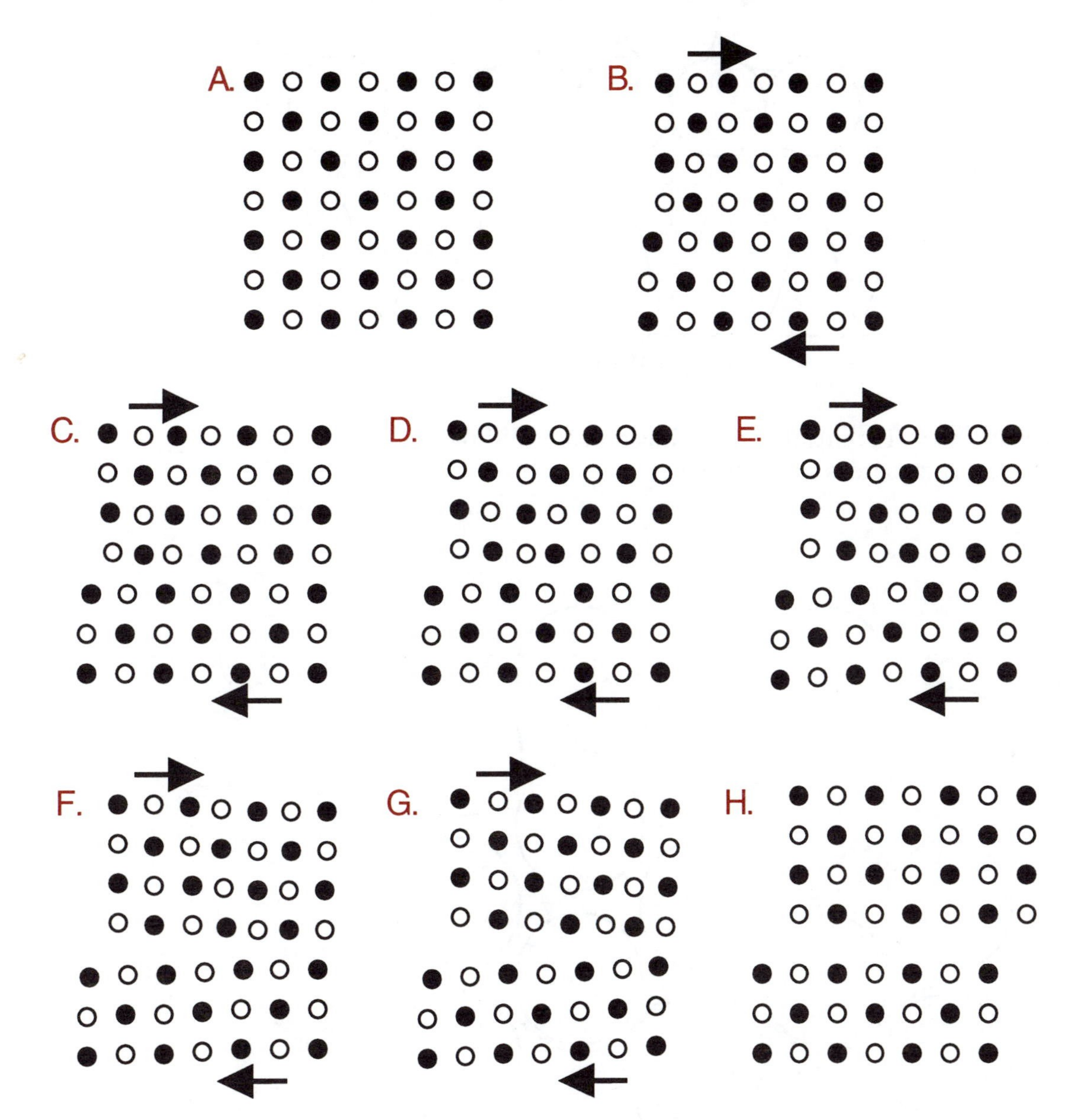

Figure 6.1. The effect of mechanical stress on an ionic crystal; filled and open circles correspond to cations and anions, respectively (or vice-versa). As a shearing force (displaced opposing forces indicated by the arrows) is applied, part of the crystal begins to slide. This sliding brings like charges into close contact, and the repulsion causes the layers to begin to separate. This separation of layers generates cracks and ultimately results in fracture of the crystal. (Compare Figure 6.8, which shows the effect of mechanical stress on a metallic crystal.)

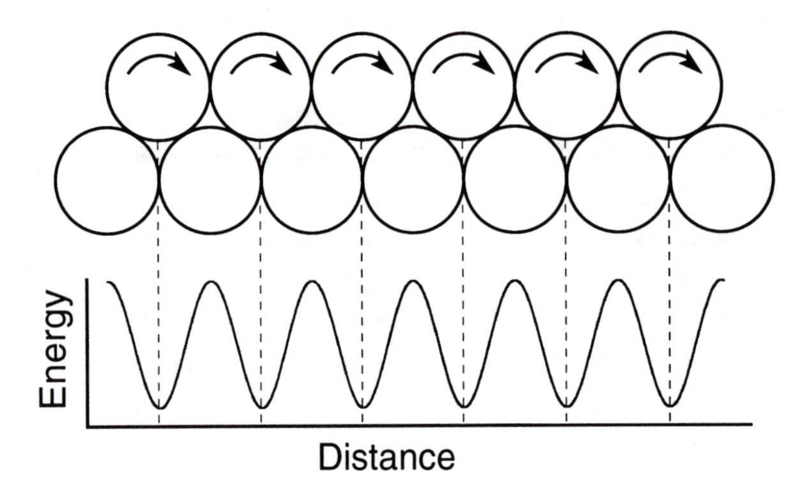

Figure 6.2. Potential energy diagram for sliding one row of spheres (the top row) across another row.

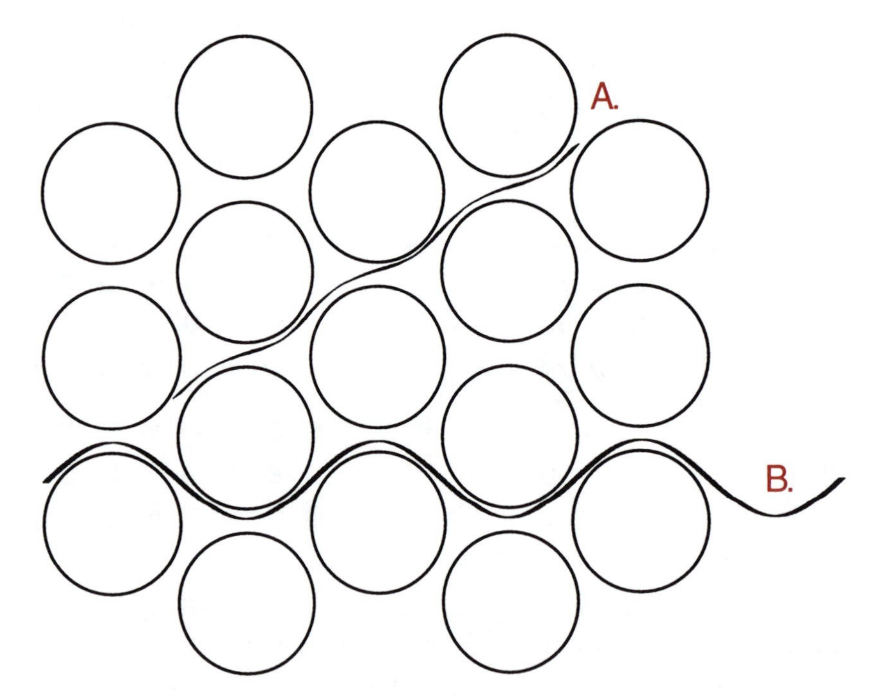

Figure 6.3. Slip plane motion between (A) close-packed planes and (B) non-close-packed planes. Distortion is much greater in the latter case.

Demonstration 6.1. Ductility and Fracturing

Materials

A piece of 10-gauge Cu wire (house wiring from a hardware store)
Solid surface on which to hammer
Wafer or chunk of silicon
Rock salt, sold for water softeners, etc. (or transparent NaCl crystal) (*see* Supplier Information)
Paddle spatula or razor blade
Hammer
Overhead projector and screen
Bubble raft
Goggles and gloves

Procedure

- Flatten a piece of copper wire with a hammer. For viewing by a classroom audience, the change in shape can be seen by comparing the initial and final outlines of the sample when it is placed on an overhead projector.
- Model the deformation using the bubble raft *(2)* (Chapter 5): gently squeeze part of the raft against a wall of the Plexiglas tray by use of the slider. This squeezing will cause it to deform, as shown in Figure 6.4A. Alternatively, use two sliders or glass slides to provide a shearing motion, as shown in Figure 6.4B, which will also deform the packing arrangement.
- While wearing goggles and gloves, flex or hammer a piece of silicon, which will cause it to shatter. **Caution: Silicon wafers, like glass, can shatter and form sharp edges when broken.**
- Fracture a sodium chloride crystal in such a way that natural cleavage planes are revealed: While wearing goggles, place the transparent crystal on an overhead projector and drive a razor blade or a paddle spatula into the plate by striking it with a pencil, screwdriver handle or similar instrument. If the blade is oriented parallel to a face of the NaCl cubic unit cell, a clean fracture occurs; if not, jagged edges are observed *(3, 4)*.

Types of Defects

The mechanism for plastic deformation shown in Figure 6.2 is an oversimplification. When scientists first began to understand that plastic deformation involves the sliding of atomic planes past one another, they assumed that the sliding occurs all at once. Only later did it become apparent that this is not the case. Reliable theoretical calculations showed that if all the bonds between two atomic planes were distorted at the same time, the stress required would be high, roughly 10^6 psi ($\sim 10^7$ kPa; $\sim 10^5$

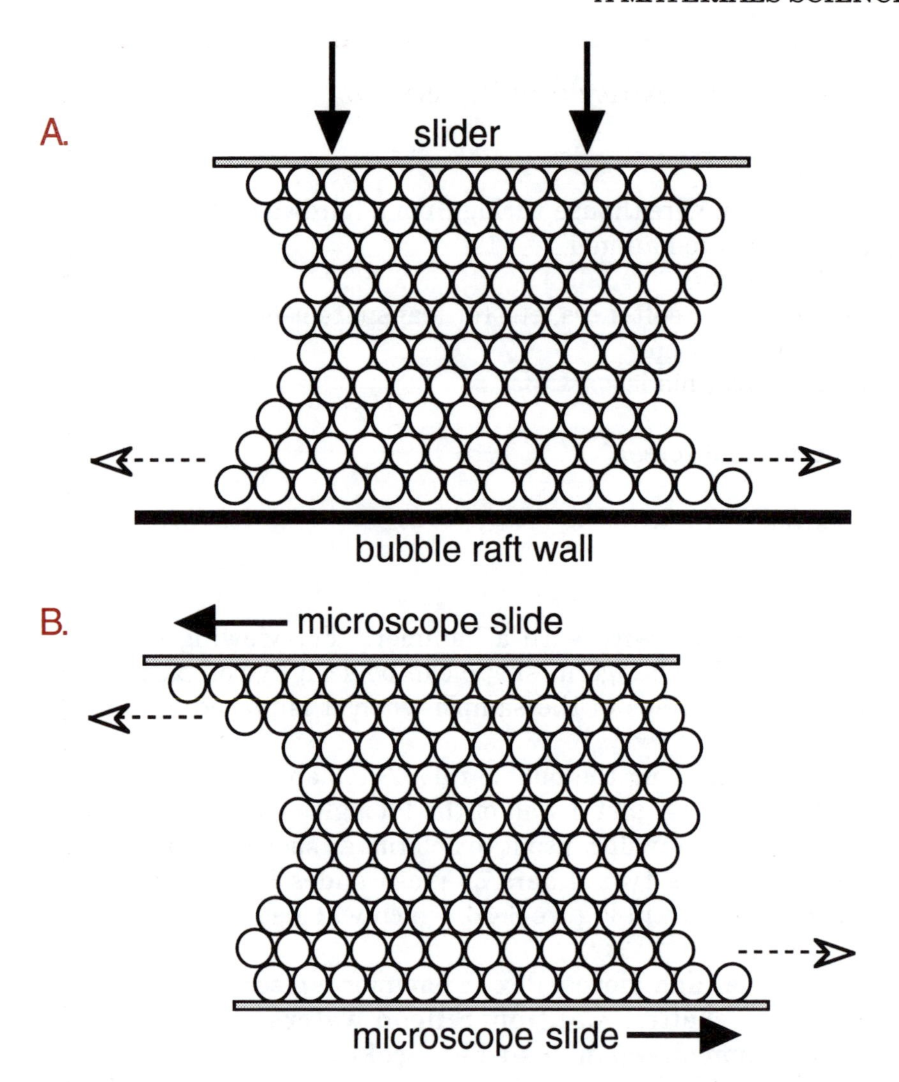

Figure 6.4. Deformations using the bubble raft (A) by squeezing the bubbles against the wall of the tray using the slider or (B) by using the shearing motion of two glass slides.

atm) for high-purity iron *(5)*. These calculations presented a direct conflict with experimental evidence, which showed that the stress required to deform pure metals is actually much lower, on the order of a few tens or hundreds of pounds per square inch. The explanation for the discrepancy lies in a type of one-dimensional crystal defect, the dislocation, first postulated in the 1930s *(6)*.

Defects in crystals are imperfections in the regular repeat pattern of the crystal and may be classified in terms of their dimensionality. A point defect is a defect that occurs at some point in a crystal and involves only one atom or interstitial site in the crystal. Examples include vacancies (atoms missing from the regular array of atoms that forms the crystal) and substitutional impurities (impurity atoms that occupy atom sites in the crystal). Interstitials are also point defects in which atoms occupy the voids between the normal atomic positions. Figure 6.5A illustrates these three types of point defects. Point defects are effectively zero-dimensional defects because they do not extend through the crystal. The concentration of vacancies is governed by chemical equilibria, as illustrated in Chapter 9.

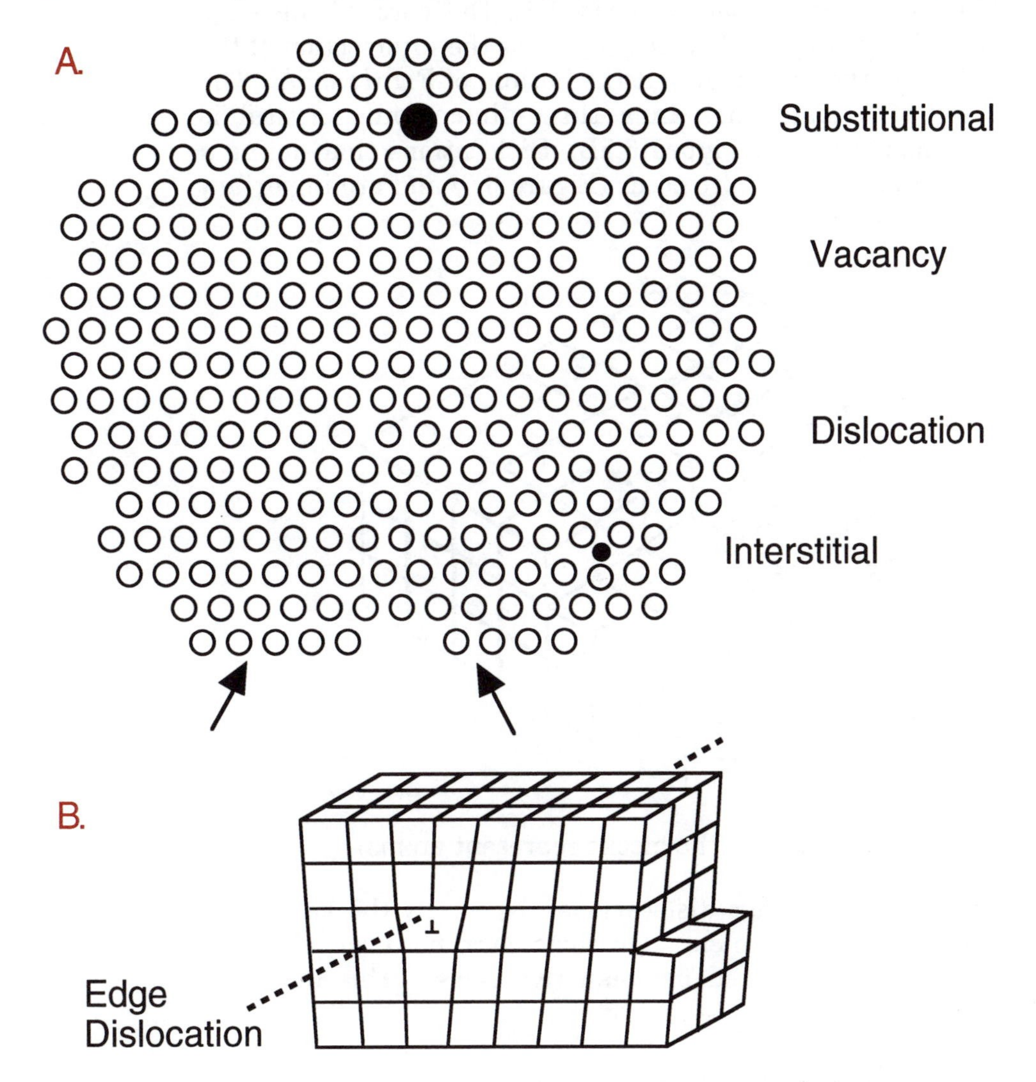

Figure 6.5. A: Crystalline defects include point defects such as the vacancy, interstitial impurities, and substitutional impurities; and extended imperfections such as dislocations. (To more easily see the dislocation, sight along the arrows while viewing from an oblique angle to the page.) B: A schematic representation of a three-dimensional crystal containing an edge dislocation, emphasizing the dislocation as a line defect. ⊥ designates the core of the dislocation.

A dislocation is a line defect; it runs somewhat like a string through a crystal. A two-dimensional representation of a dislocation is shown in Figure 6.5A in which a crystal is represented by a two-dimensional array of circles similar to the bubble raft. When viewed at an oblique angle along the directions indicated by the arrows, extra lines of atoms are seen in the top half of the figure. These lines of atoms correspond to extra planes of atoms in a three-dimensional crystal. Although the dislocation is recognizable from what appears to be an extra half-plane of atoms inserted into the structure, the line that terminates this extra half-plane is important. This terminating line is located in the core of the edge dislocation; that is, in the highly distorted region at the intersection of the directions indicated by the two arrows at the bottom of Figure 6.5A. The linelike character of a dislocation in a three-dimensional crystal is indicated by the dashed line in Figure 6.5B. Dislocations that define a spiral shape (screw dislocations) also occur (Figure 6.6) and are important in crystal growth mechanisms.

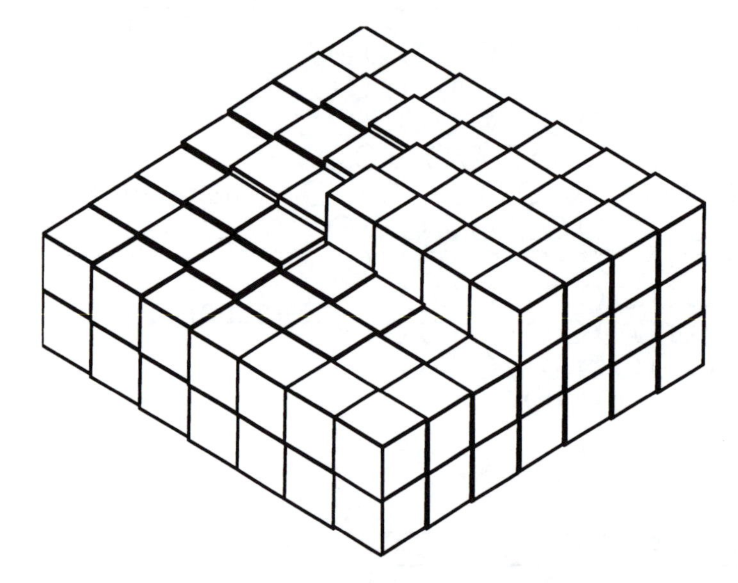

Figure 6.6. A screw dislocation (the blocks represent atoms).

An ear of corn can exhibit edge dislocations. Normally the kernels of corn grow in parallel rows on the ear. On some ears, however, an extra partial row of kernels is inserted between two rows. This defect is illustrated in Figure 6.7.

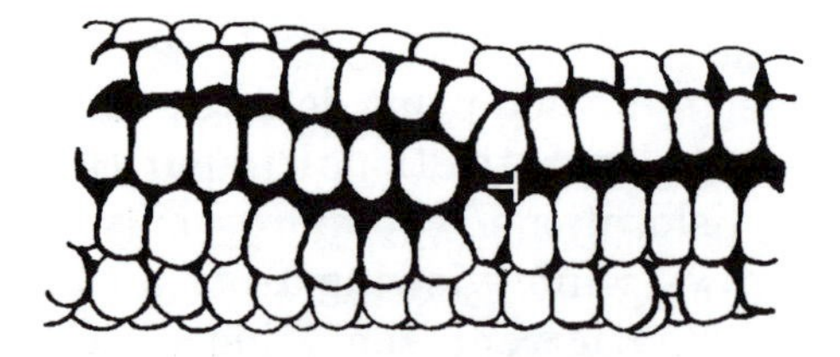

Figure 6.7. A dislocation in an ear of corn. ⊥ designates the core of the dislocation. (From reference 7.)

The presence of a dislocation relaxes the requirement that entire planes of interatomic bonds must distort and break simultaneously for plastic deformation to occur. Instead, plastic deformation can accompany the motion of a dislocation through a crystal. The motion of a dislocation from left to right through a crystal is illustrated in Figure 6.8.

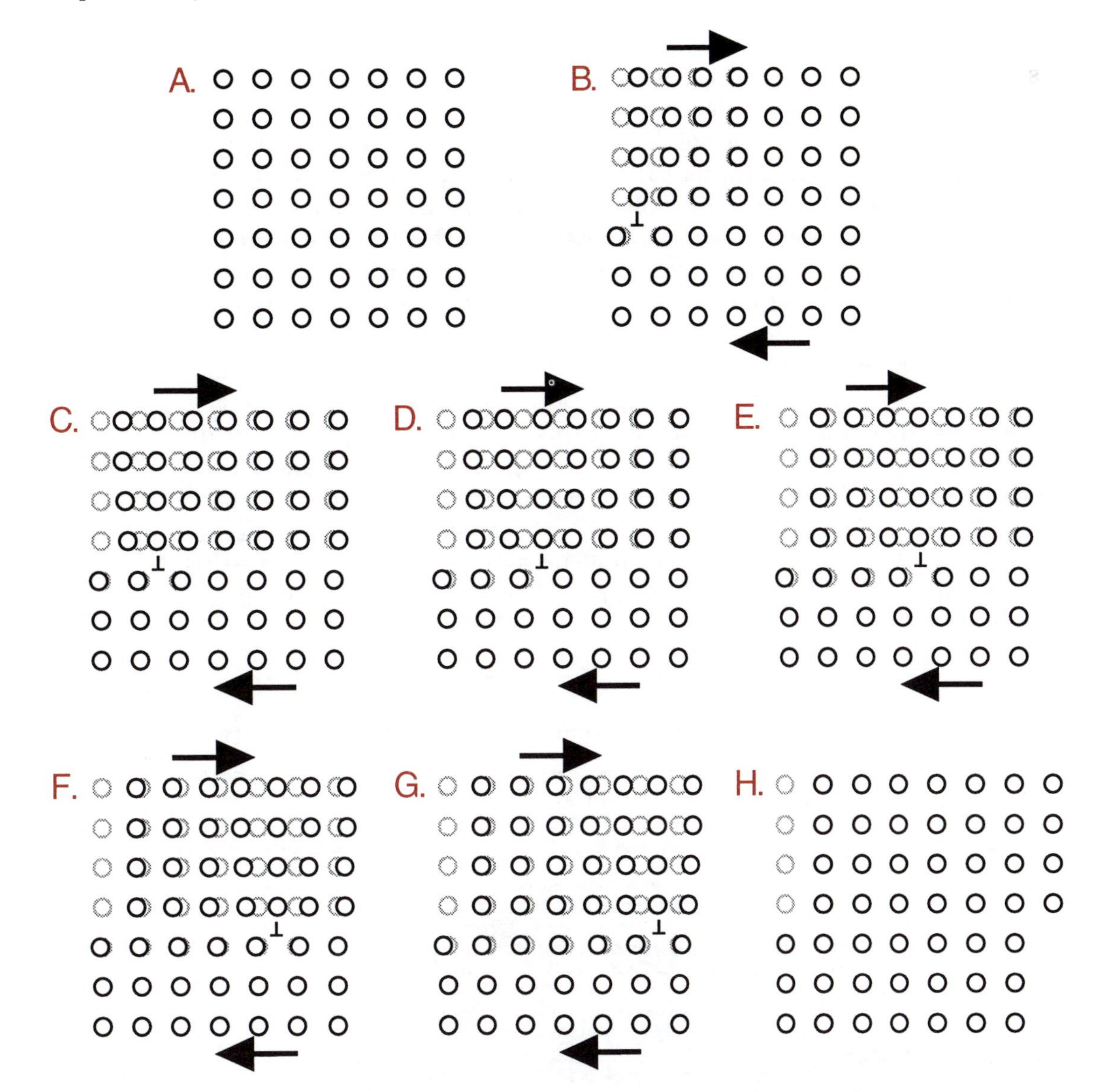

Figure 6.8. When a crystal is subjected to a shearing force (represented by bold arrows), dislocations move along crystallographic slip planes. Atoms above the slip planes shear away from their original neighbors below the slip plane, then bond with a new set of neighbors to restore the internal structure of the crystal. ⊥ designates the core of the dislocation. As the core moves through the crystal, the identity of the "extra" half plane of atoms changes. Each circle in this figure represents a line of atoms going into the page. Light circles represent the original positions of atoms.

As the dislocation moves and becomes associated with new half-planes, bonds on one side of the dislocation are broken and bonds on the other side are formed. As a consequence of the motion of the dislocation, the top half of the crystal slips to the right by one plane of atoms, yet the entire set of bonds between the planes that slip need not be broken simultaneously. Thus the motion of dislocations provides a low-energy mechanism for plastic deformation that is based on sequential shifts in atomic position. Shifts in atomic positions can occur at the speed of sound in the solid; a demonstration of this effect is described later.

Metals, like other crystalline solids, are usually polycrystalline: they are composed of many small crystals called grains (Figure 6.9). The grains within a metal usually have random orientations with respect to one another and are separated by grain boundaries, which can be regarded as two-dimensional or planar defects. (The surface of a crystal would be another example of a two-dimensional defect.) Every grain boundary is an array of dislocations. Along a grain boundary, the atomic packing is imperfect and the energies of atoms at a grain boundary are higher than those of atoms within the grains. In addition, atoms at grain boundaries are more reactive than those within grains. As a consequence, grain boundaries can often be seen using a microscope to view a polished, then chemically etched, flat metal surface. Etching with an acid removes material from grain boundaries more rapidly because of the higher energy of these sites. The patchy appearance of brass doorknobs is due to this effect, as the acids in perspiration can etch away atoms at the grain boundaries.

Laboratory. An experiment that makes use of the reactivity of alpha-particle-generated defects in polymers to detect radon gas in the home is presented in Experiment 6.

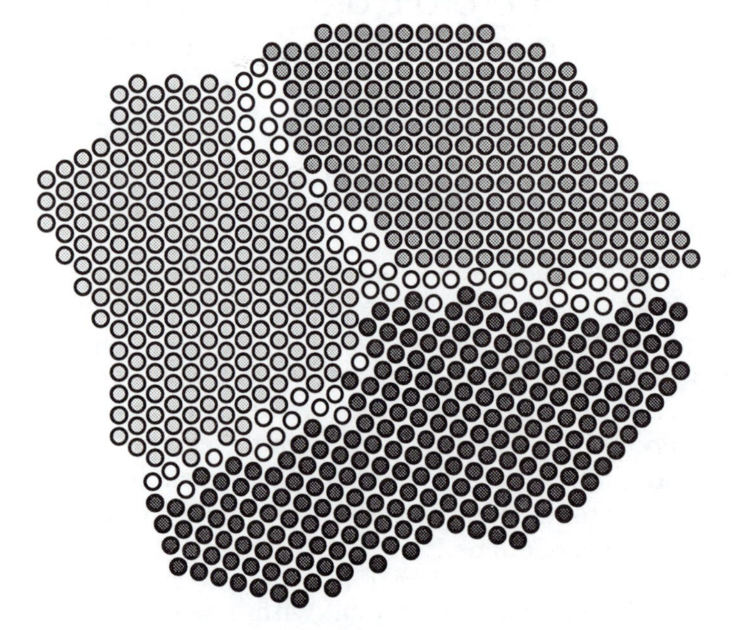

Figure 6.9. A two-dimensional representation of grains in a polycrystalline material; different shading is used for atoms in different grains. Grain boundaries are higher in energy than the interiors of the grains because of less efficient packing.

Plastic or irreversible deformation of crystalline substances occurs primarily by the movement of dislocations within grains. Dislocations are the carriers of deformation, because the passage of a dislocation through a crystal results in the relative motion of one part of a crystal past the other. Mechanical stress can cause dislocations to move through the grain until they are trapped or annihilated. Unlike vacancies, the concentrations of dislocations are not governed by chemical equilibria.

Demonstration 6.2. Dislocations and Other Defects

Materials

Bubble raft setup (*see* Chapter 5)

Procedure

- Prepare a bubble raft as described earlier.
- Pop individual bubbles with a dry finger or pencil tip to make vacancies.
- Stress the bubble raft by compressing or expanding it. Push the slider gently toward one end of the tray to compress the bubble raft, and pull the slider gently in the opposite direction to stretch it. Be careful not to the scratch the bottom of the tray.
- The bubble raft illustrates a variety of dynamic processes. As the bubble raft is forming, it undergoes reconstruction: Grains form, grow, and disappear. Point defects, dislocations, and grain boundaries can all be easily observed in a two-dimensional bubble raft model. However, because the bubble raft is a two-dimensional model, the dislocation looks more like a point (zero-dimensional) defect, and the grain boundary looks like a line (one-dimensional) defect. Many parts of the raft have a very regular hexagonal arrangement of bubbles. However, there are always a number of imperfections. When the raft is stressed, dislocations zip through the grains, moving to the grain boundaries and causing plastic deformation. As the raft is repeatedly expanded and compressed, dislocations are created and annihilated. Vacancies present in the bubble raft do not move as the raft is deformed. Vacancies are frequently annihilated, however, when they interact with dislocations that pass nearby.
- A static model illustrating many of the same concepts of structure and microstructure can be made with a sealed plastic tray of metal BB pellets. A description of the construction of this BB board is currently in press and will be available from ICE Publications, Department of Chemistry, University of Wisconsin—Madison, Madison, WI 53706.

In some metals, the motion of atoms during plastic deformation occurs at the speed of sound in the solid, which is typically on the order of several thousand meters per second or about an order of magnitude faster than the speed of sound in air. Three metals, zinc, indium, and tin, exhibit a snapping sound when rods of the metals are bent at room temperature, due to the dislocation-assisted motion of their atoms. Because the rate at which the atoms move exceeds the speed of sound in air ("Mach 1"), an audible click results.

Demonstration 6.3. Hearing Atoms Move: Breaking the Sound Barrier *(8)*

Materials

Zinc, tin or indium rods (Aldrich, 6-mm diameter; 100 mm long)
Microphone (for large audiences)

Procedure

- Bend any of the rods. A snapping or cracking sound will be heard. The sound can be made audible in a lecture room by placing a microphone near the rod as it is being bent and bringing the microphone's sensitivity to the optimum level (this should be tried in advance).

Variations

- If only short pieces of metal are available, the ends may be placed in metal or rigid plastic tubes to provide the extra leverage needed to bend them.
- Because the ability to move planes of atoms depends on the defects in the solid, which eventually become tangled, and thereby hinder atomic motion (*see* the description of work hardening, later), the experiment cannot be repeated indefinitely. Of the three metals, the zinc rods can be bent the most times without fracture and still produce the snapping sound. (For example, more than 200 students have bent a zinc rod and the sound was still audible.) Both the tin and the indium rods can be heated to about 130 °C for 24 hours in an oven to anneal the samples (*see* later), which will restore the acoustical response to bending.

Work Hardening and Annealing

The movement of dislocations is the key to plastic deformation; therefore, strengthening a metal, that is, increasing its resistance to deformation, requires either eliminating dislocations or preventing them from moving. In practice, removing all of the dislocations from a crystal or growing a crystal without dislocations is practically impossible except for very small crystals. Therefore, metallurgists go to great lengths to find clever ways of pinning dislocations (preventing the motion of dislocations) in order to strengthen metals. Dislocations are often pinned by other defects in the crystal. New dislocations are created during deformation and become pinned by the initial dislocations. The effect cascades as the material is repeatedly deformed, and dislocations collect in tangles in the metal.[1] This buildup of pinned dislocations leads to the hardening of the metal in a process known as work hardening.

An analogy for picturing dislocations and how they produce work hardening is to think of moving a heavy rug across a floor: The rug represents a plane of atoms that is to be moved relative to another plane of atoms, the floor. If the rug were to be lifted slightly at one end and dragged, a great deal of energy would be required for the movement, analogous to the aforementioned calculation of the substantial energy cost associated with breaking all of the bonds between planes of atoms simultaneously in order to move one plane of atoms past the other.

On the other hand, if a wrinkle is present in the rug (Figure 6.10A), it serves as a kind of dislocation in facilitating the movement of the rug: The wrinkle can be pushed to the edge, resulting in a net movement of the entire rug. However, at any given time only a small portion of the rug actually moves; and, consequently, relatively little energy cost is associated with the motion.

Extending the analogy, work hardening is akin to having multiple tangled wrinkles in the rug. As Figure 6.10B shows and experience verifies, if multiple wrinkles meet at various angles, it is difficult to eliminate them: The movement of one wrinkle is "pinned" by another, preventing movement of the rug. Similarly, tangled dislocations make it difficult to move atoms in a metal by creating a kind of "atomic gridlock" and effectively hardening the metal.

A work-hardened metal can be softened again by annealing (heating) at high temperatures, generally at temperatures above one-half of the melting point on the absolute temperature scale. As a piece of metal is heated, the increased thermal vibrations allow atoms to rearrange and go to lower energy states by diffusion, thereby reducing the number of dislocations present.

---

[1]The density of dislocations stored in a work-hardened metal can become huge, on the order of 10^6 km of line dislocations per cubic centimeter of material.

A. Dislocation

B. Work hardening

Figure 6.10. A rug as a model for movement and pinning of dislocations in a metal. A: If a wrinkle is present in a rug, the wrinkle may be pushed through it so that the rug moves across the floor. In this way the rug can move across the floor without requiring that the entire rug move at once; this is similar to the movement of a plane of atoms with respect to a second plane in the plastic deformation of a metal. B: If two or more wrinkles in the rug meet, it is more difficult to move them. The presence of one "pins" the other. The collection of pinned dislocations leads to the hardening of a metal in a process known as work hardening.

Demonstration 6.4. Work Hardening *(9, 10)*

Materials

Solid copper wire (10 gauge or thicker, used for 120-V house wiring and available at a hardware store)
Weight (~700 g works well. If a hole is drilled into the handle of a hammer, it can serve as the weight.)
Meker burner or torch
Tongs
Ring stand
18-gauge copper wire (optional)

Procedure

- Bend the wire. The more you bend it, the more difficult it will be to bend.
- Heat a piece of the wire red hot by holding it with tongs in the flame of a Meker burner. Let it cool slowly on the iron plate of a ring stand. The heat-treated (annealed) piece will be easy to bend at first, although continued bending will again work-harden the copper.

Variation 1

- Prepare a 9-inch-long piece of 10-gauge copper wire by heating the wire red hot in the flame of a Bunsen burner (annealing) while holding it with tongs and letting it cool slowly on the iron plate of a ring stand.
- Hold one end of the cool wire.
- Hang a ~700-g weight on the other end of the wire. The annealed wire should bend under the applied weight.
- Work-harden the wire by grasping it at its two ends and bending it many times. Show that the wire now supports the weight without bending.
- If time permits, anneal the wire by heating as described above to show that hanging the weight on its end again causes the wire to sag.

Variation 2

- Take a long piece of copper wire (about 2 feet of 18-gauge wire) and pull it two or three times over a nail or other small diameter object, as if the wire were like the rope of a pulley. The wire will bend much more easily on the first pass than on the third pass.
- Heat a section of the strained wire red hot in a Bunsen burner, and let it cool slowly. The treated wire will again bend easily at first.

Hardening of Alloys: Steel

To work-harden a metal by deformation is one way to strengthen it. However, metallurgists often rely on more sophisticated means of hardening and take advantage of the many solid phases—both equilibrium and metastable—present in binary and multicomponent alloys. The fact that the metallic bond is not particularly selective allows metal atoms of different kinds to intermingle and form alloys over wide ranges of composition. Changing the composition on an atomic scale can also increase the strength of a metal.

The presence of dissolved impurities in a metal makes it more difficult for dislocations to move, especially if the sizes of the atoms of the impurity and host metal differ significantly. (Recall from Chapter 3 that if the atomic radii of two elements differ by more than ~15% the solubility of a substitutional impurity element in a host metal is very low.) The resulting hardening of the metal is called solid solution hardening (*see* Figure 6.11). For example, dissolution of tin in copper produces bronze, which is harder than either pure copper or pure tin. The difference in the size of the atoms roughens the slip plane on the atomic scale, making it harder for dislocations to move.

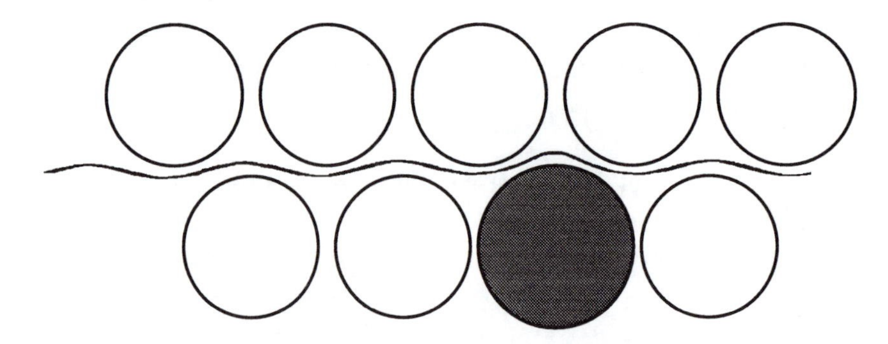

Figure 6.11. The addition of impurity atoms (shaded circle) causes slip planes (wavy line) to roughen and thereby increases the hardness of the alloy relative to the pure metal.

Precipitation hardening occurs when the room-temperature solubility of an impurity is exceeded, and a fine-grained precipitate forms throughout the metal. For example, Cu dissolves in Al at high temperatures, but its lower solubility as the solid solution cools results in precipitation of grains of $CuAl_2$. Second phases can precipitate at the grain boundaries, where the imperfect nature of the atomic packing can accommodate them more easily.

Two demonstrations that will be described provide striking illustrations of how phase transformations (also discussed in Chapter 9) and the solubility of carbon are used for strengthening in carbon steels *(11)*. Pure iron melts at 1538 °C. Between room temperature and the melting point, however, iron undergoes several allotropic phase transformations, from a body-centered cubic (bcc) phase, called α–iron or ferrite, at room temperature, to a face-centered cubic (fcc) phase, called γ–iron or austenite, at temperatures between about 900 and 1400 °C (Table 6.1). At temperatures above 1400 °C, iron transforms to a bcc structural type that is generally not an important feature in the heat treatment of steels.

Austenite, the fcc structure of iron, has the ability to dissolve much more carbon than ferrite, the bcc structure of iron. [The maximum solubility of carbon in austenite is 2.11 wt % at about 1148 °C, compared to the maximum solubility of carbon in ferrite of only 0.02 wt % at 727 °C *(11)*.]: Although the structure of ferrite is more open in terms of the total volume of interstices or voids between iron atoms, the structure of austenite contains fewer but larger interstitial sites and is able to dissolve larger amounts of carbon as an interstitial defect in the crystal.[2] The carbon atoms do not dissolve well in ferrite, because the carbon atom is too big to fit in the interstices.

Table 6.1 Phases of Iron Important in Steel

Phase	Structure	Stable Temp. Range	Carbon Solubility
α-Iron (ferrite)	bcc	under 900 °C	0.02 wt % at 727 °C
γ-Iron (austenite)	fcc	900–1400 °C	2.11 wt % at 1148 °C

Many of the important heat treatments of steels involve phase transformations across the austenite–ferrite phase boundary. When steel, such as the type used to make piano wire (containing about 1 wt % carbon), is slowly cooled below the austenite–ferrite phase-transition temperature, the excess carbon, which is no longer soluble in ferrite, precipitates as iron carbide (Fe_3C).[3] This process requires the carbon to diffuse moderate distances (~ 0.1–1 μm) through the austenite. The presence of iron carbide precipitates serves to strengthen the steel by pinning dislocations. The more carbon there is in the steel, the more iron carbide there will be, and generally, the stronger the steel will be (but it will also be more brittle).

The kinetics of the phase transformation depend strongly on temperature, a fact that allows careful tailoring of the size distributions

---

[2]Recall from Chapter 5 that a body-centered cubic structure has a packing efficiency of 68%, and a face-centered cubic structure has a packing efficiency of 74%. The fcc structure has octahedral holes that can contain a sphere of size $0.414r$ (r is the radius of the atoms forming the face-centered cube) and tetrahedral holes with radius $0.225r$. The bcc structure, on the other hand, contains a distorted octahedral hole formed by the four atoms on the face of a cube, and the two centered atoms in the unit cells that share that face. Though the distance from the center of this hole to an atom on the face is $0.633r$, the distance from the center of the hole to the atoms in the centers of the unit cell is $0.154r$. Consequently, only very small spheres will fit here. Also, slightly enlarged tetrahedral holes in the bcc structure have radii of $0.291r$. Thus, the fcc structure has larger void spaces because of the octahedral holes, even through it is a more densely packed structure. *See also* reference 12.

[3]Iron carbide (Fe_3C) is a metastable phase, meaning that it is unstable with respect to decomposition to pure iron and graphite. However, in most alloys, this decomposition is so sluggish that it is unimportant. In certain cast irons the carbon is present as graphite, and in these alloys, the formation of graphite has been catalyzed with the addition of an element like silicon.

and types of second phase precipitates within steels. If the steel is quenched from high temperature by rapid cooling in a water bath, the driving force for the austenite–ferrite phase transformation becomes very large. However, at the same time, the steel is cooled so quickly that the carbon has insufficient time to concentrate by diffusion into the Fe_3C precipitates. The result is a new phase, called martensite, which is essentially a distorted form of the bcc ferrite structure, distorted because it contains large amounts of carbon trapped during the transformation.

Martensite, when it contains large amounts of carbon, is a very hard material, but the large distortions of the crystal lattice that occur during the phase transformation produce high internal stresses that promote fracture. Therefore, martensite can be very brittle and can shatter like glass when dropped if it has not been treated by a process called tempering. In tempering, the sample is gently heated after it has been quenched by the sudden cooling. Tempering the steel allows the carbon to diffuse short distances and allows some of the internal stresses to relax, yielding a greater resistance to fracture.

Demonstration 6.5. Annealing, Hardening, and Tempering of Bobby Pins *(13)*

Materials

Four bobby pins
Tongs
Bunsen burner or torch
Iron ring stand (or other heat-proof surface)
Medium-sized beaker of water

Procedure

- Obtain four bobby pins; one pin will be left untreated for reference.
- Holding the ends of a bobby pin with crucible tongs, heat the bend of the pin red hot in a Bunsen burner flame (annealing), then allow it to cool slowly on a heat-proof surface such as the iron plate of a ring stand. This bobby pin loses its springiness and stays open when stretched.
- Heat the bend of another bobby pin red hot, but drop it into cold water to cool quickly. This bobby pin is quite brittle and snaps into pieces when bent.
- Heat a third bobby pin red hot, and drop it into cold water. Hold the now-cool pin above a flame and gently heat (tempering) until an iridescent blue coating forms and then let the pin cool slowly. This bobby pin will be as springy as the original.

A more elaborate experiment using steel piano wire can also be done as a lecture demonstration. The ostensible objective of the demonstration is to make a useful spring from a piece of piano wire.

Demonstration 6.6. Annealing, Hardening, and Tempering of Piano Wire

Materials

Two (2–foot) strands of 16.5-gauge, single-strand piano treble wire (Available in music stores or see Supplier Information)
A 1-foot length of 1-inch-diameter wooden dowel
Variable-voltage transformer (Variac)
Power cord with two large, insulated alligator clips
Bucket of water

Procedure

- Wrap the piano wire around the wooden dowel as though trying to shape it into a coiled spring. As received, the piano wire is very resistant to bending and will not retain this shape, making it unsuitable for serving as a spring.
- Connect a variable-voltage transformer (maximum current 10 A) to a heavy power cord that terminates with two large, insulated alligator clips. This device will be used to pass a current through the steel piano
- Connect one alligator clip to each end of the steel piano wire. Set the variable-voltage transformer for 20–30 V and maintain the current flow just until the wire turns red hot. **Disconnect** the wire from the variable-voltage transformer, wait about 30 seconds, and cool the wire by immersing it in a bucket of water. **CAUTION: Avoid touching the bare wire while the variable-voltage transformer is on and current is passing through the wire! And do not immerse the wire in the water unless the variable-voltage transformer is disconnected!**
- Again, wrap the piano wire around the wooden dowel, trying to shape the wire into a coiled spring. Do not cross the strands of wire as you wind the wire around the dowel. This time, the wire retains the shape of a spring when the wooden dowel is removed.
- Test the springiness of the now-coiled wire by pulling gently on one end and observing the resulting behavior. The wire retains the shape of a spring, but the spring is not very elastic.
- To harden the spring, connect the alligator clips to it, one clip on each end, and again heat the wire until it turns red hot, using the variable-voltage transformer setting of 20–30 V. Try not to let adjacent coils touch each other during this heating process. Disconnect the variable-voltage transformer from the wire, and *immediately* quench the wire in a bucket of cold water.

- Remove the alligator clips and test the springiness by pulling on one end of the coiled spring. The wire should break under a minimum of force.
- Heat a second piece of wire, anneal it, shape it into a spring and harden it as described. After the spring has been quenched, remove it from the water and connect the alligator clips to each end. Starting with the variable-voltage transformer set to less than 10 V, heat the wire gently (tempering). Avoid heating the wire to the point where it turns red hot. Turn the variable-voltage transformer off, wait about 30 seconds, and then cool the spring by immersing it in water.
- Remove the alligator clips. After tempering, the coiled spring has more elasticity than the untreated spring and loses much of the brittleness of the hardened spring. If the spring is neither brittle nor elastic, the temperature reached during tempering was probably too high.

It is valuable to summarize the processes discussed so far and to emphasize that although the details may vary for a given metal or alloy, the microscopic phenomena are universal. Specifically, the movement of dislocations in a metal causes plastic deformation. Bending a piece of copper wire leads to the creation of new dislocations, which become pinned and then pin other dislocations in a process known as work hardening. Bobby pins and piano wire are examples of materials that have been heavily work-hardened. Precipitates, as are found in steel and almost all other alloys, are used as hardening agents because they pin dislocations. To a lesser degree, solid solution impurities also pin dislocations and strengthen metals. Heat treatments are used to tailor the mechanical properties of a metal, but how a heat treatment affects the properties depends on the particular alloy system as well as the details of how the heat treatment is performed. For example, when the work-hardened copper was heated, dislocations were removed, which resulted in a net softening effect. In the steels described herein, the various heat treatments take advantage of the phase chemistry and the transformation kinetics of the steel; a variety of phases and a range of mechanical properties are seen.

Defects in Ionic Compounds

Ionic compounds also have defects. Like metals, ionic crystals exhibit grain boundaries and dislocations. However, ionic compounds differ from metals in that cations and anions are present, and their presence introduces some new possibilities for defects. Several types of defects particularly important to the physical properties of ionic compounds (such as ionic conduction) will be discussed here.

Schottky defects are most common in the alkali halides and involve pairs of vacant cation and anion sites. It has been estimated that at room temperature only one in 10^{15} pairs of cations and anions is vacant in sodium chloride. This means that a 1-mg grain of salt still has about 10,000 Schottky defects *(14)*.

A Frenkel defect is one in which a cation (or anion) is located in an interstitial site that is normally empty. In AgCl, which has the sodium chloride structure, some of the silver ions may be located in interstitial tetrahedral holes, rather than the usual octahedral sites. Because electrical neutrality must be maintained in the crystal, there will be an equal number of vacant Ag^+ octahedral sites as well. Figure 6.12 illustrates Frenkel and Schottky defects for a general ionic compound with the formula MX.

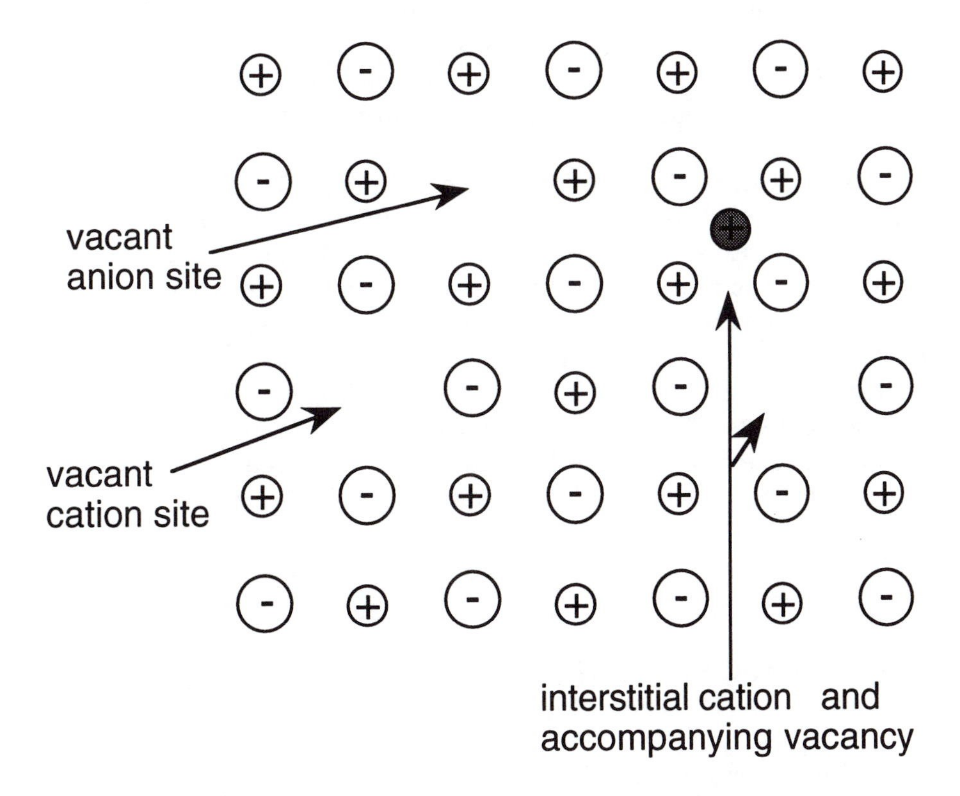

Figure 6.12. Defects in ionic compounds. Schottky defects are pairs of cation–anion vacancies and are represented by the empty sites in crystal positions normally occupied by a cation and an anion. A Frenkel defect is illustrated by the shaded atom in an interstitial site while a normally occupied site is vacant as a consequence.

Another type of defect involves substitution for some of the ions in the crystal. In sodium chloride, for example, impurity atoms such as potassium may substitute for the sodium and bromide may substitute for the chloride. [In general, if the substituting atoms are about the same size, the overall packing arrangement is maintained and a solid solution may form (*see* Chapter 3).] A different effect, however, is observed when a divalent cation such as calcium is substituted for the sodium. Such substitutions require charge compensation to maintain the overall electrical neutrality of the crystal. One way to accomplish charge compensation is that for each calcium substitution in the lattice, a second sodium is also removed from the crystal. This means that doping with calcium should serve to increase the number of Na^+ vacancies in the crystal. (This type of defect is intentionally introduced in a solid-state oxygen sensor that operates in catalytic converters. *See* Chapter 8 for details.)

Ionic compounds in the solid state are generally considered to be electrically non-conducting. However, in certain cases ionic solids act as conductors and are receiving consideration as battery electrolytes and oxygen sensors. Even sodium chloride and other classical ionic compounds conduct electricity at temperatures below their melting points. Unlike metals, which show conductivity by motion of electrons, ionic compounds are thought to be conductive due to ion mobility mediated by crystal defects. (In some cases the compounds have channels through which ions can move.)

Ionic conduction often involves the motion of the cations, presumably because of their smaller size, and can occur by a process in which a cation moves from an occupied site into a vacancy. This kind of cation motion will always leave behind a vacancy, and a second cation can move into the newly created vacancy, a third cation can move into the vacancy created by motion of the second, and so on. This process is believed to be the mechanism of conduction in sodium chloride (*see* Figure 6.13), which has predominantly Schottky defects. Substitution of a divalent cation for pairs of monovalent cations in the crystal should also serve to increase the conductivity by this mechanism, because the number of cation vacancies would be increased.

F-Centers in Salts: Spectroscopy of Defects

Certain types of point defects are readily created in alkali halides by ionizing radiation. The best understood of these point defects is the F-center, after *farbe*, the German word for color *(15–17)*. The nature of the F-center has been elucidated by various spectroscopic techniques and is described as an electron that is trapped at a halide vacancy in the crystal. This state is shown pictorially in Figure 6.14 for KCl. The absorption spectrum of the trapped electron gives rise to a deep coloration of the crystal, a beautiful deep purple in KCl. The color observed is due to the absorption of a photon by the trapped electron and excitation from the ground state of the F-center to an excited state. A related defect is also responsible for the colors of smoky quartz and amethyst. [These minerals

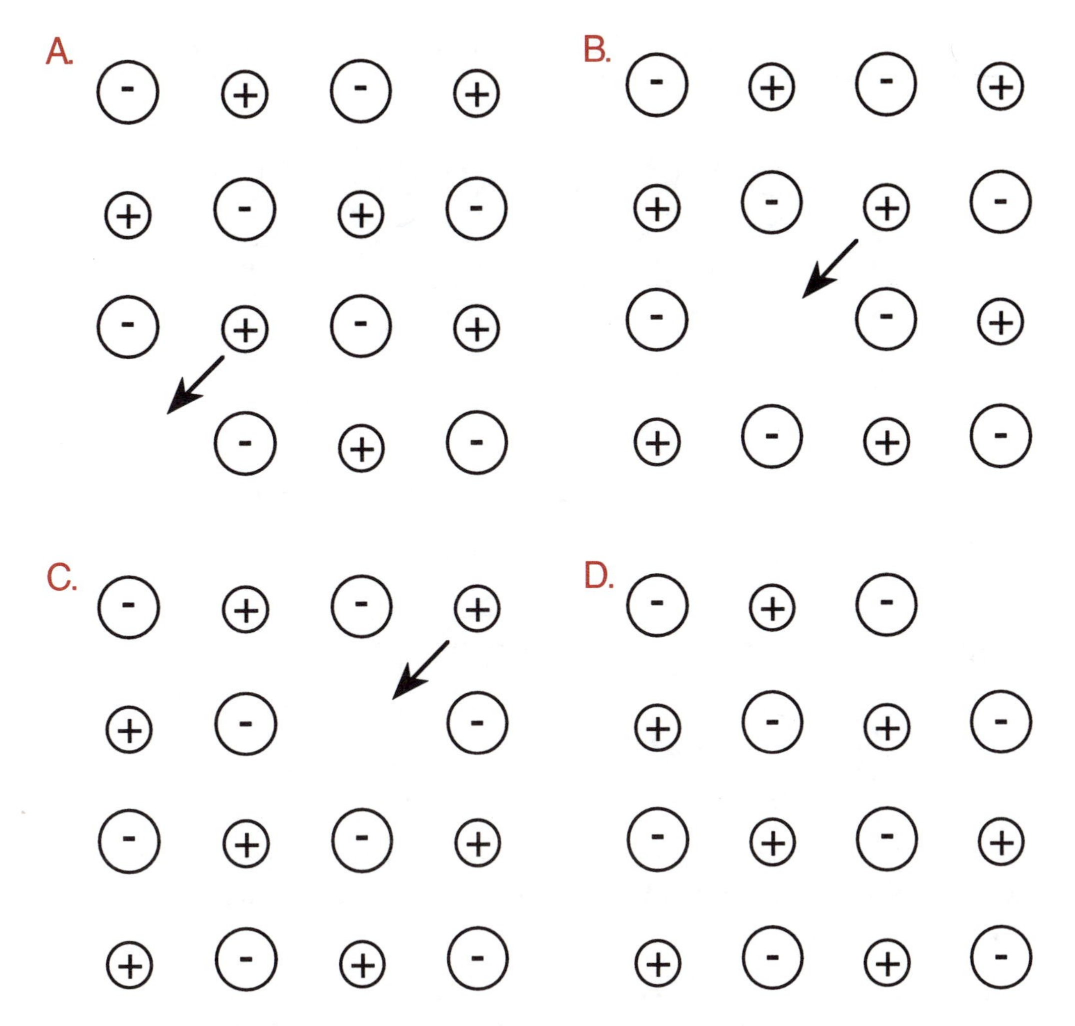

Figure 6.13. Conductivity in an ionic crystal. The sequence A–D shows how cation migration can occur by a series of movements of cations into crystal vacancies.

are predominantly SiO_2. However, the substitution of very small amounts of Al^{3+} (smoky quartz) and Fe^{3+} (amethyst) for Si^{4+} (about one substitution for every 10,000 Si atoms) in the crystal coupled with ionization of an electron from an oxygen produces the color center. As noted in Chapter 5, substitution of a trivalent cation for silicon requires the presence of one more unit of positive charge in order to maintain electrical neutrality. This unit is generally thought to be a proton located in an interstitial site *(15)*.]

The exact mechanism for F-center formation is not known; however, various models have been postulated *(18)*. One model is that high energy radiation (for example, X–rays or gamma rays) interacts with the alkali halide, causing a halide ion to lose an electron, producing a halogen atom. The electron released in the process moves freely throughout the crystal until it encounters a halide vacancy. Strong electrostatic forces trap the electron at this site, as the electron is surrounded by six positively charged alkali metal ions. Also, a halogen atom can interact with an adjacent halide ion to form a molecular-type defect such as Cl_2^-, which is the electron-deficient complement to the F-center.

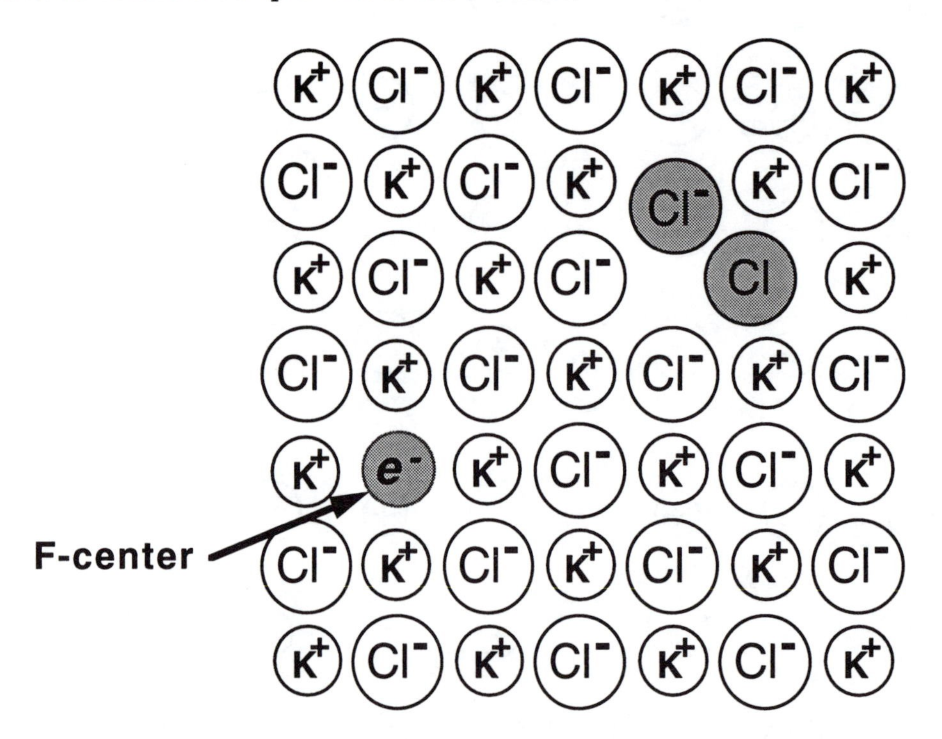

Figure 6.14. An F-center is an electron trapped in a halide vacancy. Other defects such as the formation of Cl_2^- species (indicated by the shaded pair of spheres) occur for charge balance.

F-centers can also be produced by heating the alkali halide in alkali metal vapor. In this case, the alkali metal ionizes, producing a cation, which diffuses into the crystal, and a free electron, which is trapped at a halide vacancy. The color observed in the crystal is characteristic of the crystal and does not depend on the alkali metal vapor used. Thus, potassium chloride crystals become purple whether they are heated in the presence of potassium vapor or sodium vapor.

The observation of F-centers in alkali halides arises from stoichiometric deviations on the order of one electron for every 10,000 halide ions. Thus, the effect on the optical properties of even a very small deviation from the ideal stoichiometry can be very large. F-centers can be readily prepared in alkali halides by excitation with a Tesla coil, commonly used to detect leaks in glass vacuum lines.

Demonstration 6.7. F-Centers *(19)*

Materials

Crystals of KCl, NaCl, RbCl, or KBr (Optical grade random crystal

cuttings of NaCl, KCl, and KBr can be obtained at an economical price. *See* Supplier Information.)
Small side-arm flask and stopper
Vacuum pump or oil diffusion pump
Tesla coil

Procedure

- Evacuate a small flask containing KCl or other alkali halide crystals using a good mechanical vacuum pump (10^{-3} mm Hg) or, better yet, by using a high-vacuum line with an attached oil diffusion pump (10^{-5} mm Hg or better). **CAUTION: Make sure that the flask has no defects. Small cracks may cause it to shatter under vacuum. Conduct the evacuation behind an explosion shield.**
- Direct the discharge from a Tesla coil at the KCl through the wall of the flask for several minutes, until the salt turns a bright purple (*see* Figure 6.15). This technique also yields an orange-brown coloration in NaCl, and a light blue coloration in RbCl or KBr, which persists for hours to days depending on the salt.

Variations

- If the crystals are sealed in a glass container under high vacuum, the vacuum will persist and the color can be regenerated when needed by the discharge of a Tesla coil. This demonstration works on both powdered salts or large crystals. However, one advantage of using large, optical grade crystals is that the absorption spectra of the F-centers can be readily measured with a standard spectrometer.

Figure 6.15. F-center production in an evacuated flask by a tesla coil.

Spectroscopy

After coloration of a large crystal, the absorption spectrum of the F-center can be obtained via transmission in a UV–visible spectrometer. The absorption spectra obtained from color centers in NaCl, KCl, and KBr are shown in Figure 6.16. The peaks are generally broad, but exhibit a clear trend in the variation of the band maximum (λ_{max}) with the size of the halide vacancy (as estimated by the lattice parameter, the length of an edge of the cubic unit cell): As the size of the vacancy increases, λ_{max} increases. This trend holds for color centers that have been observed in all of the alkali halides *(20)*.

Figure 6.17 is a plot of the F-center absorption energy versus the lattice parameters of the salts, which have cubic unit cells. The maximum absorption energy, *E*, is inversely related to the size of the vacancy, modeled as a cubic box:

$$E \sim a^{-1.8}$$

where *a* is the cubic lattice parameter (the length of the edge of the cubic unit cell) for a given salt.[4]

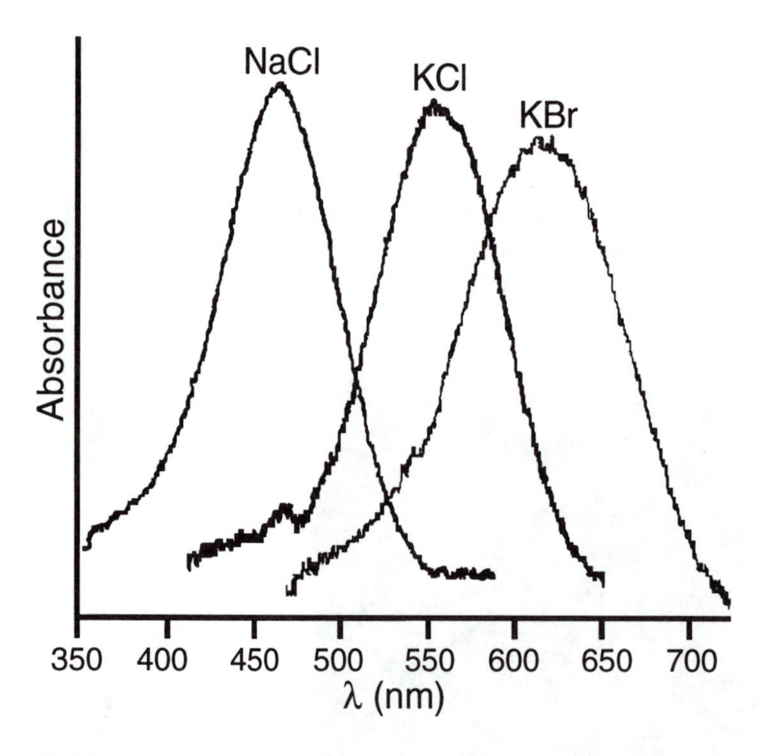

Figure 6.16. Absorption spectra from F-centers in NaCl, KCl, and KBr at 298 K. These spectra were obtained in air.

[4]This result is similar to the result of the particle-in-a-box problem in which the energy is inversely related to the square of the length of the box.

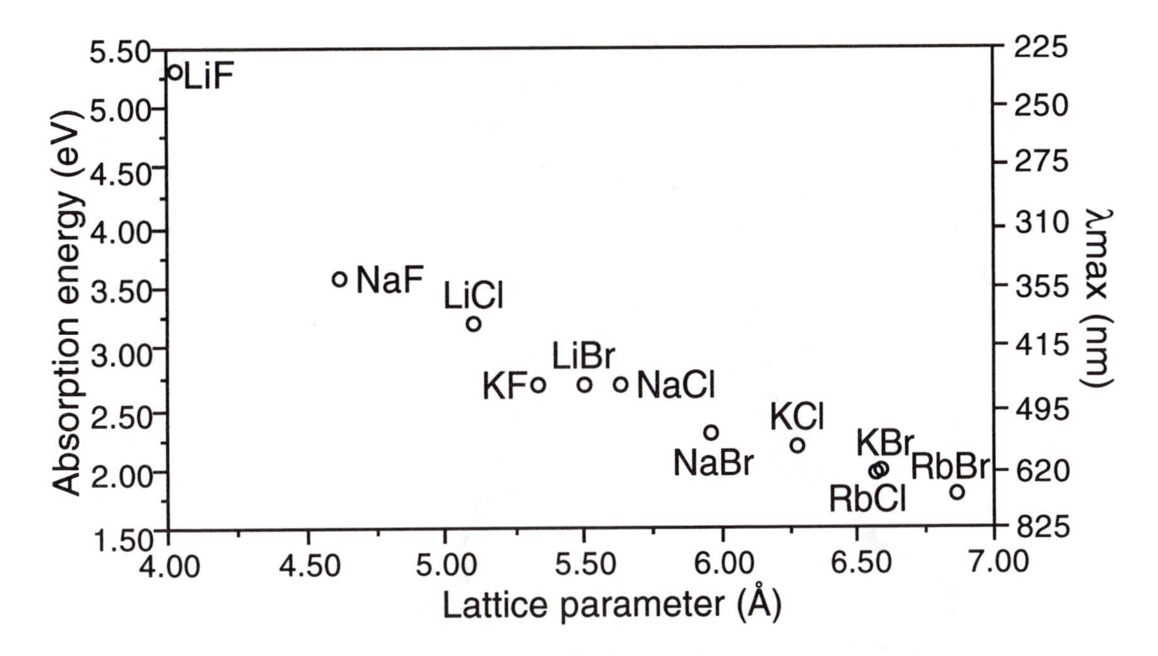

Figure 6.17. Plot of color F-center absorption maxima as a function of the lattice parameter at 298 K.

The effects of temperature and pressure on the size of the halide vacancy suggest that as the crystal is cooled or squeezed and the ions move closer together, the absorption peak should shift to higher energies. If a colored KCl crystal is cooled in liquid nitrogen, the purple color observed at room temperature changes to a deep pink color at 77 K. However, complex changes occur in the spectral distribution that have been attributed to an equilibrium between several different types of defect structures *(21)*. With increasing pressure, the absorption maxima of F-centers in NaCl and KCl shift to higher energies as the crystal contracts *(22)*.

Another method of changing the size of the halide vacancy is to prepare solid solutions (Chapter 3). Solid solutions of $K_{1-x}Rb_xCl$ can be prepared by heating a mixture of KCl and RbCl above the melting point and then cooling. F-centers produced in these mixed crystals have absorption maxima that lie between pure KCl and pure RbCl *(23)*.

The F-centers produced by the Tesla coil method bleach slowly at room temperature. The kinetics of this decay can be measured on a spectrophotometer by monitoring the absorption over a period of time. Figure 6.18 shows the decay of the F-center absorption in KBr over a period of 2 hours. This particular salt bleaches faster than either NaCl or KCl, allowing for the observation of a larger change in the absorption within a reasonable time frame. Analysis of the absorption-versus-time data does not fit a simple rate law. The decay probably occurs through several different pathways.

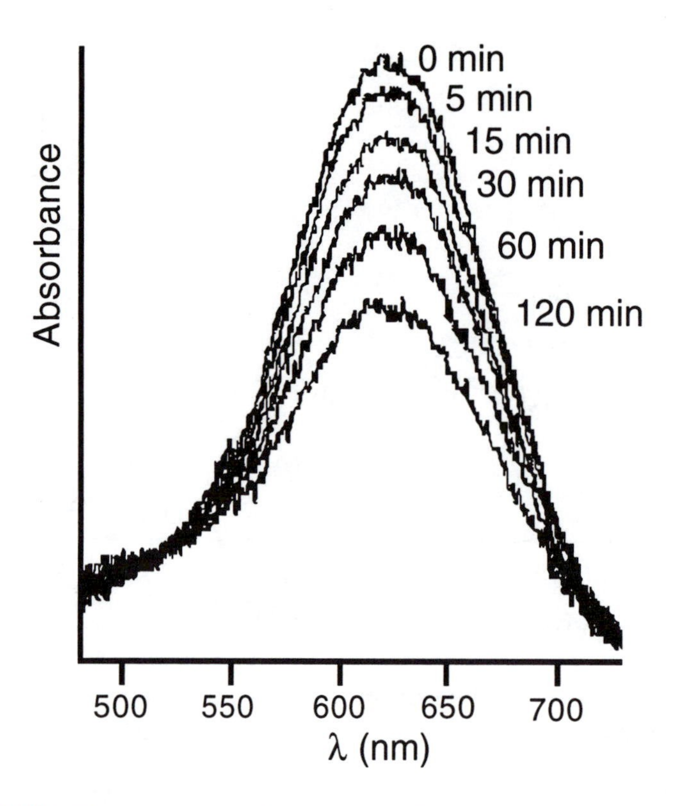

Figure 6.18. Plot showing the decay in F-center absorption of KBr at 298 K over a period of time.

F-Center Luminescence

The energy that is stored in the F-center can be released through thermoluminescence, the process of using heat to generate luminescence from a compound *(24)*. A demonstration of this phenomenon requires salt crystals with a much higher concentration of F-centers than those produced using a Tesla coil. Salts that have been irradiated with gamma rays are deeply colored as a result of a high concentration of F-centers. These samples have been irradiated but are not radioactive.

Demonstration 6.8. Thermoluminescence from F-Centers

Materials

Hot plate

Samples of irradiated, alkali halides containing high concentrations of F-centers. Irradiated NaCl, which has a brownish color, is commercially available. (*See* Supplier Information.)

Spatula

Procedure

- Turn a hot plate to the high setting (190 °C or higher).
- When the hot plate is sufficiently warm, darken the lights in the room and sprinkle a spatula full of irradiated NaCl on the hot plate. As the salt hits the hot plate, bright orange light can be seen emanating from the crystals at the surface of the hot plate.
- If samples of irradiated LiCl are sprinkled on the hot plate, the observed luminescence is bluish-white.
- When the room lights are turned back on, the salt on the hot plate is no longer colored, but has returned to its characteristic white color.

Recalling how the F-center was formed, the general phenomenon of thermoluminescence can be explained in the following way: *(25, 26)* The thermal energy provided by the hot plate allows the trapped electron of the F-center to be released from the halide vacancy trap and thereby recombine with the electron-deficient defects, such as Cl_2^- (*see* Figure 6.14), releasing part of the energy in the form of light. The thermoluminescence seen in this demonstration is in the visible region of the spectrum, but the energy required to form F-centers in NaCl and LiCl is much larger (*see* Band-Gap Energies for Ionic Solids, Chapter 7). Therefore, the emission observed is only part of the excess energy of the trapped electron. The remainder of the energy must be lost either in photons outside of the visible region or in nonradiative processes.

Finally, it is worth speculating on the reactivity of these defects. The dissolution of irradiated salt crystals in water may produce hydrated electrons that go on to react further in solution. A radiochemistry laboratory experiment described in the *Journal of Chemical Education (27)* investigates the chemical reactivity of F-centers upon dissolution in water and the exploitation of this reactivity to obtain quantitative information on the concentration of F-centers produced in irradiated NaCl.

References

1. Drexler, K. E. *Nanosystems;* Wiley: New York, 1992.
2. Bragg, L.; Nye, J. F. *Proc. R. Soc. London* **1947**, *190*, 474–481 and reprinted in Feynman, R. P.; Leighton, R. B.; Sands, M. *The Feynman Lectures on Physics*; Addison Wesley: Reading, MA, 1964; Vol. 2, Chapter 30, p 10.
3. Sprackling, M. T. *The Plastic Deformation of Simple Ionic Crystals*; Academic Press: New York, 1976; p 20.
4. Holden, A.; Singer, P. *Crystals and Crystal Growing*; Anchor Books: Garden City, NY, 1960.
5. Dieter, G. F. *Mechanical Metallurgy*, 3rd ed.; McGraw-Hill: New York, 1986, p 119.
6. Taylor, G. I. *Proc. R. Soc., London,* **1934**, [A], *145,* 362.

7. Nabarro, F. R. N. *Theory of Crystal Dislocations*; Clarendon Press: Oxford, England, 1967; p 590.
8. Feynman, R. P.; Leighton, R. B.; Sands, M. *The Feynman Lectures on Physics*; Addison-Wesley: Reading, MA, 1964; Vol. 2, Chapter 30, pp 8–10.
9. Brown, B. *200 Experiments for Boys and Girls*; Collins: Cleveland, OH, 1973, p 24.
10. Guinier, A. *The Structure of Matter: From the Blue Sky to Liquid Crystals*; Edward Arnold: London, 1984.
11. For a thorough discussion of the iron–iron carbide phase diagram, *see* Long, G. J.; Leighly, H. P., Jr. *J. Chem. Educ.* **1982**, *59*, 948–953.
12. Hume-Rothery, W.; Raynor, G. V. *The Structures of Metals and Alloys*; The Institute of Metals: London, 1954; p 58.
13. *Chemistry Can Be Fun*; Institute for Chemical Education. University of Wisconsin—Madison: Madison, WI, 1984.
14. West, A. R. *Solid State Chemistry and Its Applications;* John Wiley and Sons: New York, 1984; p 321.
15. Nassau, K. *The Physics and Chemistry of Color*; Wiley: New York, 1983; pp 184–190.
16. Greenwood, N. N. *Ionic Crystals, Lattice Defects, and Nonstoichiometry*; Butterworths: London, 1968; pp 170–175.
17. Levy P. W. In *Encyclopedia of Physics*, 2nd ed.; Lerner R. G.; Trigg, G. L., Eds.; VCH Publishers: New York, 1991; pp 162–171.
18. Royce, B. S. H. *Prog. Solid State Chem.* **1967**, *4*, 213–243.
19. "Tested Demonstrations in Chemistry;" Alyea, H. N.; Dutton, F. B., Eds.; Journal of Chemical Education: Easton, PA, 1965, p 166.
20. Markham, J. J. *F-Centers in Alkali Halides*; Academic Press: New York, 1966; pp. 7–12.
21. For more details, *see* Markham, pp 13–22.
22. For more details, *see* Markham, pp 23–26.
23. Gnaedinger, R. J., Jr. *J. Chem. Phys.* **1953**, *21*, 323–330.
24. Davison, C. C. *Chem 13 News*, November 1989, p 4.
25. Hill, J. J.; Schwed, P. *J. Chem. Phys.* **1955**, *23*, 652–658.
26. Bonfiglioli, G.; Brovetto, P.; Cortese, C. *Phys. Rev.* **1959**, *114*, 956–960.
27. Lazzarini, A. F.; Lazzarini, E. *J. Chem. Educ.* **1983**, *60*, 519–521.

Additional Reading

- Gordon, J. E. *The New Science of Strong Materials*; Princeton University Press: Princeton, NJ, 1984.
- Guinier, A. *The Structure of Matter: From the Blue Sky to Liquid Crystals*; Edward Arnold: London, 1984.
- Holden, A.; Singer, P. *Crystals and Crystal Growing*; Anchor Books: Garden City, NY, 1960.

Metals

- Cottrell, A. H. *Sci. Am.* **1967**, *217(3)*, 90–100.
- Guinier, A.; Jullien, R. *The Solid State: From Superconductors to Superalloys*; Oxford University Press: Oxford, England, 1989, pp 171–241.
- Ashby, M. F; Jones, D. R. H. *Engineering Materials: An Introduction to Their Properties and Applications*; Pergamon: Oxford, England, 1980.

- Hayden, H. W.; Moffatt, W. G.; Wulff, J. *The Structure and Properties of Materials*; Mechanical Behavior; Wiley: New York, 1964; Vol. 3.
- Moffatt, G. W.; Pearsall, G. W.; Wulff, J. *The Structure and Properties of Materials*; Structure; Wiley: New York, 1964; Vol. 1.

Acknowledgments

The section "Defects in Metals" was based in part on an article in press, *J. Chem. Educ.* (Geselbracht, M. J.; Penn, R. L.; Lisensky, G. C.; Stone, D. S.; Ellis, A. B.).

We thank Art Dodd, Eric Hellstrom, John Perepezko, Don Stone, and Frank Worzala, University of Wisconsin—Madison, Department of Materials Science and Engineering for their assistance.

We thank Candace Davison at the Penn State Reactor, who irradiated the sample (film-canister size) of LiCl.

Exercises

1. An important difference between a vacancy and a dislocation is that if you sketch a regular polygon around the vacancy in Figure 6.5 by traveling along atoms, the polygon will close. The same polygon will not close around a dislocation. Try it, using a regular hexagon to encircle each defect.
2. Calculate the density of defect-free iron from the information in Appendix 5.6. What would the density be if 0.10% of the iron sites were vacant? How much would the mass of 1.000 cm^3 of defect-free iron differ from the mass of the iron with 0.10% vacancies?
3. Why might scanning tunneling microscopy give more information about the structure of defects in a metal than X-ray diffraction?
4. How many Schottky defects are present in a 1-mg crystal if 1 in 10^{15} pairs of cations and anions are missing? (Hints: NaCl has a fcc structure with a unit cell edge length of 5.64 Å and the density of NaCl is 2.16 g/cm^3.)
5. Assume that F-centers can be generated in the following compounds that have the NaCl structure. Using information in Figure 6.17 and Appendix 5.6 for unit cell dimensions, estimate the approximate wavelength at which each would absorb.
   a. NaI
   b. LiI

6. Find and label the types of defects in the representation of a plane of atoms shown below.

A two-dimensional array of atoms showing defects.

7. Choose the correct ending to the following statement. If 1% of the Na^+ ions in a crystal of NaCl are replaced with Mg^{2+} ions, the required overall electrical charge neutrality of the crystal can be maintained if
   a. another 1% of the Na^+ ions are replaced by Al^{3+} ions.
   b. another 1% of the Na^+ ions are replaced by Ca^{2+} ions.
   c. another 1% of the Na^+ ions are replaced by Li^+ ions.
   d. another 1% of the Na^+ ions are removed from the crystal and their positions left vacant.

8. How does the movement of a caterpillar involve a kind of dislocation?

9. Consider the seats in an auditorium or coliseum, in which the number of seats in a row increases radially as you go from the stage or playing field to the back of the room or field, respectively. Make a sketch that shows the problem of lining up seats in columns, one behind the other, in registry. How do aisles serve as dislocations to address this problem?

10. Estimate the edge length of the cubic unit cell (lattice parameter) and the absorption band maximum (wavelength in nm) for an F-center in a $K_{0.5}Rb_{0.5}Cl$ solid solution.

11. How would the absorption properties of a physical mixture of equal moles of powdered KCl and RbCl differ from those of a powdered solid solution of $K_{0.5}Rb_{0.5}Cl$ if F-centers are produced in both samples using a tesla coil?

12. Describe the close-packed slip planes in a sample of an hcp metal like Zn.

Chapter 7

Electronic Structure of Crystalline Solids: Bands and Periodic Properties

Isolated atoms, as well as molecules and ions that consist of small numbers of atoms, usually have characteristic, well-separated, discrete energy levels that define the electronic structure of these species. The simplest such species, the isolated hydrogen atom, is a staple of introductory presentations on electronic structure and spectroscopy. In extended crystalline solids, however, a vast number of atoms interact electronically with one another. Their overlapping atomic orbitals lead to delocalized orbitals that encompass the entire solid, and enormous numbers of delocalized orbitals are packed into a relatively small energy range.

Figure 7.1 illustrates the electronic structure of a bulk extended solid. This energetically nearly continuous set of electronic states is called a band and accounts for many of the common macroscopic properties of bulk solids. In this chapter, we will specifically discuss examples of electrical, thermal, and optical properties that result from the presence of bands in solids. Furthermore, we will show that bands arise irrespective of whether the bonding in the extended solid is principally metallic, covalent, or ionic. Although this chapter will focus on three-dimensional extended solids, bands also can arise in two-dimensional layered solids, like graphite, and in one-dimensional solids built from chains, like the conducting polymer polyaniline.

2725–X/93/0187$13.25/1

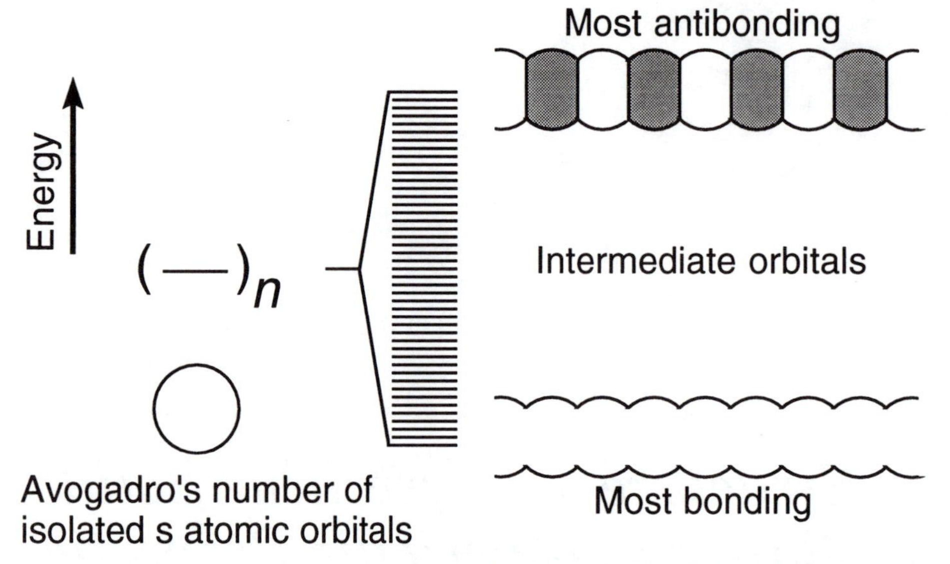

Figure 7.1. In a solid sample, s-type atomic orbitals can combine to give many delocalized orbitals that are very close in energy. The collection of these orbitals forms a nearly continuous set of energy states called a band. The top of the band is the totally antibonding combination formed, in this case, from s atomic orbitals of alternating signs, which are represented by the shaded and unshaded areas. The bottom of the band is the totally bonding combination formed by combining all of the s atomic orbitals with the same sign.

Bands in Metals

Bands result from the overlap of atomic orbitals in a solid, much as molecular orbitals form in molecular systems. In the molecular orbital theory of diatomic molecules, molecular orbitals can form when an atomic orbital from one atom overlaps with an atomic orbital from the second atom to give a lower-energy bonding molecular orbital and a higher-energy antibonding molecular orbital, each of which is delocalized over both atoms. Application of this approach to larger molecules leads to an increase in the number of molecular orbitals, because one molecular orbital is generated for each atomic orbital that is included in the model. A crystal of a metal can be regarded as an enormously large molecule for which the number of molecular orbitals, or energy levels, is on the order of the number of atoms in the solid.

The general picture of how bands arise (shown in Figure 7.1 for an s band, i.e., a band formed exclusively from combinations of s orbitals) can be fleshed out with illustrations from the alkali metals and alkaline earth metals (Figure 7.2). The relatively nondirectional bonding in the alkali metals (bcc, coordination number 8) and alkaline earth metals (hcp or bcc, coordination numbers 12 or 8) make the use of unhybridized orbitals a logical choice to describe the formation of bands in these metals. Thus, bands of interest will in general be those derived from the overlapping valence orbitals of atoms constituting the solid, for example the 3s orbitals of Na. Core orbitals like the 1s, 2s, and 2p orbitals of Na do not overlap as significantly. Much as the more stable, core energy levels that are filled with electrons are ignored in simple molecular orbital treatments of valence electronic configurations, bands derived from core orbitals typically have less significance in characterizing the chemical and physical properties of solids and will also be ignored.

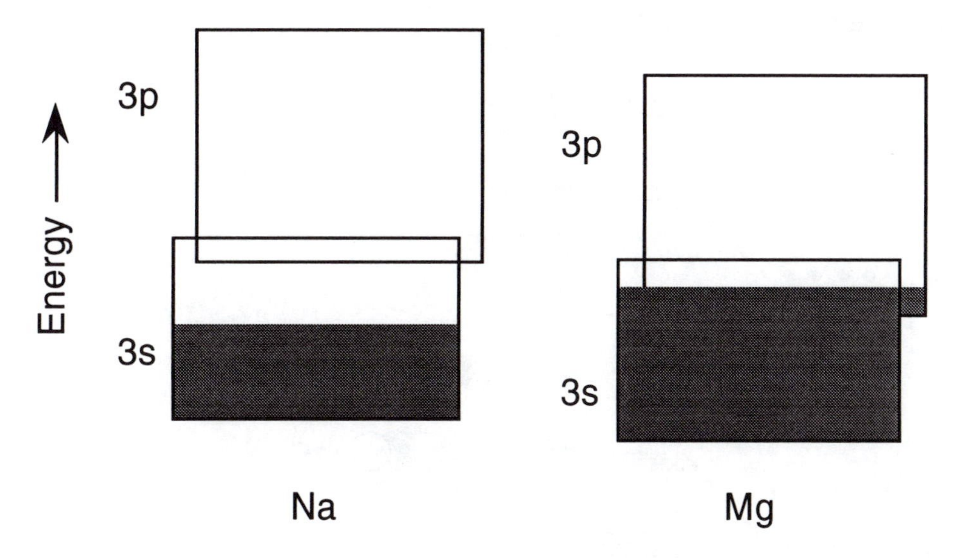

Figure 7.2. Band structure for sodium and magnesium. The 3s and 3p bands are offset for clarity. Shaded regions correspond to energy levels within the band that are occupied by electrons. (The overlap in energy between the 3s and 3p orbitals permits some of the valence electrons of magnesium to occupy energy levels in the 3p band.) Band widths and overlap are not to scale.

Figure 7.3 shows the spreading of the valence 3s orbitals of Na into a band as increasing numbers of Na atoms are combined to make the solid. The right side of the diagram presents a more molecular orbital (MO)-based view, showing the algebraic combinations of these atomic orbitals that correspond to the delocalized orbitals of the band; the most energetically stable (most bonding) combination has the fewest nodes between atoms, and the most unstable (most antibonding) combination has the most nodes between atoms, with the intermediate orbitals varying in numbers of nodes between these two extremes. For a piece of sodium with Avogadro's number, N_A, of atoms, there must be N_A energy levels within the 3s band, reflecting the conservation of 3s orbitals.

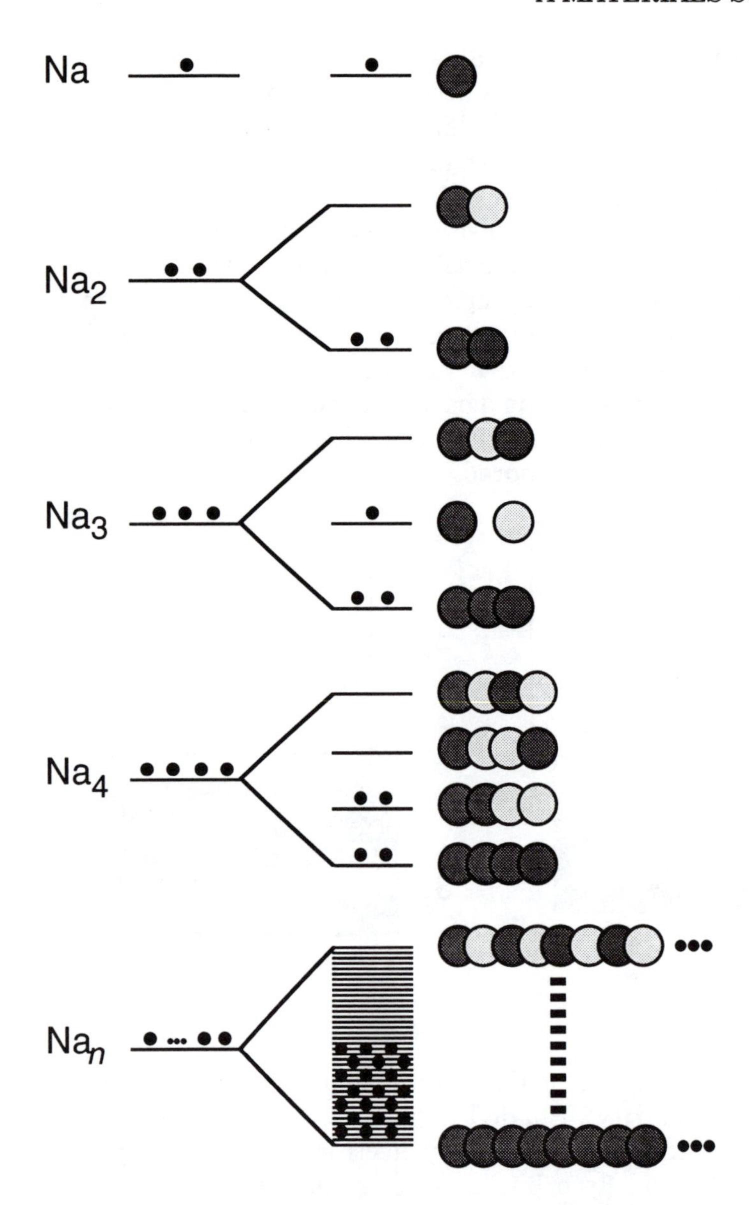

Figure 7.3. Atoms, such as Na(g), and small molecules, such as Na_2(g), Na_3(g), and Na_4(g), have well-separated energy levels. In crystalline solids, large numbers of atoms interact to produce bands of delocalized orbitals that have a relatively narrow distribution of energies. The ideas are illustrated for combinations of the 3s orbitals of sodium atoms, leading to a 3s band. The shading of orbitals (large circles) is described in Figure 7.1; the small dark circles represent electrons.

The energy spread of the band reflects the atomic overlap. As the overlap between orbitals increases, the most bonding orbital of the band is stabilized to a greater extent, the most antibonding orbital of the band is destabilized to a greater extent, and a relatively wide band (one with a large spread in energy) results. Similar considerations are applied to the MO picture of bond formation found in general chemistry texts for diatomic molecules. In contrast, if there is little orbital overlap, the energy levels of the orbitals that are combined to form the band are little affected, and a narrow band results. The net overlap of the 1s, 2s, and 2p core orbitals in Na is very small, and bands derived from these orbitals are considerably narrower than the band derived from the 3s orbitals.

The number of electrons must be preserved in the band diagram. The Pauli principle limits the population of each delocalized orbital comprising the band to two spin-paired electrons, just as would be the case for orbitals in isolated atoms, ions, or molecules. Thus, the 1s, 2s, and 2p bands are completely filled with electrons, because each such orbital of each atom contributing to these bands had its full complement of two electrons. In contrast, the 3s band is exactly half full, because each contributing Na atom has a $3s^1$ or half-full electronic configuration.

For Mg, an isolated 3s band would be filled with electrons, reflecting the individual $3s^2$ configuration of individual Mg atoms. But Mg illustrates the important property that bands can overlap in energy: Although the 3p band of magnesium would be predicted to have no electrons in it, because individual Mg atoms do not have 3p valence electrons, Figure 7.2 shows that the bottom electronic states of the 3p band are lower in energy than the upper states of the 3s band. The electronic states of the 3p band are thus energetically accessible to electrons of the 3s band, because electrons can occupy lower energy states in the 3p band. This feature is crucial to describing electrical conductivity in Mg, as will be described.

Electrical Conductivity of Metals

The conventional description of metallic bonding, as metal ions surrounded by a sea of mobile electrons, is consistent with the band picture sketched. Given the large coordination numbers of metals, the number of valence electrons is insufficient for the kind of electron sharing needed for covalent bonding. The similar electronegativities of the atoms involved precludes the electron transfer characteristic of ionic bonding. The band picture for Na suggests that the solid could be regarded as a bcc crystal of Na^+ ions surrounded by freely moving valence electrons. One plane of atoms in such a crystal is sketched in Figure 7.4.

An important property for comparing the electrical conductivities of various solids and liquids is their carrier concentration, the number of charged carriers (electrons or ions) per unit volume. Carrier concentrations are typically expressed in density units of carriers per cubic centimeter. Assuming all of the 3s band electrons are mobile, Na has the enormous carrier concentration of roughly 10^{22} electrons/cm^3.[1] For perspective, pure water, for which hydrogen ions and hydroxide ions would

---

[1]In a metal, only those electrons that are near the top of the filled energy levels of an unfilled band (*see* later) are actually charge carriers. When an electric field is applied, all electrons gain energy, but most simply shift from the level that they occupied to a level that was previously occupied. As such, these electrons provide no net contribution to the conductivity. The remaining carriers near the top of the

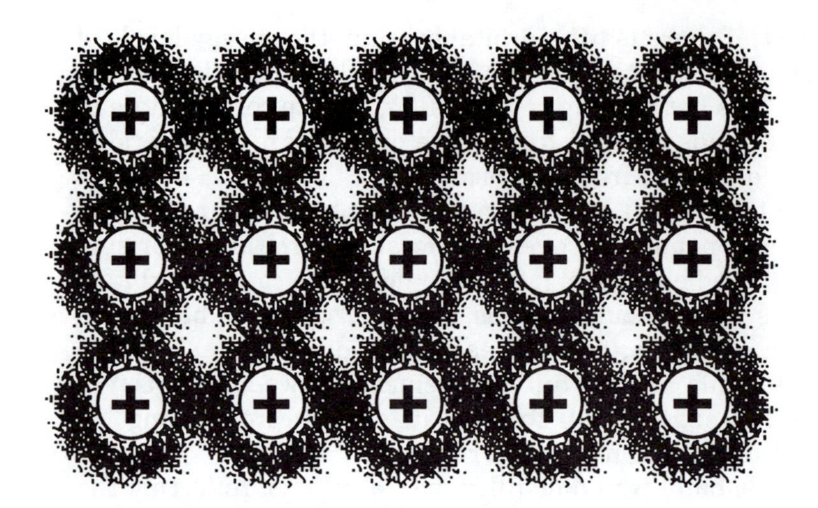

Figure 7.4. Bonding in metals can be viewed as a sea of delocalized and mobile valence electrons acting as a kind of glue to hold the metal cations together.

be the charge carriers, would have a carrier concentration of only 10^{14} ions/cm^3, while a 1 M NaOH aqueous electrolyte would have a carrier concentration of about 10^{21} ions/cm^3.[2]

For a solid to be a good electronic conductor, a property we associate with metals, a high carrier concentration of electrons is a necessary but not sufficient condition. The two requirements for good conductivity by electrons are (1) the presence of a band, whose delocalized orbitals provide a kind of electronic highway through the solid; and (2) an electronic population of the band that corresponds to its being only partially filled, typically from about 10 to 90% of its capacity.

The second requirement can be appreciated from the limiting cases of a completely empty and completely filled band. A completely empty band cannot contribute to electrical conductivity, because it has no electrons to produce the net flow of charged particles that represents an electrical current. A completely filled band is unsuitable for electrical conduction because it has no vacant orbitals for the electrons to jump into to achieve a net physical motion of charge carriers in a particular direction. But if the band is only partially filled, electrons are easily promoted into the vacant, energetically accessible orbitals of the band by absorbing thermal energy or energy from an electric field. These unpaired electrons are then free to move under the influence of an electric field such as that produced by a battery, or to be swept along by a magnetic field as in a generator.

---

filled energy levels move to previously unoccupied levels and increase the conductivity *(1)*. Consequently, in sodium metal, approximately 1% of the electrons in the 3s band contribute to the conductivity.

[2]Although the number of charge carriers in sodium metal and sodium hydroxide solution is comparable, the ions, being more massive than electrons, are less mobile. Thus, the conductivity, which depends on both carrier concentration and mobility (a quantitative measure of how the charged particles respond to an electric field), is much lower in 1 M NaOH solution.

Demonstration 7.1. Electron Flow in a Band

Materials

Three clear bottles with caps, one filled with sand, one half-filled with sand, and one empty

Procedure

- Tip each bottle to about a 45° angle from vertical and watch for the flow of sand. The tilting of the bottle represents the applied potential from a battery. If the bottle is empty or completely filled, no sand moves. Only if the bottle is partly filled can sand flow toward the lower side; moreover, only the sand near the top moves, reflecting the fact that only electrons near the top of the filled energy levels contribute to conductivity.[1]

With its half-filled 3s band, Na is an excellent conductor of electricity. A half-filled s band is a general characteristic of the alkali metals because of the ns^1 electronic configurations they have in common. The alkaline earth metals owe their conductivity to the overlap of their p bands with their s bands, as noted in Figure 7.2 for Mg. Overlapping bands provide the same kind of mechanism for conductivity obtained from a single partially filled band: An enormous number of energetically proximate energy levels are accessible through absorption of energy from an electric field. As one moves through the rest of the metallic elements in the periodic table, additional atomic orbitals are found in the valence shells, and overlap of some combination of bands derived from s, p, d, and f orbitals will usually occur to provide the partial filling of bands needed for electrical conductivity.

Conductivity may be increased by adding electrons to an empty band or removing electrons from a filled band. This increase can be demonstrated by using the layered solid graphite (Chapter 5). Graphite is a poor metal (a semimetal) with only a little overlap between the highest energy filled band and neighboring higher energy empty band (contrast Mg, with substantial band overlap). The conductivity of a graphite sample can be substantially enhanced by oxidation with bromine. Oxidation removes electrons from the filled band. (The bromide ion that is generated by the reaction intercalates or moves into the spaces between the layers of the solid, as shown in Figure 7.5.)

Graphite intercalation by electron donors, such as alkali metals, or by electron acceptors, such as bromine or fluorine, is also possible. With donors, electrons are added to the higher energy band, partially filling it and also increasing the conductivity relative to pure graphite. Thus, reaction with either oxidants or reductants increases the conductivity of graphite, and enables it to reach metallic values.

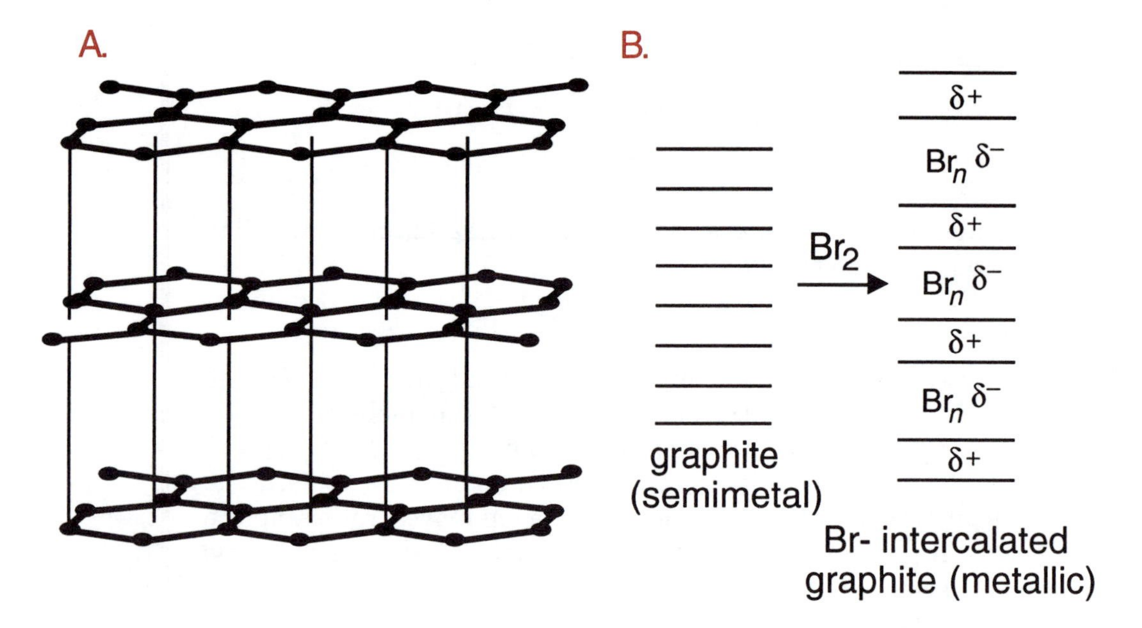

Figure 7.5. A: The layered structure of graphite. B: The oxidative intercalation of bromine into graphite removes electrons from the highest energy filled band in the graphite sheets (the sheets are represented by the parallel lines), while simultaneously forcing them further apart because of the intercalation of bromide ions between the layers.

Demonstration 7.2. Oxidative Intercalation of Graphite

Materials

- Two glassy carbon electrodes or graphite rods (1/4-inch diameter)*
- Teflon tubing (1/4 inch I.D.) about 3 inches long (or a Teflon cylinder with a 1/4-inch hole drilled lengthwise) with a second hole or notch cut into the center—*see* Figure 7.6
- Medium-sized beaker or crystallizing dish (or a glass tube with a hole in center large enough to contain the Teflon tube)
- 1–2 g of crystalline graphite powder, 200 mesh (Ultra Carbon F works)*
- Ring stands and clamps
- 10% (v/v) Br_2 in MeOH in dropper bottle
- Elemental bromine (optional)
- 6-V battery
- 1–2-Ω light bulb*
- Wires and alligator clips to connect the parts in series
- Ohmmeter and leads

Procedure

The intercalation reaction is carried out in a section of Teflon tubing containing the graphite powder. To construct the apparatus,

- Sandwich about 5 mm of crystalline graphite powder between two glassy carbon rods in the Teflon tubing, as shown in Figure 7.6. Apply enough pressure to the graphite so that the resistance between the two rods is 10–30 Ω.
- Place this apparatus in a glass tube with a hole in the top. (The purpose of this tube is to contain excess bromine solution. Alternatively, the apparatus may be arranged so that it is held over a beaker or a crystallizing dish.) Clamp the electrode assembly in a horizontal position. Connect it in series via alligator clips to a 6-V battery and a light bulb with a resistance of 1–2 Ω. The light bulb should glow faintly or be dark, showing that the graphite powder is resistive.
- If you are using the outer glass tube, slide the glass tube off the Teflon tubing to expose the center of the tube. With a razor blade, cut a notch in the Teflon tubing to expose a small area of the graphite powder. (If a thicker Teflon cylinder is used, the notch should be precut.) Place the glass tube over the Teflon tube so that the hole in the glass is aligned with the notch in the Teflon.
- Introduce several drops of 10% v/v Br_2 in methanol solution directly into the notch, and after a few seconds the light bulb will begin to glow brightly.

CAUTION: Bromine is volatile, corrosive, and highly toxic. If a fume hood is available, elemental bromine may be used to produce a more dramatic conductivity change. In either case, the demonstrator should wear gloves. The change in conductivity is sufficient in either case to light the bulb. A solution of sodium thiosulfate can be used to decompose any unreacted bromine and to wash the apparatus upon completion of the demonstration.

The intercalation process swells the individual graphite particles, and increases the pressure inside the Teflon tubing. This improves electrical contact between particles, and thus further lowers the resistance.

Although not explicitly demonstrated here, this experiment can lead easily into a discussion of the anisotropy of layered solids. Ask students how they think the anisotropy affects the electrical and mechanical properties of graphite. What differences in conductivity might be expected using a single-crystal (*see* Chapter 6) of graphite?

*See Supplier Information

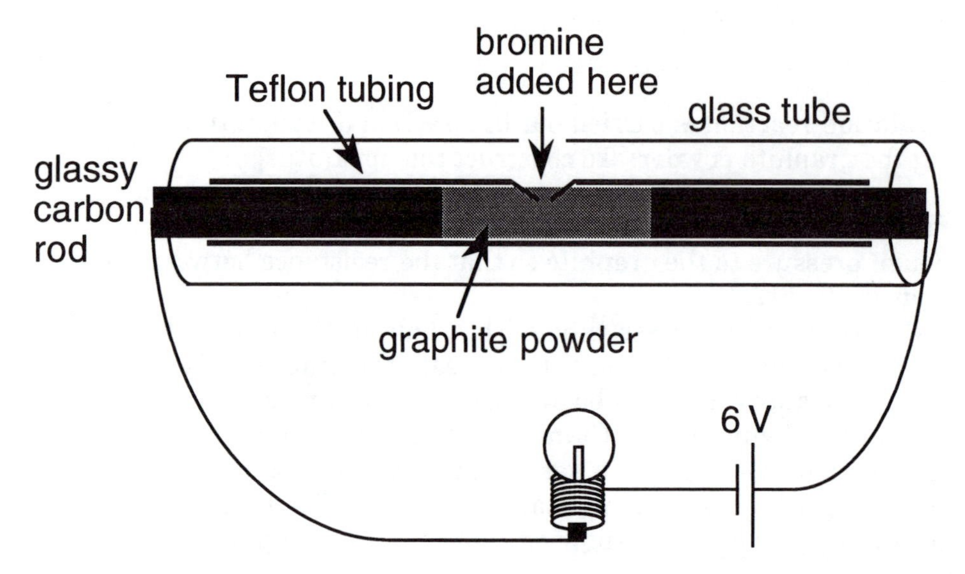

Figure 7.6. The conductivity apparatus for demonstrating the intercalation of bromine into graphite.

> **Laboratory.** An experiment that, through chemical reduction, converts WO_3 to H_xWO_3 with an accompanying enhancement in electrical conductivity is described in Experiment 8.

Solids are generally classified as metals, semiconductors, and insulators, depending on their ability to conduct electricity. Among materials, metals have a relatively high electrical conductivity, σ (sigma). Alternatively, we can speak of the electrical resistivity, ρ (rho), which is the reciprocal of conductivity ($\sigma = 1/\rho$).[3] Conductivity is an unusual property in that it spans roughly 30 orders of magnitude—one of the largest ranges known for a physical property! An electrical conductivity scale encompassing a variety of common substances is shown in Figure 7.7.

How are new metals—be they combinations of metallic elements (alloys), ceramics, or conducting polymers—recognized as such? By the variation of how strongly they resist the motion of electrons as a function of temperature. If a substance exhibits metallic conductivity, its resistivity increases (and its conductivity decreases) as the temperature increases. In contrast, the resistivity of semiconductors and insulators (discussed later) decreases with increasing temperature. By way of illustration, the resistivity of tungsten doubles (the conductivity decreases by a factor of 2) between 750 and 1500 K. For silicon, the prototypical semiconductor, the

---

[3]The experimentally measured electrical resistance of a wire, R, is related to the resistivity, ρ, by the equation $R = \rho L/A$, where L is the length of the wire and A is the cross-sectional area. Consequently, resistivity values have the units of ohm-(distance) and are usually reported in ohm-centimeters. Conductivity has the units of reciprocal ohm-centimeters, which is equivalent to mho per centimeter and siemens per centimeter.

resistivity decreases by a factor of 100 (the conductivity increases 100-fold) over the same temperature range *(3)*. The trends are illustrated in Figure 7.8.

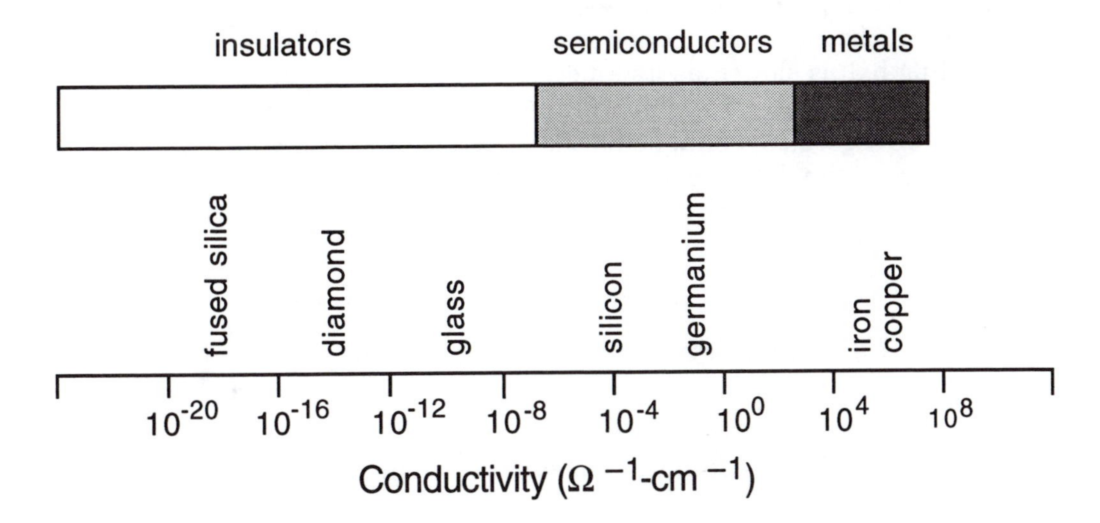

Figure 7.7. The entire range of conductivities of solids spans roughly 30 orders of magnitude. (Adapted from reference 2.)

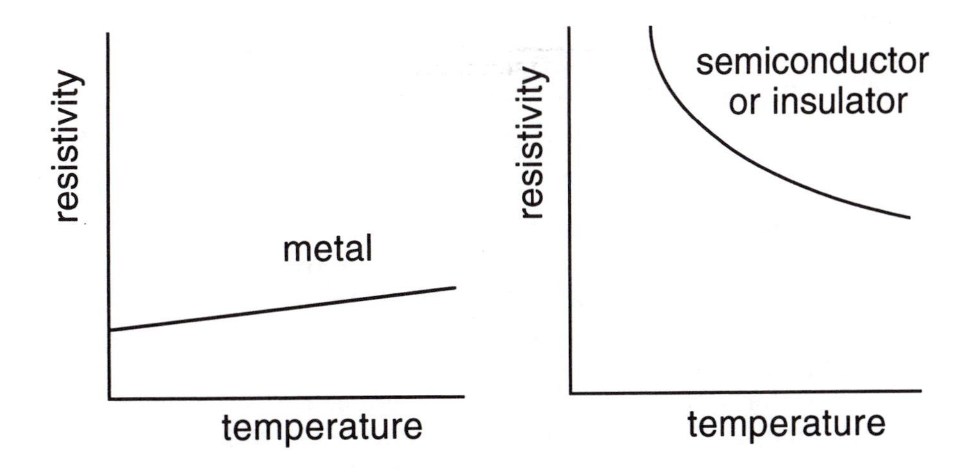

Figure 7.8. The resistivities of metals and semiconductors or insulators show different behavior when the temperature is changed. As the temperature is increased, the resistivity of a metal increases slightly, but the resistivity of a semiconductor or insulator decreases dramatically. The relative size of the effect is not shown to scale in this figure.

The electrical resistance of a solid is analogous to mechanical friction: As electrons move through a material under the influence of an electric field (away from the negative terminal and toward the positive terminal of a battery), they occasionally are scattered by the vibrating atoms of the crystal, by defects, or by impurity atoms in the solid. In this manner, part of the kinetic energy of the electrons' motion is converted into heat, just as the motion of a rolling ball is slowed as its kinetic energy is converted into heat through friction. Increasing the temperature causes the atoms in the crystal to vibrate more vigorously, enhancing the effectiveness with which electrons are scattered and leading to modest increases in electrical resistance. (The decrease in resistivity in semiconductors and insulators is discussed in a subsequent section.)

Demonstration 7.3. Electrical Resistance

Materials

Nichrome wire [22 gauge (0.025-inch diameter), 14 inches long]
Wax
9-V battery
Battery snap with two leads and alligator clips
Overhead projector with a plastic protection sheet (optional)

Procedure

- Prior to the demonstration, heat a container of wax until the wax is just above its melting point.
- Coil the middle section of the nichrome wire around a thin dowel, glass rod, or pencil such that you have about 10 turns that are about 3/8 inch in diameter. (Make sure that the coils do not cross.) This will leave a tail about 1 inch long on each end. Cool the wire in a freezer.
- Dip the coils of the wire rapidly into the heated wax and withdraw the wire immediately. (If the wire is precooled in a freezer, a satisfactory coating will result after one dip. To get even more wax on the wire, cool it in the freezer before dipping a second time.)
- Allow the wax-covered wire to cool.
- When it is time to perform the demonstration, connect the wax-coated piece of nichrome wire to the alligator clips attached to the wires from a battery snap. Press the battery snap onto the terminals of the 9-V battery. Be ready to disconnect the nichrome wire quickly. **CAUTION: Do not connect the wire for more than a few seconds, as the battery may become hot!** When electricity is passed through the wire, the wax melts and drips onto the plastic protection sheet, an effect that can be displayed on an overhead projector.

  Both the wire length and gauge are important variables that must be controlled to achieve a high enough temperature to melt the wax.

Variations

Electrical resistance can be demonstrated in a number of other ways. Toasters, electric stove burners, and light bulbs can be used to demonstrate resistive heating. Memory metal can also be employed: When a sample of memory metal is resistively heated, it will return to a shape that it has been trained to remember. (*See* Chapter 9 for a description of memory metal and the details of the demonstration.)

All of these demonstrations show that wire possesses some resistance to current flow, which results in the conversion of electrical energy to thermal energy or heat.

Demonstration 7.4. Temperature Dependence of Resistance in a Metal

Materials

Two choke coils or inductors (Mouser Electronics; *see* Supplier Information)
Hacksaw
Liquid nitrogen in a small Dewar flask
Ohmmeter

Procedure

- Use a hacksaw to remove a portion of the outer casing of one of the choke coils to show the inside. The coil contains 150 m of tightly wound Cu wire in a plastic container.
- Connect the two ends of the other choke coil to the leads of an ohmmeter and measure the resistance. A value of about 150 Ω will be found at room temperature.
- Plunge the choke coil into the liquid nitrogen. **CAUTION: Liquid nitrogen is extremely cold. Do not allow it to come into contact with skin or clothing, as severe frostbite may result. Wear gloves when transferring and using liquid nitrogen.** The resistance will be found to drop to about 15 Ω, as the metal cools. (Do not place a coil that has been cut open into liquid nitrogen, as it will crack.)

Something like a choke coil is needed for this experiment because metals are such good conductors: To obtain significant resistance values, the metal needs to be as long and thin as possible, because, for a given material, electrical resistance increases with length and decreases as the wire thickness increases.[3] The analogy can also be made that the flow of electricity is like the flow of water through pipes: Less resistance to flow obtains for pipes with larger openings and shorter lengths.

Thermal Conductivity of Metals

Partially filled bands lead not only to good electrical conductivity, but also to good thermal conductivity. If one end of a metal rod is heated, the other end will soon become warm through a process called thermal conduction, wherein heat is transferred without physically displacing atoms: At the heated end, atoms vibrate more rapidly, and the electrons in the partially filled conduction band easily absorb thermal energy (moving into the thermally accessible, unfilled energy levels of the band) and thereby increase their kinetic energy. This increased kinetic energy leads to more frequent and more energetic collisions with atoms throughout the solid, increasing their vibrational energy and effectively transferring the added thermal energy along the length of the rod. Demonstrations 7.5 and 7.6 illustrate the high thermal conductivity of metals, and Table 7.1 presents values for various materials.

Demonstration 7.5. Thermal Conductivity in the Classroom

Materials

A student desk or chair that has wooden and metal parts (or any other sources of wood and metal that are in the room)

Procedure

- Ask the students to touch the wood and the metal. Which feels cooler?

  This is a simple demonstration of the relatively high thermal conductivities of metals. A wooden lecture room chair having a metal frame can be a particularly effective example. Both the wood and metal are at thermal equilibrium and thus at room temperature. However, the metal will feel cooler to the touch than the wood because the free electrons in the partially filled band can rapidly remove thermal energy from our bodies and thereby make the metal feel cool.

Demonstration 7.6. A More Quantitative Measure of Thermal Conductivity

Materials

6-inch long, 1/4-inch-diameter rods made of various materials, including aluminum, copper, brass, wood, iron, and Plexiglas. (These rods are available from hardware or hobby stores.)
Styrofoam cup
Hot water (~150 mL)
Elbow macaroni noodles
Butter or margarine
A clock or watch capable of measuring time in seconds

Procedure

- Dip one end of each rod in butter or margarine to provide a sticky surface that will hold a piece of macaroni in place. Place a piece of macaroni on one end of each rod to be studied.
- Then place all of the rods in a Styrofoam cup with the end holding the macaroni and margarine up, as shown in Figure 7.9.
- Add hot water to the cup and begin timing. As the heat travels by conduction along each rod, the margarine melts and the macaroni falls off; relative rates of thermal conductivity are thus defined. The rate in this case is 1/time. A plot of typical data, shown in Figure 7.9, reflects the far greater thermal conductivity of metals.

Laboratory. Students can measure cooling curves for hot water in various containers. A container such as a Styrofoam cup is filled with hot water, and the temperature is monitored as a function of time. The experiment can be repeated with different container materials, amounts of water (varying surface-to-volume ratios), or initial temperatures of water. The rate of cooling should depend on the thermal conductivity of the container. If students are given thermometers, this can be done as a take-home lab. A variety of variables need to be constrained. This is a simple enough experiment that it is an opportunity to let students set up their own experimental procedure, and no further directions are included in this book.

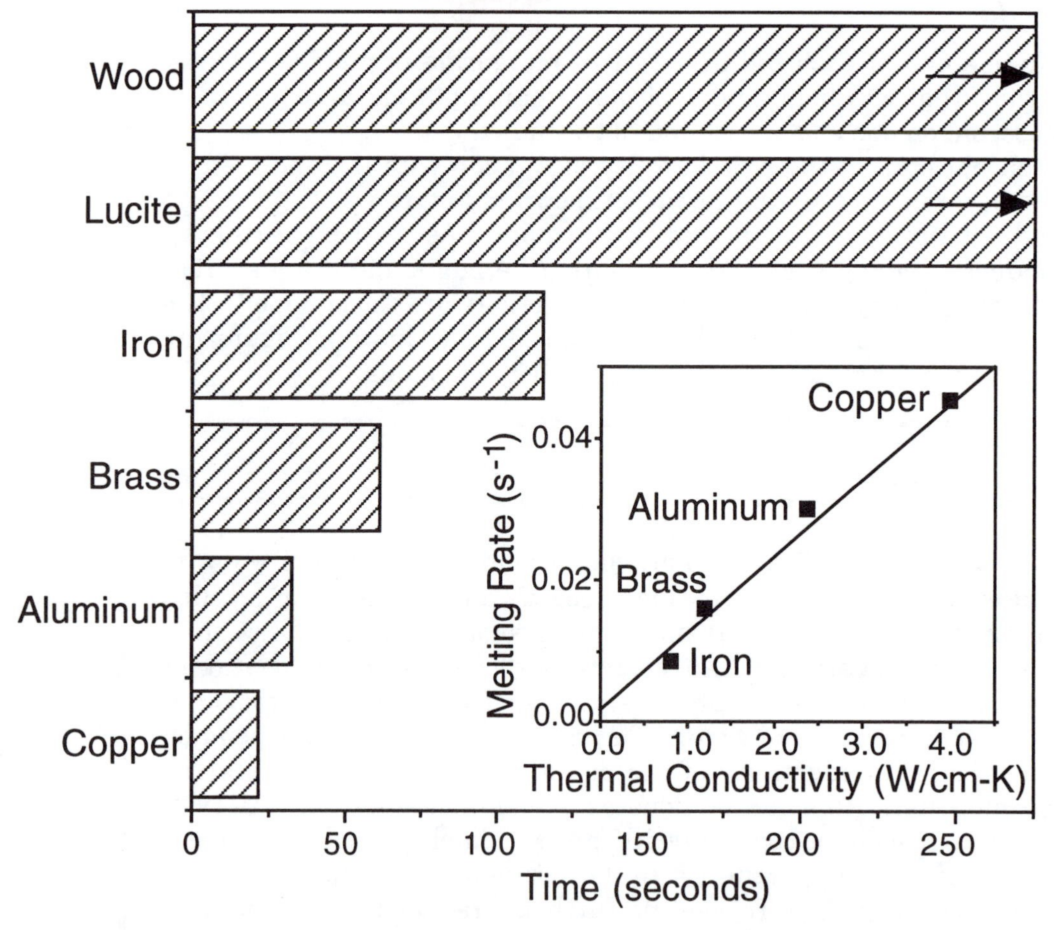

Figure 7.9. Test of thermal conductivity described in Demonstration 7.6. Typical data are shown, as is a plot of the thermal conductivity of the rod types (from Table 7.1) versus the relative melting rate (the reciprocal of the amount of time it took for the noodle to fall off the rod).

Table 7.1. Thermal Conductivities of Various Substances

Material	Thermal Conductivity (watt/cm-K)
Air	0.00026
Glass wool	0.00042
Corkboard	0.00043
Carbon tetrachloride	0.0010
White pine	0.0011
Oak	0.0015
He	0.001520
Water	0.0061
Glass	0.0072–0.0088
Hg	0.083
Concrete	0.0086–0.013
SiC	0.090 (100 °F)
NaCl	0.092 (0 °C)
ZnS (zinc blende)	0.264 (0 °C)
Al_2O_3	0.303 (100 °C)
Pb	0.353
Cs	0.359
MgO	0.360 (100 °F)
Rb	0.582
Fe	0.804
Li	0.848
K	1.025
C (graphite)[a]	1.1–2.2
Zn	1.16
Brass	1.2
Na	1.42
Mg	1.56
Be	2.01
BeO	2.20 (100 °F)
Al	2.37
Au	3.18
Cu	4.01
Ag	4.29
C (diamond)[b]	9.9–23.2

NOTE: All values are at room temperature unless otherwise noted.

[a] Value is dependent on the impurities in graphite and on the orientation of graphite, being larger in the direction parallel to the layers of carbon atoms.

[b] Value is highly dependent on impurities and defects.

SOURCE: Adapted from reference 4.

Optical Properties of Metals

Certainly one of the most characteristic features of clean metals is their shininess, or high reflectivity for visible light and other wavelengths of electromagnetic radiation. We often overlook the fact that they are also opaque. This effect is also interpretable in terms of the band structures of these solids. Figure 7.10 illustrates the kind of electronic transitions that can be expected for a metal with its partially filled band. Absorption can take place over enormous regions of the electromagnetic spectrum because of the nearly continuous nature of the electronic states within the band. The vertical arrows in Figure 7.10 represent electronic transitions ranging from very low energy (short arrows), well below the visible spectrum (infrared radiation, microwaves, radio waves, etc.), to well above the visible spectrum (long arrows), into the UV region and beyond.

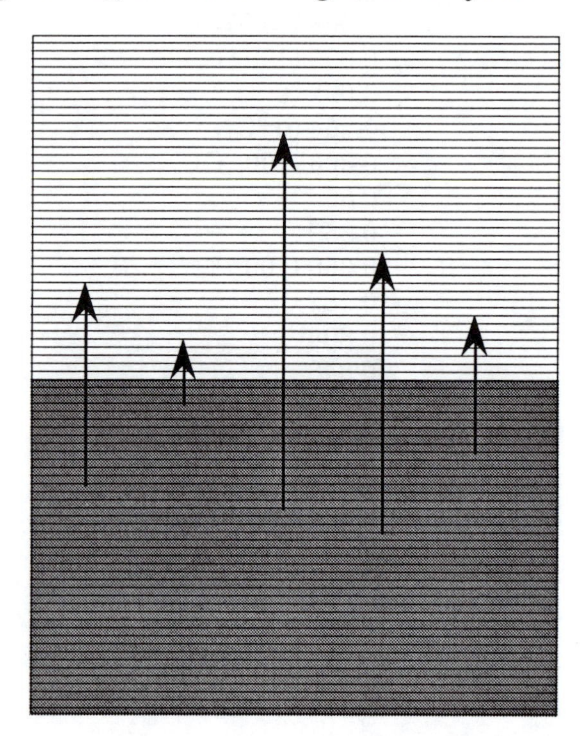

Figure 7.10. Possible electronic transitions in a half-filled band of a metal.

It may seem somewhat paradoxical, but metals are both strong absorbers and strong reflectors of electromagnetic radiation. Since metals are electrical conductors, absorption of light (an electromagnetic wave) causes the many free electrons in the solid to oscillate at the frequency of the radiation and to re-radiate at the same frequency. The re-radiated light exits from the metal (as reflected light) and also cancels the incident light propagating into the metal. Figure 7.11 shows that the reflectivity of visible light from various metals typically exceeds 90%.

The colors we associate with certain metals like Cu and Au arise from nonuniform reflectivity across the visible spectrum. For Cu, in particular, the band structure leads to higher absorption and reflectivity in the red, orange, and yellow portions of the visible spectrum (Figure 7.11), and thus to the reddish color associated with the metal. The intensity of absorption and reflectivity in general is a complex function of selection rules (certain electronic transitions are favored, as they are in the spectroscopy of atomic and molecular systems) and of the density of electrons and electronic states that can be involved in the transitions.

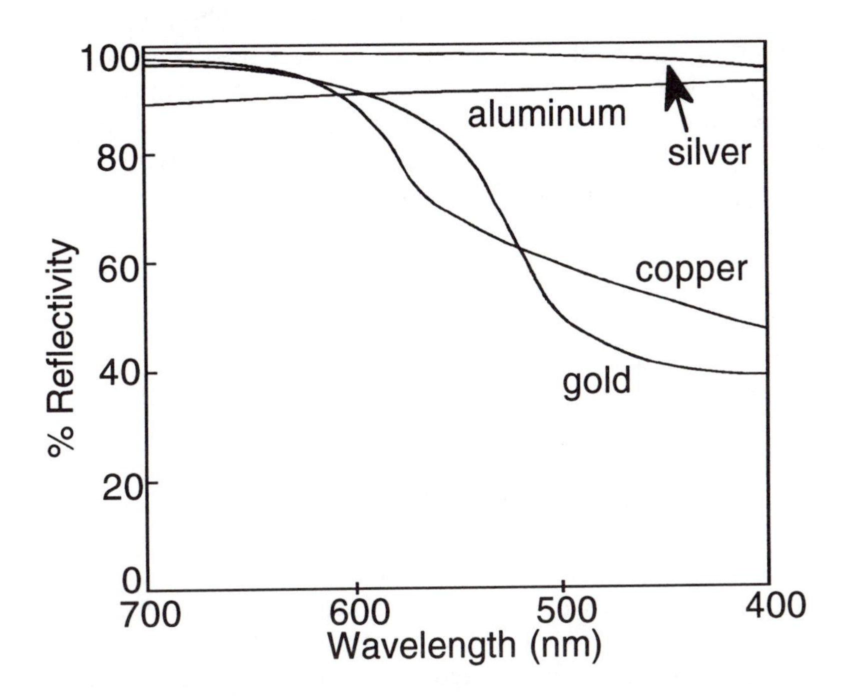

Figure 7.11. Reflectivity as a function of wavelength in the visible region of the spectrum for several metals. The color of gold and copper arises from nonuniform reflectivity across the visible region of the spectrum. (Adapted with permission from reference 2. Copyright 1983 John Wiley and Sons.)

Demonstration 7.7. Metallic Reflectivity

Mirror-quality thin films of metals like Ag and Cu can be plated onto glass surfaces by the chemical reduction of Ag^{+} and Cu^{2+} ions. A silver mirror can be prepared for viewing in a large lecture hall by reduction of silver nitrate. These demonstrations have been described elsewhere *(5, 6)*. If a flat region of glass is available for the deposition, a low power laser beam can be reflected off the surface before and after deposition to highlight the metal's reflectivity. **CAUTION: Take care to ensure that the viewers will not inadvertently look directly at the beam where it emerges from the laser. Clamp or tape the laser in a fixed position so that the beam will not accidentally shift. Aiming a laser at a person or looking directly at a source of laser light can be harmful to the eye!**

Covalent Insulators and Semiconductors

Prototypical electrical insulators and semiconductors, like diamond and silicon, are found in the nonmetals portion of the periodic table. Many of these solids have the diamond or zinc blende structures discussed in Chapter 5 that feature tetrahedral bonding geometries. This more directional, covalent bonding can be described with bonds derived from the same sp^3 hybrid orbitals commonly introduced in general chemistry courses to discuss methane and related tetrahedral molecules. The families of extended solids that exhibit tetrahedral bonding provide a wealth of illustrations of periodic properties.

Localized and Delocalized Bonding Pictures

Before introducing a delocalized bonding model to describe these solids (their band structure), trends in many of the properties to be described can be discussed by using exclusively a simple localized bonding picture. In particular, periodic trends in atomic orbitals and atomic electronegativity can be used to predict trends in bond length, bond strength, and energies of electronic transitions associated with these extended solids. These correlations will be noted throughout the remainder of this chapter, in Chapter 8 and in Experiment 7 as localized descriptions. Thus many interesting properties of families of semiconductors and insulators can be described in introductory courses without recourse to bands, if desired. The delocalized model discussed in the next section will make evident some of the limitations of this localized approach.

Tetrahedral Solids

Given the technological importance of tetrahedral solids like diamond and silicon, it is not surprising that their electronic structures have been intensively investigated. A qualitative approach that will be presented here is based in part on a treatment in Cox's book *(7)*.

The methane molecule is a good starting point for describing band formation in extended tetrahedral solids. On the left of Figure 7.12 is shown a carbon atom that has been hybridized to yield four equivalent sp^3 hybrid orbitals, directed to the corners of a tetrahedron. The four hybrid orbitals conserve the original total number of valence 2s and 2p orbitals, and each holds one electron, representing the original total number of valence electrons. At the right of Figure 7.12 are the valence 1s orbitals of the four hydrogen atoms. These are combined with the four carbon sp^3 hybrid orbitals to give four bonding orbitals that are stabilized in energy relative to the atomic orbitals of carbon and hydrogen; and four antibonding orbitals that are destabilized. All eight valence electrons can be accommodated by the four bonding orbitals, representing the four sigma bonds between carbon and each of the four hydrogen atoms. (A similar picture would result for carbon tetrachloride, using, for example, one 3p orbital from each of the chlorine atoms to combine with the carbon sp^3 hybrid orbitals.)

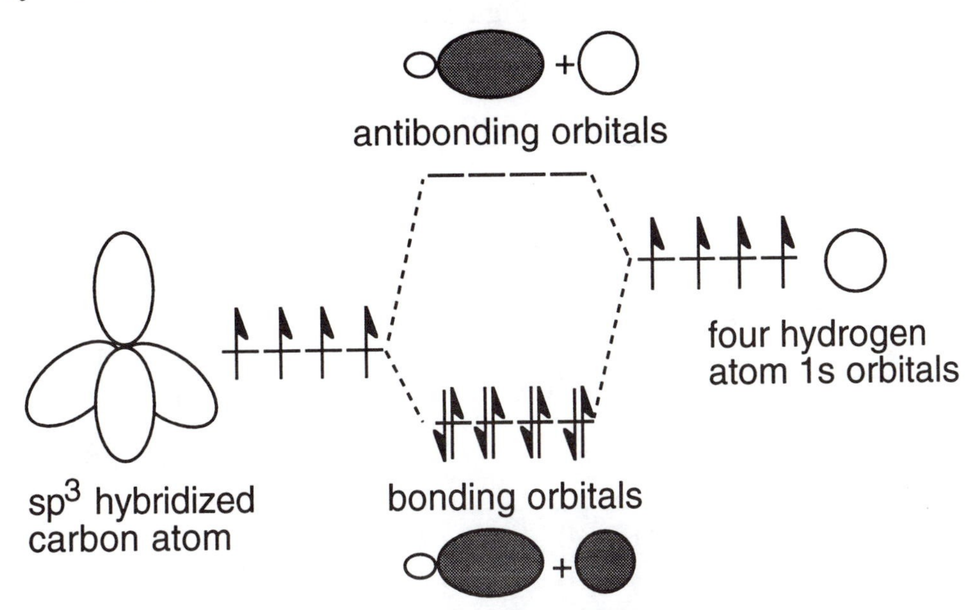

Figure 7.12. A simplified molecular orbital diagram for CH_4. The left side of the diagram shows the four sp^3 hybrid orbitals on the carbon atom (the smaller back lobes are omitted for clarity) and the 1s orbitals of the four hydrogen atoms are shown on the right. Bonding and antibonding combinations of the two sets of orbitals produce two sets of discrete energy levels. The lower set is completely filled by the eight valence electrons.

The situation in the diamond allotrope of carbon is analogous to that for methane except that the orbitals on the right side of the diagram become sp^3 hybrid orbitals of the four adjacent carbon atoms that are bonding to the central carbon atom, (Figure 7.13A.) Once again the same kind of splitting between the four bonding orbitals and four antibonding orbitals occurs. However, other, weaker orbital interactions occur, too. For example, a hybrid orbital on the central carbon atom has some overlap with the hybrid orbitals on the adjacent carbon atoms that are not pointed directly at it, as shown in Figure 7.13B. These interactions and still weaker ones with atoms yet farther away from the central atom cause the bonding orbitals to broaden into one band, the valence band, and the antibonding orbitals to broaden into a second, higher energy band, the conduction band.

Again, the total number of orbitals in the bands preserves the original number of atomic orbitals. For a large number, N, of atoms in the solid, with each atom using four valence orbitals for bonding (the 2s and three 2p orbitals), $2N$ orbitals will make up the lower energy valence band, and $2N$

orbitals will make up the higher energy conduction band. Each atom has four valence electrons. These 4*N* valence electrons of the solid can be entirely accommodated by the 2*N* orbitals of the valence band in accord with the Pauli principle's limitation of two spin-paired electrons per orbital. The result, at a temperature of absolute zero, is a filled valence band and an empty conduction band, shown in Figure 7.13.

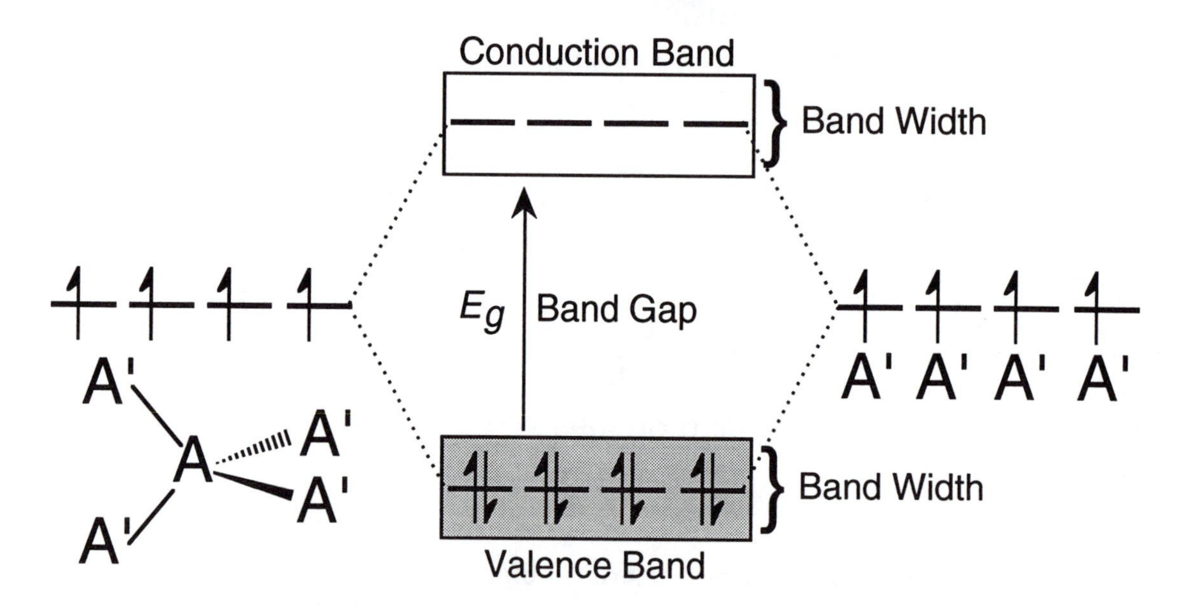

Figure 7.13A. Schematic bonding picture of the Group 14 tetrahedral solid diamond. The four equivalent hybrid orbitals on the central carbon atom are shown on the left, and one overlapping hybrid orbital from each of the adjacent carbon atoms, A', is shown on the right. These orbitals are combined to produce a set of four bonding orbitals and four antibonding orbitals (horizontal lines in the center of the picture). Secondary interactions of these orbitals (*see* part B of this figure) lead to band formation: The bonding orbitals broaden into the so-called valence band, which is completely filled with electrons (shaded) at 0 K, through the interactions of all the atoms in the solid. The antibonding orbitals broaden into the so-called conduction band, which is empty of electrons (unshaded) at 0 K.

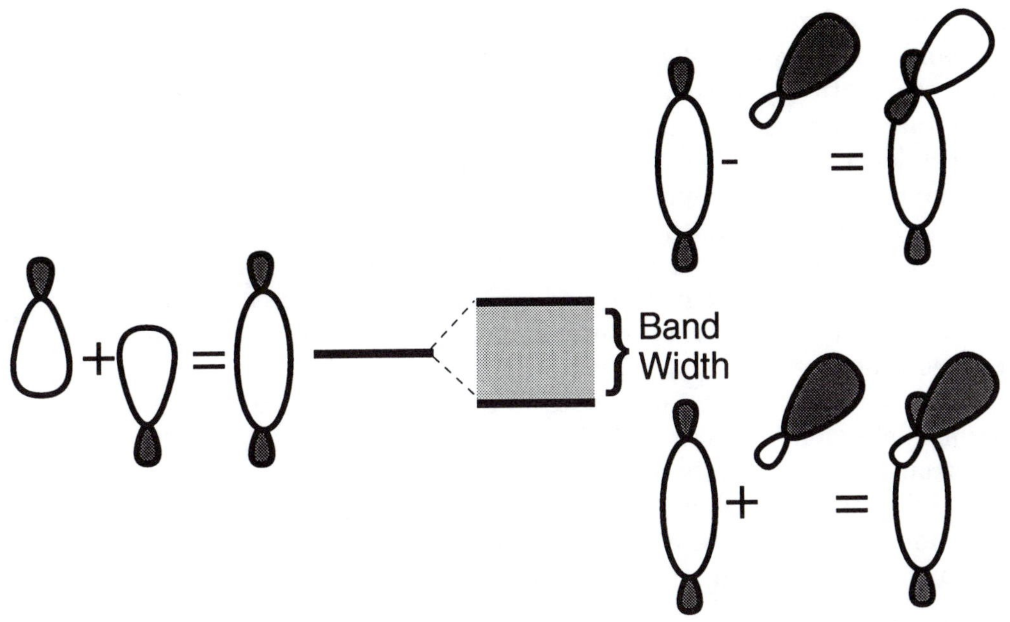

Figure 7.13B. A representation of the kind of secondary orbital interactions that cause broadening of the bonding and antibonding orbitals into bands. The bonding orbital on the left side of the diagram (shown as formed from hybrid orbitals on adjacent carbon atoms A and A' in Figure 7.13A) can have positive or negative overlap with the other hybrid orbitals of these atoms, leading to slightly stabilized (bottom combination on the right-hand side) and destabilized (top combination on the right-hand side) energy levels, respectively; shading is used to represent a change in sign. The collective effect of all of these interactions is to produce the set of tightly spaced energy levels that defines the valence band and its band width. Analogous interactions involving the antibonding orbitals cause broadening into the higher energy conduction band.

Electrical and Optical Properties of Insulators and Semiconductors Having the Diamond Structure

The energy separation between the top of the valence band and the bottom of the conduction band is a fundamental property of the solid, called its band gap, E_g, and it reflects the structure and bonding in the solid. The bands and band gap arise from the periodic, repeating arrangement of atoms in crystalline solids. The magnitude of the band gap is derived from two factors: the strength of the interaction that leads to the separation of the bonding from the antibonding orbitals (roughly, the separation in energy between the centers of the two bands shown in Figure 7.13A), and the strength of the secondary interactions that lead to the energy spread, the band width, of each band (Figure 7.13B). Both will increase as the internuclear distance decreases and the orbital overlap increases. The terms insulator, semiconductor, and metal are used to classify solids based on the size of the band gap (which is essentially zero for a metal) and the effect of the band-gap energy on electrical conductivity.

Figure 7.14 schematically illustrates how insulators and semiconductors differ from metals. In an electrical insulator, diamond being an excellent example, the valence band is essentially filled with electrons, and the conduction band is essentially empty. The band gap of an insulator is large (an arbitrary definition is that E_g > 3 eV ~ 300 kJ/mol; this energy corresponds to a photon of wavelength λ < 400 nm), and the relatively small amount of thermal energy available at room temperature (kT, where k is Boltzmann's constant and T is absolute temperature, is ~0.025 eV ~ 2.4 kJ/mol) ensures that few electrons will be promoted thermally across the band gap where they could then contribute to electrical conductivity. A semiconductor has a smaller band gap (E_g < 3 eV ~ 300 kJ/mol; λ > 400 nm), and electrons are more easily thermally promoted into the conduction band. The metal shown in Figure 7.14 corresponds to a case such as that discussed earlier for Mg, where a partially filled band arises from the energy overlap of a filled with an unfilled band.

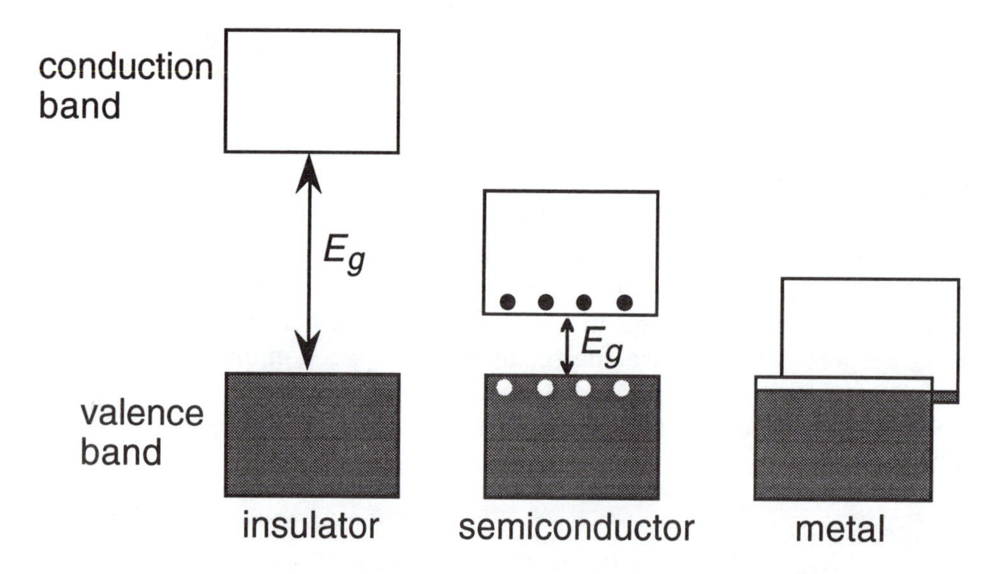

Figure 7.14. Schematic band-structure diagrams for an insulator, a semiconductor, and a metal. The band gap, E_g, shown as the double-headed arrow, is the separation between the top of the valence band and the bottom of the conduction band. The size of the band gap decreases in passing from an insulator to a semiconductor to a metal. Electron–hole pairs are shown for a semiconductor as filled circles in the conduction band (electrons) and open circles in the valence band (holes; *see* Chapter 8).

An important "litmus test" for distinguishing metals from semiconductors and insulators is the temperature dependence of electrical resistivity. As noted, metals have a resistivity that increases slowly with temperature, owing to increased scattering of the free electrons by crystal atoms with temperature. This effect occurs in semiconductors as well but is usually completely swamped by another effect: exponentially increasing numbers of charge carriers with temperature. The carrier concentrations of insulators and semiconductors at thermal equilibrium are dependent on the magnitude of the band gap *(8)*, E_g, as given by exp $(-E_g/2kT)$ [or as exp $(-E_g/2RT)$ on a molar basis; *see* Chapter 8]. Thus, the carrier concentrations are typically orders of magnitude below those of metals (a value of $\sim 10^{10}$ carriers/cm^3 in Si at room temperature, for example, compared to $\sim 10^{22}$ carriers/cm^3 in Na), but the resistivities drop rapidly with increasing temperature, as shown in Figure 7.8. (A related plot that shows the number of charge carriers as a function of reciprocal temperature for Si, Ge, and GaAs appears in Figure 8.5.)

In the Group 14 solids that crystallize with the diamond structure (C, Si, Ge, and α-Sn; *see* Chapter 5), E_g progressively decreases down the group. Table 7.2 lists the Group 14 solids, C through α-Sn, along with their cubic lattice parameters (the length of the side of the cubic unit cell), estimated bond-dissociation energies (D_0), and band-gap energies (300-K values). The size of the unit cell scales with interatomic distance in the solid and follows the expected periodic trend: the increase in atomic radius with atomic number is reflected in larger unit cells. The trend in dissociation energy reflects the greater orbital overlap obtainable as the Group 14 atoms become smaller, which leads to shorter, stronger bonds.

The trend in band gaps has to be analyzed with caution because, as noted, several factors contribute to it. The dominant effect in causing E_g to increase up the group is the increased orbital overlap with smaller internuclear separation, which causes a greater splitting between the centers of the bonding and antibonding bands derived from the sp^3 hybrid orbitals. Although the band widths do change—they increase somewhat up the group for, in part, the same reason of better overlap with smaller internuclear separation—this effect appears to be weighted less strongly in this series.[4]

Diamond has exceptionally strong covalent bonds that result from excellent orbital overlap; these bonds make diamond the standard against which the hardness of materials is measured. The large band gap of ~5.5 eV (580 kJ/mol) makes diamond a good electrical insulator. The optical properties of diamond can also be explained by the band gap, which gives the threshold energy for an electronic transition from the valence band to the conduction band. This transition would be analogous to the lowest energy electronic transition in a molecule, wherein an electron in the highest occupied molecular orbital (HOMO) is promoted to the lowest unoccupied molecular orbital (LUMO) by absorbing a photon whose energy corresponds to the energy difference in these two levels. In the extended

---

[4]The interplay of factors affecting the band gap is seen by noting that the widths of the valence and conduction bands for the Group 14 solids are directly proportional to the initial energy separation between the valence s and p orbitals, $(E_s - E_p)$ *(11)*. The energy difference, $(E_s - E_p)$, drops from ~8.5 eV (820 kJ/mol) for carbon to ~6.6 eV (640 kJ/mol) for tin, a result suggesting that the widths of the valence and conduction bands are getting smaller as the Group 14 atoms become larger. Had there been no change in the separation between the bonding and antibonding orbitals down the group (the separation between the centers of the bands), the band gap would have increased down the group!

solid, the many filled electronic states below the valence-band edge and the many empty states above the conduction-band edge means that electronic transitions and absorption occur for a broad range of energies above the band-gap energy, as shown in Figure 7.15A. A generic absorption spectrum for insulators and semiconductors is sketched in Figure 7.15B and illustrates this onset of electronic absorption at the band-gap energy. For diamond, the E_g value of 5.5 eV (λ = 230 nm) is in the ultraviolet portion of the electromagnetic spectrum rather than in the visible portion, which ranges from violet light at ~3.1 eV (λ = 400 nm), to red light at ~1.7 eV (λ = 730 nm), making diamond transparent to visible light.

Table 7.2. Periodic Properties of the Group 14 Solids Possessing the Diamond Structure

Element	Lattice Parameter , Å[a]	D_0, kJ/mol[b]	E_g , eV (λ, nm)[c]
C	3.57	346	5.5 (230)
Si	5.43	222	1.1 (1100)
Ge	5.66	188	0.66 (1900)
α-Sn	6.49	146	< 0.1 (12,000)

[a] Repeat distance, a, of the cubic unit cell at 300 K, obtained from X-ray diffraction data. These values are known to greater precision, but are rounded off here for simplicity and are taken from reference 9.
[b] Bond-dissociation energy was determined from heats of atomization data from reference 10.
[c] Band-gap energy at 300 K is from reference 9.

On the basis of the explanation given for the high thermal conductivities of metals, diamond, with its paucity of free electrons, might be predicted to have poor thermal conductivity. However, diamond actually has a remarkably high thermal conductivity because of a second mechanism for transferring heat that is based on the quantized high-energy vibrations of atoms in the solid, called phonons (*see* Chapter 2, Heat Capacities). Diamond is sometimes said to have a "stiff" crystal structure because of the strong carbon–carbon bonds. The phonon mechanism also imparts high thermal conductivities to many ceramic materials like Al_2O_3 and BeO.

In descending Group 14 to silicon, the cubic unit cell's lattice constant (the length of an edge of the unit cell) expands to 5.43 Å. The reduced orbital overlap is reflected in Si–Si bonds that are longer and weaker than the C–C bonds of diamond and in a substantially smaller band gap of ~1.1 eV (110 kJ/mol). Electrons are more easily promoted across the band gap by thermal energy, and silicon is a semiconductor. Furthermore, each electron that is thermally excited across the gap helps conductivity in two ways: by helping to partially fill the conduction band and by helping to partially empty the valence band (a concept that will be elaborated upon in more detail in Chapter 8, using a formalism by which missing valence-band electrons are treated as positively charged particles called holes). The small band gap of silicon corresponds to a threshold wavelength for absorption of ~1100 nm, below the visible spectrum. Thus, crystalline silicon absorbs light from the entire visible spectrum and appears black.

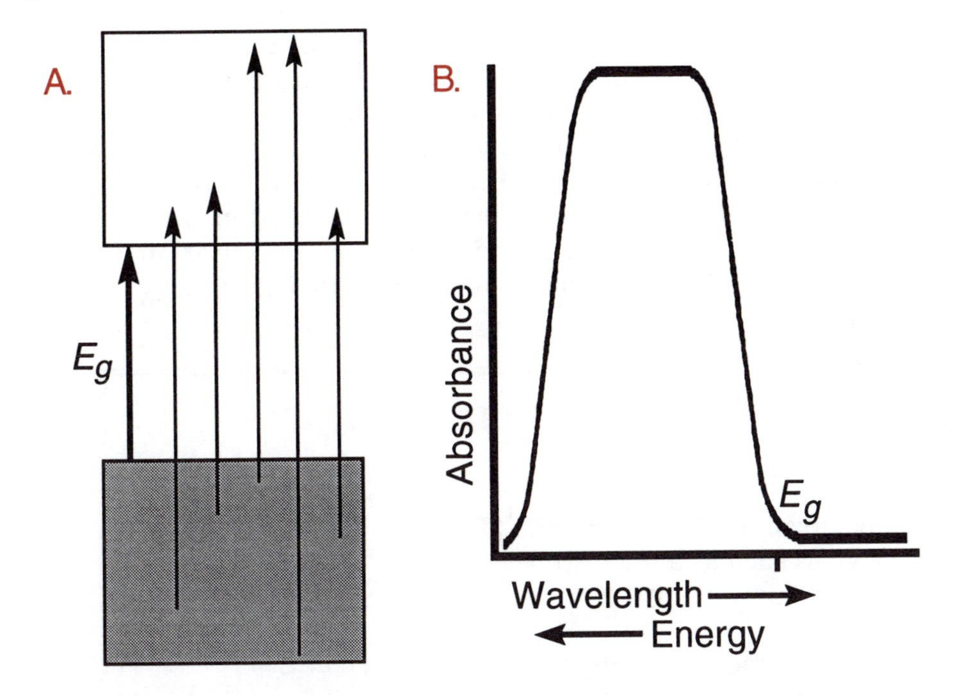

Figure 7.15. A: A band diagram showing possible transitions of electrons from the valence band to the conduction band. The bold arrow represents the minimum energy for the transition and corresponds to an electron that is promoted from the top of the valence band to the bottom of the conduction band. This energy corresponds to E_g. The longer arrows correspond to absorption of photon energies greater than E_g. B: Sketch of a generic absorption spectrum of a semiconductor. The absorbance increases sharply as the energy of light approaches the band gap of the semiconductor. Light with greater energy than the band gap is absorbed because it corresponds to a transition from an individual orbital in the valence band to another orbital in the conduction band. This feature makes some semiconductors useful as optical cutoff filters. Absorption will decline again for photons whose energies exceed the combined energies of the band widths and the band gap.

Only a modest increase in unit-cell size occurs in passing from Si to Ge (5.66 Å), presumably reflecting the interposition of the transition elements. The Ge band gap, 0.66 eV (64 kJ/mol), is slightly reduced from that of Si; absorption begins for wavelengths less than ~1900 nm, giving Ge a black color. A larger change in unit-cell size is seen in passing to α-Sn. This phase of tin, which exhibits the diamond structure, is stable only below room temperature (Chapter 9). The lattice constant of α-Sn is 6.49 Å, and the solid has a band gap of < 0.1 eV (< 9.6 kJ/mol; λ > 12,000 nm). In this case the valence and conduction bands are close to merging. The final member of the Group 14 elements is lead, which does not have the diamond structure. However, if Pb were to adopt this structure, it has been predicted *(12)* to be a metal having a filled s band and partially filled p band.

In a localized bonding picture, promoting an electron across the band gap can be thought of crudely as freeing one of the electrons from the directional covalent bonds in the crystal, making it a mobile charge carrier. The increasing likelihood of this happening down the Group 14 extended solids has been correlated with the trends in bond length, bond strength and atomic ionization potentials.[5] Both thermal excitation and photoexcitation can cause this ionization to occur; photoexcitation corresponds to absorption of lower photon energies as the energy requirement declines.

Demonstration 7.8. The Hardness of Diamond

Materials

Diamond scribe (available from Aldrich, Z22554-1)
Small piece of glass such as a microscope slide
Overhead projector
Penny
Piece of aluminum metal

Procedure

- Use the diamond scribe to scratch a piece of glass. This scratching can be shown on an overhead projector and contrasted with the inability of other materials like a penny or aluminum to mar the glass surface.

Demonstration 7.9. Photoconductivity of CdS

Materials

CdS photocell (obtained from Mouser or Radio Shack; *see* Supplier Information). CdS has the wurtzite structure, rather than the zinc blende structure.
Ohmmeter (ideally with an audible conductivity test function)
Light source such as an overhead projector or sunny window

Procedure

- Connect the photocell to the ohmmeter.
- Measure the resistance of the device with and without exposure to a light source to note the effect of light-induced promotion of electrons across the band gap. A multimeter with an audible conductivity test feature is an effective way to demonstrate this in a classroom.

---

[5]Correlations of these parameters with band gap have all been analyzed and their limitations discussed. *See* reference 13.

Electrical and Optical Properties of Insulators and Semiconductors Having the Zinc Blende Structure

Isoelectronic–isovalent principles are compellingly illustrated with the tetrahedral extended solids.[6] The total valence-electron count of elements having the diamond structure is preserved in compounds of AZ stoichiometry having the zinc blende structure (also called "sphalerite," after the mineral ZnS that possesses the structure), which is identical to the diamond structure except that two different kinds of atoms replace the single kind of atom in diamond (Chapter 5).

Trends in periodic properties can be used to explain the band gaps in the group of elements that symmetrically flank the Group 14 elements in the periodic table. In Figure 7.16, complementary AZ pairs are indicated with similar shading. All A atoms are tetrahedrally coordinated exclusively to Z atoms; and all Z atoms are tetrahedrally coordinated exclusively to A atoms. Solids such as Ge, GaAs, ZnSe, and CuBr are isostructural and can be considered to be isoelectronic: For example, in passing from Ge to GaAs, half of the Ge atoms are replaced by Ga atoms having one fewer valence electron, and the other half are replaced by As having one additional

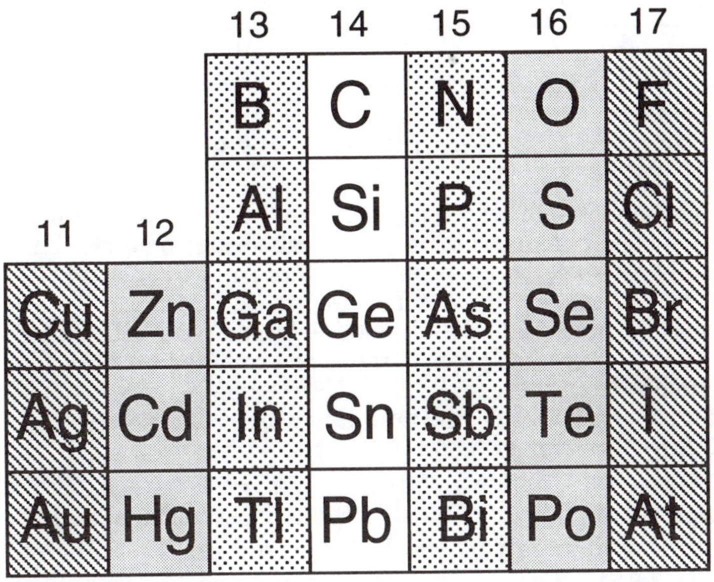

11	12	13	14	15	16	17
		B	C	N	O	F
		Al	Si	P	S	Cl
Cu	Zn	Ga	Ge	As	Se	Br
Ag	Cd	In	Sn	Sb	Te	I
Au	Hg	Tl	Pb	Bi	Po	At

Figure 7.16. Portion of the periodic table emphasizing the formation of AZ solids that are isoelectronic with the Group 14 solids. Complementary pairs are indicated with similar shading; for example, Ge, GaAs, ZnSe, and CuBr.

---

[6]As used here, isoelectronic species are those that preserve the number of valence electrons in orbitals with the same principal quantum number. The ions Na^{+} and Mg^{2+} are considered to be isoelectronic with each other. The solids Ge, GaAs, and ZnSe have the same average number of valence electrons per atom and are also considered to be isoelectronic. Isovalent species are those that have the same number of valence electrons per atom, but the electrons are in orbitals with different principal quantum numbers. Thus, carbon and silicon are isovalent.

valence electron. In ZnSe, the Zn atoms that replace half of the Ge atoms have two fewer valence electrons, and the Se atoms that replace the other half have two additional electrons. Thus, the total electron count is preserved in both instances. (This trend is like playing the musical composition *Chopsticks* on the periodic table: A central "note" is struck on the table, then the two flanking columns are struck together, followed by the next two adjacent flanking columns, etc.)

In Table 7.3, the cubic unit-cell lengths (lattice parameters), the Pauling electronegativity differences, and the band-gap energies at 300 K are presented for Ge, GaAs, ZnSe, and CuBr. Structurally, in passing from Ge to CuBr, the lattice parameter is essentially constant at 5.67 ± 0.02 Å. At the same time, however, the bonding in this series acquires an increasingly ionic contribution: the electronegativity differences increase from 0.0 to 0.9 along this series. As shown in Figure 7.17 and Table 7.3, band gaps monotonically increase with the difference in electronegativity from Ge to CuBr.[7]

Table 7.3 Periodic Properties of a Family of Isoelectronic, Tetrahedral Semiconductors

Material[a]	Cubic Unit-Cell Parameter, Å[b]	$\Delta\chi$[c]	E_g, eV (λ, nm)[d]
Ge	5.66	0.0	0.66 (1900)
GaAs	5.65	0.4	1.42 (890)
ZnSe	5.67	0.8	2.70 (460)
CuBr	5.69	0.9	2.91 (430)

[a] The indicated materials have either the diamond structure (Ge) or the zinc blende structure of AZ stoichiometry in which all A atoms are bonded to four Z atoms and all Z atoms are bonded to four A atoms in a tetrahedral geometry. The solids are listed sequentially as isoelectronic semiconductors from Groups 14, 13 and 15, 12 and 16, and 11 and 17 in the periodic table (*see* Figure 7.16).
[b] Repeat distance, *a*, of the cubic unit cell at 300 K, obtained from X-ray diffraction data. These values are known to greater precision, but are rounded off here for simplicity and are taken from reference 9.
[c] Difference in Pauling electronegativities for the interatomic bonds in the crystal.
[d] Band-gap energy at 300 K from reference 9.

The trend in band gaps of this family of isoelectronic, isostructural solids is reflected in their colors, which can be predicted based on the color of transmitted light (Figure 7.18). Both Ge and GaAs have band gaps below the visible spectrum and look black. With a band gap of 2.7 eV (460 nm), ZnSe absorbs light of higher energy than indigo light and thereby acquires a yellow appearance. Pure, powdered CuBr looks white, because of its absorption onset at 2.9 eV (430 nm), which is nearly out of the visible region of the spectrum.

---

[7]Because of its d-orbital bonding contributions, CuBr does not fit some of the trends in properties along the isoelectronic series as well as the other members and could be dropped from the series at the instructor's discretion. *See* reference 13 and reference 7. Reference 13 also discusses limitations on the correlation between band gap energies and electronegativity differences along this isoelectronic series (Sections 6.4.1 and 6.4.4).

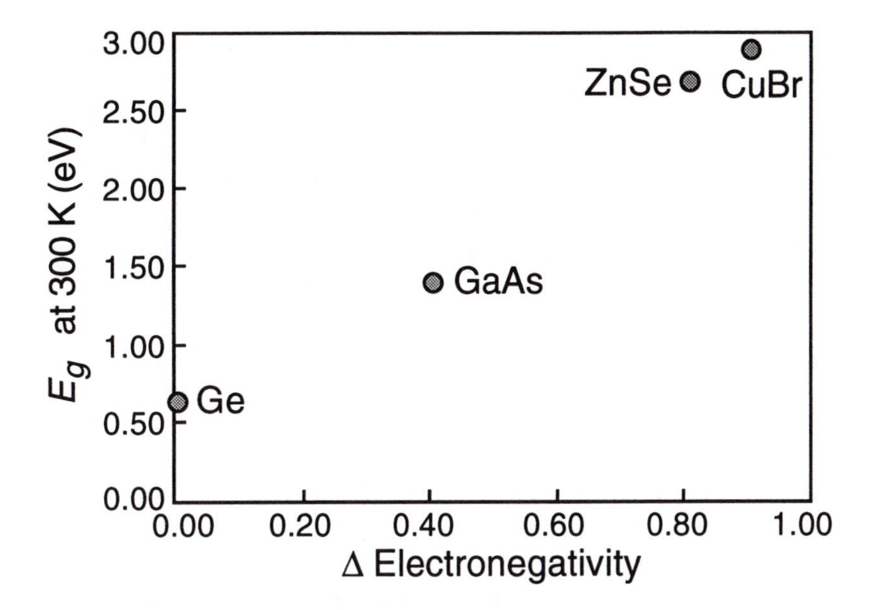

Figure 7.17. Plot of the band-gap values from Table 7.3 versus the difference in Pauling electronegativities in the same table. The unit-cell edge length is roughly constant for these materials.

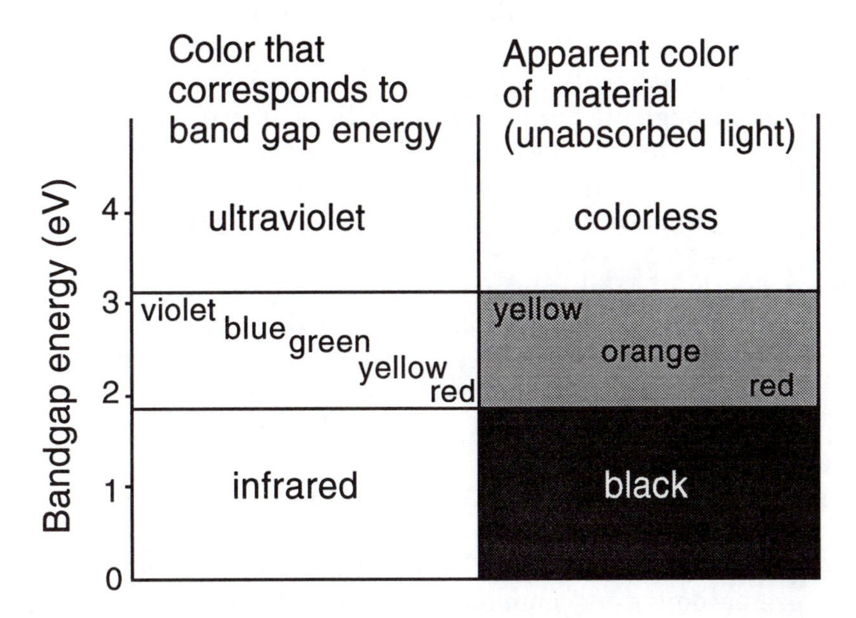

Figure 7.18. Diagram showing the relationship between band-gap energy and color. Substances with band gaps larger than the energies of visible light appear colorless, while those with band gaps below the visible region of the spectrum absorb all visible light and appear black. Materials with band gaps in the visible region of the spectrum will absorb some light above the band gap but not light below the band gap (Figure 7.15). The observed colors will be related to the colors of the unabsorbed light. (Adapted with permission from reference 2. Copyright 1983 John Wiley and Sons.)

Solid Solutions Having the Zinc Blende Structure: Tunable Band Gaps

The ability to form solid solutions between solids having the same zinc blende structure and possessing atoms of comparable size (Chapter 3) is of great technological importance. Rather than being restricted to the band gaps corresponding to the AZ stoichiometries obtainable with the elements in the periodic table, solid solutions provide a chemical means for tuning band-gap energies.

The zinc blende semiconductors GaAs and GaP constitute one of many pairs of solids that can be combined to yield solid solutions, symbolized for this family as GaP_xAs_{1-x} ($0 \leq x \leq 1$). As noted in Chapter 3, these are disordered solid solutions with the stoichiometric fractions x and $1 - x$ denoting the probabilities that an atom on a Group 15 site in the zinc blende crystal structure is P or As, respectively.

The smaller atomic radius of P relative to As leads to a smaller unit-cell constant for GaP (5.45 Å) relative to GaAs (5.65 Å). The unit-cell constants of the GaP_xAs_{1-x} solid solutions follow Vegard's law, which means that they are a weighted average of the unit cell constants of the components, that is, they vary linearly with the value of x in the formula. Thus, as shown in Figure 7.19, GaP_xAs_{1-x} solid solutions have unit-cell constants that are continuously tunable between 5.45 and 5.65 Å. There is little difference in the electronegativities of P and As (various scales rank both as having larger electronegativity), suggesting that ionic bonding contributions are similar throughout the GaP_xAs_{1-x} solid solution series. The increase in band gap from GaAs (1.4 eV; λ = 890 nm) to GaP (2.3 eV; λ = 540 nm) , Figure 7.19, appears to correlate most strongly with the reduction in internuclear distance and enhanced orbital overlap with increasing phosphorus content.

In contrast, another family of solid solutions, $Al_xGa_{1-x}As$, features an essentially constant cubic unit cell size of 5.658 ± 0.004 Å throughout the series *(14)*, but increasingly ionic bonding as Ga atoms are replaced by less electronegative Al atoms. Increasing ionic character is reflected in the increase in band gap observed in passing from GaAs (1.4 eV; λ = 890 nm) to AlAs (2.1 eV; λ = 590 nm), Figure 7.20.

The tunability of the band gap in the GaP_xAs_{1-x} and $Al_xGa_{1-x}As$ families of solids is used extensively in the design of light-emitting diodes (LEDs) and diode lasers. These devices, which will be described in more detail in Chapter 8, are becoming as common as the electric light bulb. Already, they are evident in consumer electronic products ranging from indicator lights on digital clocks and microwave ovens to compact disc players, laser printers and pointers, and fiber-optic telephone transmission lines.

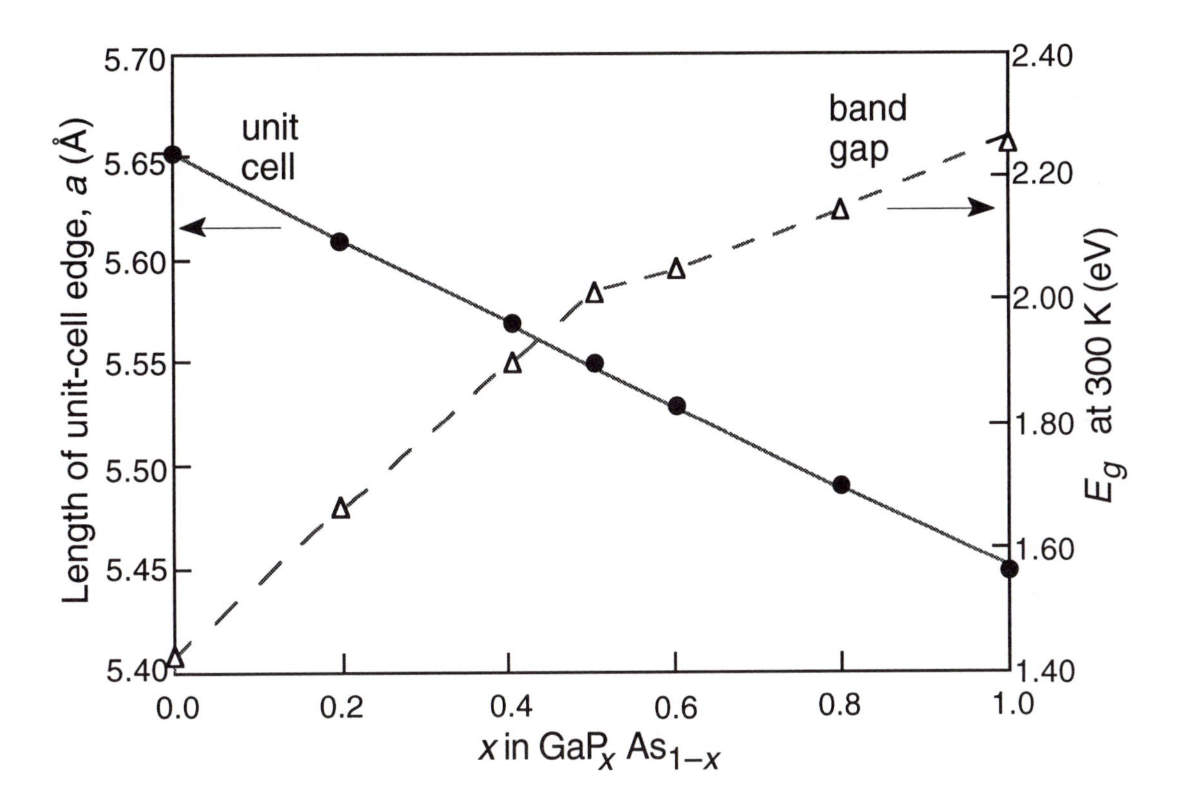

Figure 7.19. Trends in the cubic unit-cell parameter, a (Å), (filled circles), and the band gap at 300 K (open triangles) as a function of composition, x, for the solid solution series GaP_xAs_{1-x}. The kink in the band gap at $x = 0.45$ corresponds to a change from a direct band gap to an indirect band gap.[8] (Data are taken from references 14–16.)

[8]When photons are absorbed or emitted, both energy and momentum in the solid must be conserved. For GaAs, As-rich GaP_xAs_{1-x} ($0.00 \le x \le 0.45$) and Ga-rich $Al_xGa_{1-x}As$ ($0.00 \le x \le 0.38$) samples, momentum associated with absorption or emission of a photon can be conserved exclusively by production or recombination of an electron–hole pair, respectively. Such solids, called "direct band gap" materials, absorb light strongly at the band-gap energy and emit band-gap energy light with high efficiency. In contrast, GaP, P-rich GaP_xAs_{1-x} ($0.45 \le x \le 1.00$), and Al-rich $Al_xGa_{1-x}As$ ($0.38 \le x \le 1.00$) samples, cannot conserve momentum during an electronic transition at the band-gap energy unless the absorption or emission is accompanied by a change in crystal momentum, corresponding to absorption or emission of quantized vibrations of the crystal atoms called phonons. This process is less likely, and these "indirect band gap" materials absorb light more weakly and yield less efficient radiative recombination. This feature makes them inferior LED materials, unless they are doped with an impurity atom, which permits an efficient emissive transition by relaxing these rules.

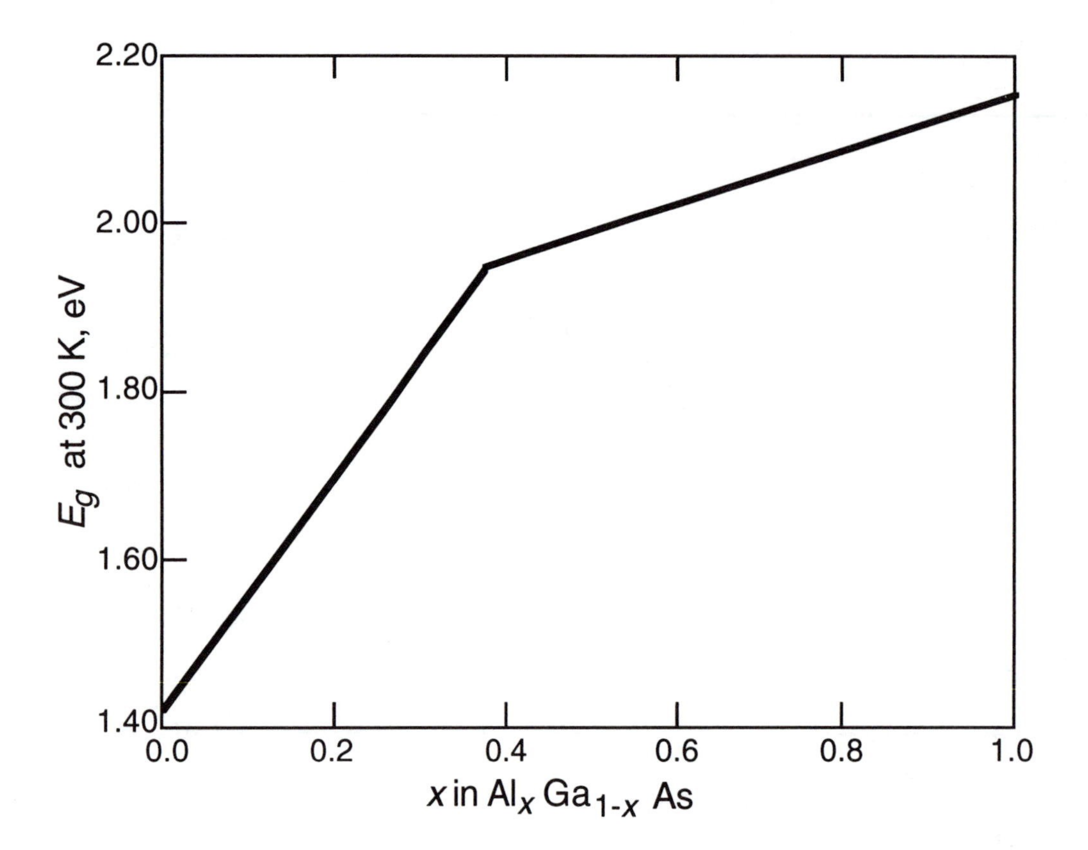

Figure 7.20. Trend in the band gap at 300 K as a function of composition, x, for the solid solution series $Al_xGa_{1-x}As$. The kink in the band gap at $x = 0.38$ corresponds to a change from a direct band gap to an indirect band gap.[8] (Data are taken from reference 17.)

The color of light emitted by these devices is often roughly the band-gap energy. This feature illustrates the basic spectroscopic principle that the energy of an electronic transition is controlled by the difference in energy of the levels between which the electron jumps: electrons can be promoted across the band gap in LEDs and diode lasers by absorbing energy from an applied electric field (a battery, e.g.). The subsequent return of the electrons from the conduction band back to the valence band releases roughly the band-gap energy (Figure 7.21A). This return is often described as a recombination of the electron in the conduction band with an absent electron in the valence band (treated as a positively charged particle called a "hole"; *see* Chapter 8) and can occur both nonradiatively (the energy is released as heat through the quantized vibrations of atoms in the crystal called "phonons") and radiatively, producing photons of roughly band gap energy. (The simplistic localized picture corresponding to this process is that the energy needed to remove an electron from a bond in the solid can be released as a photon upon the return of the electron to the bond. As the energy associated with this process increases, so does the corresponding photon energy.) These electro-optical devices are optimized to release as much of the energy as possible as photons.

Emission spectra of LEDs are relatively narrow, as sketched in Figure 7.21B. The reason for this narrowness is that even though electrons can be created at energies well above the conduction-band edge (and holes with which the electrons will recombine at energies well below the valence-band edge), the electrons and holes will rapidly come to the band edges by losing their excess energy as heat, which is given to the crystal (holes "rise" to the valence band edge to reach lowest energy; *see* Chapter 8 and Demonstration 8.1).

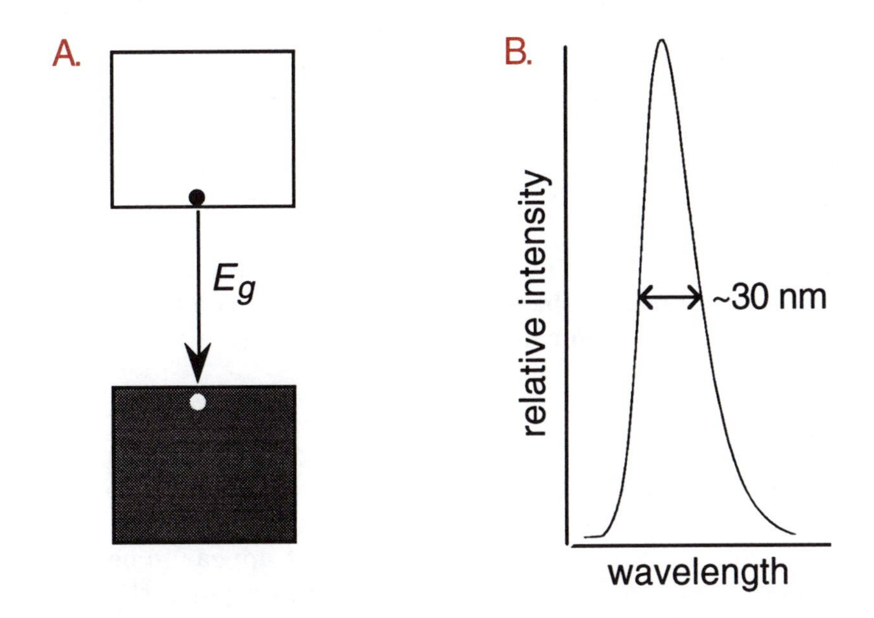

Figure 7.21. A: The return of an electron from the conduction band to the valence band can release a photon with an energy roughly equal to the band-gap energy. The electron can also return without the release of light through atomic vibrations in the crystal (phonons). B: A typical emission spectrum corresponding to the electronic transition shown in A.

Demonstration 7.10. Tunable Band Gaps Based on Solid Solutions

Materials (For electronic parts, *see* Supplier Information)

GaP_xAs_{1-x} LEDs, shown schematically in Figure 7.22A
1-kΩ resistor
9-V battery
Battery snap
LED socket
1-MΩ resistor (optional)
Liquid nitrogen
foam cup

Voltmeter
Soldering gun and solder
Magnifying lens or Micronta 30× microscope (Radio Shack)

Procedure

- Build the circuit shown in Figure 7.22B.
- Plug the available LEDs into the circuit and note the color emitted. The photon energies corresponding to the colors observed can be taken to be a rough estimate of the band gaps of the LEDs.
- To show simultaneously the range of colors that are possible in the solid solution series, you can wire together several LEDs to form a strip of red, orange, yellow, and green lights. The circuit diagram is shown in Figure 7.22C, and the assembly directions are given in Appendix 7.1.
- Use a magnifying lens or the Radio Shack Micronta hand-held microscope to see and appreciate the small size of the semiconductor chips, which is on the order of 1 mm^2.

Variation

- Another rough estimate can be made in the following way: Place a large resistor (1 MΩ) in the circuit instead of the 1–kΩ resistor so that little current passes. Plug one of the LEDs into the circuit. With the large resistor in the circuit, the LED may not appear to be lit at room temperature. Connect the voltmeter and measure the voltage drop across the LED. If the measured voltage is more than 3 volts, then the electrical leads on the LED should be reversed.
- Cool the LED in liquid nitrogen. **CAUTION: Liquid nitrogen is extremely cold. Do not allow it to come into contact with skin or clothing, as severe frostbite may result. Wear gloves when transferring and using liquid nitrogen.** Measure the voltage drop across the LED while it is cold. The measured voltage corresponds roughly to the energy (in electron-volts) of the band gap.[9] If more than 3 volts is measured, then current is not flowing through the LED, perhaps because contraction of the plastic case has broken the electrical contact; raise the LED to just above the liquid nitrogen level. Repeated dunking in liquid nitrogen seems to damage some LEDs.

---

[9]Poor correlations may to some extent reflect the fact that some LEDs contain impurity atoms (called "dopants;" *see* Chapter 8) to enhance the intensity of the emitted light. Emission from the dopant-containing LEDs occurs below the band-gap energy (the dopants introduce electronic states within the band gap that are involved in the recombination mechanism that produces light) and complicates correlations of band gaps with the composition of the host semiconductor.

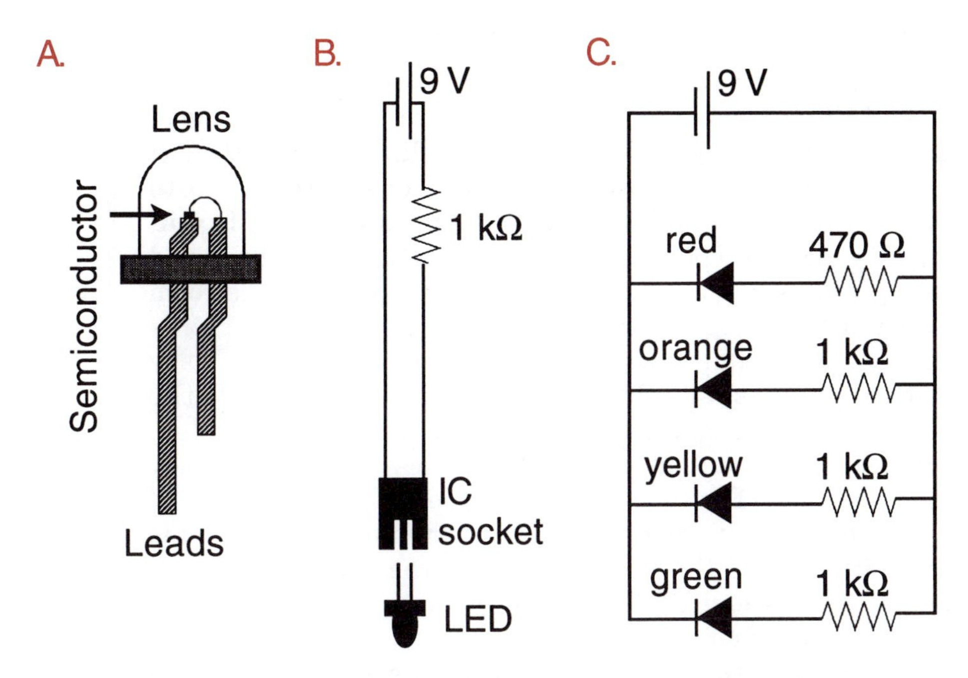

Figure 7.22. Experimental setup for the LED experiment. A: Schematic picture of a commercially available LED, showing the location of the semiconductor chip. B: Circuit diagram for a single LED. C: Circuit diagram for the reference strip of LEDs.

A visually stunning demonstration is provided by plunging an LED into liquid nitrogen. Most impressive in a darkened lecture hall is to take a commercially available $GaP_{0.40}As_{0.60}$ LED that is shown to emit weak red light at room temperature and to cause its emission to become bright orange as it is cooled in liquid nitrogen.

Demonstration 7.11. The Effect of Temperature on LED Emission

Materials

$GaP_{0.40}As_{0.60}$ (red) or $GaP_{0.65}As_{0.35}$ (orange) LED (Mouser or Radio Shack, but the flat-topped LEDs from DigiKey work especially well–*see* Supplier Information)

Circuit containing 1 kΩ resistor as described in previous demonstration

Styrofoam cup

Liquid nitrogen

Procedure

- Prepare the LED circuit as described in Figure 7.22B and insert an LED. The LED has a red or orange color at room temperature.

- After lighting the LED, darken the lecture hall.
- Immerse the LED into a Styrofoam cup filled with liquid nitrogen. **CAUTION: Liquid nitrogen is extremely cold. Do not allow it to come into contact with skin or clothing, as severe frostbite may result. Wear gloves when transferring and using liquid nitrogen.** The light intensity becomes substantially brighter and lights the cup with a red-orange or orange-yellow glow for the room temperature red and orange LEDs, respectively.

The intensity enhancement and visible spectral shift can vary significantly; samples should be tested in advance of the demonstration. Some samples also glow more intensely when initially plunged into liquid nitrogen, but then lose intensity as they remain in the coolant. For such samples, if the LED is pulled out of the liquid nitrogen once it has reached its maximum intensity and is then held a little above the liquid nitrogen, the intensity can be maintained at what is essentially a somewhat warmer temperature than that of liquid nitrogen.

These particular LED sample compositions were selected because their color changes illustrate a shift of the band gap to higher energy with decreasing temperature, a common feature of most semiconductors; and greatly enhanced efficiency for light emission upon cooling. Both effects are generally found with GaP_xAs_{1-x} (Figure 7.23) and $Al_xGa_{1-x}As$ (Figure 7.24) LEDs. However, not all of the GaP_xAs_{1-x} compositions of Demonstration 7.10 permit this kind of "eyeball spectroscopy." In particular, as noted in Demonstration 7.10, LEDs that exhibit emission involving the presence of impurities (dopants) may have their spectrum dominated by those peaks, which occur at energies below the band gap and which may or may not shift in parallel with the band gap.

An increase in band gap with decreasing temperature would generally be predicted on the basis of the enhanced orbital overlap that accompanies the contraction of the crystal when it is cooled. For example, in GaP, the cubic unit-cell size shrinks from 5.451 Å at 300 K to 5.447 Å at 77 K *(18)*. (A similar effect on the unit cell size was noted in the discussion of F-centers in Chapter 6.) The increased orbital overlap is expected to contribute to an increase in the band-gap energy by increasing the separation between the centers of the valence and conduction bands, providing, as noted, that this effect is larger than the effect of the increased overlap on the widths of the two bands. For GaP, E_g increases from 2.27 eV (λ = 550 nm) at 300 K to 2.33 eV (λ = 530 nm) at 77 K.

If a spectrometer is available, the LED spectral shifts can be quantified. Clear shifts in the wavelength of emitted light to higher energy with decreasing temperature is best observed for LEDs that have a single emission peak at room temperature. LEDs with multiple emission peaks arising from the deliberate introduction of impurities into the crystal typically exhibit more complex temperature effects, Figure 7.23.[8]

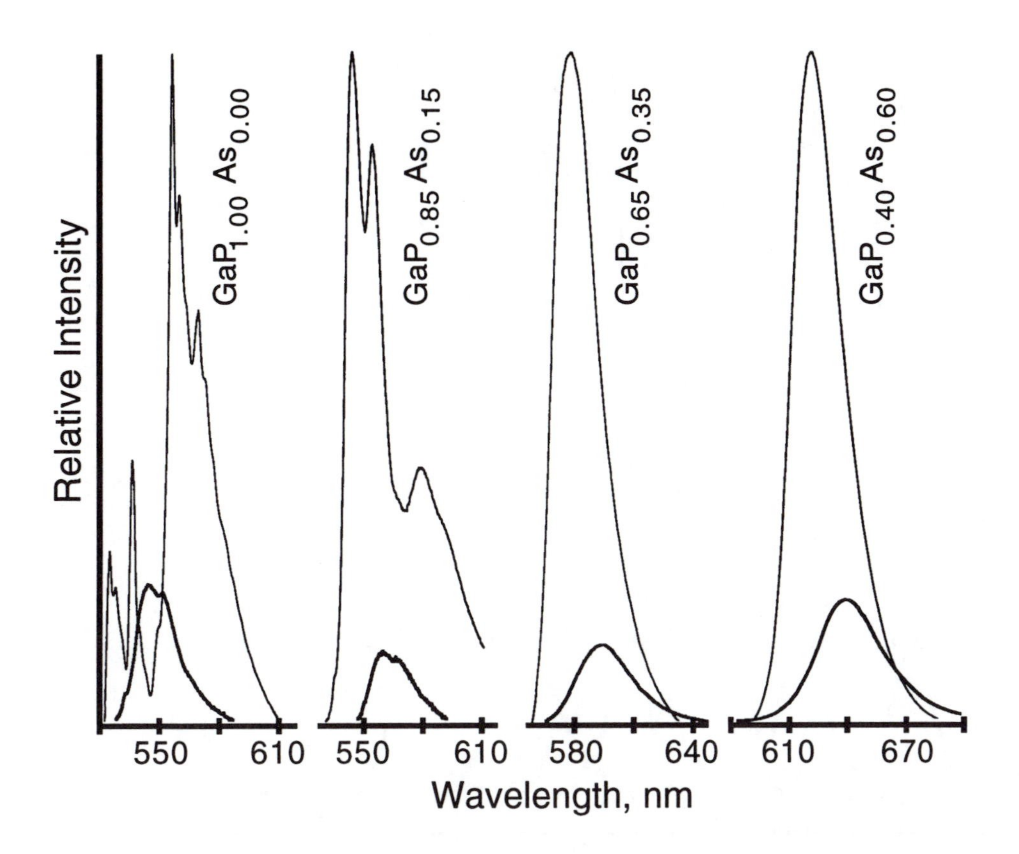

Figure 7.23. Spectra of GaP_xAs_{1-x} LEDs at room temperature (thicker lines) and at liquid nitrogen temperature. $GaP_{1.00}As_{0.00}$ and $GaP_{0.85}As_{0.15}$ are not suitable for "eyeball spectroscopy" predictions of band gap spectral shifts because of the complexity of the spectra.

The enhancement in LED intensity at low temperature reflects the competition between whether the return of the electrons from the conduction band to the valence band occurs radiatively, giving off light, or nonradiatively, giving off heat through vibrations of atoms in the crystal. At lower temperatures, the vibrations in the crystal play a less effective role (they are being "frozen out"), and lead to significant enhancements of the radiative recombination process. LED efficiency enhancements of 1–2 orders of magnitude are common in this experiment.

Laboratory. Experiment 7 has students use LEDs to study the effect of varying the chemical composition of the LED material on the wavelength of the emitted light and the effect of temperature on emitted light intensity.

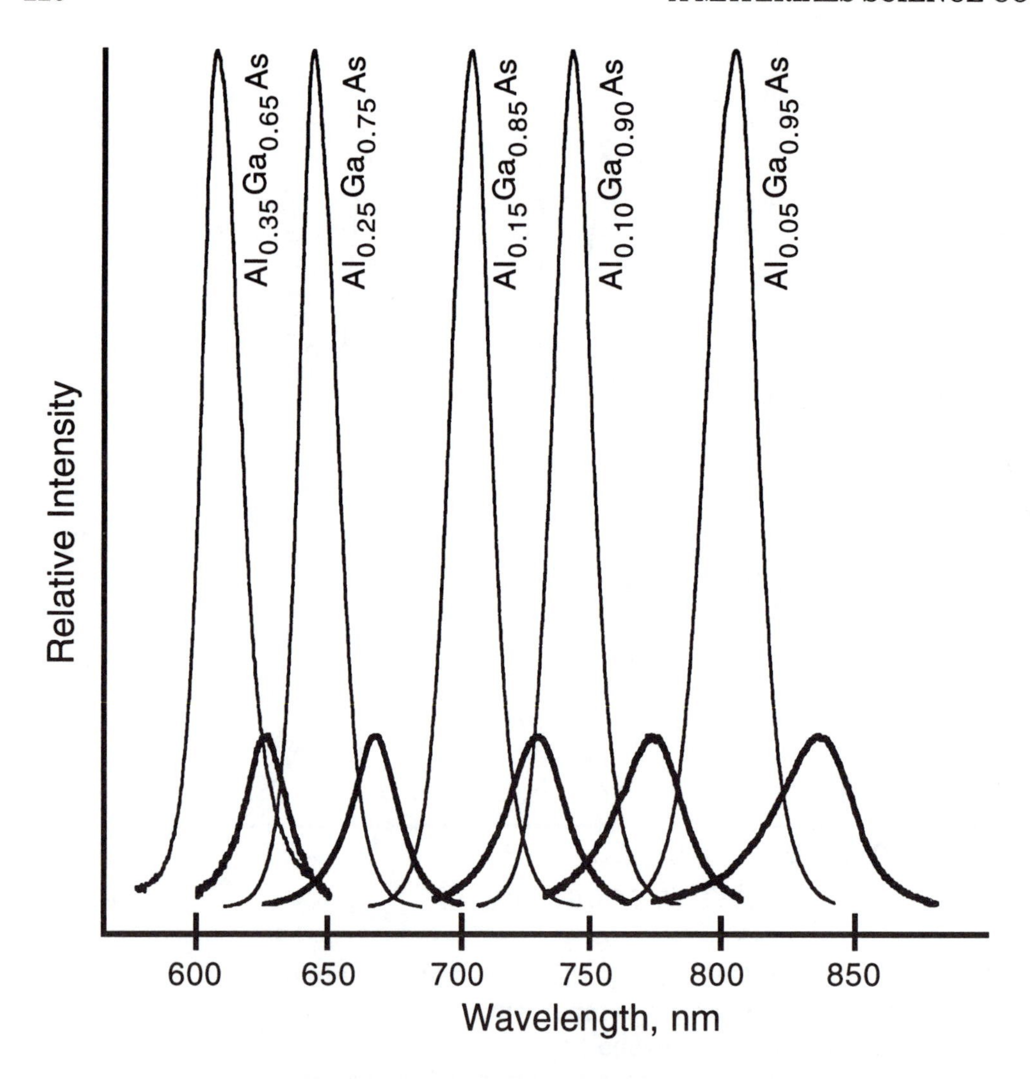

Figure 7.24. Spectra of $Al_xGa_{1-x}As$ LEDs at room temperature (thicker lines) and at liquid nitrogen temperature.

Ionic Insulators and Semiconductors

The electronic structures of extended salts may also be described with the band formalism. Sodium chloride illustrates many of the concepts used to develop the electronic structure of these solids. The large difference in Na and Cl electronegativities is reflected in the classification of this solid as ionically bonded, comprising Na^+ and Cl^- ions, as discussed in Chapter 5. Both the anion and cation are closed-shell ions; therefore, bands that are exclusively empty or full of electrons are expected, and the salts would be predicted to be insulators or semiconductors, depending on the magnitude of the band gap.

Bands in these solids reflect the importance of electrostatic interactions. A typical band diagram, that of NaCl, is shown in Figure 7.25. The valence band has primarily Cl 3p orbital character. Since Cl^- has a $3p^6$ configuration, this band is completely filled with electrons. The conduction band is derived largely from the empty 3s orbitals of Na. These 3s-derived orbitals of the conduction band are at higher energy than the 3p-derived orbitals of the valence band because of chlorine's greater nuclear charge, which stabilizes all of the atomic orbitals of the halogen relative to those of the alkali metal.

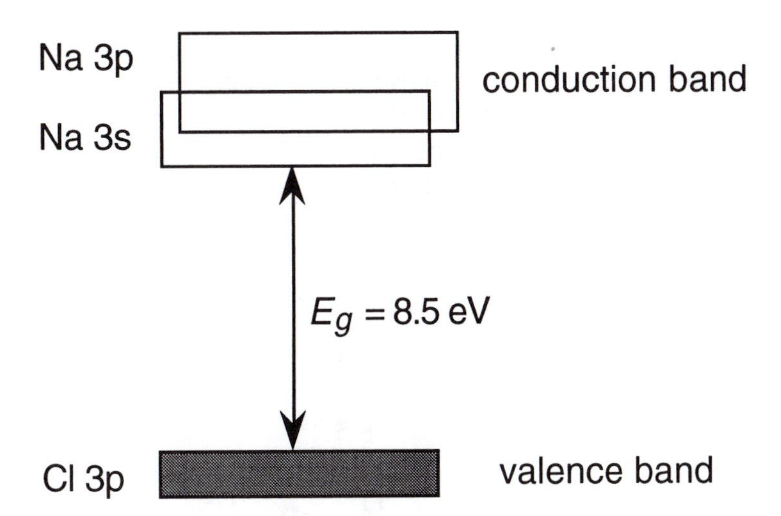

Figure 7.25. Band-structure diagram for NaCl. The valence band is primarily chlorine 3p in character. The conduction band is primarily sodium 3s in character, and there is overlap in energy with the sodium 3p band.

As shown in Figure 7.25, the two bands give rise to an enormous band gap of ~ 9 eV, nearly twice that of diamond! The large gap separating the filled valence band from the unfilled conduction band accounts for the electrically insulating nature of crystalline NaCl and for its optical transparency. In a simple localized picture, promoting an electron across the band gap is somewhat like trying to transfer an electron in the solid from Cl^- back to Na^+, an energetically costly process:

$$Cl^- + Na^+ \longrightarrow Cl + Na$$

Generally ionic compounds become conductors of electricity only at high temperatures. Under these circumstances the mobile species commonly are not electrons but ions moving through lattice defects (*see* Chapter 6).[10]

Band gaps in families of salts, as in families of covalent solids, reflect the interplay of several parameters *(19)*. Some of the families of simple salts do, however, have trends in band gaps that are due principally to

[10]In some cases, defects and impurities give rise to conductivity that is electronic in nature.

variations in electrostatic interactions, which increase with decreasing interionic distances.

Table 7.4 presents a collection of band gaps of alkali halide salts. Among the alkali halides, the largest band gap occurs for LiF, which has a value of 14 eV. These two ions are the smallest of those in the table, and their large electrostatic interaction leads to the largest splitting, in this case between a valence band that has largely 2p orbital character (from F^- ions) and a conduction band that has primarily 2s orbital character (from Li^+ ions). Band gaps are seen to either decrease or remain roughly constant as the interionic distance increases with larger cations or anions.

Salts with more covalent character, like the silver halide salts so critical to the photographic process, have smaller band gaps more typical of semiconducting materials, with absorption onsets that fall in or near the visible portion of the spectrum. Part of the reduction in band gap in these materials may be attributed to covalent interactions that widen the bands.

Table 7.4. Band Gaps for Some Ionic Compounds

Compound	Band Gap (eV)
LiF	14
LiCl	9.5
NaF	12.
NaCl	8.5
NaBr	7.5
KF	11
KCl	8.5
KBr	7.5
KI	5.8

NOTE: All of the compounds have the NaCl structure. Band gaps are approximate.
SOURCE: Adapted from reference 20.

Appendix 7.1. Assembly of LED Reference Strip

1. Slice perfboard using a band saw, cutting through lines of holes to leave strips of four holes. Use drill to slightly enlarge four holes to match diameter of battery snap wires. Spray-paint back side black. Let dry.

2. Attach LEDs. The plastic case goes on the black side. Leads stick through to solder side. All LEDs must be oriented the same way; for example, put the short lead on the left. Bend left leads over, flush to board and along horizontal row of holes. Solder left leads.

3. Connect one side of the LEDs by bending the lead from the top LED flush to the board and along the vertical row of holes. Solder right leads and trim indicated excess.

4. Bend resistors and position diagonally, inserting wires into outside holes. Solder left leads and trim indicated excess.

5. Take a trimmed lead and lay along the row of resistors to connect them. Solder along the outside edge. Trim excess resistor leads from the black side.

6. Attach battery snap by threading leads through the perfboard, verifying correct polarity to light LEDs. Solder.

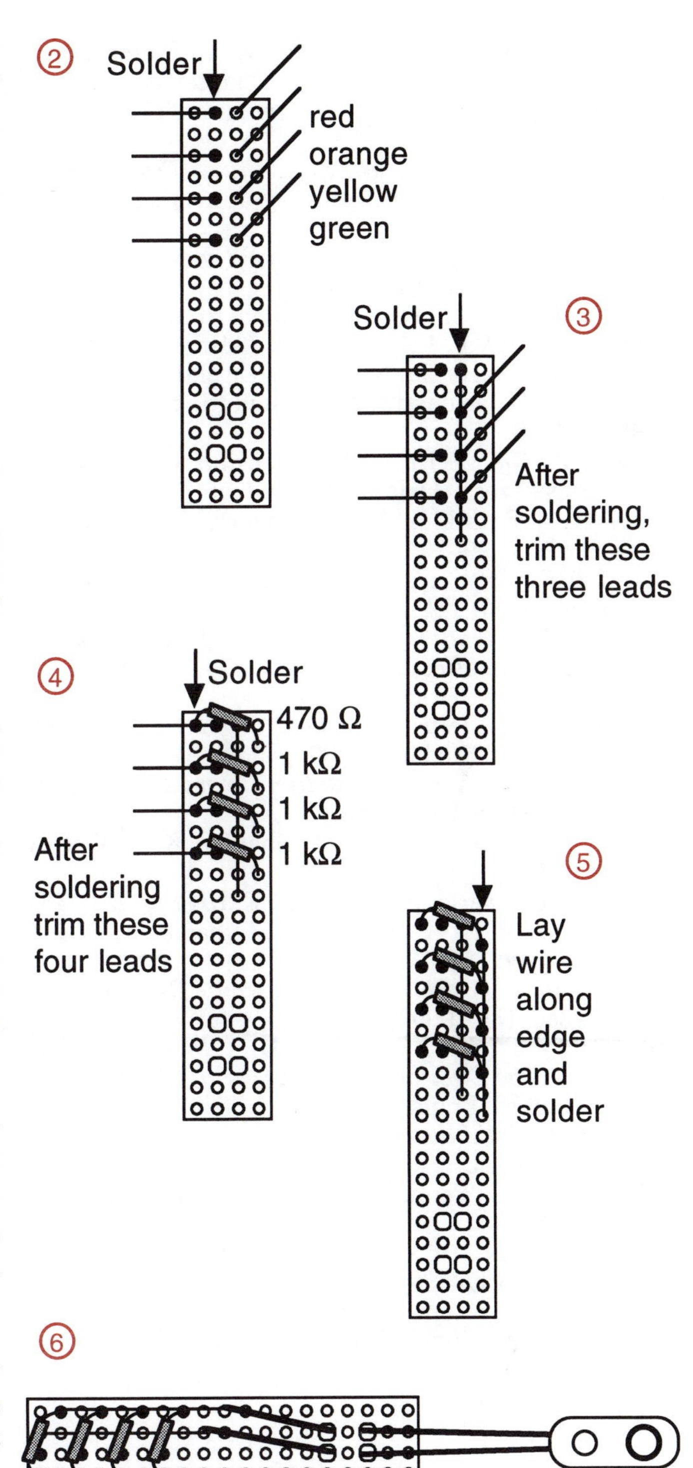

References

1. Honig, J. M. In *Modern Aspects of Solid State Chemistry*; Rao, C. N. R., Ed.; Plenum Press: New York, 1970; Chapter 20.
2. Nassau, K. *The Physics and Chemistry of Color*; John Wiley and Sons: New York, 1983.
3. Guinier, A.; Jullien, R. *The Solid State: From Superconductor to Superalloys*; International Union of Crystallography, Oxford University Press: Oxford, England, 1989; p 102.
4. *CRC Handbook of Chemistry and Physics*, 68th ed.; Weast, R. C., Ed.; CRC Press: Boca Raton, FL, 1987.
5. Shakhashiri, B. Z. *Chemical Demonstrations;* University of Wisconsin Press: Madison, WI; Vol. 4, 1992, p 240. (Also included in this volume are demonstrations of Cu, Ni, Zn and Cr plating.)
6. For a simple demonstration of Cu plating, *see* Herrmann, M. S. *J. Chem. Educ.* **1992**, *69*, 60.
7. Cox, P. A. *The Electronic Structure and Chemistry of Solids*; Oxford University Press: Oxford, England, 1987, Chapters 3 and 4.
8. Shriver, D. F.; Atkins, P. W.; Langford, C. H. *Inorganic Chemistry*; Freeman: New York, pp 99–100.
9. *Landolt-Bornstein New Series*; Madelung, O., Ed.; Springer-Verlag: Berlin, Germany, 1987; Vol. III, p 22a.
10. Huheey, J. E. *Inorganic Chemistry*, 3rd ed.; Harper and Row: New York, 1983.
11. Harrison, W. A. *Electronic Structure and the Properties of Solids*; W. H. Freeman: New York, 1980.
12. Cox, P. A. *The Electronic Structure and Chemistry of Solids*; Oxford University Press: Oxford, England, 1987; p 55.
13. Levin, A. A. *Solid State Quantum Chemistry*; McGraw-Hill: New York, 1977; p 154 and Section 6.4.
14. Thompson, A. G.; Cardona, M.; Shaklee, K. L.; Wooley, J. C. *Phys. Rev.* **1966**, *146*, 601–610.
15. Mathieu, H.; Merle, P; Ameziane, E. L. *Phys. Rev.* **1977**, *B15*, 2048–2052.
16. Staumanis, M. E.; Krumme, J. P.; Rubenstein, M. *J. Electrochem. Soc.* **1977**, *114*, 640–641.
17. Streetman, B. G. *Solid State Electronic Devices,* 3rd ed.; Prentice Hall: Englewood, NJ, 1990; p 60.
18. Deus, P.; Voland, U.; Schneider, H. A. *Phys. Stat. Sol.* (a) **1983**, *80*, K29–K32.
19. Cox, P. A. *The Electronic Structure and Chemistry of Solids*; Oxford University Press: Oxford, England, 1987; Chapter 3.1 has a more detailed discussion.
20. West, A. R. *Solid State Chemistry and Its Applications*,; John Wiley and Sons: New York, 1984.
21. Lisensky, G. C.; Penn, R.; Geselbracht, M. J., Ellis, A. B. *J. Chem. Educ.* **1992**, *69*, 151–156.
22. Baumann, M. G. D.; Wright, J. C.; Ellis, A. B.; Kuech, T.; Lisensky, G. C. *J. Chem Educ.* **1992**, *69*, 89–95.
23. Cox, P. A. *The Electronic Structure and Chemistry of Solids*; Oxford University Press: Oxford, England, 1987; Chapter 6.

Additional Reading

- Cox, P. A. *The Electronic Structure and Chemistry of Solids*; Oxford University Press: Oxford, England, 1987.
- *Semiconductors*; Hannay, N. B., Ed.; Reinhold: New York, 1959.
- Streetman, B. G. *Solid State Electronic Devices* 3rd ed.; Prentice Hall: Englewood, NJ, 1990.

Acknowledgments

This chapter is based on references 21 and 22 and conversations with Thomas Kuech, University of Wisconsin—Madison, Department of Chemical Engineering; Clark Landis, University of Wisconsin—Madison, Department of Chemistry; Nathan Lewis, California Institute of Technology, Department of Chemistry; Andrew Bocarsly, Princeton University, Department of Chemistry, and Allen Adler. Demonstration 7.2 was developed by Thomas Mallouk and Steven Keller, University of Texas—Austin.

Exercises

1. Choose the correct answer. A semiconducting solid solution used in manufacturing an LED has the zinc blende structure and the chemical formula $Al_xGa_{0.25}As_yP_{0.60}$, where
   a. $x = 0.75$ $y = 0.40$
   b. $x = 0.40$ $y = 0.75$
   c. $x = 1.00$ $y = 0.00$
   d. $x = 0.25$ $y = 0.60$
2. Which of the following would be expected to have the same structure and total valence electron count as α-Sn?
   a. InSn
   b. CdSb
   c. CdTe
   d. InTe
3. Choose the correct answer. If a material has a band gap in the infrared portion of the spectrum, it will appear
   a. black
   b. red
   c. green
   d. colorless
4. Which of the following is not a characteristic of a metal?
   a. high reflectivity of light
   b. high thermal conductivity
   c. high electrical conductivity
   d. increasing electrical conductivity with increasing temperature

5. Consider the situation shown in the diagram, in which a solid has a low energy band that is partially filled with electrons (shaded region) that is separated by a large gap from an empty higher energy band. If the solid is treated with a chemical reducing agent that completely fills the lower energy band with electrons, the solid's electrical properties will become
   a. metallic
   b. semiconducting
   c. superconducting
   d. insulating

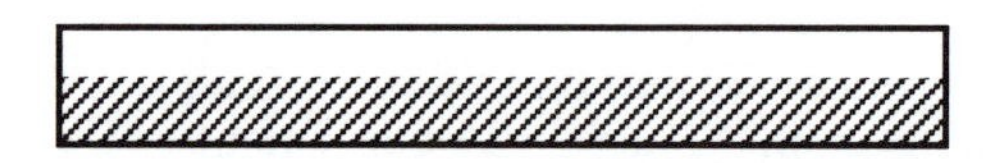

6. Choose the correct answer. A new material will be recognized as a metal if
   a. its electrical resistivity is independent of temperature
   b. its electrical resistivity increases with increasing temperature
   c. its electrical resistivity decreases with increasing temperature
   d. its electrical resistivity is zero

7. Choose the correct answer. An example of a solid that possesses the zinc blende structure is
   a. NaCl
   b. GaAs
   c. CsCl
   d. Zn

8. Calculate the carrier concentration (in carriers per cubic centimeter) in 1 M NaOH, assuming all ions present in the solution are carriers.

9. A coil of copper wire is 150 m long and has a resistance of 150 Ω. If the resistivity of copper wire is 1.67×10^{-6} Ω-cm, find
   a. the cross-sectional area of the wire.
   b. the diameter of the wire.

10. The room-temperature band gap of CdS is about 2.4 eV. What color will a crystal of CdS appear to be?

11. Suppose that you want to create a red cutoff filter (of all of the colors in the visible region of the spectrum, the filter will transmit only red light). What should the band gap be to make such a filter out of a semiconductor?

12. The color emitted from an LED is roughly its band-gap energy. What composition, if any, of GaP_xAs_{1-x} (*see* Figures 7.18 and 7.19) would you use to make an LED that emits light that is
    a. red?
    b. orange?
    c. yellow?
    Can a blue LED be made from this family of solids? Why or why not?

13. Some LED materials can be prepared by combining Ga, In, As, and P in the zinc blende structure. If the formula of one such solid is $Ga_{0.4}In_xAs_yP_{0.7}$, what are x and y equal to, and how would you interpret this formula based on the zinc blende structure?

14. Solar cell **A** uses a semiconductor with a band gap of 1 eV. Solar cell **B** uses a semiconductor with a band gap of 2 eV. Either cell is thick enough to absorb all the light above its band gap. Both cells can work simultaneously with one on top of the other to operate more efficiently in a small space. But which cell has to be on top (first to receive the sunlight) for both to work simultaneously and why?

15. A compound has the following band diagram, with the shading representing occupancy by electrons: The solid has a completely filled lower energy band and a partially filled higher energy band.

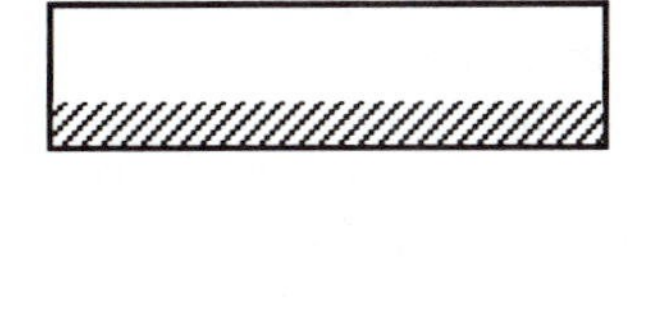

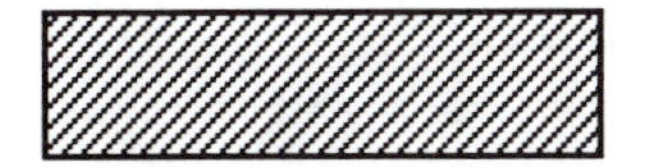

a. Predict the electrical properties of this solid on the basis of its band diagram.
b. What would happen to the electrical properties of this compound if all of the electrons were removed from the conduction band and why? Would you use a chemical oxidizing agent or reducing agent to accomplish this and why?

16. The band gaps of $GaAs_{1-x}P_x$ solid solutions are continuously tunable from about 2.3 eV (GaP) to 1.4 eV (GaAs) at room temperature. The band gap increases linearly with x up to about $x = 0.45$ where the band gap is 2.0 eV. Assuming that the wavelength of emitted light from the solid solutions is at about the band-gap energy, calculate the value of x (i.e., what solid solution would you prepare) to obtain emission at 7000 Å from an LED.

17. Name two solids with the zinc blende structure that are isoelectronic with α-Sn, and predict how their band gaps will compare to that of α-Sn.

18. Suggest a two-element (binary) compound that is isoelectronic with diamond and that might rival it in hardness.

19. Explain why $CdSnP_2$ has the same valence electron count as GaAs.

20. Draw figures similar to those in Chapter 5, Figure 5.12A or 5.12B , for the contents of the planes in the unit cell at $z = 0, 0.25, 0.5,$ and 0.75 for the semiconductors Si and GaP.

21. Explain why the partially filled band in sodium metal is exactly half filled.

22. Explain why the conductivity of a semiconductor is strongly temperature dependent.

23. Compare the electrical conductivities of semiconductors at room temperature and liquid helium temperature.

24. Which contain partially filled bands and why: Mg, Si, NaCl?

25. Why does the reaction of a small amount of bromine with graphite convert the semiconductor to a metallic conductor, but the reaction of bromine with sodium converts the metal into an insulator?

26. Consider the molecule polyacetylene, which consists of an array of the repeat unit $(-CH{=}CH-)_n$ attached end to end and extending to infinity.
   a. Draw a portion of the molecule (ignoring the ends of the molecule).
   b. What is the hybridization about each carbon? How many electrons on each carbon are used to form the sigma bond network in the molecule?
   c. What orbital is left over on each carbon? How many electrons are left over per carbon? They will form the pi bond network on the molecule. Draw a second (resonance) structure for polyacetylene. What effect will this have on the structure of the molecule?
   d. These remaining orbitals and electrons can be considered to form a one-dimensional band in polyacetylene. Use a rectangle to represent the band. Would the band be full? Empty? Half full? Explain. If polyacetylene had this structure, would it be a metal, semiconductor, or insulator? Explain.
   e. For reasons that will not discussed here (*see*, for example, reference 23), the structure of polyacetylene contains alternating short and long bond lengths and has the band structure shown below. On the basis of this information, is polyacetylene a metal, semiconductor, or insulator? Explain.

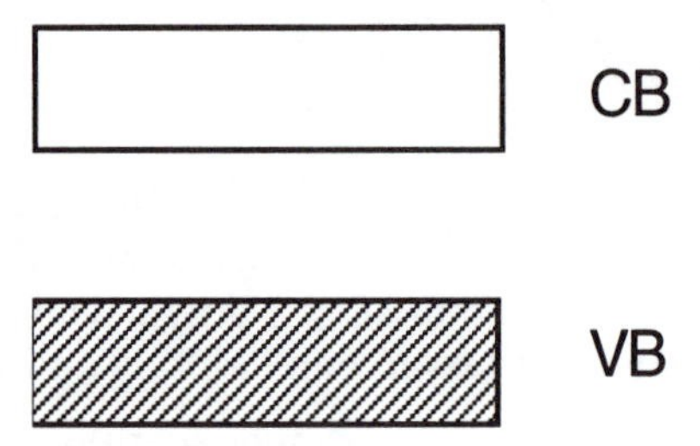

   f. On the basis of this band diagram, what would happen to the conductivity if polyacetylene were allowed to react with a small amount of an oxidizing agent such as Br_2? What would happen to the conductivity if polyacetylene were allowed to react with a small amount of a reducing agent such as Li?

27. Another compound being used for LEDs is derived from the chalcopyrite structure.

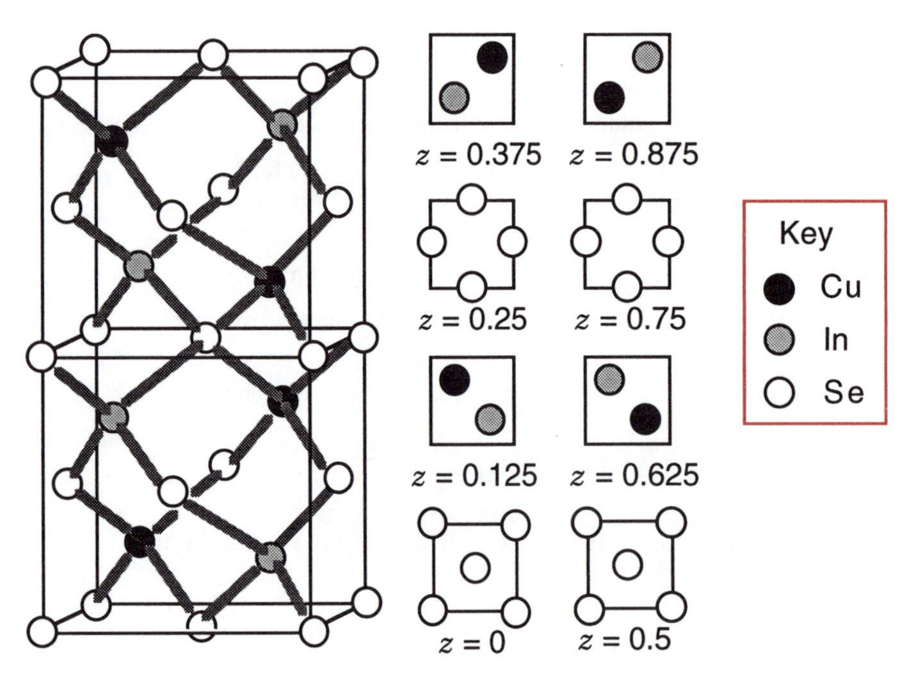

a. To what empirical formula does this compound correspond?
b. How is this structure related to the zinc blende structure?
c. Show that this structure is isovalent with the zinc blende family of semiconductors used in LEDs.

28. How would the absorption and emission spectra of a powdered 1:1 GaP and GaAs physical mixture differ from the spectra that would be seen for a $GaP_{0.5}As_{0.5}$ solid solution? (Other combinations that the instructor could use include $CdS_{0.5}Se_{0.5}$ and $Al_{0.5}Ga_{0.5}As$.)

29. Sketch the most bonding and most antibonding orbitals in a band formed by p_z orbitals in a one-dimensional solid where the atoms lie on the *z* axis.

30. The luminescence spectra of a crystal of CdSe are shown as a function of sample temperature. (Luminescence spectra are obtained by exciting the sample with light of energy greater than the band gap, and the spectrum obtained is of the light emitted by the sample as its excited electrons return from the conduction band to the valence band as in Figure 7.21.)
a) Why does the peak maximum shift as the sample is heated?
b) Why does the intensity decrease as the sample is heated?

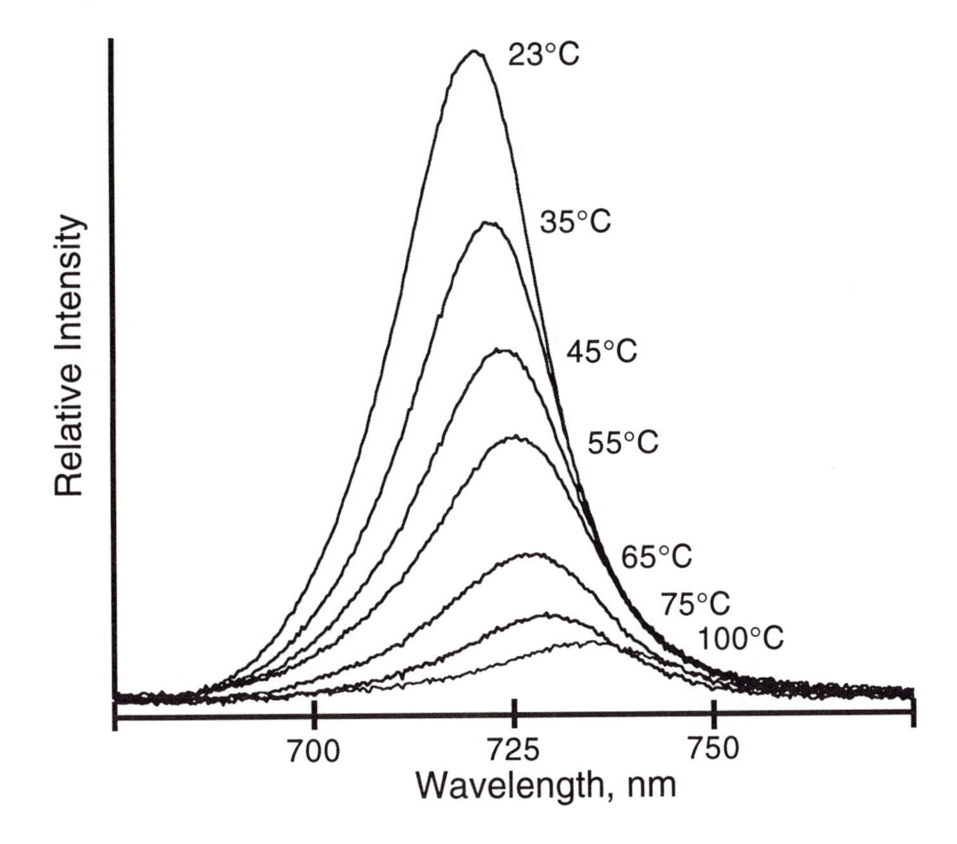

31. Estimate the shift in band gap energy (in eV) between 300 and 77 K for $Al_{0.05}Ga_{0.95}As$ from the data in Figure 7.24. (Any of the other compositions in the figure may also be used.)

Chapter 8

Chemical Equilibrium: Acid–Base and Redox Analogies in Solids

The basic principles that apply to equilibria in solution can be used to gain insight into the behavior of important classes of solids like semiconductors and ion conductors. In this chapter we introduce the notion that traditional, wet acid–base equilibria and Nernstian redox chemistry have counterparts in the solid state. For example, doping aluminum into silicon is analogous to dissolving HCl in water, and a light-emitting diode is analogous to an electrochemical concentration cell.

The electronically and ionically conducting solids that form the basis for many of these analogies illustrate not only chemical equilibria, but the application of several other concepts, including the solutions to the Schrödinger equation for the hydrogen atom and periodic properties. Moreover, the chemical principles that govern these solids are readily connected with common devices like solar cells, light-emitting diodes, diode lasers, gas sensors used in catalytic converters, and ion sensors.

Electrons and Holes

One way of describing conductivity in a semiconductor such as silicon is to use a localized bonding picture (Figure 8.1A). In this model, the focus is on the bonds that bind each silicon atom to four other silicon atoms. Because the silicon–silicon bonds are relatively strong, the electrons are held fairly tightly and are not mobile. However, a few electrons can be excited out of the covalent bonds by absorption of thermal energy (or by light of at least band-gap energy; *see* Chapter 7), giving rise to small electrical conductivities near room temperature. A representation is shown

2725–X/93/0237$13.80/1

in Figure 8.2, in which the arrows show electrons (e^-) that have been freed to roam throughout the silicon crystal.

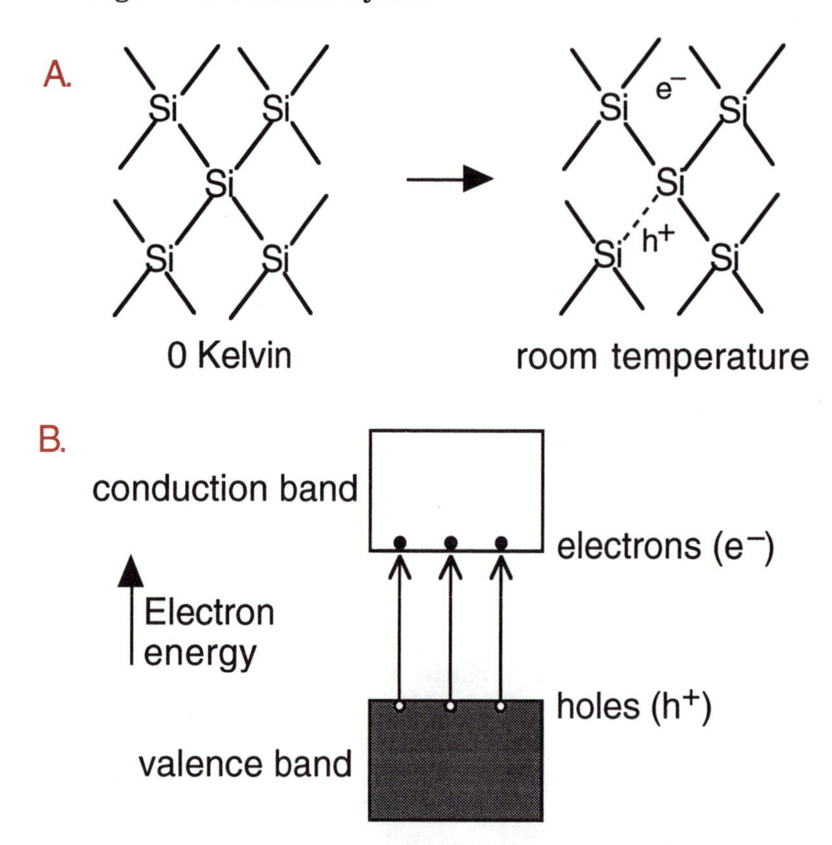

Figure 8.1. A: A localized bonding picture of a semiconductor. In the absence of heat or light, the valence electrons are localized in two-electron Si–Si bonds. However, absorption of heat or light energy produces some mobile electrons (e^-) and an equivalent number of mobile holes (h^+). The one-electron bond resulting from creation of the electron–hole pair is indicated with a dashed line. B: The band structure of a semiconductor using a delocalized bonding picture. Electrons (filled circles) may be excited from the valence band by absorption of light or thermal energy. Holes (open circles) are left behind in the valence band.

Mobile electrons and holes (*see* next paragraph) are collectively called *carriers* because they are the species responsible for the conductivity of a semiconductor. They "carry" the electric current in the semiconductor.

Equally important in this representation are the one-electron bonds resulting from the creation of the mobile electrons. One of the most powerful formalisms in discussing solids is to treat missing electrons as particles called "holes"; they are symbolized by h^+. The mobility of the holes in this localized bonding picture is illustrated in Figure 8.2, which shows electrons from adjacent two-electron bonds moving into the holes: this movement restores one-electron bonds to two-electron status, but makes an equal number of what were two-electron bonds become new one-electron bonds. This process thus conserves the number of holes and effectively moves the holes throughout the crystal.

The positively charged nature of the hole is appreciated by imagining the effect of a voltage that is applied to the sample, as shown in Figure 8.2: the negatively charged electrons that are actually moving would be attracted to the positive terminal, but this is equivalent to thinking of the holes as moving toward the negative terminal. This can be modeled in the classroom as shown in Figure 8.3.

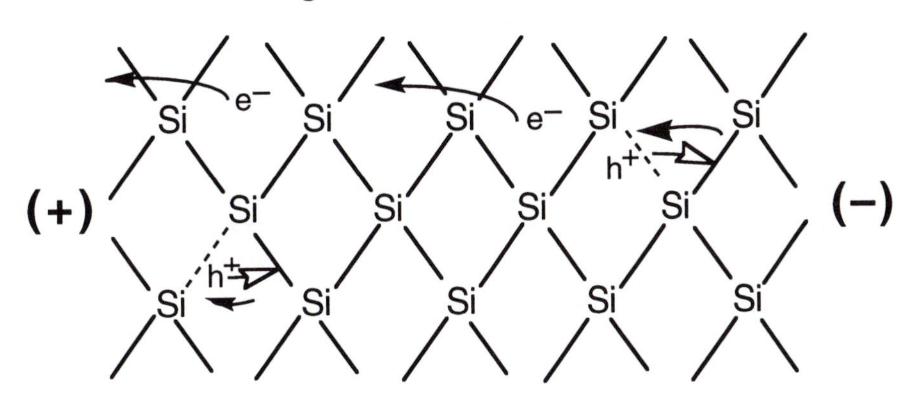

Figure 8.2. If a voltage is applied to a crystal of semiconducting material, the electrons migrate toward the positive terminal and the holes migrate toward the negative terminal. Two kinds of electron motion are illustrated. Excited electrons (e^-) move through the crystal toward the positive terminal. In addition, an electron in a bonding pair can jump into a one-electron bond (hole). This results in net motion of the electrons (dark arrows) and holes (open arrows) in opposite directions.

Figure 8.3. Hole mobility can be modeled by students. Five students are electrons and one empty chair is a hole. Having each student move one chair to the right (dark arrows) produces a result that is equivalent to the empty chair moving to the left (open-tipped arrows).

Another way of describing the conductivity in silicon is to use a delocalized bonding picture (Figure 8.1B). In Chapter 7, we noted that semiconductors are characterized by a filled valence band that is separated by a band gap from an unfilled conduction band at higher energy. Ambient thermal energy promotes a few energetic electrons across the band gap, and this process becomes more favorable with an increase in temperature and a decrease in band gap. Each electron promoted to the conduction band becomes a mobile electron and results in an electron missing from the valence band; a missing electron is also treated as a mobile hole in the valence band. To compare with the localized bonding picture, the promotion of electrons can be thought of as removing electrons from the delocalized bonding orbitals (the valence band) connecting the atoms in the crystal.

The energy scales of electrons and holes are inversely related. By convention, in energy diagrams the energies of electrons increase vertically upward on the page (Figure 8.1B). In contrast, hole energies decrease in the upward direction, giving rise to the analogy that holes are like bubbles in that they float or rise to reach lower energy. This effect can be demonstrated as described in Demonstration 8.1.

Demonstration 8.1. Hole Energies and Conductivity

Materials

Stoppered test tube, nearly filled with glycerine

Procedure

- Invert the test tube.
- A bubble will slowly move from the bottom to the top of the test tube. By way of analogy, holes are like bubbles in a fluid. They are the absence of something, in this case the absence of electrons. The shift in contents can be explained as a movement of the bubble or a movement of the fluid.

  The entire tube may also be viewed as a band, with the fluid being electrons and the air being holes. Inverting the tube animates the phenomenon that electrons (fluid) move to the bottom of the band (tube) to achieve lowest energy, or equivalently, holes (air) move to the top of the band (tube) to reach lowest energy.

Demonstration 8.2. A Classroom Model of Hole Conductivity

Materials

A row of chairs or desks with students in all but one of them

Procedure

- Arrange a row of students so that the empty seat is at one end of the row (see Figure 8.3).
- Have an instructor or student stand at each end of the row, with one of these two individuals representing the positive terminal of a battery (this is the person next to the empty seat) and the other the negative terminal.
- Instruct the student in the seat next to the empty one to move into it. Then tell the next student to move into the newly empty one. Repeat until all students have moved by one seat.

  In this model, the students are analogous to valence-band electrons. As the students shift seats, moving toward the individual who represents the positive terminal, it can be pointed out that, simultaneously, the empty seat is a hole that is moving toward the negative terminal. A more extensive analogy in which the entire classroom is regarded as a band (and a nearby room as an adjacent band) recently was published *(1)*.

The Autoionization Analogy

A solid like silicon can be regarded as a medium, somewhat analogous to water. Just as autoionization in water produces H^+ and OH^- ions, which can be regarded as mobile defects in the solvent, the analogous autoionization process in silicon produces a mobile electron and a mobile hole. In the delocalized bonding picture, promotion of electrons produces conduction band electrons, e^-, and valence-band holes, h^+, in equal numbers, just as dissociation of water molecules produces protons and hydroxide ions in equal numbers. In the localized bonding picture, the equations for autoionization in the two media are given by equations 1–2:

$$\mathrm{H:O:H} \rightarrow \mathrm{H^+} + \mathrm{OH^-} \qquad (1)$$

$$-\overset{|}{\underset{|}{\mathrm{Si}}}:\overset{|}{\underset{|}{\mathrm{Si}}}- \;\rightarrow\; -\overset{|\,\oplus}{\underset{|}{\mathrm{Si}}}\cdot\overset{|}{\underset{|}{\mathrm{Si}}}- \;+\; e^-$$

$$\text{or} \qquad (2)$$

$$-\overset{|}{\underset{|}{\mathrm{Si}}}:\overset{|}{\underset{|}{\mathrm{Si}}}- \;\rightarrow\; h^+ \;+\; e^- \quad \text{where} \quad -\overset{|\,\oplus}{\underset{|}{\mathrm{Si}}}\cdot\overset{|}{\underset{|}{\mathrm{Si}}}- \quad \text{is } h^+$$

In equation 2, the symbol $-\overset{|}{\underset{|}{Si}}\cdot\overset{\oplus\ |}{\underset{|}{Si}}-$ represents the one-electron silicon–silicon bond—the "hole"—that results from ionization of a bonding electron. Its shorthand notation, h^+, reflects the fact that, as noted, the hole is mobile and can move away from the site of its generation.

The process of dissociation in water is countered by the very high likelihood that when H^+ and OH^- ions encounter each other in aqueous solution, they will recombine to make water molecules. A similar process, the recombination of electron–hole pairs, occurs in the solid. As shown in Figure 8.4A, when an itinerant electron encounters a one-electron silicon–silicon bond, a hole, the normal two-electron bond is restored. When a mobile conduction-band electron recombines with a mobile valence-band hole, both carriers are annihilated and no longer contribute to electrical conductivity in the solid. The delocalized bonding view of the same event is shown in Figure 8.4B:

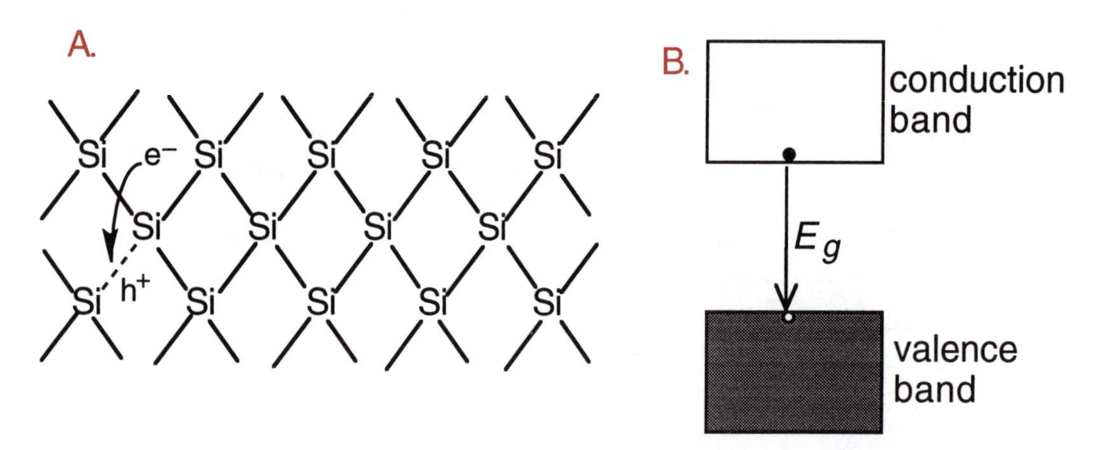

Figure 8.4. Recombination of the electron–hole pairs annihilates both carriers and releases energy that is roughly equivalent to the band-gap energy. This process may be viewed (A) as a free electron encountering a one-electron bond, a hole (localized picture); or (B) as an electron falling from the conduction band back to the valence band (delocalized picture).

With silicon, as with water, the rates of dissociation and recombination are equal at thermal equilibrium in the dark (light having greater than band gap energy creates additional electron–hole pairs; *see* Chapter 7.) The processes in equations 1 and 2 may be replaced by an equilibrium double arrow.

How are these processes quantified? In water, the equilibrium constant K_W is used as a measure of the extent of ionization at equilibrium,

$$K_W = [H^+][OH^-] \tag{3}$$

At room temperature K_W is 1×10^{-14} M^2, which corresponds to 4×10^{27} ions/cm^6, when concentrations are given as numbers of ions per cubic centimeter. The concentrations $[H^+]$ and $[OH^-]$ are, of course, equal to one another and to the square root of K_W, yielding values for them of ~1×10^{-7} M or ~6×10^{13} ions/cm^3 at room temperature.

The origin of the small value of K_W (corresponding to a large, positive value of $\Delta G^o = \Delta H^o - T\Delta S^o$; $K = e^{-\Delta G^o/RT}$, where ΔG^o, ΔH^o, and ΔS^o are the standard free energy, enthalpy, and entropy for a reaction with equilibrium constant K) is noteworthy: The dissociation reaction is endothermic with a strongly positive ΔH^o value (the exothermic nature of the reverse reaction, the neutralization reaction, is commonly demonstrated in discussions of thermochemistry) and has a substantial negative ΔS^o value, reflecting the fact that water molecules order themselves in solvation spheres around these ions. The autoionization that does occur at room temperature is due to the high entropy of a *dilute* solution of protons and hydroxide ions.

In extending the analogy between water and a semiconductor, measurements of electrical conductivity can be used to establish concentrations, in units of reciprocal cubic centimeter, of valence band holes, denoted p or $[h^+]$, and conduction band electrons, denoted n or $[e^-]$. In the pure solid "solvent," $p = n$, and the product of the two is also governed by chemical equilibrium:

$$K = [h^+]\,[e^-] = p \times n \tag{4}$$

The value of K, the equilibrium constant for the autoionization of silicon at room temperature, has been determined to be about 2×10^{20} cm^{-6}, meaning that $p = n = K^{1/2} \sim 1 \times 10^{10}$ cm^{-3}. These values of n and p correspond to a very small degree of autoionization in silicon (eq 2), as can be seen by comparing them to the density of atoms in silicon, which is on the order of 10^{22} cm^{-3}: ionization occurs only for about one silicon–silicon bond in 10^{12}! Room temperature values of K for several other semiconductors are given in Table 8.1.

Table 8.1 Values of K at 300 K for Semiconductors

Semiconductor	K (cm^{-6})[a]	Band Gap
GaAs	4.0×10^{12}	138 kJ/mol (1.43 eV)
Si	2.2×10^{20}	107 kJ/mol (1.11 eV)
Ge	6.2×10^{26}	66 kJ/mol (0.68 eV)

[a]Equilibrium constant for autoionization.
SOURCE: Data are taken from references 2 and 3.

As with water, the dissociation is endothermic, and what little autoionization there is in the semiconductors of Table 8.1 arises because of the entropy increase associated with the creation of a dilute "solution" of electrons and holes; there is no counterpart in semiconductors analogous to the ordering of solvent molecules that occurs in the aqueous autoionization reaction.

The direction of energy flow in the semiconductor autoionization reaction can be appreciated by considering the reverse reaction: Recombination of electrons and holes (Figure 8.4), like the neutralization reaction in water, yields energy, which can be released in the form of heat and/or light (*see* the discussion of light-emitting diodes (LEDs) in Chapter 7 and following herein). Table 8.1 shows that a decrease in the band-gap energy, a measure of the energy cost associated with autoionization, corresponds to an increase in K (K is proportional to $e^{-E_g/RT}$; the band gap E_g is an internal energy).

The endothermicity of the semiconductor autoionization reaction is also established from the dependence of the equilibrium constant or carrier concentration ($n = p = K^{1/2}$) on temperature. Figure 8.5 makes a direct comparison between the autoionization equilibria of water and the semiconductors silicon, germanium, and gallium arsenide by showing the value of the carrier concentration ($[H^+] = [OH^-]$ for water; $n = p$ for the semiconductors) as a function of reciprocal temperature. The negative slopes demonstrate that these are endothermic reactions that can be used as illustrations of Le Chatelier's principle: Increased temperature will shift the autoionization equilibria to the right, as the system tries to relieve the stress from the additional thermal energy by creating more charge carriers.

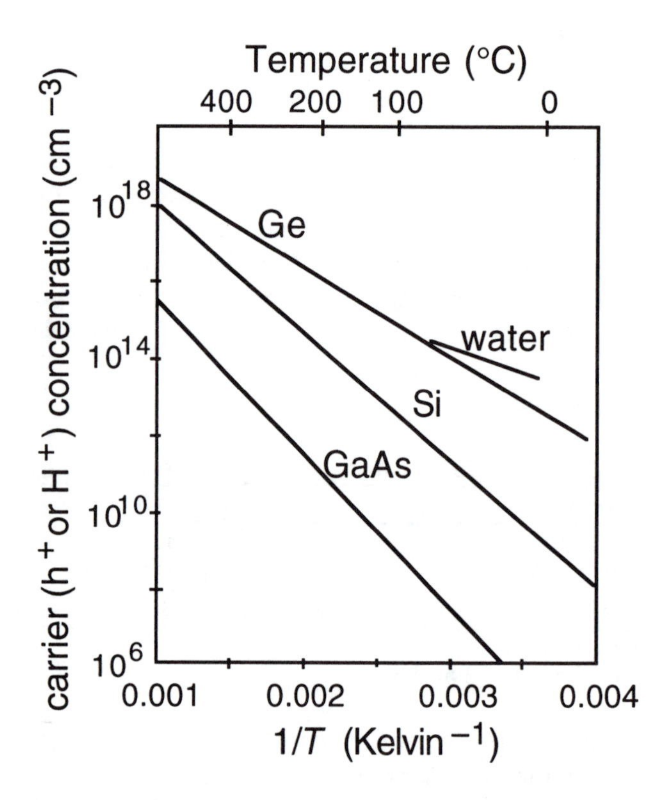

Figure 8.5. A comparison of semiconductor hole concentration, p, and $[H^+]$ in pure water versus $1/T$. (Adapted from references 2 and 3.)

Doping with Electron Donors and Acceptors

Given the analogy between pure water and pure semiconductors as equilibrium-governed solvents, it is natural to wonder whether the relative concentrations n and p can be altered. Bases and acids are routinely added to water to increase [OH^-] or [H^+], respectively, at the expense of the other: The analogous behavior in semiconductors involves adding atoms having additional valence electrons relative to the host crystal atoms or adding acceptor atoms having fewer valence electrons relative to the host crystal atoms.

Deliberately adding impurity atoms, called "dopants," to substitute for atoms that normally comprise the crystal can often be accomplished by incorporating the dopant atoms during the growth of the solid, or afterward, by heating the solid in the presence of a volatile source of the dopant; diffusion of the dopant into the material is enhanced at high temperatures. Because of the small degree of autoionization, often only very small amounts of the dopant, on the order of parts per million, are needed to significantly alter the electrical properties of the solid.

The simplest cases of control of p and n involve covalent tetrahedral semiconductors like silicon. In choosing a dopant that might increase the concentration of conduction-band electrons n at the expense of the concentration of valence-band holes p (recall eq 4, that $p \times n = K$, a constant at a given temperature), a logical strategy is to choose a dopant that is about the same size as the atom to be replaced, so it readily fits into the structure, and that has an additional valence electron. Inspection of the periodic table suggests that phosphorus and arsenic would be good choices.

The localized bonding model of semiconductors may be used to explain the effects of doping. The Lewis structure of silicon doped with a phosphorus atom, shown in Figure 8.6A, suggests qualitatively why phosphorus is a donor atom. The extra valence electron of phosphorus is easily ionized, because it is not needed to bond the phosphorus atom to its four neighboring silicon atoms in the tetrahedral geometry of the atoms in the crystal. Instead, these electrons are free to roam throughout the solid, increasing n. The reaction may be generally written as

$$D \rightarrow D^+ + e^- \quad (5)$$

where D represents a donor atom in a semiconductor. A mobile electron and an immobile cation are produced by each ionization event.

In the delocalized band picture, this ease of ionization is described by the introduction of a donor energy level near the conduction-band edge. (A quantitative estimate of the energy of the donor level relative to the conduction-band edge is given in the next section.) Consequently, thermal energy is sufficient to promote a large fraction of these donors' excess electrons into the conduction band. This situation is illustrated in Figure 8.6B.

In a similar manner, atoms like aluminum or gallium, having one fewer valence electron than silicon but of similar size, are logical acceptor-type dopants for increasing the valence-band hole concentration p at the expense of n. Again, the localized bonding picture may be used to explain the effects of doping with aluminum. In the localized picture, Figure 8.6C, an aluminum atom, with three valence electrons, fits into the silicon lattice by making three two-electron bonds to each of three neighboring silicon atoms; but must make a one-electron bond to its fourth silicon neighbor. This one-electron bond can be converted to a two-electron bond by an electron from a neighboring two-electron silicon–silicon bond, leading to a mobile hole. The reaction may be written

$$A \rightarrow A^- + h^+ \tag{6}$$

where A represents an acceptor atom in a semiconductor. Thus, a mobile hole and immobile anion result from the ionization process.

In the delocalized picture, the unfilled orbitals on aluminum lie just above the filled valence band. As such, thermal energy is sufficient to promote electrons out of the filled valence band into these energy levels, simultaneously producing valence-band holes. This process is illustrated in Figure 8.6D.

The Hydrogen Atom Model for Estimating Dopant Ionization Energies

Figures 8.6A and 8.6C suggest a simple model for quantitatively estimating the ionization energies of dopants. The extra valence electron of a phosphorus-donor dopant atom that is not used for covalent bonding is held to the phosphorus atom through electrostatic attraction in much the way that the electron of a hydrogen atom is held to the nucleus; in both cases, ionization corresponds to freeing the electron and leaving behind a center of positive charge. For the hydrogen atom, allowed energies for the electron, determined from the Schrödinger equation (or from the earlier Bohr model), are commonly presented in introductory courses to illustrate quantization. The models may be applied to give a rough estimate of dopant ionization energies in semiconductors, as well.

Equation 7 describes the allowed energies for the hydrogen atom, where e is the charge on an electron, m is the mass of an electron, ε_o is the vacuum permittivity, ε_r is the relative permittivity of the medium ($\varepsilon_r = 1$ for a vacuum), n is the principal quantum number, and h is Planck's constant.

$$E = \frac{-e^4 m}{8\,\varepsilon_o^2\,\varepsilon_r^2\,h^2\,n^2} = \frac{-2.18 \times 10^{-18}\ \text{J}}{n^2} \tag{7}$$

In most introductory chemistry texts this equation appears as $E = -(\text{constant})/n^2$, because in the hydrogen atom, all of the terms except n are constants and can be combined into one constant. Equation 8 is the equation for the Bohr radius, the most probable distance from the nucleus for finding the electron.

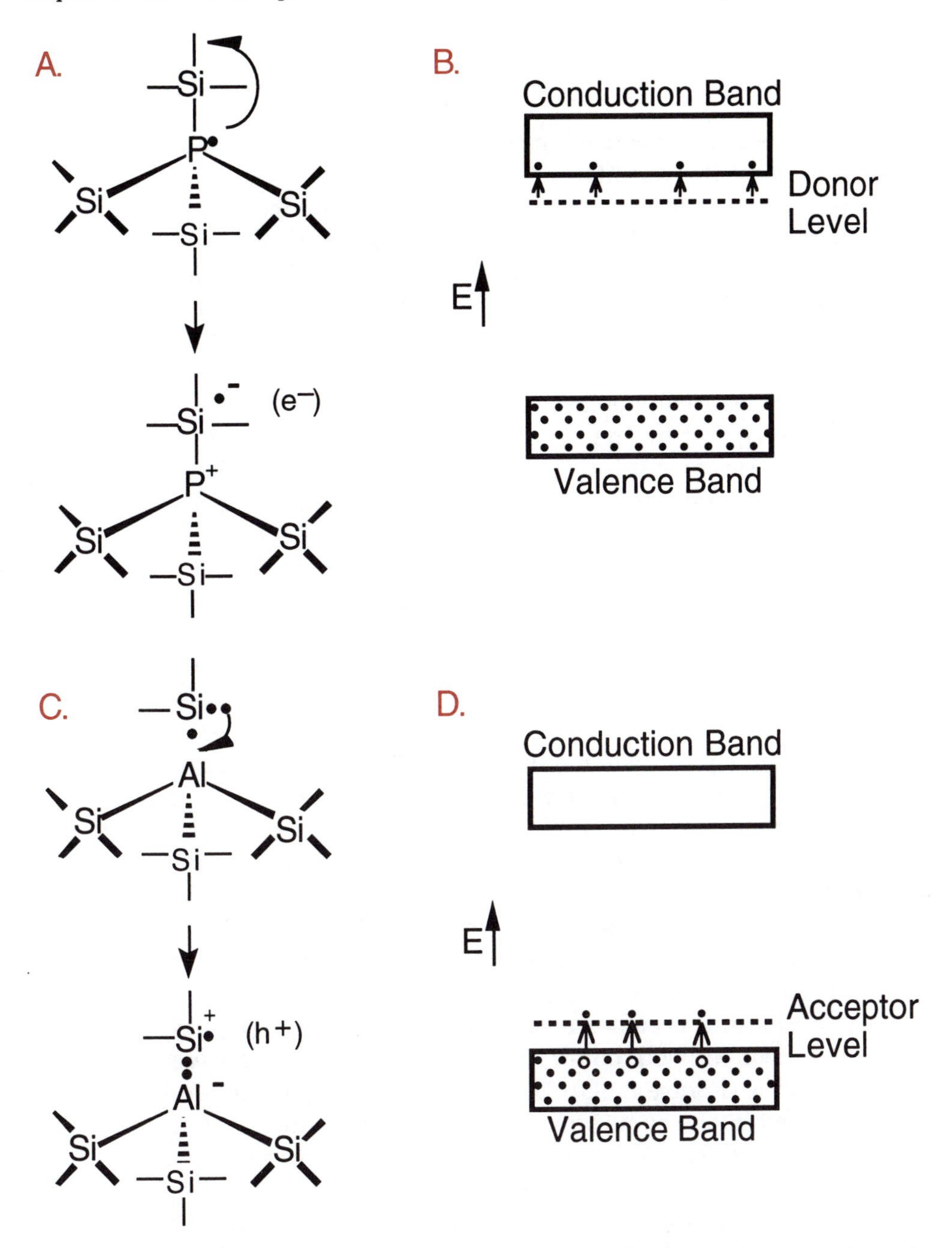

Figure 8.6. A: A localized bonding picture showing the effect of adding phosphorus to a silicon crystal. B: The addition of a donor atom such as phosphorus introduces electrons in donor levels that are close to the conduction band. As such, the electrons (filled circles) are easily promoted into the conduction band. C: A localized bonding picture showing the effect of adding aluminum to a silicon crystal. D: The addition of aluminum to a silicon crystal introduces acceptor levels that lie just above the valence band. Thermal energy is sufficient to promote electrons from the valence band into these levels, creating holes (open circles) in the valence band.

$$r = \frac{n^2 h^2 \varepsilon_o \varepsilon_r}{\pi e^2 m} \quad (8)$$

This equation usually appears as $r = a_o n^2$ (after collection of the constants into a_o, the radius of the first Bohr orbit).

The hydrogen-atom model can be applied to this doped semiconductor with two modifications: First, unlike the vacuum assumed in the space between the proton and electron of the hydrogen atom, the solid has intervening atoms that shield the electron from the nucleus. This shielding reduces the attraction of the nucleus for the electron by the square of the dielectric constant of the medium (the dielectric constant is a measure of how easily the medium is electrically polarized). In a vacuum, ε_o is used (ε_r = 1); in a solid, ε_o is multiplied by the relative dielectric constant of the solid. (The relative dielectric constant, ε_r, is on the order of 12 for silicon.) This approach leads to an approximately hundred-fold reduction in the electrostatic attraction. Second, the kinetic response of electrons to electrical forces indicates that electrons behave as though they have a mass, called the "effective mass," that is usually somewhat smaller than the value of their mass in vacuum. Consequently, m^*, the effective mass of the electron, replaces m in equation 8. With these two changes, essentially the same equations can be used for this donor-doped semiconductor as for the hydrogen atom.

The combined effect of these two modifications is that instead of requiring 13.6 eV to remove an electron from a hydrogen atom in a vacuum (or 10.5 eV to remove an electron from a phosphorus atom in vacuum), a phosphorus atom doped into a silicon crystal has an ionization energy that is far lower, by more than a factor of 200; the measured value is ~0.044 eV.

The case of aluminum doped into silicon is equivalent, except that a positively charged hole is to be ionized from a negatively charged acceptor: When an electron from a neighboring silicon–silicon bond is trapped at an aluminum atom, the aluminum becomes a center of fixed negative charge with a one-electron silicon–silicon bond, the positively charged hole, electrostatically bound to it (Figure 8.6C). The allowed energy levels and Bohr radius are calculated in the same way, and a similar result is found. Experiments show that only 0.057 eV is needed to ionize the holes provided by aluminum atom dopants in silicon. Moreover, the first Bohr radius typically expands from 0.53 Å in hydrogen to tens of angstroms, spanning many atoms, for common semiconductor dopants *(4)*. Ionization energies of common donors and acceptors in silicon are summarized in Figure 8.7.

Figure 8.7 shows the placement of Group 15 donors like phosphorus near the conduction-band edge and of Group 13 acceptors like aluminum near the valence-band edge. The small amount of energy needed to ionize these dopants means that at room temperature the amount of thermal energy available, ~0.025 eV, is sufficient to ionize a substantial fraction of them.[1]

[1]Effects due to interactions among dopants at common doping levels can effectively reduce the band gap and bring the dopant levels even closer to the band edge.

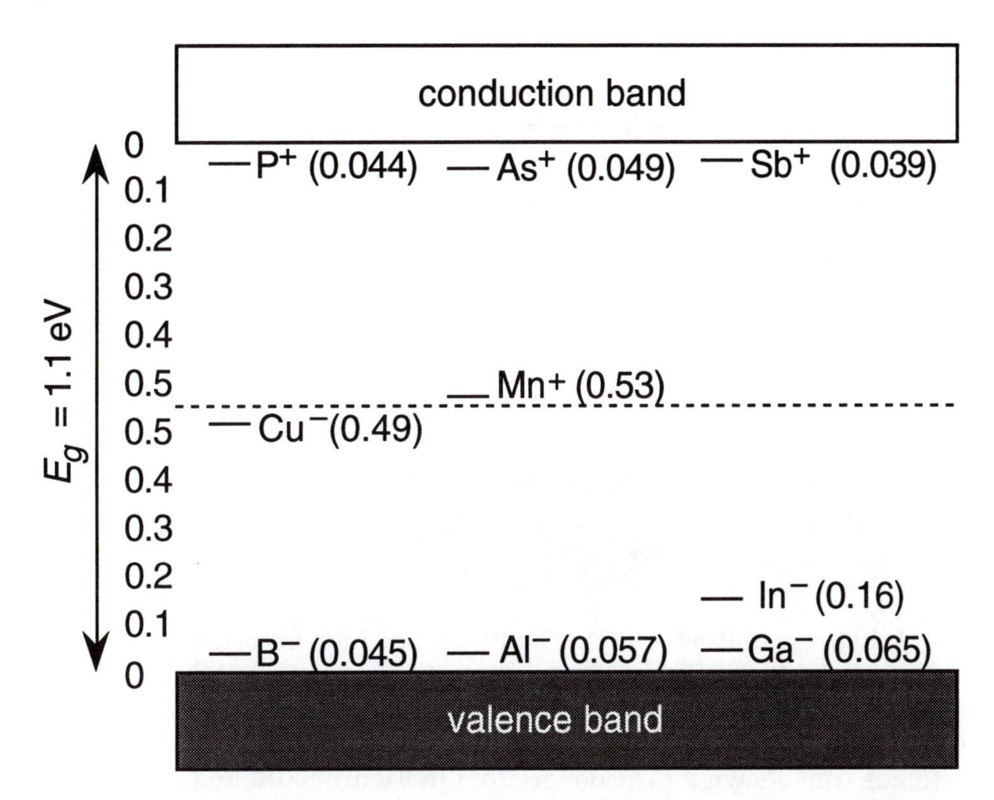

Figure 8.7. Energy levels for isolated impurities in silicon, with the ionization energies in parentheses. The energies are measured from the nearest band edge and are reported in electron volts. Donors are indicated with a plus charge and acceptors with a minus charge, their charges after ionization leading to conduction-band electrons and valence-band holes, respectively. (Adapted from reference 5.)

The celebrated absorption and emission spectra of the hydrogen atom, which arise from electronic transitions between the quantized electronic energy levels of the atom, also have counterparts in doped semiconductors. Similar energy levels in the band-structure diagram for semiconductors, are shown in Figure 8.8 for silicon doped with donor atoms. The energies of the ground state of the dopant (the state corresponding to the principal quantum number n = 1; this is the state shown in diagrams like Figure 8.7) and its excited states (n = 2, 3, etc.) are located relative to the conduction-band edge, which represents n = infinity and corresponds to ionization. Because of the minimal thermal energy needed to ionize the dopants, extremely low temperatures are required to observe these transitions spectroscopically *(6)*.

Figure 8.8. A hydrogen-atom analogy may be used to describe dopants in semiconductors. Superimposed on the normal band diagram for a donor-doped semiconductor are several discrete energy levels that correspond to excited states for the donor's bound electron. The ground state is that shown for the donor atom in Figure 8.6B, and promotion of the donor's electron into the conduction band is equivalent to ionization. In this figure, the separation between the energy level of the donor atom and the conduction band has been expanded to show the energies of the excited states more clearly. Like the hydrogen atom, a doped semiconductor would be expected to show narrow spectral bands due to electronic transitions between discrete energy levels; these bands can be observed at very low temperatures. A similar diagram could be drawn for an acceptor-doped semiconductor.

Dopants for Compound Semiconductors

For compound semiconductors like GaAs, the same principles involving dopant size and valence-electron count can be utilized to alter the solid's electrical properties. For example, substitution of a comparably sized Group 16 element like Se or Te for As leads to an n-type material, because the dopant is a donor with one extra valence electron relative to As. Substitution of a comparably sized Group 12 element like Zn for Ga produces p-type material, because the Zn atom, with two valence electrons, is an acceptor relative to Ga. An interesting case is provided by Group 14 dopants, which would be predicted to serve as donors if they substitute for Ga and as acceptors if they substitute for As. Both types of situations occur: in GaAs, C is found to be an acceptor and Sn is found to be a donor.

The Semiconductor as an Acid–Base System

In addition to water, many other solvents autoionize with the formation of cationic and anionic species that are regarded as the strongest acidic and basic species that exist in the medium. For example, NH_3 dissociates into NH_4^+ and NH_2^- ($K \sim 10^{-26}$); and H_2SO_4 into $H_3SO_4^+$ and HSO_4^- ($K \sim 10^{-4}$). A species that increases the characteristic cation concentration of the solvent is regarded as an acid, and a species that increases the characteristic anion concentration as a base. Scales analogous to the pH scale of water can be constructed with the neutral point occurring where the concentrations of the cation and anion are equal.

Applying this kind of definition to a semiconductor like silicon, the mobile hole can be regarded as the acid species and the mobile electron can be regarded as the basic species. Dopants can be classified on the basis of the way they affect the relative concentrations of these electrons, n, and holes, p. In the sections that follow, the analogy will be made that the holes correspond to like-charged protons (h^+ is like H^+); and that the electrons correspond to like-charged hydroxide ions (e^- is like OH^-). An alternative analogy that emphasizes the correspondence of electrons in the semiconductor with protons in water—the species that are physically transferred in the two media—can also be made.

Effect of Doping on Conductivity

We can elaborate on the semiconductor–acid–base analogy by considering species in both media that strongly dissociate. The species in semiconductors that best mimic strong acids and bases, which fully ionize when placed in water, are acceptor and donor dopants whose energy levels lie very near the valence- and conduction-band edges, respectively (Figure 8.7). These so-called "shallow dopants" can be ionized to a substantial extent by thermal excitation at room temperature: Electrons are excited from the donor orbitals into the conduction band to produce carriers in the conduction band; or from the valence band into the acceptor orbitals to create holes in the valence band.

The behavior of silicon illustrates why small concentrations of dopants can have an enormous effect on electrical conductivity, which depends on the density of charge carriers (Chapter 7). As noted, pure silicon has equal but small numbers of electrons and holes carrying charge in the conduction and valence bands, respectively, with $n = p = \sim 1 \times 10^{10}$ cm^{-3}. If only one in a million silicon atoms is replaced with a phosphorus atom, then at temperatures at which most of the donors are ionized, the value of n leaps to $\sim 10^{17}$ cm^{-3}, roughly a ten-million-fold increase in charge carrier density and electrical conductivity over pure silicon.[2] Because the chemical equilibrium must satisfy equation 4, the value of p must drop by the same factor, to ~10^3 cm^{-3}. Semiconductors for which $n > p$ are called n-type semiconductors and are analogous to alkaline solutions for which $[OH^-] > [H^+]$.

Similarly, if aluminum atoms had been introduced at the part-per-million concentration level and were largely thermally ionized, p would be ~10^{17} cm^{-3} and n would drop to ~10^3 cm^{-3} ($= K/p$). Semiconductors for which $p > n$ are called p-type and are analogous to acidic solutions for which $[H^+] > [OH^-]$.

In semiconducting materials with an imbalance in n and p, whichever particles are in larger concentration are called majority carriers and those in smaller concentration are minority carriers. Thus in n-type materials, electrons are majority carriers and holes are minority carriers, with the reverse the case in p-type materials. The inversely related values of n and p can also be regarded as an analogy for the common ion effect: addition of a donor shifts the equilibrium position of equation 4 so that the concentration of valence-band holes is suppressed; addition of an acceptor likewise suppresses the concentration of conduction-band electrons.

Analogies to Weak Acids and Bases

In addition to aluminum or phosphorus, a variety of other atoms can enter a silicon crystal and introduce energy levels within the band gap in accord with their ionization properties. Figure 8.7 shows, for example, that copper acts as an acceptor and introduces energy levels at 0.49 eV above the valence-band edge. For this so-called "deep acceptor," thermal energy at room temperature can fill only a small fraction of these states with electrons from the valence band. Thus, copper somewhat increases the hole concentration in the valence band, but by less than if the copper atoms reacted completely to give Cu^- and an equal number of holes. This process is analogous to the ionization of a weak acid in water: the proton concentration is increased but by much less than the stoichiometric number of moles of acid added.

Similarly, manganese serves as a deep donor, characterized by a state lying 0.53 eV below the conduction-band edge. This state is relatively far from the conduction band edge; at room temperature few of the manganese atoms will be thermally ionized to Mn^+ and an equal number of conduction-band electrons. This reaction is analogous to the behavior of a weak base in water, which increases the concentration of hydroxide ions but by much less than the stoichiometric number of moles of the base added.

---

[2]This number can be calculated by determining the number of atoms per cubic centimeter of silicon (density times Avogadro's number divided by the atomic mass of Si) and then dividing the value obtained by 10^6. The latter factor represents the substitution of one phosphorus atom per million silicon atoms and assumes that they are all ionized. The value obtained is 5×10^{16} charge carriers/cm^3, which was rounded to ~$10^{17}/cm^3$ in the text.

The Fermi Level and Its Analogy to pH

Irrespective of what species have been added to water, their influence on the solution is described in terms of the concentration of hydrogen ions through the pH value (related to the thermodynamic chemical potential): Both pure water and solutions with a pH of 7 are described as neutral; all solutions with pH less than 7 are acidic and those with pH greater than 7 are alkaline.

The analogous property of interest in characterizing electronic materials is the concentration of electrons. This concentration is reflected in the so-called Fermi level, E_f (the thermodynamic electrochemical potential), which is defined as the energy at which the probability of finding an electron is 1/2, i.e., below the Fermi level it is more likely that the electronic states are occupied with electrons, and above the Fermi level it is more likely that they are not. For a metal at room temperature, the Fermi level occurs at roughly the energy of the highest occupied level in the partially filled band at 0 K[3] (Figure 8.9A).

The position of the Fermi level relative to the band edges provides the analog of the pH scale associated with aqueous solutions. A pure semiconductor for which $n = p$ will have its Fermi level in about the middle of the band gap[4] (Figure 8.9B), an analogy to pure water having a pH of 7, in the middle of the common pH range. As acceptors are added to a neutral semiconductor, increasing the concentration of holes and reducing the concentration of conduction band electrons (making $p > n$) , the Fermi level will shift toward the valence-band edge (Figure 8.9C), reflecting the fact that it is less likely that electrons will be found occupying the higher energy levels of the conduction band. This is analogous to adding acid to a neutral solution, moving the pH to values <7. Conversely, as donors are added to a neutral semiconductor, enhancing the concentration of conduction-band electrons (making $n > p$) and the likelihood that electrons will be found occupying the higher energy levels of the conduction band, the Fermi level will shift toward the conduction-band edge (Figure 8.9D), just as adding base to a neutral solution moves the pH to values >7.

---

[3]At 0 K, the electrons will be paired in the lowest energy levels of the solid, and the Fermi level will occur at the top of the occupied energy levels: there is a sharp break between the occupied (probability of occupancy is one) and the unoccupied (probability of occupancy is zero) energy levels at 0 K. As the temperature increases, some of the electrons will be promoted to higher energy levels by absorbing thermal energy, causing the occupation probability to change less abruptly, but leaving the energy at which the occupancy probability is 1/2 (the Fermi level) at about the same energy.

[4]Because of the statistical origin of the Fermi level, there need not be an occupiable energy level at the Fermi level. Occurrence of the Fermi level near the middle of the band gap reflects the fact that the probability of occupancy is very high in the valence band and very low in the conduction band. The exact position of the Fermi level depends on several factors, including the effective masses of the electrons and holes; *see* reference 7.

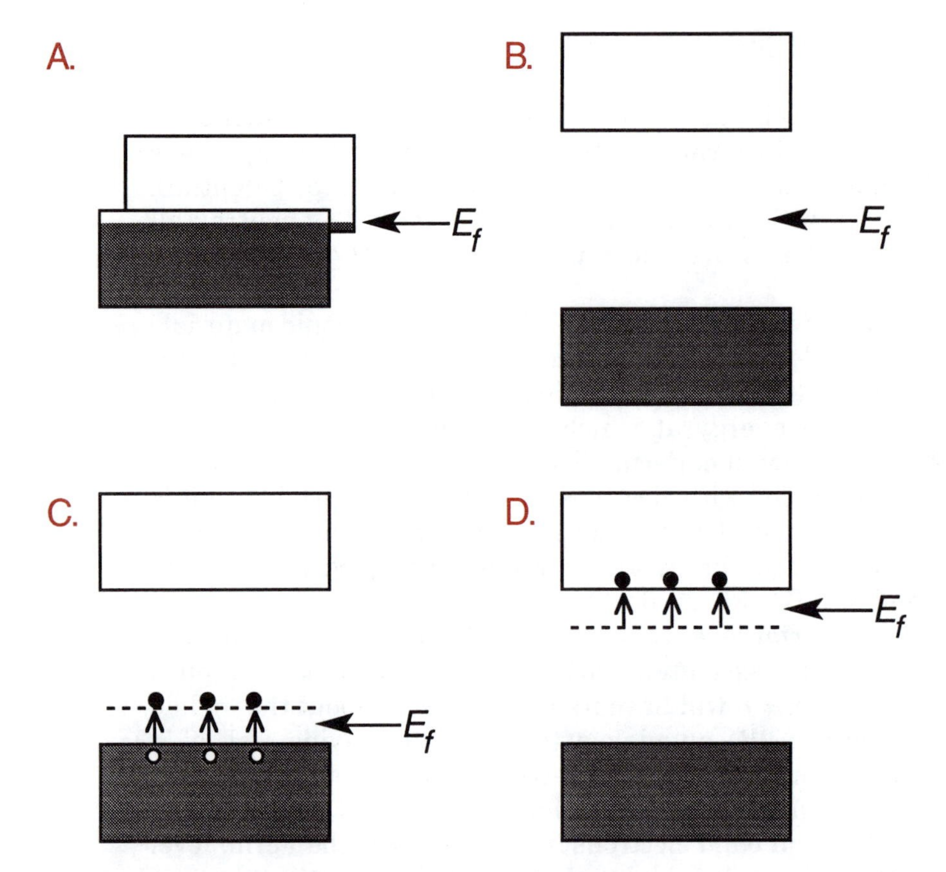

Figure 8.9. The position of the Fermi level (E_f) for (A) a metal, (B) an undoped semiconductor, (C) a p-type semiconductor, and (D) an n-type semiconductor.

Extent of Ionization

The pH of aqueous solutions and the Fermi level in semiconductors play analogous roles in determining the extent of ionization in the two media. Figure 8.10 encapsulates the analogy. For a weak acid, HA, the fraction of molecules in the protonated form depends on the relative values of pK_a and pH: as pH is lowered relative to the pK_a of the acid, a progressively larger fraction of molecules will be in the protonated form, HA; conversely, as pH increases to values exceeding pK_a, a progressively larger fraction of these molecules will be in the deprotonated form, A^-. When pH = pK_a, equal quantities of the protonated and deprotonated species are present. These relationships are easily seen by rearranging the equilibrium constant expression to: $K_a / [H^+] = [A^-] / [HA]$; or, $pH - pK_a = \log \{ [A^-] / [HA] \}$.

Similarly, for a weak acceptor whose ionization energy is E_I (*see* Figure 8.7), the Fermi level position E_f will dictate the percentage of acceptors that are ionized (formally A^-) or un-ionized (formally A): As the Fermi level moves toward the valence-band edge (analogous to moving to low pH), progressively more of the acceptors will be un-ionized; as the Fermi level moves toward the conduction-band edge (analogous to moving to high pH), progressively more of the acceptors will be ionized. When the Fermi level is at the acceptor energy level, there will be roughly equal numbers of ionized and un-ionized acceptors.[5]

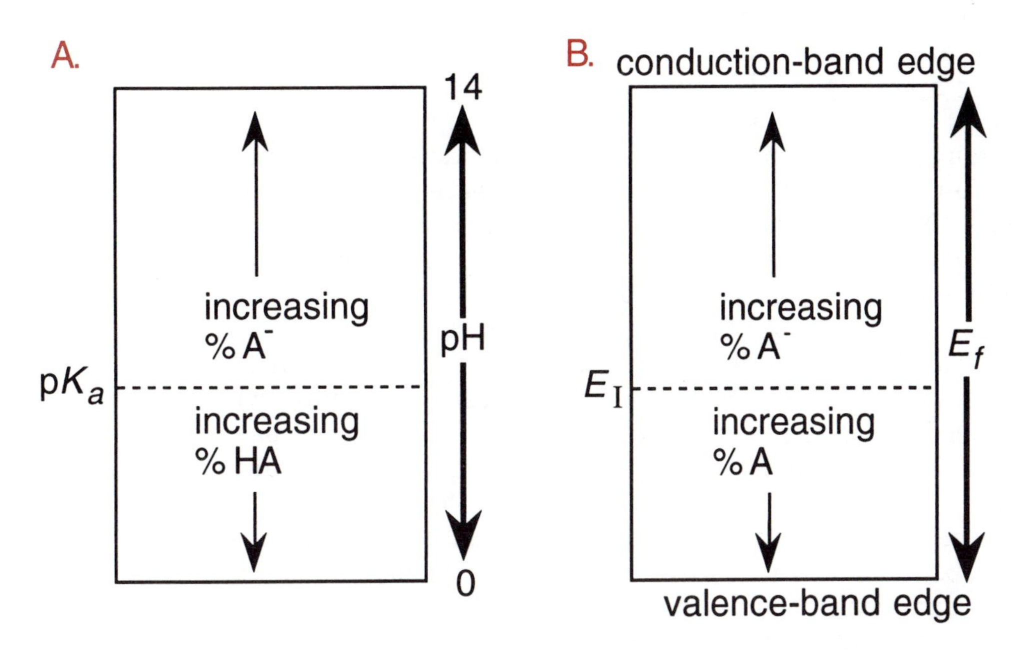

Figure 8.10. A: The relative amounts of HA and A^-, the protonated and deprotonated forms of a weak acid, can be varied by changing the pH relative to pK_a. B: Similarly, the relative amounts of A and A^-, the un-ionized and ionized forms of a weak acceptor, respectively, can be varied by changing the Fermi level, E_f, relative to the ionization energy level of the acceptor, E_I. The separation of the band edges is equal to the band gap, E_g.

Similar analogies can be established for weak bases in water and weak donors in the semiconductor: the ratio of protonated to deprotonated base, BH^+/B, will increase with decreasing pH; and the ratio of ionized to un-ionized donor, D^+/D, will increase as the Fermi level moves toward the valence band.

Neutralization Reactions and Buffers

Neutralization reactions can be established in the solid just as they are in aqueous solution and followed by shifts in the Fermi level in the same way that pH changes would be used to chart the course of titrations in water. As an example, consider a sample of p-type, aluminum-doped silicon with a concentration of acceptors, N_a. As noted, Al readily accepts electrons, a behavior making it analogous to a strong acid, for which the Fermi level will lie near the valence-band edge, as shown in Figure 8.9C.

[5]The 1:1 ionized:un-ionized concentration analogy is not exact at this energy because of the spin of the electron. *See* reference 3.

The concentration of p should thus be ~N_a, just as [H^+] is taken as the concentration of a strong acid in water.

If the sample is now heated in phosphorus vapor to achieve uniform incorporation, ionizable electrons from these donor phosphorus atoms that are introduced at concentration N_d can combine with these valence-band holes, reducing p to ~ $(N_a - N_d)$ and shifting E_f toward the conduction band. When $N_d = N_a$, the equivalence point has been reached, $n = p$, and E_f should be roughly in the middle of the band gap, analogous to a neutral aqueous solution. Additional phosphorus will increase n and continue to decrease p [$n \sim (N_d - N_a); p \times n = K$] and shift E_f closer to the conduction band, just as adding base past the equivalence point would move the pH into the alkaline regime. A graph of the position of E_f versus $|(N_a - N_d)|$, plotting the course of the so-called "compensation" process, is shown in Figure 8.11. The shape is similar to a strong-acid–strong-base titration. Of course, this same sequence can be run in reverse, beginning with a phosphorus-doped silicon sample to which aluminum is added: In this case, the Fermi level begins near the conduction band and moves toward the valence band.

Similar compensation reactions can occur between a strong donor and weak acceptor, between a strong acceptor and a weak donor, and between a weak acceptor and weak donor. Analogous to titrations between strong bases and weak acids, and between strong acids and weak bases, a buffer region is established: just as pH is roughly constant over large changes in the strong-acid or strong-base concentration because of the interconversion of conjugate weak acid–base forms, the Fermi level is constant over large changes in strong-acceptor or strong-donor concentrations because of the interconversion of conjugate ionized–un-ionized donor or acceptor forms (relative concentrations of ionized and un-ionized forms).

As in aqueous solution, these relations can be understood in light of the equilibria that are involved. Addition of strong donors in the titration of a weak (deep) acceptor A, such as Cu in Figure 8.7, will cause formation of the ionized acceptor, A^-, rather than releasing the donor's electron to the conduction band; the analogy can be made to titrating a weak acid like HF with hydroxide ion:

$$HF + OH^- \rightarrow H_2O + F^- \tag{9}$$

$$A + e^- \text{(from strong donor)} \rightarrow A^- \tag{10}$$

Similarly, addition of strong acceptors in the titration of a weak (deep) donor D, such as Mn in Figure 8.7, will cause formation of the ionized donor, D^+, rather than increasing the concentration of holes in the valence band; the aqueous analogy is to titrating a weak base like NH_3 with strong acid:

$$NH_3 + H^+ \rightarrow NH_4^+ \tag{11}$$

$$D + h^+ \text{(from strong acceptor)} \rightarrow D^+ \tag{12}$$

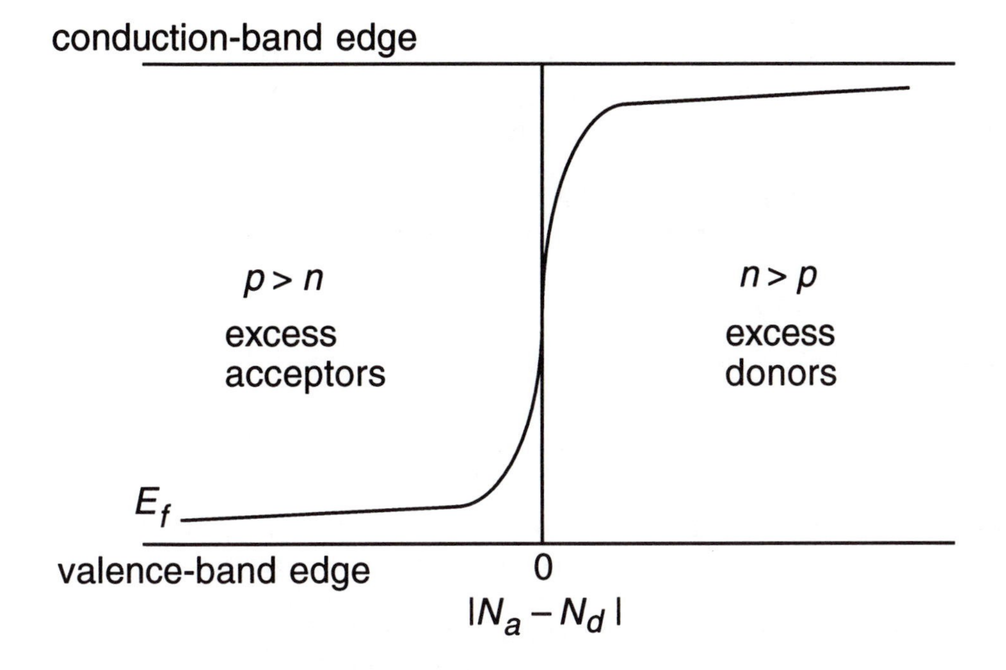

Figure 8.11. A plot of the position of the Fermi level (E_f) as a function of doping. The left side of the diagram corresponds to a p-doped semiconductor, with N_a equal to the number of acceptor atoms. In p-type semiconductors the Fermi level is just above the valence band (*see* Figure 8.9C). Electrons provided by the addition of small amounts of donor atom impurities (N_d) will annihilate some of the valence-band holes in what can be thought of as a neutralization reaction in the semiconductor. The net number of acceptors remaining will be $|N_a - N_d|$. As a result, the Fermi level shifts toward the conduction band. When the number of donors equals the number of acceptors, the "equivalence point" is reached, and the Fermi level is roughly halfway between the valence and conduction bands. As more donor atoms are added to the semiconductor, the Fermi level continues to rise. When the number of donors has exceeded the number of acceptors present initially, the semiconductor has become essentially n-type. This process is analogous to the titration of a strong acid with a strong base. The x-axis in this plot, $|N_a - N_d|$, is analogous to the volume of titrant added. (Adapted from reference 8.)

The p–n Junction: A Solid-State Concentration Cell

If acidic and alkaline solutions are mixed, the neutralization reaction is favorable and rapid. Alternatively, it is possible to exploit the difference in concentration of H^+ of these two solutions to create an electrical output by means of a concentration cell. As shown in Figure 8.12, if a couple like

H^+–H_2 is established in the two solutions (hydrogen gas at 1 atm is bubbled over Pt electrodes immersed in the solutions), and the two solutions are connected by a salt bridge, a potential difference (voltage) can be measured. The cell can be written as $H_2 | NaOH(aq) || HCl(aq) | H_2$. For solutions that are 1 M in acid and base, a voltage of 0.83 V is realizable. This value is determined from the Nernst equation at 298 K by treating the system as a concentration cell for the reaction

$$[H^+]_{\text{acid side}} \rightarrow [H^+]_{\text{base side}} \tag{13}$$

The Nernst equation for this concentration cell may be written as

$$E = E^o - \frac{RT}{nF} \ln Q = \frac{-2.3\,RT}{F} \log \frac{[H^+]_{\text{base side}}}{[H^+]_{\text{acid side}}} \tag{14}$$

where R is the universal gas constant; T is absolute temperature; n is the moles of electrons; F is Faraday's constant, and Q is the reaction quotient. Because $[H^+]_{\text{acid side}} > [H^+]_{\text{base side}}$, E is positive. Of course, if the output energy is recovered as electricity by allowing current to pass, both solutions will eventually have equal values of $[H^+]$, corresponding to pH 7, and the chemicals must be replenished to continue the cell's operation.

If a p-type semiconductor is again regarded as analogous to an acidic solution and an n-type semiconductor to an alkaline one, their juxtaposition as in Figure 8.13 also represents a concentration cell, because the concentrations of electrons, the Fermi levels, in the two phases (high in the n-type material; low in the p-type material) will seek to come to equilibrium by the reaction

$$[h^+]_{\text{p-type side}} \rightarrow [h^+]_{\text{n-type side}} \tag{15}$$

or

$$[e^-]_{\text{n-type side}} \rightarrow [e^-]_{\text{p-type side}} \tag{16}$$

The "built-in" voltage that develops between these two phases that are brought together at what is called a p–n junction is given by the same kind of Nernstian expression, reflecting the analogous equilibrium considerations:

$$E \text{ (in volts)} = \frac{-2.3\,RT}{F} \log \frac{[h^+]_{\text{n-type side}}}{[h^+]_{\text{p-type side}}} = \frac{-2.3\,RT}{F} \log \frac{[e^-]_{\text{p-type side}}}{[e^-]_{\text{n-type side}}} \tag{17}$$

One means for preparing a p–n junction has been intimated in connection with the neutralization (compensation) experiment (Figure 8.11). If a sample of n-type Si is exposed to a high concentration of aluminum vapor for a short time, the surface region of the solid can be made to become p-type by virtue of now having a higher acceptor concentration than donor concentration. The bulk sample is still n-type, however, because the aluminum atoms are not given time to diffuse into

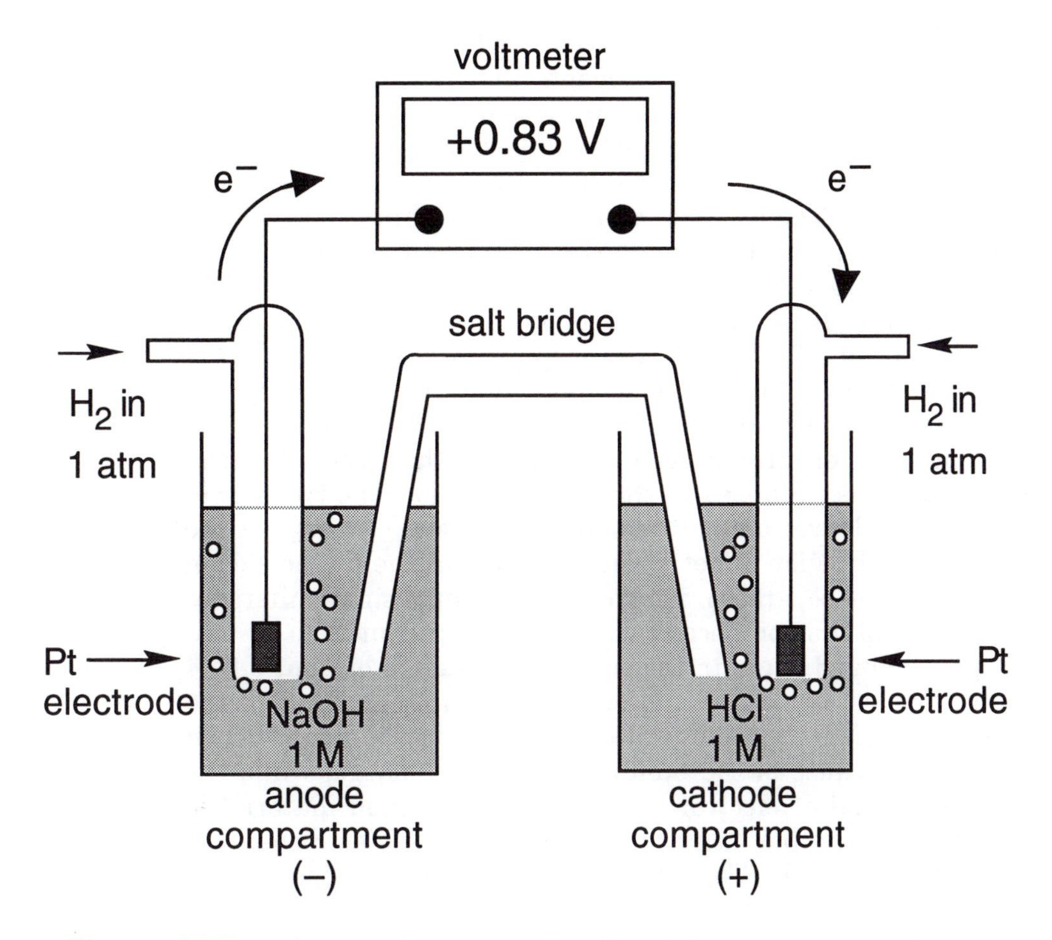

Figure 8.12. An electrochemical cell. The right compartment is the standard hydrogen electrode (1 M HCl, 1 atm H_2) and the left compartment contains 1 M NaOH and 1 atm H_2. This cell will produce a potential of +0.83 V at 298 K and may be thought of as a concentration cell in which the right compartment has $[H^+] = 1$ M and the left compartment has $[H^+] = 1 \times 10^{-14}$ M.

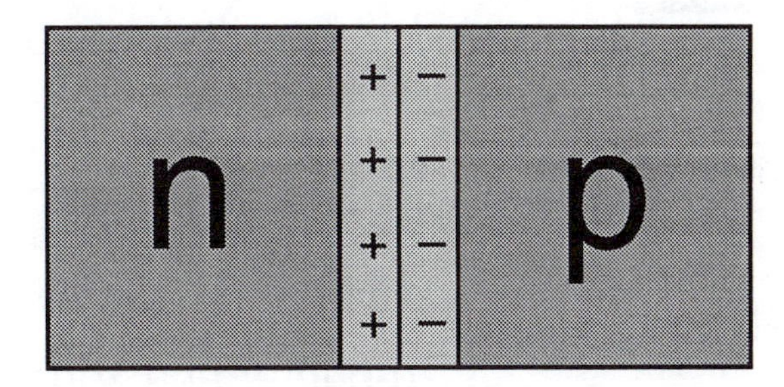

Figure 8.13. A p–n junction. Electrons spontaneously flow from the n-type side to the p-type side until equilibrium is reached. This leaves a small region with a net positive charge on the n-type side of the junction and a region with a net negative charge on the p-type side, opposing the additional net flow of charge across the junction.

the bulk at the elevated temperature employed. And at room temperature, diffusion is sufficiently slow for these dopants that it can be neglected. This process will not give an atomically abrupt p–n junction but rather a narrow region over which the conductivity will gradually change from n-type to p-type. With modern growth methods based on chemical vapor deposition (CVD) and molecular beam epitaxy (MBE), junctions that are virtually atomically abrupt can now be prepared (*see* Chapter 10).

The electronic properties of the p–n junction provide an explanation for the operation of LEDs, diode lasers, and photovoltaic (solar) cells. As shown in Figures 8.13 and 8.14, when a junction is formed between an n-type and p-type semiconductor, electrons flow from the n-type semiconductor (at higher electron concentration) to the p-type semiconductor (at lower electron concentration) until the electrochemical potential (Fermi level) is equalized in the two materials. The transfer of charge at the junction leaves a small region of net positive charge at the n-type side and a small region of net negative charge at the p-type side of the junction. The charge separation internally builds up a small difference in potential energy, and thus produces a voltage across the junction.

The bending of the band edges in the equilibrium sketch of Figure 8.14 reflects the distribution of the internal or "built-in" voltage through the junction. The electrochemical potential of electrons in the donor-doped material is lowered because of the buildup of positive charge, while the electrochemical potential of electrons in the acceptor-doped material is raised because of the buildup of negative charge, until the electrochemical potentials have equilibrated. Then there is no further net transfer of charge across the interface.

If a circuit is now prepared by making low resistance (ohmic) contacts with metals, one to the n-type material and one to the p-type material, a variety of energy conversion schemes can be realized once a voltage source or light source is introduced into the circuit.

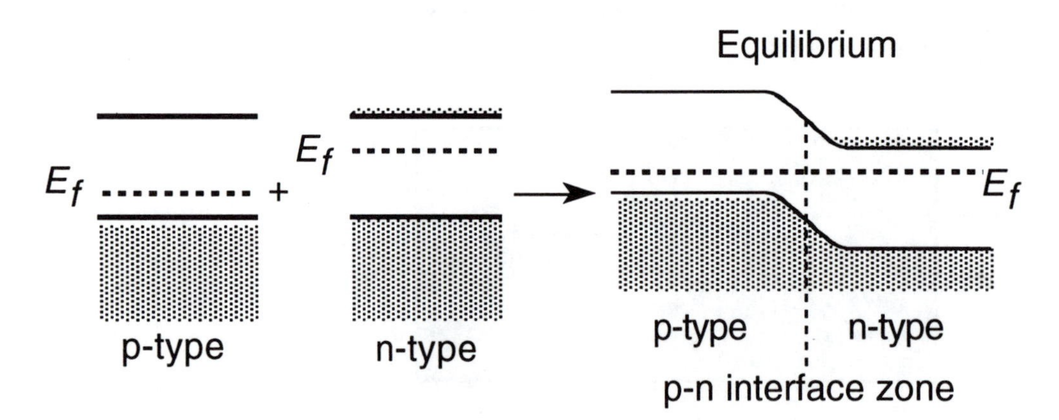

Figure 8.14. A band-energy diagram of a p–n junction. The two semiconductors are in equilibrium with each other and must have their Fermi levels at the same energy. The band-gap energy does not change throughout the solid. Only the top portion of the valence band and the bottom portion of the conduction band are shown.

Biasing the p–n Junction

The introduction of an external voltage source into the circuit (called biasing) to induce a current disturbs the equilibrium situation just described. Substantial current flow through the junction requires that electrons in the conduction band move from the side with more electrons to the side with fewer electrons. Similarly, holes in the valence band must move from the side with more holes to the side with fewer holes. Thus, electrons must be given enough energy to move over the potential energy barrier from the n-type side to the p-type side, while holes must be forced to move under the barrier from the p-type side to the n-type side during current flow (recall from Demonstration 8.1 that hole energies increase downward).

If the voltage applied to the n-type side is negative relative to that applied to the p-type side, the voltage is referred to as forward bias. The energy barrier for electrons and holes to flow through the junction is decreased relative to the equilibrium energy barrier, Figure 8.15A, and substantial current can flow through the junction. Under reverse bias the opposite situation occurs. The applied voltage adds to the internal voltage, Figure 8.15B, making current flow even more difficult. This is the essence of diode behavior: The current increases exponentially with forward bias and is negligible with reverse bias, as sketched in Figure 8.15C.

Under forward bias, the electrons reaching the p-type side and the holes reaching the n-type side are minority carriers in these regions. They can recombine with the majority carriers that are in abundant supply (Figure 8.16). This recombination results in luminescence (emitted light), as electrical energy is converted to light by the p–n junction. A p–n junction used to produce light is called a light-emitting diode, LED, Figure 8.17.

The color of the luminescence can be controlled by varying the band gaps of the junction materials, as noted in Chapter 7. Employing semiconductors that are solid solutions is a particularly effective way to continuously tune the band gap (*see* Chapter 7). Semiconductors with the formulas GaP_xAs_{1-x} ($x = 1.00 - 0.40$) can be used to tune the bandgap from 1.88 eV (red light) to 2.23 eV (green light) *(9)*.

Laboratory. A laboratory that explores observed properties such as the color, wavelength and energy of light, and excitation voltage for a series of compound semiconductors appears as Experiment 7 of this book.

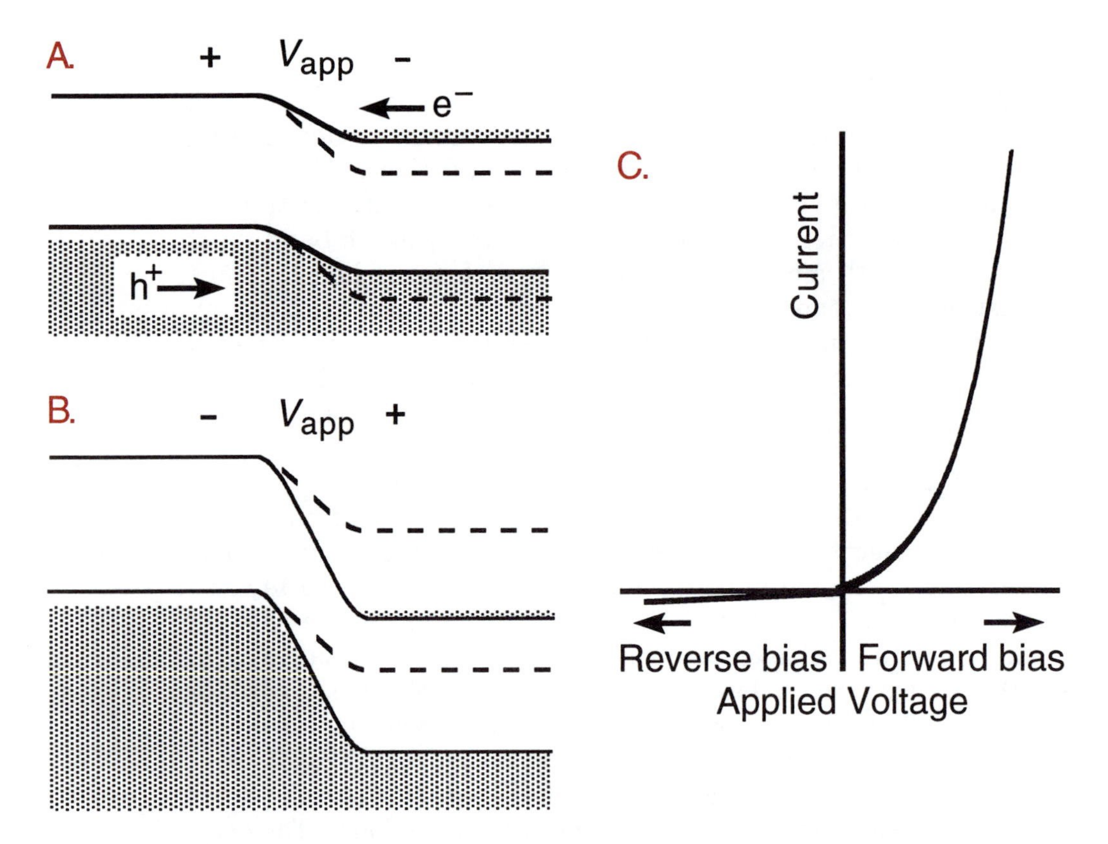

Figure 8.15. A: A p–n junction under forward bias conditions. B: A p–n junction under reverse bias. The dashed lines in A and B show the equilibrium band bending when no bias is applied, Figure 8.14. C: Current through a p–n junction as a function of applied voltage, V_{app}.

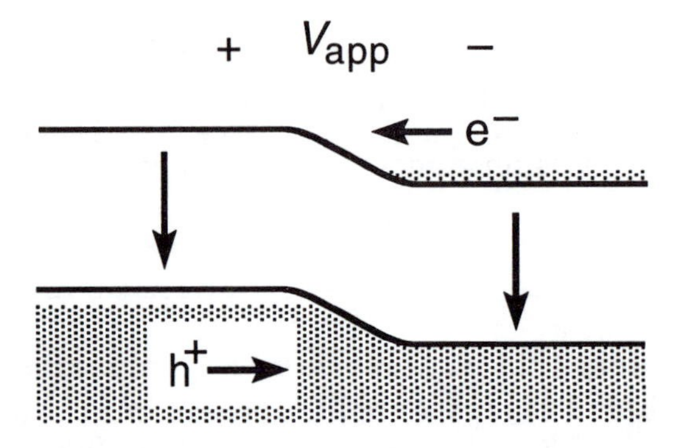

Figure 8.16. A p–n junction acting as a light-emitting diode (LED) under forward bias conditions. As electrons in the conduction band migrate toward the positive potential applied at the p-type side, they can return to the valence band (recombine with the abundant holes) and emit light of the band-gap energy. As holes in the valence band migrate toward the negative potential at the n-type side, they can recombine with the abundant electrons and emit light of the band-gap energy.

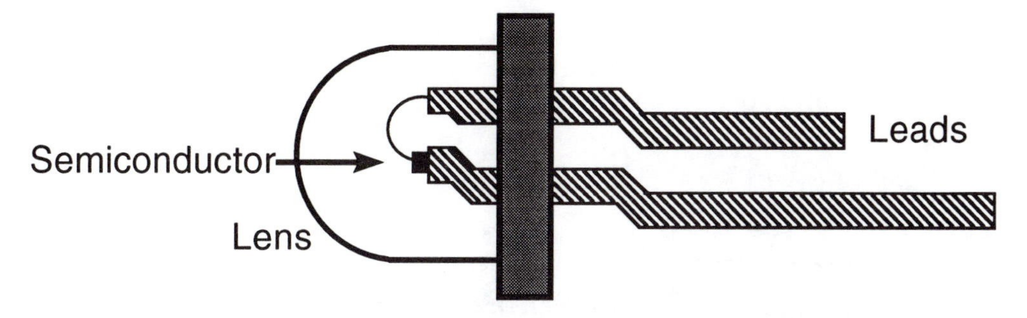

Figure 8.17. Schematic diagram of an LED.

Demonstration 8.3. Biasing and Dissection of an LED

Materials

LED (*see* Supplier Information for Chapter 7)
second LED with the cover removed (prepared before class)
A few milliliters of dichloromethane
LED circuit (containing a 1-kΩ resistor) described in Chapter 7 (*see* Figure 7.22B)
Magnifying lens or Micronta 30× microscope (Radio Shack)
Microscope connected to a television camera (optional)

Procedure

- Connect the LED to the battery in the two possible ways. The LED will light in one way (forward bias) and not in the other (reverse bias).
- Before class, place the second LED in dichloromethane. **CAUTION: Avoid inhalation of the vapors. Perform in a fume hood.** The plastic cover will slowly dissolve.
- Using a magnifying lens, observe that the semiconducting material is very small (~1 mm^2) and rests on a metal plate, which acts as the electrical contact for one side of the p–n junction. The electrical connection between the top of the chip (the other side of the p–n junction) and the electrical lead is made by a very thin piece of gold wire. This setup is shown in Figure 8.17. You may show the LED to the entire class at once using a microscope connected to a television camera. This equipment is often available in biology departments.
- Forward bias the LED by connecting it to a battery. If electrical contact through the gold wire is maintained, the LED will still light with its plastic lens removed.

It is possible to prepare a diode laser from a semiconductor p–n junction by making the structure in the shape of a laser cavity and electrically stimulating the recombination of such large numbers of carriers that lasing action results (Figure 8.18) *(10)*.

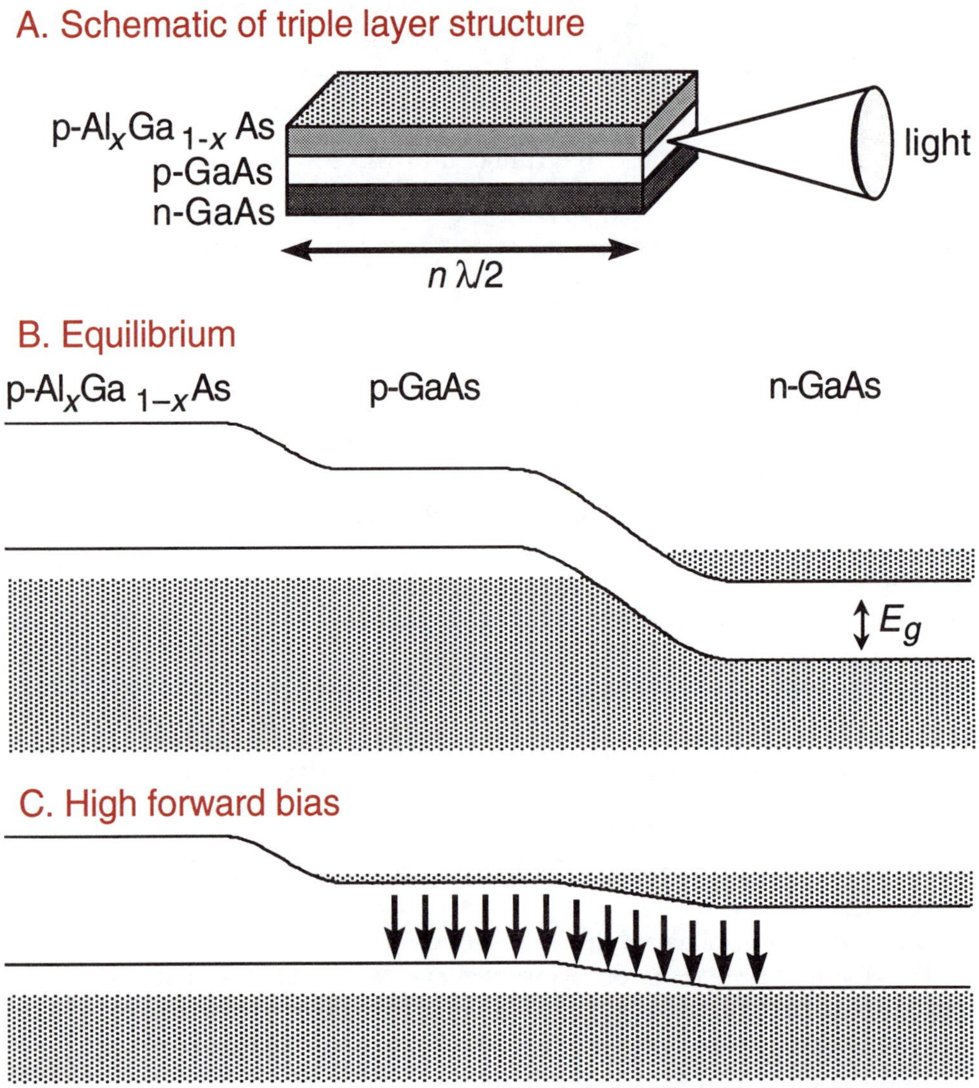

Figure 8.18. A: Schematic of a diode laser. The light is emitted from the p–n junction area. The cavity length itself must be a half-integral number of wavelengths of the light emitted. B: Confinement of the electron–hole pairs within the active lasing area of the solid is accomplished by bounding the active region with another semiconductor whose band gap is larger than that of the first semiconductor. In this case, a solid solution of $Al_xGa_{1-x}As$ is used as the larger band-gap material, and GaAs is used as the narrower band-gap material. C: Under high forward bias, this construction forces electrons to stay within the GaAs region, where they recombine with holes that are present, and emit light in the process.

Finally, if light having an energy above the band gap of the semiconductor shines on the junction, the electron–hole pairs created in the junction region are separated by the electric field therein, as shown in Figure 8.19. This separation leads to an electrical current (at maximum efficiency, each photon absorbed moves one electron through the electric circuit), the basis of the photovoltaic cell in which the energy of light is converted into electricity by this remarkable solid-state concentration cell.

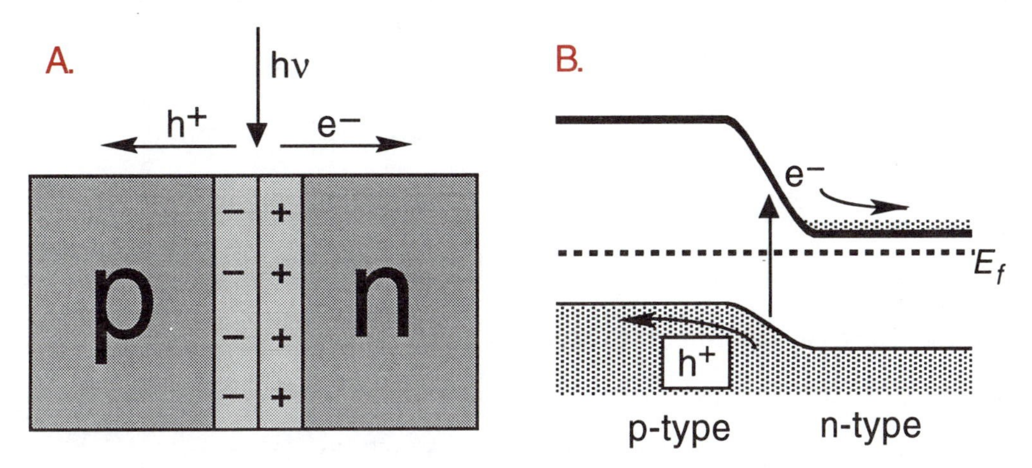

Figure 8.19. Two views of a photovoltaic cell based on the p–n junction. A: If light above the bandgap energy falls on the junction, the semiconductor will absorb the light. Electrons and holes will be generated and move toward opposite ends of the junction. The positive and negative signs represent the charges in the junction region resulting from the formation of the junction. B: A band diagram of the p–n junction. As light above the band-gap energy is absorbed in the junction region, electrons will be promoted from the valence band to the conduction band, leaving holes in the valence band. The slope of the band edge corresponds to the built-in electric field that causes the electrons to migrate toward the right (the n-type side) while the holes move toward the left (the p-type side) to create an electrical current. The magnitude of the current is determined by the absorbed light intensity: each photon can provide one electron to the current flow, but not all of the electrons may be collected. For example, defects in the solid may promote the recombination of the photogenerated electron–hole pairs, and thus lead to smaller electric currents.

Demonstration 8.4 can be used to illustrate how light absorption depends on solution path length and concentration, and thereby to discover or verify the Beer–Lambert law that is a cornerstone of spectroscopy. Use of a diode laser pointer as the light source and a solar cell based on a silicon diode as the light detector demonstrate the versatility of the p–n junction.

Demonstration 8.4. Counting Photons Using a Solar Cell

Materials

Diode laser (670 nm, 5 mW); other lasers (He–Ne, e.g.) and solutions can be substituted for those used here (*see* Supplier Information for Chapter 4)
Solar cell, approximately the size of the laser beam (Radio Shack); instructions for reducing the size of the solar cell are given in Laboratory Experiment 9
Two wires
Digital multimeter
Five or six Plexiglas cells (12 cm × 11 cm × 3.5 cm or 5.0 in × 4.5 in × 1.5 in)
$CuSO_4 \cdot 5\,H_2O$
Red food coloring
Darkened room or a cardboard box to serve as a light shield

Procedure

- Solder two wires to opposite sides of the solar cell in order to accommodate leads from the digital multimeter for measurements of current (*see* Experiment 9).
- Make a 0.01 M solution of Cu^{2+} by dissolving 2.5 g of $CuSO_4 \cdot 5\,H_2O$ in 1 L of water. Add a few drops of sulfuric acid if the solution is cloudy.
- Align the laser to strike the solar cell, allowing space in between for interposing as many of the Plexiglas cells as will be used. It is best to clamp the laser and solar cell in place, and to fix the laser in the "on" position. (See the diagram in Experiment 9.) **Caution: Do not look into the laser; its beam can damage the eye!** Try to set up the system so that there is only a small background current (minimal ambient light).
- *Uncorrected* relative numbers of photons transmitted by the copper solutions can be obtained by recording the solar cell current as a function of path length (here, the number of identical solution cells interposed in the path of the laser beam). Typical values are shown as in Figure 8.20A for zero to five solution cells. The fraction of light transmitted by a given solution or set of solutions is the current measured in their presence divided by the current measured in their absence. If time permits, a more accurate measurement can be made by correcting for reflective losses at the air–Plexiglas and Plexiglas–solution interfaces: the fraction of light transmitted is the current measured in the presence of the copper solutions divided by the current measured when the same cells are filled with water instead of cupric ion solution.

As each solution cell is added, providing a linear increase in path length, the same fractional decrease in transmitted light should be observed. Thus a semilogarithmic plot of the current versus the number of solution cells should be linear, as shown in Figure 8.20B.

- Make a 0.05 M cupric ion solution and dilute to 0.04 M, 0.03 M, 0.02 M, and 0.01 M concentrations. This step can either be done ahead of time or as part of the presentation of the demonstration. Again, add a few drops of sulfuric acid if the solutions are cloudy.
- Measure the current transmitted by each solution to estimate the fraction of transmitted light as a function of concentration. Data can again be corrected as described. A plot of the current versus cupric ion concentration or a semilogarithmic plot of the current versus cupric ion concentration can be prepared (Figures 8.20C and D). As was found for increases in path length, linear increases in concentration will decrease the transmitted light by the same fraction. Collected data should be well fit by the Beer–Lambert law: $-\log_{10}$ (fraction of light transmitted) = absorbance = absorptivity × concentration × path length
- Fill a cell with water and place enough red food coloring in it to make the solution color extremely dark. Despite the intense absorption in the visible spectrum, there should be little effect on the intensity of the transmitted laser light, a result illustrating the lack of absorbance by this solution at the monochromatic laser wavelength of 670 nm. Conversely, if a cell containing the visibly very pale cupric ion solution intercepts the beam, a significant decline in current will be seen, corresponding to the much stronger absorptivity at this wavelength of the cupric ions.

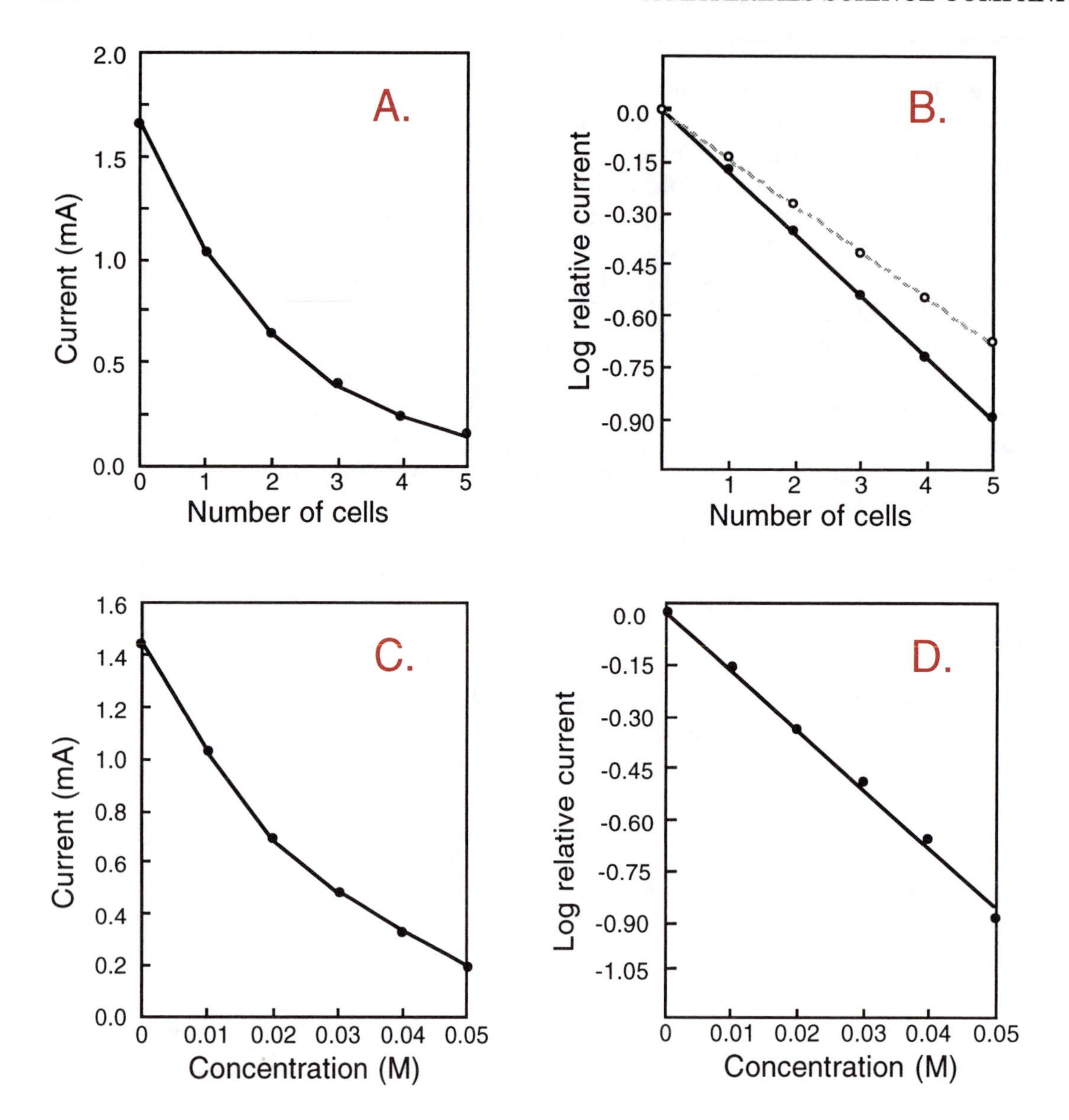

Figure 8.20. Data from Demonstration 8.4. A: Plots of photocurrent (uncorrected; see text) versus the number of cells, each of which contains 0.01 M copper sulfate. B: Plots of the logarithm of the relative photocurrent versus the number of cells used. The closed circles are for uncorrected data, while open circles represent data that have been corrected for reflective losses (*see* text). C: Plots of photocurrent versus copper sulfate concentration using a single cell. D: Plots of the logarithm of relative photocurrent versus the copper sulfate concentration. The same result was obtained with or without the correction for reflective losses.

The analogies between semiconductors and aqueous solution chemistry are summarized in Table 8.2.

Table 8.2 Parallels Between Aqueous Solution Chemistry and Semiconductors

Chemical Reaction	Aqueous Example	Silicon Example
Solvent autoionization	$H_2O \rightarrow H^+ + OH^-$ $K_w = [H^+][OH^-]$ $[H^+] \approx 10^{14}$ ions/cm^3	$Si_{(crystal)} \rightarrow h^+ + e^-$ $K = [h^+][e^-] = p \times n$ $[h^+] \approx 1.5 \times 10^{10}$ cm^{-3}
Strong acid–acceptor, strong base–donor	$HCl \rightarrow H^+ + Cl^-$ $NaOH \rightarrow Na^+ + OH^-$	$Ga \rightarrow h^+ + Ga^-$ $As \rightarrow As^+ + e^-$
Weak acid–acceptor; base–donor	$CH_3COOH \rightleftarrows H^+ + CH_3COO^-$ $NH_3 + H_2O \rightleftarrows NH_4^+ + OH^-$	$Cu \rightleftarrows h^+ + Cu^-$ $Mn \rightleftarrows Mn^+ + e^-$
Common ion effect	adding base (acid) suppresses the concentration of H^+ (OH^-)	adding a donor (acceptor) suppresses the concentration of h^+ (e^-)
Concentration cell[a]	$H_2 \mid NaOH \mid\mid HCl \mid H_2$ $E = \frac{-2.3RT}{F} \log \frac{[H^+]_{base\ side}}{[H^+]_{acid\ side}}$	—[n-type | p-type]— $E = \frac{-2.3RT}{F} \log \frac{[h^+]_{n\text{-}side}}{[h^+]_{p\text{-}side}}$

[a]Cell is treated as a concentration cell $[H^+]_{acid\ side} \rightarrow [H^+]_{basic\ side}$.
SOURCE: Adapted from reference 11.

Ionic Concentration Cells

Oxygen Sensors

Another kind of solid-state concentration cell is employed to control the ratio of air to exhaust gas in catalytic converters *(12)*. It is based on ZrO_2 that has been doped with CaO. Crystals of ZrO_2 doped with CaO have the fluorite structure (Chapter 5): the zirconium atoms form a face-centered cubic arrangement and the oxygen atoms fill all eight tetrahedral holes in the structure. If some of the Zr^{4+} is replaced by Ca^{2+}, oxygen vacancies must be created in order to maintain electroneutrality. The presence of vacancies in the tetrahedral holes now allows migration of oxide ions through these vacancies, and CaO-doped ZrO_2 acts as an ionic conductor in the solid state.

The oxygen sensor shown in Figure 8.21 acts in the following way: Oxygen molecules from the exhaust gas are adsorbed by platinum electrodes in the cell and converted into oxygen atoms. If the concentration of oxygen atoms in the other compartment (air, which has a constant oxygen content) is different, the oxygen atoms will have a tendency to migrate through the doped ZrO_2 solid as oxide ions. On the side with the higher oxygen content the following reactions are occurring:

$$O_2\,(g) + Pt\,(s) \rightarrow 2O\text{ (on the Pt surface)} \tag{18}$$

$$O\text{ (on the Pt surface)} + 2e^- \rightarrow O^{2-}\text{ (migrating through }ZrO_2) \tag{19}$$

The reverse reactions are occurring on the side with the lower oxygen concentration. (This behavior is analogous to the reactions $2\,H^+ \rightarrow H_2$ and $H_2 \rightarrow 2\,H^+$ in a concentration cell based on the standard hydrogen reaction.) The concentration on one side is known, so the measured potential difference can be related to the unknown oxygen concentration through the Nernst equation.

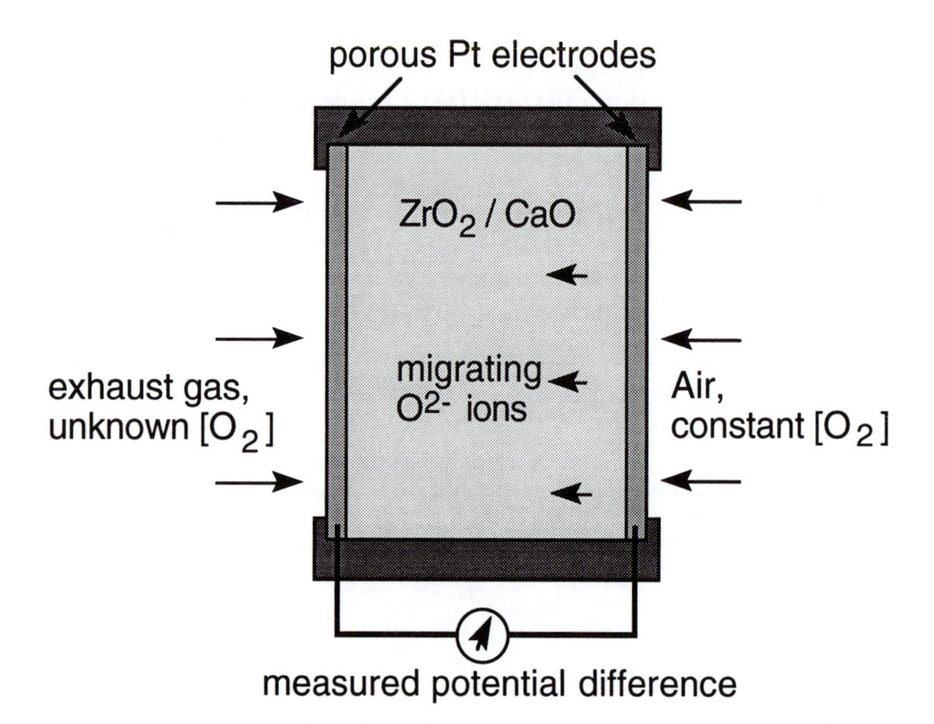

Figure 8.21. An oxygen sensor based on CaO-doped ZrO_2. Oxide ions are created at the platinum electrodes from O_2. If the concentration of O_2 is different in the two compartments, a potential develops. If the concentration in one compartment is known, the concentration in the other can be calculated using the Nernst equation. (Adapted with permission from reference 12.)

The Fluoride-Selective Electrode

Crystals of lanthanum fluoride (LaF_3) have provided the basis for a fluoride selective electrode, first developed in 1966 *(13, 14)*. Pure LaF_3 is conductive because of the mobility of the fluoride ions through the crystal; however, in the commercially available fluoride electrode, the samples are doped with EuF_2. Doping with a divalent cation creates fluoride vacancies for reasons described in the previous section and further increases the conductivity. A slice from a single crystal of the doped material serves as the electrode membrane.

At the interfaces of the membrane, ionization creates a charge on the membrane surface that depends of the fluoride concentration in solution, as follows:

$$LaF_3(\text{solid}) \rightleftarrows LaF_2^+(\text{solid}) + F^-(\text{solution}) \tag{20}$$

and the cell potential becomes

$$E_{\text{cell}} = \text{constant} - \frac{2.3RT}{nF} \log a_{F^-}, \tag{21}$$

where the constant depends on the reference electrodes and a_{F^-} is the activity of fluoride ions in solution.

Commercially available fluoride electrodes are very selective, with only the hydroxide ion interfering, and then only above a pH of 8. Below pH 5, protonation of fluoride to form hydrogen fluoride, to which the electrode does not respond, also causes error in total fluoride determinations. The LaF_3 fluoride-selective electrode has found applications in the analysis of minerals, fertilizers, drinking and waste water, teeth and bones. Solid-state electrodes are available that sense the other halides, CN^-, S^{2-}, SCN^-, Cd^{2+}, Cu^{2+}, Pb^{2+}, and Ag^+.

Laboratory. Experiment 12 involves the synthesis and study of Cu_2HgI_4, a compound that is an ionic conductor.

References

1. Lomax, J. F. *J. Chem. Educ.* **1992**, *69*, 794–795.
2. Streetman, B. G. *Solid State Electronic Devices*, 3rd ed.; Prentice Hall: Englewood, NJ, 1990; p 79.
3. *Semiconductors*; Hannay, N. B., Ed.; Reinhold: New York, 1959; p 196.
4. Ashcroft, N. W.; Mermin, N. D. *Solid State Physics*; Saunders College: Philadelphia, PA, 1976.
5. Streetman, B. G. *Solid State Electronic Devices*, 3rd ed.; Prentice Hall: Englewood, NJ, 1990; p 108.

6. *Semiconductors*; Hannay, N. B., Ed.; Reinhold: New York, 1959; p 463.
7. Streetman, B. G. *Solid State Electronic Devices*, 3rd ed.; Prentice Hall: Englewood, NJ, 1990; p 71.
8. Smith, R. A. *Semiconductors*, 2nd ed.; Cambridge University Press: New York, 1978; p 91.
9. Lisensky, G. C.; Penn, R.; Geselbracht, M. J.; Ellis, A. B. *J. Chem. Educ.* **1992**, *69*, 151–56.
10. Baumann, M. G. D.; Wright, J. C.; Ellis, A. B.; Kuech, T.; Lisensky, G. C. *J. Chem. Educ.* **1992**, *69*, 89–95.
11. Hannay, N. B. *Int. Sci. Technol.* October 1963; p 65.
12. Shriver, D. F.; Atkins, P. W.; Langford, C. H. *Inorganic Chemistry*; W. H. Freeman: New York, 1990; pp 590–591.
13. Skoog, D. A.; Leary, J. J. *Principles of Instrumental Analysis*; Saunders: Philadelphia, PA, 1992; p 499.
14. Koryta, J. *Ion Selective Electrodes*, Cambridge University Press: New York, 1975; p 99.
15. Fonash, S. J. *Solar Cell Device Physics*; Academic Press: New York, 1981; p 26.
16. Fonash, S. J. *Solar Cell Device Physics*; Academic Press: New York, 1981; p 3.

Additional Reading

- Ashcroft, N. W.; Mermin, N. D. *Solid State Physics*; Saunders College: Philadelphia, PA 1976.
- *Semiconductors*; Hannay, N. B., Ed.; Reinhold: New York, 1959.
- Streetman, B. G. *Solid State Electronic Devices,* 3rd ed.; Prentice Hall: Englewood, NJ, 1990.

Acknowledgments

This chapter is based on discussions with Thomas Kuech, University of Wisconsin—Madison, Department of Chemical Engineering; Norman Craig, Oberlin College, Department of Chemistry; Nathan Lewis, California Institute of Technology, Department of Chemistry; Andrew Bocarsly , Princeton University, Department of Chemistry; Eric Hellstrom, University of Wisconsin—Madison, Department of Materials Science and Engineering; and Allen Adler.

Exercises

1. At room temperature, which pure semiconductor would give rise to more electron–hole pairs, Ge or Si? GaAs or Si? Explain your answers based on Table 8.1.

2. Use Figure 8.5 to answer the following questions.
   a. Estimate the value of the autoionization equilibrium constant K for Si at 300 °C.
   b. From the temperature dependence of all of the substances shown, is autoionization a heat-absorbing or heat-releasing process? How do you know?
   c. At 20 °C, roughly how many times more positively-charged carriers does water have than GaAs?

3. Sketch a plot of Fermi level versus added dopant for a situation in which
   a. P is added as a dopant to In-doped Si
   b. Al is added as a dopant to Sb-doped Si

4. Show that the voltage in the cell shown in Figure 8.12 is 0.83 eV using either
   a. standard reduction potentials , or
   b. a concentration cell calculation

5. Determine whether the following are n- or p-type semiconductors:
   a. CdS in which silver substitutes for some cadmium atoms
   b. CdS in which bromine substitutes for some sulfur atoms

6. For the element Ge, which has the diamond crystal structure, the autoionization equilibrium constant at 300 K is 6×10^{26} cm^{-6}.
   a. What are the values of n and p at 300 K? Which elements might serve as dopants to increase n; to increase p. Why?
   b. If a donor is added that makes $n = 3 \times 10^{18}$ cm^{-3}, what is the corresponding value of p?
   c. The band gap for Ge is 0.7 eV. Sketch the band diagram, and indicate on it the positions of strong and weak acceptors and donors. Indicate where the Fermi level is for the three cases of $n = p$, $n > p$, and $n < p$.
   d. Compare the K value with that of Si (2×10^{20} cm^{-6}) and account for the difference qualitatively.
   e. Characterize In and Sb as donors or acceptors, explain why they have this characteristic in Ge, give the equations for their ionization, and describe their effect on the electron and hole concentrations.
   f. Predict what happens to the Fermi level if you begin with only In-doped Ge and then add Sb until the Sb concentration is much higher than the In concentration. What aqueous reaction is this like?
   g. When gold, Au, is doped into Ge, it is a weak acceptor, with an energy level of 0.2 eV. Where would it be placed on the band diagram and why?

7. Consider a solar cell (photovoltaic cell) made of GaAs, which has a band gap of 1.4 eV at room temperature.
   a. A rough absorption spectrum for GaAs is shown in Figure 8.22A. Explain its qualitative shape.
   b. A p–n junction is made from GaAs by starting with Se-doped GaAs (Se substitutes for As) and heating the solid in an atmosphere of Zn, which substitutes for Ga. Explain why this makes a p–n junction. With which semiconductor containing only a single kind of atom is GaAs isoelectronic?
   c. Each photon absorbed by a GaAs solar cell creates one conduction-band electron and one valence-band hole that contribute to electrical current in the solar cell. The efficiency of a solar cell (defined as electrical energy produced divided by the photon energy absorbed) is greatest for photons having roughly the band-gap energy. Why?
   d. Shown in Figure 8.22B is the solar output at the earth's surface. Why is GaAs a good choice for a solar cell material?

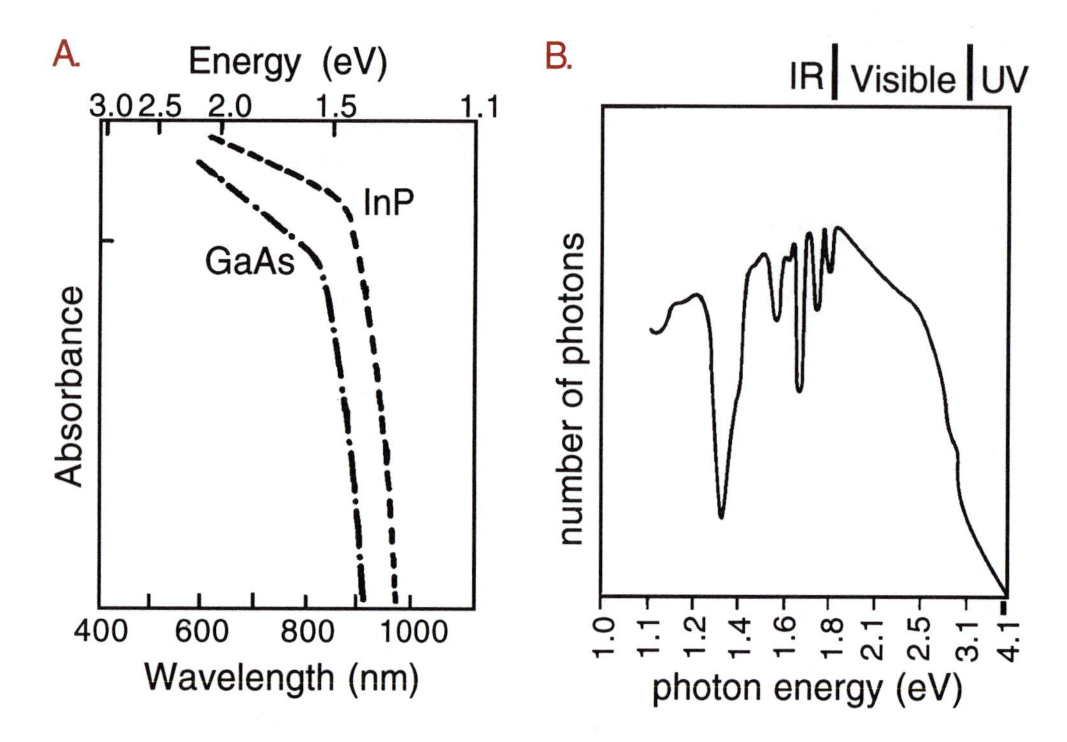

Figure 8.22. A: Absorption spectra of GaAs and InP semiconductors. B: The solar energy spectrum as measured at the earth's surface under average weather conditions. (Adapted with permission from references 15 and 16.)

8. Explain why the semiconductivity of pure silicon is strongly temperature-dependent while that of doped silicon is much less so.

9. An electrochemical cell is to be used to run a LED.
   a. Sketch all the components of an electrochemical cell that consists of Ag and Mg electrodes and suitable 1 M electrolytes (identify what they are).
   b. What voltage will this cell generate if the E^o values are as follows:

$$Ag^+ + e^- \rightarrow Ag \qquad E^o = 0.80\ V$$

$$Mg^{2+} + 2e^- \rightarrow Mg \qquad E^o = -2.37\ V$$

   c. If the voltmeter is removed, sketch the direction in which electrons will spontaneously flow in the wire and describe the net chemistry that will take place in each cell compartment.
   d. An InP LED is connected between the electrodes. Does it matter which side of the p–n junction is connected to which electrode and, if so, how do you connect it to make it glow?
   e. Suggest dopants that will substitute for In to make the p-type region of the LED and for P to make the n-type region. Explain your choices.
   f. Use the sketch shown in Figure 8.22A to predict the energy of the LED's light output, and explain your prediction.
   g. Discuss how long the LED can remain lit by the electrochemical cell by analyzing qualitatively what happens to the cell voltage over time. What energy conversions are taking place?

10. The ionization of a weak donor might be written as $D \rightarrow D^+ + e^-$; or as $D + h^+ \rightarrow D^+$. Interpret these two equations with an aqueous analogy.

11. In this chapter it is stated that doped semiconductors will give rise to narrow spectra like that of the hydrogen atom. Calculate the energies for the following transitions in phosphorus-doped silicon. For each calculated energy, determine in what region of the electromagnetic spectrum these transitions will fall. (Take $0.3m_{e-}$ for the effective mass of the electron and $\varepsilon_r = 12$.)
   a. $n = 4$ to $n = 3$
   b. $n = 4$ to $n = 1$
   c. $n = 3$ to $n = 2$
   d. These spectra are often obtained at liquid helium temperature, 4.2 K. What is the ambient thermal energy at this temperature (kT, or RT on a molar basis) and how does it compare to the energies associated with these transitions?

12. Which of the following represents a junction between p-type ZnSe (band gap 2.7 eV) and n-type GaAs (band gap 1.4 eV) at equilibrium? The dashed line is the Fermi level.

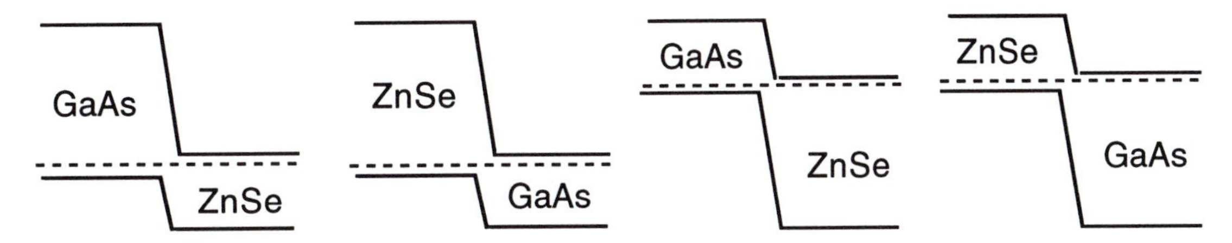

13. In problem 12, which color light will pass through both semiconductors without being absorbed?
   a. ultraviolet light of 3.0 eV
   b. green light of 2.4 eV
   c. red light of 1.7 eV
   d. near-infrared light of 1.1 eV

14. If zinc is doped into germanium, it can contribute two holes to the valence band. Explain. Can you make an analogy to a particular kind of aqueous acid?

15. Using the Nernst equation, calculate the voltage that could be obtained from a silicon p–n junction where the majority carriers in the n- and p- type regions have a concentration of 10^{17} cm^{-3} ($K \sim 10^{20}$ cm^{-6}).

16. A 5-mW diode laser with output at 670 nm is directed at a solar cell.
   a. To how many moles of photons/second does this power correspond (1 W = 1 J/s)?
   b. If all of these photons are absorbed by the solar cell and each photon produces one electron in the solar cell circuit, how much current in amperes (1 A = 1 coulomb/sec) would be measured ?

Chapter 9

Applications of Thermodynamics: Phase Changes

Phase changes include interconversions of solids, liquids, and gases, as in, for example, ice melting to liquid water and solid carbon dioxide (dry ice) subliming into the gaseous state. There are many technologically important classes of phase changes, however, that occur exclusively in the solid state.

In this chapter, after a qualitative discussion using Le Chatelier's principle to interpret the effects of temperature and pressure on equilibria involving solids, three types of solid-state phase changes will be presented. The first, involving NiTi "memory metal," is representative of a class of phase changes called martensitic transformations, wherein changes in temperature and pressure (mechanical stress) can cause a reversible shift of atomic positions in the crystal that leads to remarkable changes in mechanical properties. The second, based on the $YBa_2Cu_3O_{7-x}$ superconductor, highlights a phase change characterized by a pairing of electrons that results in striking changes in electromagnetic properties. Finally, temperature-induced changes in ionic conductivity and optical properties arise from a solid-state phase change in Cu_2HgI_4 that involves shifting from an ordered arrangement of cations to a disordered arrangement, a so-called "order-disorder" phase change.

Additional solid-state phase transformations include the conversions between hexagonal close-packed and face-centered cubic (fcc) structures in metals (Chapter 5) and interactions of electrons in magnetic garnets (*see* Experiment 13).

2725–X/93/0277$13.80/1

Effects of Temperature and Pressure on Equilibria: Illustrations of Le Chatelier's Principle with Solids

Qualitative discussions of chemical equilibrium typically focus on the effects of concentration and temperature on gaseous and solution equilibria and of pressure or volume changes on gaseous equilibria. Solids often are discussed in connection with heterogeneous equilibria such as solubility equilibria, where the point is made, seemingly paradoxically at first, that solids can be ignored so long as they are present!

In fact, in addition to the electronic and ionic-based equilibria described in Chapter 8, solids engage in a variety of pressure- and temperature-dependent equilibria that enrich the qualitative predictions afforded by Le Chatelier's principle. A practical example involving nickel–titanium memory metal appears later in this chapter, but some general illustrations involving phase changes are given first.

The free energy, G, of a pure material can be used to predict which phase of the substance is favored at different combinations of pressure and temperature: At a given pressure and temperature, the phase having the lowest free energy will be the most stable phase. Le Chatelier's principle provides a means for predicting pressure and temperature effects: Equilibria of endothermic reactions shift to products with an increase in temperature, and an increase in pressure shifts the equilibrium to favor the denser phase. The free-energy changes corresponding to the predictions reflect entropy and volume considerations that are discussed briefly in Appendix 9.1.

The equilibrium involving liquid water and ice (eq 1) illustrates both temperature and pressure effects. As commonly written,

$$\text{reaction energy} + \underset{\text{less dense}}{H_2O_{(s)}} \rightleftarrows \underset{\text{more dense}}{H_2O_{(l)}} \qquad (1)$$

Le Chatelier's principle predicts that heat added to an ice–water mixture at equilibrium will convert some of the ice to liquid water.

For the same reaction, there is also a volume change, because ice is less dense than water, as illustrated by the flotation of ice in water. The conversion of water to ice can produce mechanical work, as is sometimes spectacularly illustrated by the volume increase accompanying freezing in an "ice bomb" *(1)*. In general, the direction of a chemical equilibrium that yields a volume increase can be used to perform mechanical or pressure–volume "PV" work.[1]

---

[1]The actual magnitude of mechanical energy effects is generally much smaller than thermal energy effects: even when large pressures are involved, they are often coupled to small volume changes so that the product of pressure and volume is a modest energy change.

Le Chatelier's principle predicts that if external pressure is applied to ice, the system will respond by shifting the equilibrium to favor the phase having greater density, which is liquid water. The ability to ice skate is partially attributable to the fact that pressure from the ice skate blade can help reduce friction by pushing the equilibrium toward formation of liquid water, and thus effectively lower the melting point of H_2O. References 2 and 3 present further discussion of the effects of the ice–water equilibrium applicable to skating and Demonstration 9.1.

A material with a similar property is elemental gallium, which melts at about 30 °C. Like H_2O, gallium expands on freezing. In principle, it should be possible to skate on a gallium rink. Of course, skate blades that do not react with the liquid metal would be needed!

Demonstration 9.1. The Effect of Pressure on the Melting Point of Ice

Materials

Iron pipe (1-inch diameter) and a rubber stopper (the length of pipe is not critical but a 1-foot-long piece works well)
Two ring stands and clamps
28-gauge copper wire (1-foot length)
Two 1-kg weights
Two foam blocks with notches cut in them

Procedure

- Prepare a cylinder of ice by placing a stopper in one end of the pipe, adding water until the pipe is full, and placing this container upright in a freezer. The ice may be removed from the pipe by passing warm water over the outside.
- Place the ring stands about 8 inches apart. Clamp the foam blocks and position the cylinder of ice in the grooves of the foam like the crossbar of a goalpost.
- Tie a 1-kg weight to each end of the wire. There should be about 10 inches of wire between the two weights.
- Hang the wire and weights over the piece of ice so that one weight will be on each side of the cylinder. The wire will begin to cut through the ice as the pressure from the mass of the wire and weights melts the ice.
- After 2–3 minutes the wire will have passed through the ice and the weights will fall to the floor. (You may want to place rags on the floor to lessen the noise when the weights fall.) The water above the cut will have refrozen.
- Show that the ice cylinder can be lifted off the support in one piece after the wire passes through and the pressure is removed.

The same principles can be applied to transformations occurring exclusively in the solid state. For all of the examples that follow, external pressure can be used to convert the less dense to the more dense phase. A typical illustration is the conversion of bcc to fcc metals:

$$\underset{\text{less dense}}{\text{bcc}} \rightleftarrows \underset{\text{more dense}}{\text{fcc}} \tag{2}$$

The bcc packing efficiency is only 68%, and the fcc packing efficiency is 74%, so that an increase in pressure favors the close-packed fcc structure (Chapter 5).

A second example is the pressure-induced conversion of graphite to diamond, a method by which synthetic diamonds are made at high temperature. Pressure can be used to overcome the unfavorable enthalpy and entropy changes associated with this transformation. The Solid-State Model Kit can be used to build both graphite and diamond and to directly compare their packing efficiencies.

$$\text{reaction energy} + \underset{\text{less dense}}{\text{graphite (s)}} \rightleftarrows \underset{\text{more dense}}{\text{diamond (s)}} \tag{3}$$

[As noted in Chapter 1, there are now ways to produce diamond films, relatively inexpensively, by chemical vapor deposition methods (Chapter 10)].

An example of a phase change with historical significance is the conversion of less dense α-tin ("gray tin"; cubic diamond structure, Chapter 5) to denser β-tin ("white tin"; a tetragonal structure having a unit cell with a square base and rectangular sides), which occurs a little below room temperature.

$$\text{reaction energy} + \underset{\text{less dense}}{\alpha\text{-tin (s)}} \rightleftarrows \underset{\text{more dense}}{\beta\text{-tin (s)}} \tag{4}$$

In early European cathedrals, organ pipes made of tin were found to swell and then to disintegrate in cold weather *(4)*. The origin of this so-called "tin disease" is the phase transformation of tin: Low temperatures shift the equilibrium toward α-tin, and the lower density of this phase results in swelling and ultimately disintegration from the relatively large shifts in atomic position that are demanded by the phase transformation.

The phase changes that salts and minerals undergo are also important in understanding geological processes. The transformation of sodium chloride, for example, from the rock salt to the cesium chloride structure, equation 5, can be accomplished at 298 K at high pressure (~10^5 atm) *(5)*:

$$\underset{\text{less dense}}{\text{NaCl (rock salt structure)}} \rightleftarrows \underset{\text{more dense}}{\text{NaCl (cesium chloride structure)}} \tag{5}$$

An important feature of solid-state phase changes that distinguish them from many other phase changes is that superheating and supercooling are extremely common, being more the rule than the exception. Recall that solid-to-liquid phase changes like ice to liquid water occur abruptly at a well-defined temperature and pressure for pure samples. However, liquids can be supercooled, freezing substantially below their normal melting point, if a nucleation site for the growth of ice crystals is unavailable.[2] Similarly, liquids can be superheated, boiling above their normal boiling point, if a nucleation site for bubble formation is absent.

In solid-state phase changes, superheating and supercooling are common, because the new phase generally has to nucleate and grow within the phase that is initially present. The shifts in atomic position that are required for this growth are relatively slow, particularly if diffusion is required at temperatures near and below room temperature (*see* Chapter 10). Ultimately, the energy requirement for the growth of the new phase is met by shifting the temperature and/or pressure enough beyond the point of thermodynamic equilibrium through superheating or supercooling that sufficient chemical free energy is available to complete the structural conversion.

Nickel–Titanium Memory Metal

A captivating property of the alloy NiTi is its ability to "remember" a shape into which it has previously been fabricated *(7)*. The processes underlying the shape-memory effect illustrate links between the structure, microstructure, and composition of NiTi (the stoichiometry of the solid can vary by a few percent from 1:1 Ni:Ti) and general thermodynamic principles of phase transformations.

Samples of NiTi readily undergo a martensitic phase transformation, a kind of ballet on the atomic scale: atoms in solids like NiTi can make subtle orchestrated shifts in their positions, which, because little energy is involved, can occur in the vicinity of room temperature and below. In any martensitic phase transformation the high-temperature phase is called austenite, and the low temperature phase is called martensite, regardless of the crystal structures of the phases. This nomenclature was originally derived from the martensitic phase transformation in steel (Chapter 6).

Of the many alloys that exhibit the shape-memory effect, *(8, 9)* the "memory metal" comprising nickel and titanium atoms is one of the most accessible and dramatic. Memory metal is sometimes called nitinol, which is short for **ni**ckel **ti**tanium **N**aval **O**rdnance **L**aboratory and which acknowledges the site of its discovery in 1965 *(10)*. The relatively low cost of the alloy and its ready availability make it ideal for lecture demonstrations and laboratory experiments in general chemistry courses. In addition to illustrating the martensitic phase change, the ability of NiTi to remember its shape under certain experimental conditions makes it an

[2]The related phenomenon of supersaturation can be demonstrated by using sodium acetate solution or sodium thiosulfate solutions. *See* reference 6.

entry point for discussing “smart” materials. “Smart” materials have the capability to sense changes in their environment and respond to the changes in a preprogrammed way. These new high-tech solids are being used in a variety of artistic, medical, and engineering applications.

The Structural Cycle of Shape Memory

As described in Demonstration 9.2, thin, straight wires and rods of NiTi display an astonishing, counterintuitive property: they can be distorted into a variety of shapes while in their low-temperature phase at room temperature, then, upon gentle warming into the high-temperature phase, the samples will return to their original shape. Cooling the sample from its high-temperature phase into its low-temperature phase will not generally cause an observable change in its shape.

Demonstration 9.2. The Shape-Memory Property of Memory Metal

Materials

NiTi wire (3 inches long by 0.03 inches in diameter works well) with transition temperatures between 30 and 50 °C; *see* Supplier Information
Hot water (over 50 °C) in a crystallizing dish or small beaker (or heat gun or hair dryer)
Gloves or tongs (optional)
Tape (optional)
Overhead projector and screen (optional)
9-V battery (optional)
Battery snap with alligator clips attached to the leads (optional)
Sample of memory metal that spells “ICE” (available from the Institute for Chemical Education) (optional)
3–6-V power supply (such as a lantern battery) and leads with alligator clips (optional)

Procedure

- Place a sample of low-temperature phase martensite NiTi wire on the overhead projector to show its linear shape.
- Bend it; for a particularly dramatic demonstration, coil the wire around a finger to form a helical shape.
- Then, holding one end of the coiled wire, immerse it in hot water. (Alternatively, place it in the hot-air stream of a hair dryer. As a third option, while wearing gloves or holding the wire with tongs, place the wire in the hot-air stream of a heat gun.) As the wire is heated, it transforms into the austenite phase and straightens back

into the linear shape it had been "trained" to remember.

Variations

To demonstrate this effect in a large lecture room, place a crystallizing dish containing hot water on an overhead projector so that the straightening of the wire is readily visible. Alternatively, tape the coiled wire at one end onto the overhead projector (to prevent the wire from blowing away), and use a heat gun to straighten the wire. In a laboratory setting, students can do this experiment with their own strip of NiTi (see Experiment 10).

You may use resistive heating in lieu of hot air or water to demonstrate the shape-memory effect. For this experiment, connect a coiled NiTi wire to a 9-V battery with two alligator clips soldered to a 9-V battery snap. The current provided by the battery heats the NiTi and thereby transforms it back into the austenite phase. **CAUTION: Because of the large current drain, do not maintain the connection for more than 10 seconds. If the battery becomes hot, disconnect the wire immediately.**

Another twist on all of these experiments is to use a sample of NiTi that has been trained to spell "ICE" (Figure 9.1). After either crumpling the metal or stretching it to the point where it is no longer readable, the shape can be restored either by immersing it in water at about 50 °C, heating with hot air, or attaching alligator clips to the ends of the sample and connecting them to two 1.5-V dry cells hooked in series. A 6-V lantern battery will also work.

Figure 9.1. An illustration of the shape of a strand of memory metal that is trained to spell "ICE". (*See* Supplier Information.)

An understanding of the features of the shape-memory cycle lies in the relationship between the structures of the high- and low-temperature phases. These structures have been investigated in detail using X-ray diffraction (Chapter 4) *(11, 12)*. In Figure 9.2A the structure of austenite, the high-temperature phase, is shown from two different perspectives. The austenite phase adopts the CsCl structure (Chapter 5), shown at the left, which can be described as having Ni atoms at each corner of the unit cell cubes and Ti atoms in the centers of the cubes, or vice versa. Smaller spheres lie behind the plane of the paper in this view.

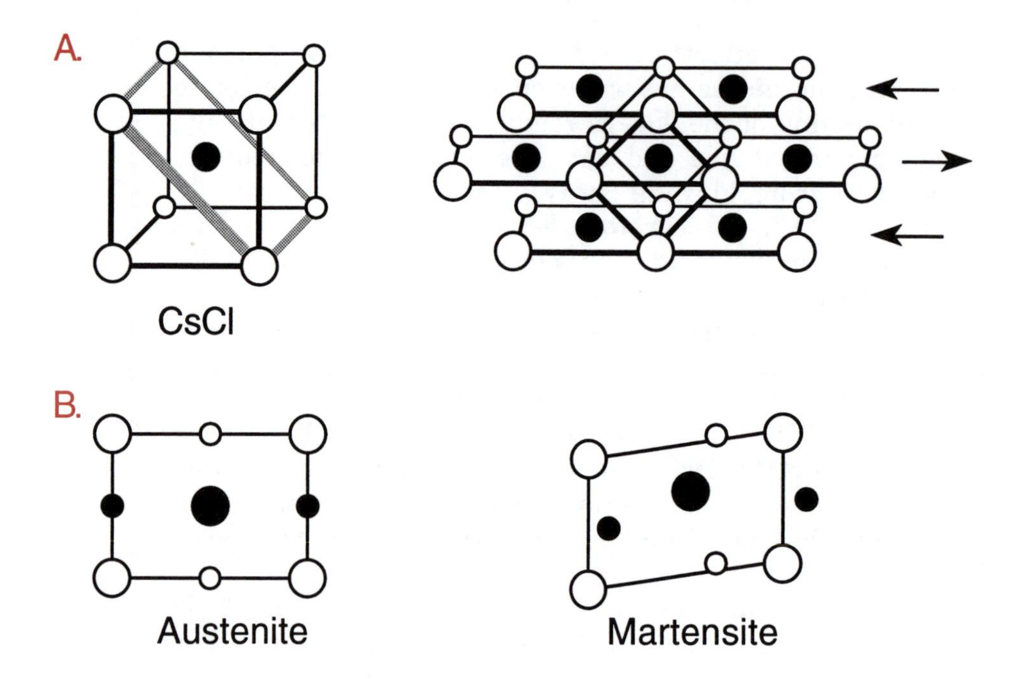

Figure 9.2. A: Two different views of the CsCl structure adopted by NiTi in the austenite phase. The common depiction is at the left, showing a cube of one type of atom with the other atom at the center of the cube. The white and black spheres represent the two kinds of atoms. The size of the spheres indicates depth; larger spheres are closer, and smaller spheres are farther away. At the right, the structure is depicted as series of stacked planes, with the cube also shown for comparison. The arrows indicate one component of the sliding of the planes that leads to changes in the atomic positions during the austenite-to-martensite phase transition. B: The structures of the austenite and martensite phases are represented by a two-dimensional projection of one of the rectangles shown in part A. Once again, the size of the spheres represents depth; the smaller spheres belong to the layer below the plane of the larger spheres. An additional shearing mechanism (not shown in A) also changes the angles in the structure of the martensite phase from 90 to 96°.

If the NiTi unit cell cube is balanced on a cube edge, the structure of austenite may also be represented as a stack of planes, as shown at the right in Figure 9.2A. This view is particularly useful in describing the relationship of the structure of austenite to the structure of martensite. As a more compact presentation of this view, the stack of planes can be reduced to a two-dimensional rectangular projection from above, with the rectangle derived from two edges and two face diagonals of the cubic unit cell (shaded outline at the left of Figure 9.2A). Two of the layers in the stack are depicted in this manner at the left of Figure 9.2B: In this view, large spheres represent the atoms in the middle layer of Figure 9.2A, and small spheres represent the atoms in the next layer above or below.

During the structural transformation from austenite to martensite upon cooling, these particular planes of atoms in the austenite structure slide relative to one another (indicated by arrows on the right of Figure 9.2A), which does not change the 90° angles, and deform by shearing, which does change the 90° angles to about 96°. (A shearing motion, illustrated in Figure 9.3, can result when two opposing forces are displaced from one another.) Comparing the two-dimensional projection of the martensite with that of the austenite (Figure 9.2B) highlights the loss of 90° angles and the offset of alternating layers relative to one another as the austenite phase transforms into the low-temperature martensite phase.

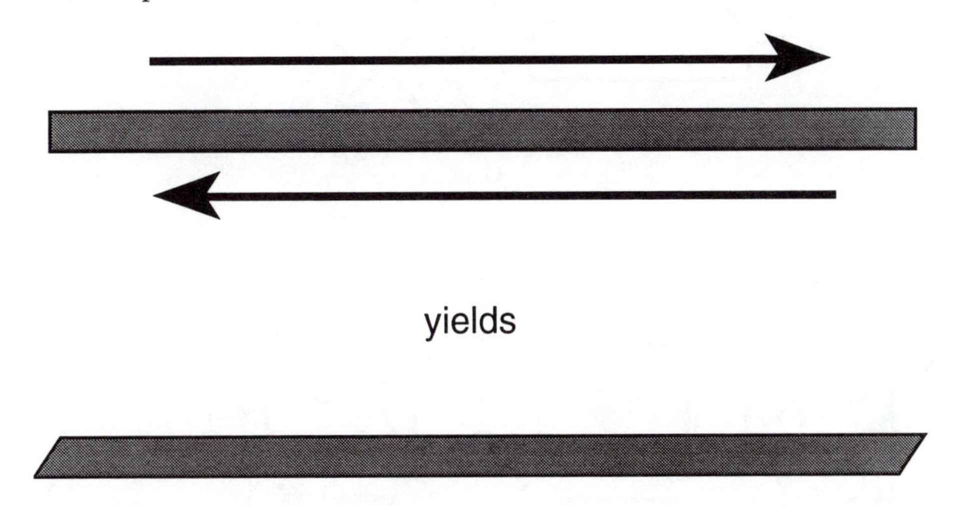

Figure 9.3. Application of displaced opposing parallel forces to an object creates shear.

Although the motions involved in the transformation from austenite to martensite are relatively simple, there are 24 different ways to carry out the transformation. To understand the origin of these 24 different so-called "variants," the directions of the two types of shearing during the transformation are shown in Figure 9.4A. The planes can shift relative to one another in each of two directions parallel to the face diagonal (marked by the dark arrows labeled a), and shear in each of two directions parallel to the cell edges (marked by pairs of arrows labeled b and c). In addition, there are six equivalent face diagonal planes in the CsCl structure, highlighted in Figure 9.4B. Each of these six sets of planes can thus shift in one of two directions and distort in one of two directions. The result is $6 \times 2 \times 2$ or 24 different ways to transform the structure into martensite.

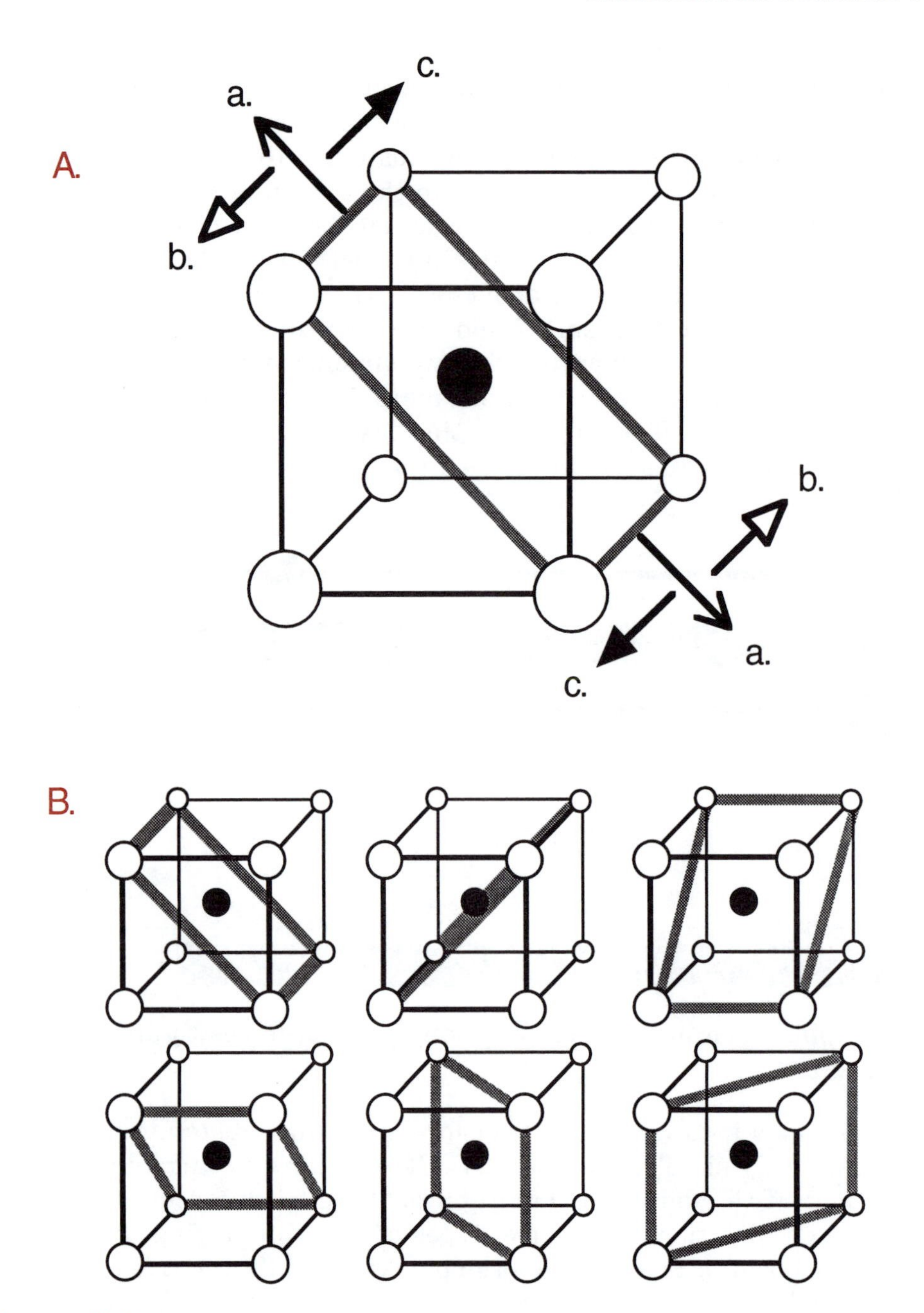

Figure 9.4. A: A total of four martensite variants may grow from each plane passing through a face diagonal in the CsCl structure. The planes may shift in the direction of either of the arrows labeled a at the edge of the face diagonal plane. This shifting will not change the 90° angles of the plane. An additional shear can occur through simultaneous motion in the directions indicated by either pair of arrows labeled b or c. These motions will destroy the 90° angles of the plane. B: Six equivalent planes pass through face diagonals in the CsCl structure. Thus a total of $6 \times 2 \times 2 = 24$ different variants may grow from the planes.

The phase transformation and shape-memory cycle is presented in Figure 9.5. At the top of the figure is the two-dimensional projection representation of the structure of austenite, a rectangle. The matrix of rectangles shown below this representation depicts how the structural units pack to fill space on a larger scale. As a sample of austenite is cooled through the phase-transition temperature, the structure transforms to martensite. To reflect the 24 different orientations of martensite, the variants, the long-range structure of martensite is represented in two dimensions in Figure 9.5 as a set of tilted parallelograms. These parallelograms can be packed together so that the overall shape does not change significantly during the phase change, and, as already noted, no observable change in shape accompanies the austenite-to-martensite transformation: The figure is meant to emphasize that the shape of the matrix of rectangles, and the matrix of tilted parallelograms on the atomic scale is roughly the same on the macroscopic scale. In fact, the density difference in the two phases is less than 0.5% (the martensite is slightly denser, discussed later) *(13)*.

The NiTi wire, as purchased, is typically in the low-temperature martensite phase at room temperature. The wire has been previously trained to remember a linear shape in the austenite phase. When the NiTi is bent or coiled in the low-temperature martensite phase, the effect on the microstructure is a reorientation of the variants corresponding to a macroscopic change in shape. This reorientation is shown in the lower left of Figure 9.5, as the inclination of most of the parallelograms to a common direction. This reversible mechanism for accommodating stress distinguishes NiTi from most metals, in which similar stress would introduce defects into the crystal structure, or cause planes of atoms to slip past each other and permanently deform the metal.[3]

In the last step of the shape-memory cycle, heat is used to transform the martensite back into the austenite phase. The atoms recover their initial positions, and the initial macroscopic shape of the sample is restored. Although several pathways back to the austenite structure exist, only the lowest energy pathway restores the initial ordered CsCl structure in which Ni atoms are surrounded exclusively by Ti atoms and vice versa. Other paths would produce higher energy structures in which Ni and Ti atoms have a mixture of Ni and Ti atoms as nearest neighbors in the crystal.

The enthalpy associated with the phase change is relatively small, as is typical for a solid-state phase transformation: The conversion of austenite to martensite is modestly exothermic with an enthalpy change on the order of only –2 kJ/mol.

---

[3]Bending a metal can result in work hardening. *See* Chapter 6.

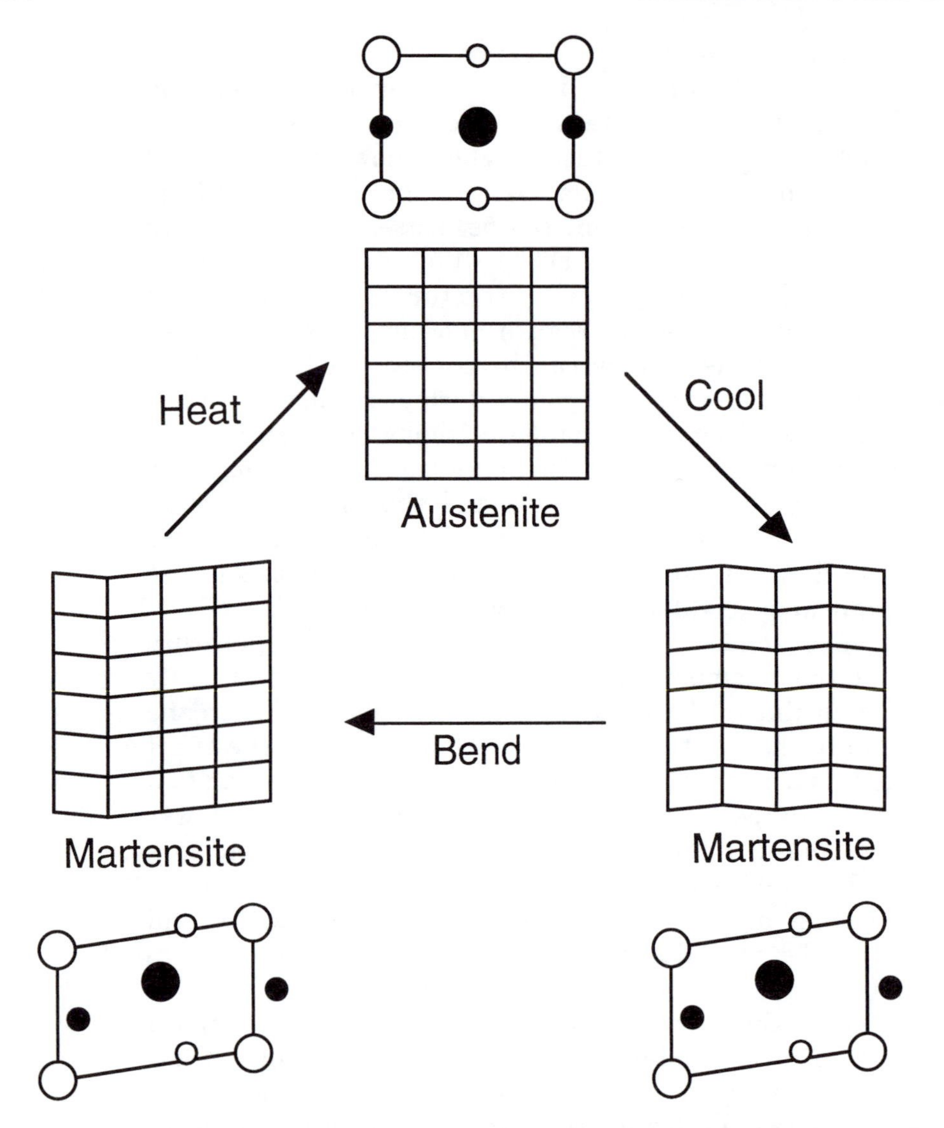

Figure 9.5. The structural features of NiTi that give rise to the shape-memory effect. The cycle starts with the NiTi in the austenite phase shown at the top of the figure. As the NiTi is cooled, going clockwise around the cycle, the transformation into the martensite proceeds. The diagonal planes slide past each other (as shown in the upper right of Figure 9.2A) and deform slightly to a parallelogram in the two-dimensional projection. In this projection, two differently oriented martensite variants, one angled to the left and one to the right, may be observed. When the martensite is bent, the variants can reorient from left to right or vice versa to relieve the stress. When the NiTi is heated, the lowest energy pathway returns the atoms to their original positions and maintains the ordering of the atoms, with nickel atoms surrounded exclusively by titanium atoms and vice versa.

Changing the "Memorized" Shape of NiTi

As purchased, samples of NiTi are polycrystalline, meaning that regions of crystallinity on the order of micrometers in size (occasionally as large as a millimeter) are separated from one another by grain boundaries (Chapter 6). Within the grains, the nickel and titanium atoms are arranged with almost perfect order. However, occasional mistakes in the packing, such as dislocations and other defects described in Chapter 6, may occur *(14)*. The nickel and titanium atoms must return to exactly the same position each time the NiTi is heated into the austenite phase, so the configuration of defects in this phase effectively pins the austenite into a given shape. To give the NiTi a new shape to remember requires substantial energy, provided, for example, by heating the NiTi in a candle flame to about 500 °C, while it is physically held in the desired shape. During the annealing (heating) process, the atoms surrounding the defects gain enough energy to relax into lower energy configurations, and this new configuration of defects effectively pins the austenite into its new shape. The gentle heating used in the shape-memory demonstration to return to the austenite phase from the martensite phase does not provide enough energy to allow the defects to readjust.

Demonstration 9.3. The Annealing of NiTi into a New Shape

Materials

NiTi wire (3 inches long, 0.03-inch diameter; *see* Supplier Information)
Candle and matches
Heat gun or hot water

Procedure

- Grasp the two ends of the wire, and place the middle of the wire in the center of the candle flame. Try to bend the wire into a V-shape. It will yield as it becomes hot, at which point it should be removed immediately from the flame.
- The wire will cool off in a few seconds by waving it in the air or blowing on it. After the wire has returned to room temperature, Demonstration 9.2 can be repeated: The wire can be coiled, but upon heating with hot water or the heat gun, it will now return to the V-shape, not the linear shape.

More complicated shapes like the "ICE" sample described in Demonstration 9.2 require a jig like that shown in Figure 9.6, because the heat treatment will initially cause the wire to try to recover its original shape before it relaxes into the new shape. The jig itself consists of a series of metal pegs firmly mounted in a metal base. The wire is wound tightly around the pegs and fastened securely at either end. The apparatus is placed in a standard box furnace for about 15 minutes at 500–540 °C. Prolonged heating should be avoided as that seems to degrade the shape-memory behavior. Again, these are the conditions that have been used to prepare the "ICE" sample; the preparation of samples such as this with fairly complicated shapes is very sensitive to temperature, oven geometry, and heating time. The conditions described here may have to be modified to form other shapes.

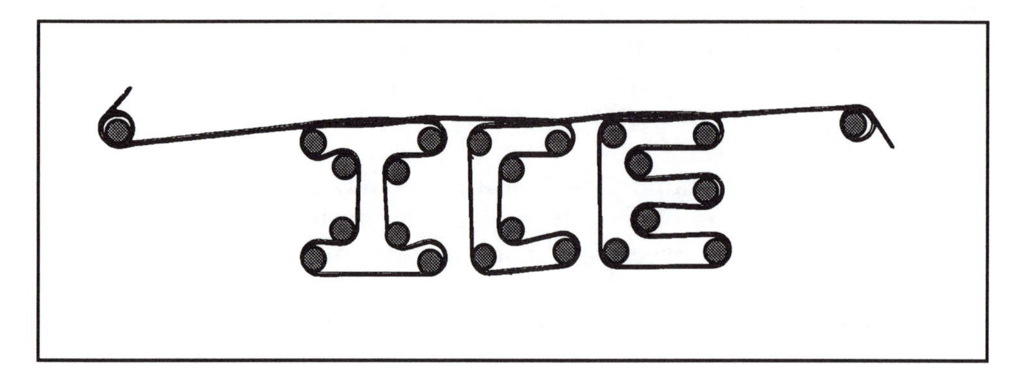

Figure 9.6. The jig used to train the memory metal that spells "ICE".

Mechanical Properties and Superelasticity of NiTi: Changing Phase with Pressure

The different structures of austenite and martensite result in considerably different mechanical properties for the two phases. The CsCl austenite structure is relatively rigid and hard. In contrast, the ability to re-orient variants of the martensite phase imparts mechanical flexibility and makes the low-temperature phase a little softer than the high-temperature phase.

Although the structural cycle just discussed is based on using temperature to interconvert the two phases of NiTi, the equilibrium is also a strong function of pressure. Under certain conditions, the austenite phase may be mechanically transformed into the martensite phase, and become elastic; that is, when the stress is removed, the martensite phase will transform back to the austenite phase and the NiTi will return to its undeformed shape. This mechanical property is sometimes known as "pseudoelasticity,"[4] or more specifically "superelasticity," and many NiTi applications are based on it.

Superelasticity may be understood as another example of the pressure-induced phase transformations discussed at the beginning of this chapter. Applying mechanical pressure to NiTi in the austenite form can cause the transformation to martensite to occur without a change in temperature.

[4]It is pseudoelasticity and not elasticity because the strain, or fractional change in length of the wire under tension or compression, is not quite linear with the amount of tension or compression that is applied.

This mechanically induced pressure, called stress, is a force exerted over an area of the material, and thus has units of pressure. Both tensile stress, in which the atoms are pulled apart (by applying tension), and compressive stress, in which the atoms are pushed together (by compression), can occur when a material is bent, as shown in Figure 9.7.

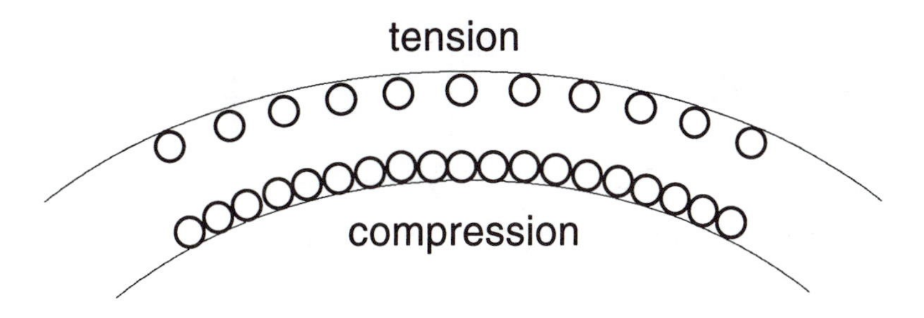

Figure 9.7. The simultaneous occurrence of tension (atoms are pulled apart) and compression (atoms are pushed together) when a metal rod is bent.

The conversion under pressure of the lower density, austenite structure to the higher density, martensite structure is analogous to the transformation under pressure of low-density, gaseous water to the denser, liquid form of water. In accord with the prediction of Le Chatelier's principle, application of pressure will favor formation of the denser martensite phase (*see* Figure 9.8), even though the difference in densities of the two phases is less than 0.5%.

Another way to visualize this relationship is with the traditional phase diagram for water that is shown in most general chemistry texts. On the pressure–temperature phase diagram, lines with positive slopes (like that representing the gas–liquid water equilibrium) indicate that an increase in pressure at a given temperature will raise the temperature at which interconversion of the two phases occurs (the boiling point, for example), and thus permit conversion of some of the gas to the liquid state. The Clapeyron equation can be used to quantify the effect.[5] Similarly, bending a sample of NiTi in the austenite form will cause an increase in the characteristic temperature at which the sample begins to be converted into the martensite phase in the region of the solid that is being compressed by the bending motion. If the increase in this temperature is large enough, calculable with the Clapeyron equation, some of the austenite will be converted to martensite. Releasing the stress causes the martensite to transform back into austenite with restoration of the original shape.

---

[5]In the case of the martensite–austenite transformation, a modified form of the Clapeyron equation may be used to determine the change in the martensite start temperature, M_S (the temperature at which the sample begins to transform into martensite; Figure 9.9), based on the applied stress, σ, and the strain, ε (the fractional change in length): $d\sigma/dM_S = -\Delta H/T\varepsilon$. The analogous equation in terms of pressure and volume changes is $dP/dT = \Delta H/T\Delta V$.

energy + $H_2O_{(l)}$ ⇄ $H_2O_{(g)}$

high density, low volume ⇄ low density, high volume

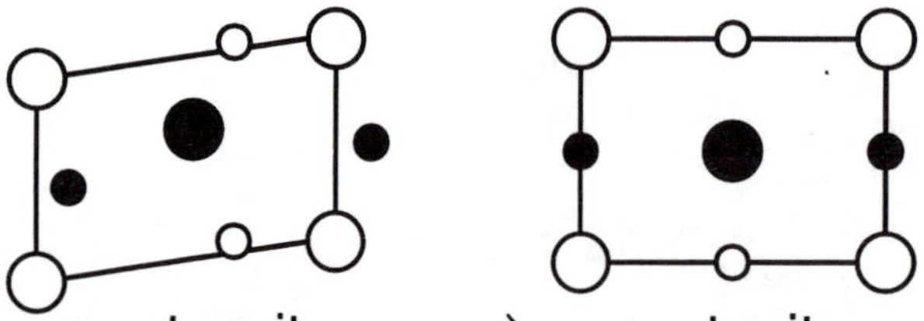

energy + martensite ⇄ austenite

high density, low volume ⇄ low density, high volume

Figure 9.8. The application of pressure will cause the austenite form of NiTi to transform into martensite at temperatures higher than those at which the transformation would occur at ambient pressure. This behavior is analogous to the observation that the boiling point of water rises under conditions of applied uniform pressure.

Thicker NiTi rods can be used to illustrate the critical role that atomic structure plays in defining mechanical properties.

Demonstration 9.4. The Mechanical Properties of Two NiTi Phases

Materials

NiTi rods (Shape Memory Applications; see Supplier Information) Rods of differing stoichiometry (2.5 inches long by 0.10 inch in diameter) are available in both the low-temperature (martensitic rods) and high-temperature phase (austenitic rods) at room temperature. The rods have slightly different Ni:Ti atomic ratios, leading to different temperatures for the phase transition, as will be discussed *(15)*.

Hot water (above 50 °C)

Liquid nitrogen in a 1-L Dewar flask

Tongs

Gloves

Heat gun

Overhead projector and screen (optional)

Procedure

- Try to bend rods that are in the two different phases at room temperature. A rod that is in the high-temperature austenite phase at room temperature will be extremely difficult to bend into a V-shape. In contrast, a rod that is in the low-temperature martensite phase at room temperature is comparatively flexible. Of course, the bent rod can be placed in very hot water (above 50 °C) or in front of a heat gun to restore its linear shape.
- Cool an inflexible rod that is *in the austenite phase at room temperature* in liquid nitrogen. **CAUTION: Liquid nitrogen is extremely cold. Do not allow it to come into contact with skin or clothing, as severe frostbite may result. Wear gloves when transferring and using liquid nitrogen.**
- Use tongs to remove the rod from the liquid nitrogen. While wearing gloves, bend it into a V-shape. As the rod warms back to room temperature, it will return to a linear shape. (The bent rod can be placed on an overhead projector to show its return to linearity to an audience.)
- Alternatively, immerse a rod that is in the martensite phase at room temperature in very hot water. It will become inflexible when hot and flexible as it cools back into the martensite phase.
- Try to scratch a rod in the martensite phase with one in the austenite phase. The end of an austenitic rod will scratch the martensite, but a martensitic rod will not scratch the surface of the austenite. This activity demonstrates the hardness of the austenite compared to the martensite.

Superelasticity can be demonstrated with a pair of eyeglasses having frames made from NiTi memory metal. The NiTi must start in the austenite phase for this application; therefore, the transition temperature must be set slightly below room temperature.

Demonstration 9.5. Superelasticity

Materials

Eyeglass frames made from NiTi memory metal (commonly available from optometrists)
Overhead projector
Liquid nitrogen and Dewar flask with a large enough mouth to accommodate the frames
Gloves
Tongs

Procedure

- Lay the eyeglass frames on the overhead projector to show the initial shape.
- Bend the frames. When bent, the frames yield. But once the stress is released, the frames snap back into their original shape.
- Cool the frames in liquid nitrogen. **CAUTION: Liquid nitrogen is extremely cold. Do not allow it to come into contact with skin or clothing, as severe frostbite may result. Wear gloves when transferring and using liquid nitrogen.** Allow them to remain there until thermal equilibrium is reached.
- Put on insulating gloves and remove the frames from the liquid nitrogen using the tongs. Bend the frames. Lay the bent frames on the bench top or on an overhead projector. Observe what happens as the eyeglasses return to room temperature. The frames stay bent if bent when cold, but return to their original shape on warming to room temperature.

Acoustic Properties of NiTi Phases

The structural differences between austenitic and martensitic NiTi affect sound propagation in the two forms of the solid. The regular structure of the austenite leads to a ringing sound when a sample is dropped or struck: a sound wave launched in the material travels relatively unimpeded through it. In the martensite phase, the boundaries between regions with different orientations (the variants) appear to act as baffles for vibrations, resulting in a more muffled-sounding thud when martensitic samples are dropped.[6] Cold-worked metals, which have a very high density of induced defects, are very good at absorbing sound.

Laboratory. The marked acoustical difference in the two phases also permits students to determine the approximate temperature of the phase transformation for the NiTi rods. A laboratory procedure is given in Experiment 10.

[6]Above about 20 K, defects in metals are the primary absorbers of sound waves traveling through metals, and boundaries between martensitic variants behave similarly. *See* reference 16.

Demonstration 9.6. Acoustic Properties of Two NiTi Phases

Materials

NiTi rods (one austenitic and one martensitic; 0.10-inch diameter, 2.5 inches long; *see* Supplier Information)
String
Stirring hot plate and stirring bar
400-mL beaker
Water
Thermometer
Ring stand and ring

Procedure

- Tie a string around a rod that is in the martensite phase at room temperature.
- Fill the 400-mL beaker with water and place it on the stirring hot plate. Add a stirring bar and begin stirring but not heating.
- Loosely tie the string to the ring of the ring stand and suspend the rod in the water so that it is not touching the side or the bottom of the beaker but is completely submerged.
- After the rod has been immersed for about 2 minutes, record the temperature of the water, untie the string from the ringstand, and immediately drop the rod onto a solid surface such as a benchtop or a cement floor. Note the sound that the rod makes when dropped (thud, ring, or intermediate).
- Return the sample of nitinol back to the water bath and begin slowly heating it. The hot plate should be on the lowest setting. Remove the rod from the water bath at 10 °C intervals and drop it, recording the sound made upon dropping the sample until it is clearly in the high-temperature phase. (A clear ring will be heard. This sound can be verified by dropping the austenite rod.)
- The characteristic acoustic signatures of the two phases of NiTi are revealed simply by dropping them. The austenite rods will ring like a bell, and the martensite samples will yield a dull thud! A sample that contains both phases (discussed later) will yield a sound that is intermediate between a ring and a thud. For best results, the rod should be dropped from a roughly reproducible initial position that is approximately parallel to the surface it will strike, and from a height that ensures an easily recognizable sound. The audio characteristics can also be measured as the bath cools back to room temperature.

Hysteresis

More quantitative measurements of the phase change in NiTi, through magnetic, electrical, calorimetric, or diffraction measurements, reveal that the heating curve may be displaced toward higher temperature from the cooling curve, as shown in Figure 9.9, a phenomenon known as hysteresis. Hysteresis reflects the fact that one solid phase needs to grow within a region of the other. Growth of the new phase creates elastic strain in the crystalline region around it, adding to the chemical free energy necessary for enlarging the grain of the new phase *(8, 17)*. Thus, no matter how slowly the conversion takes place, this thermodynamically based displacement of the heating and cooling curves will occur, although it may be very small.

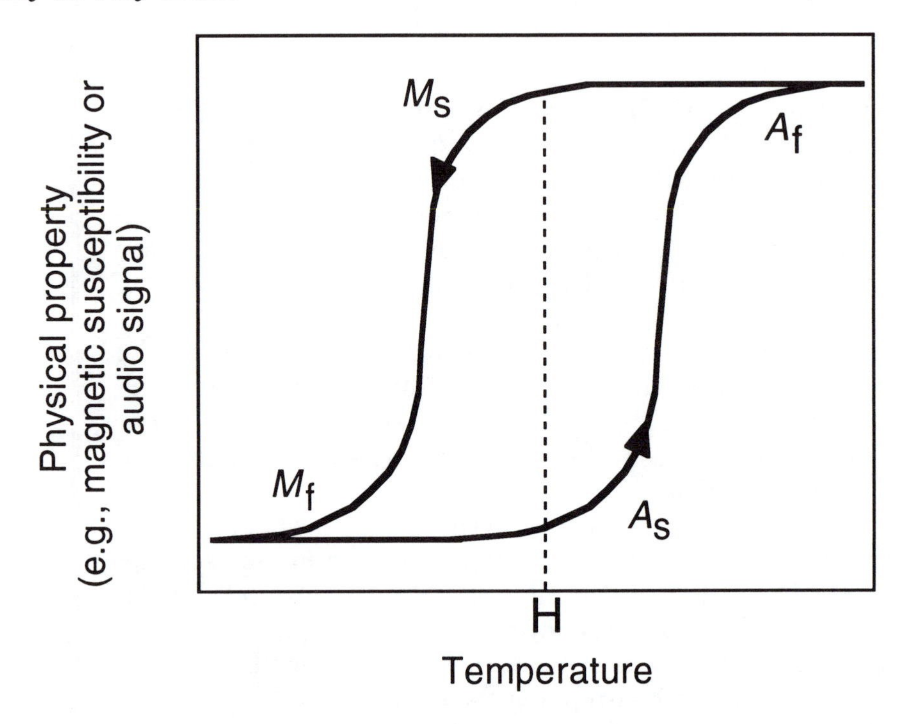

Figure 9.9. The martensitic phase transformation in NiTi is evident in a number of physical properties. For example, the general characteristics of the magnetic susceptibility as a function of temperature show four transition temperatures: A_s, the start of the martensite-to-austenite transition on heating; A_f, the end of the martensite-to-austenite transition on heating; M_s, the start of the austenite-to-martensite transition on cooling; and M_f, the end of the austenite-to-martensite transition on cooling. The heating and cooling curves are often separated by as much as 20 °C. This separation, or hysteresis, allows both the austenite and martensite to be accessible at the same temperature (marked H), depending upon whether the sample was previously heated or cooled.

The rods used to investigate the acoustic properties of the two phases may or may not exhibit hysteresis in the experiment. With some NiTi samples, an intermediate phase may grow during the austenite-to-martensite transition, the so-called R (rhombohedral) phase, which may show relatively little hysteresis in its conversions with austenite *(9, 18)*. Any samples that are found to have a significant displacement in the heating and cooling curves lend themselves to the following interesting experiment: At certain temperatures in the middle of the transition region, such as the one labeled H in Figure 9.9, the sample will thud if it is first cooled into the martensite phase, then heated to temperature H and dropped. In contrast, if the sample is first heated into the austenite phase, then cooled in the bath to temperature H and dropped, it will ring. That is, the sample phase(s) present at temperatures within the hysteresis loop depend on the thermal treatment used to reach the temperature.

Chemical Composition

NiTi has been prepared traditionally by heating the elements together above the NiTi melting point (1200–1300 °C, depending on composition). Exclusion of oxygen is critical during the synthesis because of the ease with which Ti oxidizes.

NiTi can tolerate small deviations in chemical composition around the 1:1 stoichiometry before the shape memory effect is lost. A more accurate representation of NiTi within this range is Ni_xTi_{1-x}, wherein the shape-memory effect has been observed when x lies between 0.47 and 0.51. However, the temperature at which the martensitic phase change occurs in NiTi is a sensitive function of stoichiometry. As indicated in Figure 9.10, variations of a few percent around the equiatomic point in Ni_xTi_{1-x} cause substantial changes in the temperature of the phase transformation *(15)*.

One of the reasons that the transition temperature begins to fall with increasing Ni content is that small domains of other phases begin to precipitate as the composition deviates from 1:1 ($x = 0.50$). In addition, small quantities of impurity atoms that may be present may also influence the transition temperature. Tuning of the transition temperature to as high as several hundred degrees has been accomplished by partial or complete substitution of Ni with Pd or Pt *(19, 20)*. Conversely, substitution of Ni with a few percent of Co or Fe substantially reduces the transition temperature. A discussion of the effects of adding a third transition metal may be found in reference 21. The result of this strong dependence of transition temperature upon composition is that samples of memory metal (such as the rods used in the demonstrations) can be made in either the martensite or the austenite phase at room temperature.

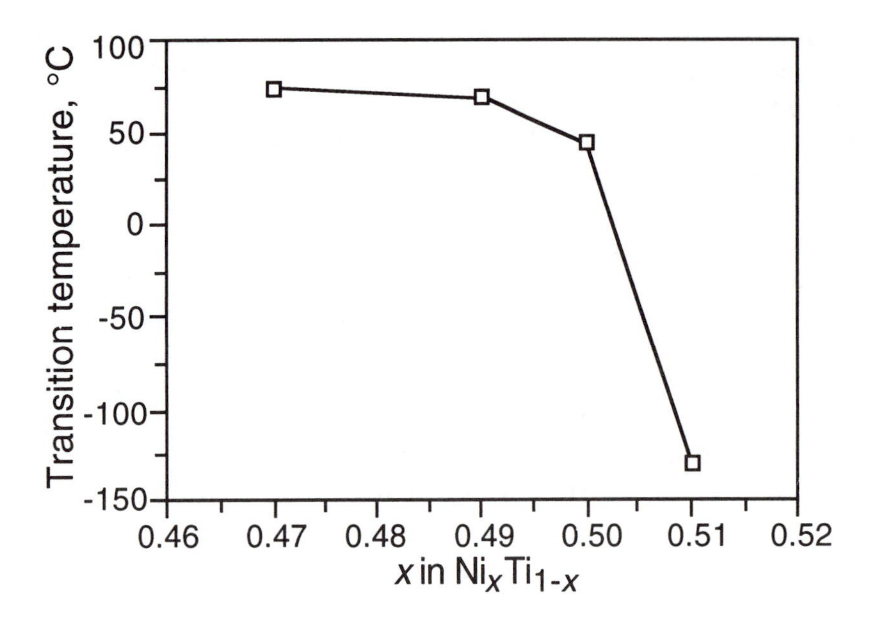

Figure 9.10. The effect of nickel concentration on the temperature of the phase transition. (Adapted from reference 18. This is a compilation of data by several researchers, who measured the composition and transition temperatures by different methods.)

Applications

Shape-memory alloys have been incorporated into a wide variety of applications. An overview of recent applications using shape-memory alloys may be found in reference 22. Some creative suggestions from undergraduates at the University of Wisconsin—Madison are presented in the following list. The shape-memory effect creates a temperature-sensitive "on–off" switch that has been used in coffee makers and scald-proof shower heads. Diesel-fueled Mercedes Benz cars have a shape-memory alloy-based valve that regulates the flow of transmission fluid in the engine as a function of temperature *(23)*. The NiTi memory metal valve controls the shifting pressure in the automatic transmission (as the temperature in the engine goes up, a higher pressure is desirable), and this pressure control in turn smoothes shifting between gears.

As shown in Figure 9.11, one NiTi spring, one steel spring, and a piston provide the switching mechanism. The springs and the piston are housed in a case with an inlet and outlet at the top and bottom. When the engine is cold, the NiTi spring is in the martensite phase, and is flexible. The steel spring provides enough force to push the piston to the left and collapse the NiTi spring. This collapse closes the flow path through the valve. As the engine warms up, the NiTi transforms into the austenite phase and remembers its expanded shape. It pushes against the piston and the steel spring, opens up the valve, and allows oil to flow through. If the NiTi is not strongly deformed, it can be switched through millions of cycles.

Suggested Uses of NiTi
Fishing hook—straightens when heated for easy removal
For spies—send ordinary looking wire, pop in hot water for message
Siding and roofing on houses—sun repairs baseball damage
Cookware—heating repairs dents
Meat thermometer—sticks out when done
Jewelry
Clamps and locks
Fire detectors
Safety device for household irons
Wind chimes made from austenitic rods
Malleable "Chemistry Action Figures" made from the martensite form
Moving sculpture
Levers

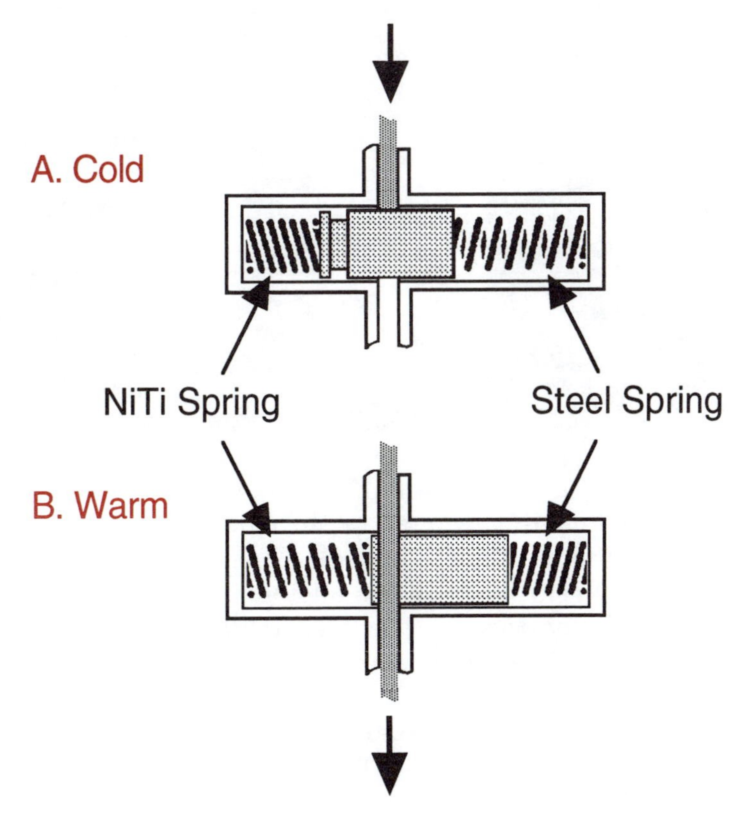

Figure 9.11. NiTi-actuated governor valve for automatic transmissions. The steel spring easily compresses the memory metal spring when it is in the low-temperature martensite phase. As the temperature increases, the NiTi spring remembers its extended shape (it transforms into the austenite phase) and becomes rigid, and thus forces the steel spring to compress. This compressed spring opens the pathway for transmission fluid through the valve. (Reprinted with permission from reference 23. Copyright 1991.)

NiTi has been used to create sculptures with moving parts. Olivier Deschamps designed the sculpture entitled "Les Trois Mains" (The Three Hands) *(24)*. When the weather is cool, the NiTi hands are in the martensite phase, and the hands are pointed down toward the center of the sculpture. If the day is warm, the NiTi transforms into the austenite phase, and the hands reach upward. Similarly, a sculpture by the same artist entitled "Espoir–Desespoir" (Hope–Despair) shows a kneeling woman with a baby on the ground in the martensite phase; upon warming, the sculpture is transformed so that the woman lifts the baby toward the sky. (Figure 9.12). An intriguing idea is to make automobile bumpers out of memory metal: The effects of an accident could be reversed simply by heating the car!

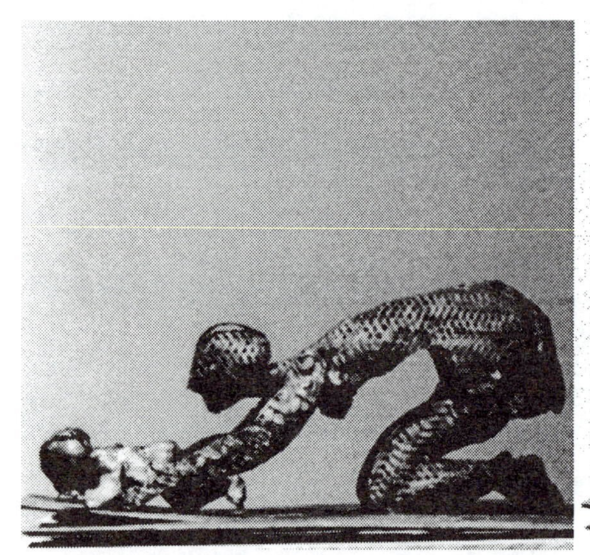

Figure 9.12. The sculpture "Espoir-Desespoir" (Hope-Despair) by Olivier Deschamps shows a kneeling woman with a baby on the ground (left photo; NiTi is in the martensite phase). Upon warming, the sculpture is transformed (NiTi is then in the austenite phase) so that the woman lifts the baby toward the sky (right photo). (Photographs courtesy of Olivier Deschamps, Paris, France.)

The most common use of NiTi is in biomedical applications for which its combination of strength, flexibility, and biocompatibility is desirable. The pseudoelastic property of NiTi is used in orthodontic braces. To straighten teeth, an "archwire" is connected across each tooth by a brace. The archwire is pulled tight, and this aligns the teeth. If a NiTi archwire in the austenite phase is used, the stress from tightening causes it to transform into the martensite phase. After tightening, the force the wire exerts as it tries to return to the austenite gradually pulls the teeth into position. The advantage of NiTi is that the wire can be pulled tighter than other types of wire, and therefore fewer trips to the orthodontist are needed. Because the force comes from the phase transformation, it is more even and continuous over time than the force from other types of wire, which tend to pull a lot at first, and then relax. Other biomedical applications for NiTi include staples that use body heat to cause the phase transformation and clamp broken bones together, and guidewires for arthroscopic surgery that are strong but flexible.

The 1-2-3 Superconductor

Superconductivity owes its discovery to a Dutch physicist, Heike Kamerlingh Onnes. In the early 1900s, Kamerlingh Onnes accomplished the then remarkable feat of liquefying helium. The ability to drop temperatures to a few kelvins (helium boils at 4.2 K at atmospheric pressure) suddenly opened a larger window for studies of thermal effects. Kamerlingh Onnes exploited his new tool by measuring the electrical resistance of metals at these very low temperatures. In 1911, while measuring the resistance of mercury, he found an astonishing result: at about 4 K, the resistivity abruptly dropped below his ability to measure it. Nor was mercury unique in displaying what appeared to be zero resistance to a flow of electrical current. Several other metallic elements exhibited the same effect at temperatures of a few kelvins. This remarkable electrical property was treated as a diagnostic for a new state of matter, the superconducting state. The temperature below which the material becomes a superconductor (this represents a kind of phase change; discussed later) is called the critical temperature, T_c. *(25)*

About 20 years later, in 1933, another surprising characteristic of superconductivity was discovered by W. Meissner and R. Ochsenfeld. They found that a superconducting material will not permit a magnetic field to penetrate its bulk, a property that has come to be called the Meissner effect.

These two diagnostics of superconductivity, resistanceless current flow and perfect diamagnetism, were appreciated even many years ago as having tremendous technological implications. But two major obstacles to implementing any new technology existed. First, extremely cold temperatures were required to achieve the superconducting state. Although liquid helium could serve as a coolant, its scarcity and processing costs made it expensive; moreover, sophisticated equipment was required to handle it. The second problem was that the superconducting state of these elemental metals was easily destroyed by the application of modest external magnetic fields or electrical transport currents; thus their use in electromagnetic applications was impractical.

These problems prompted an intensive search for new materials that would become superconducting at higher temperatures and retain their superconductivity in the presence of large magnetic fields and electrical currents. In particular, scientists who studied this phenomenon dreamed of achieving superconductivity at or above 77 K (–196 °C), in which case liquid nitrogen (boiling point, 77 K), which is cheaper and more easily handled, could be used as a coolant.

Until recently, alloys of niobium, particularly Nb–Ti, had been responsible for the greatest advances in superconductor technology. Their ability to remain in the superconducting state while supporting large electrical currents led to limited but extremely important applications such as the construction of powerful electromagnets. In 1973, superconductivity with Nb_3Ge was observed at 23 K, which until 1986 was the highest observed T_c value. Another breakthrough involved the observation of superconductivity with certain organic salts, whose structures feature one-dimensional stacks of organic moieties that can be modified by chemical substitution. These salts have required liquid-helium-range temperatures to become superconducting.

In spite of these advances, there was no evidence of superconductivity at temperatures anywhere near 77 K, even after some 70 years of investigation: More than 20 metallic elements had been found to exhibit superconductivity (Figure 9.13), but they did so only at very low temperatures near that of liquid helium *(26)*. Furthermore, most of the simple alloys or compounds based on combining two or three elements also required disappointingly low temperatures for superconductivity.

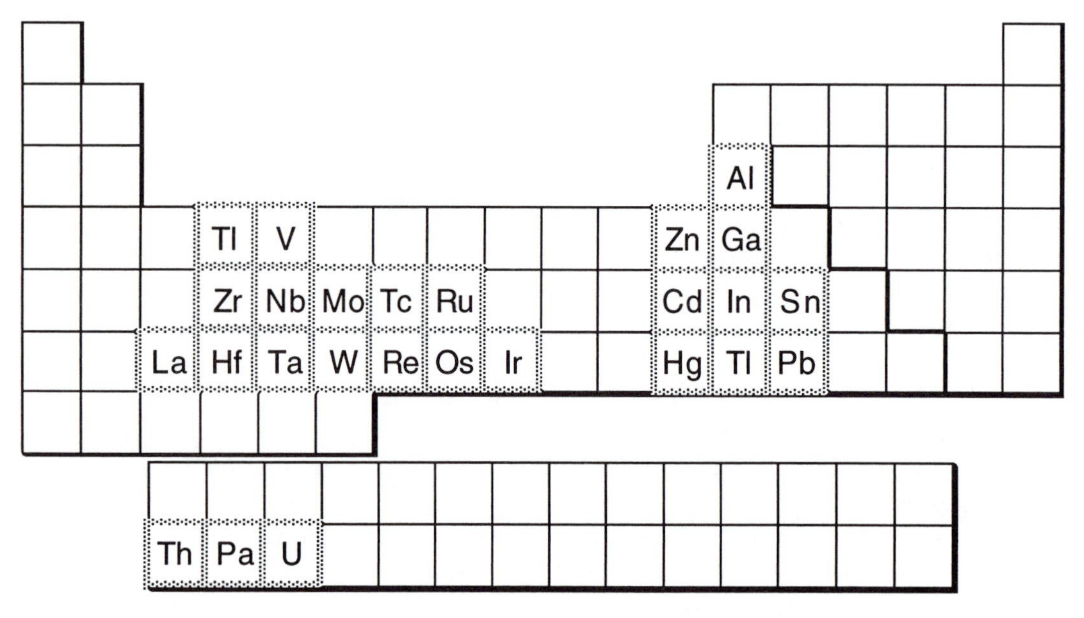

Figure 9.13. The periodic table, with elements that exhibit superconductivity at ambient pressure indicated by outlined boxes. All of these elements require temperatures of less than 10 K to attain the superconducting state *(26)*.

It was against this setting that two scientists working at the IBM facility in Zurich, Georg Bednorz and K. Alex Müller, reported a startling result in early 1986: A ceramic oxide of lanthanum, barium, and copper lost its resistance at about 30 K *(27)*. That a metallic oxide should not only exhibit superconductivity but do so at such a high temperature struck many in the scientific community as incredible. Suddenly a whole new family of materials, ceramics, was ripe for investigation, and the pace of research accelerated markedly.

In 1987, groups at the University of Houston and University of Alabama, headed by Paul Chu and M. K. Wu, Jr., respectively, announced the discovery of a related oxide that was superconducting above 77 K. This material was later determined to have the formula $YBa_2Cu_3O_{7-x}$ (commonly called "1-2-3" because of the Y:Ba:Cu atomic ratio) *(28, 29)*. Part of the charm of this high-tech material is its relative ease of preparation. Superconductivity at the temperature of liquid nitrogen is now commonplace, and efforts continue to find materials that can push superconductivity up toward room temperature. At present, however, much of the focus of research has shifted from increasing T_c to increasing the critical current density (the current density above which superconductivity is lost, often reported in amperes per square centimeter) and making the materials in usable forms such as thin films and wires.

Laboratory. Experiment 11 describes the synthesis and explores the properties of the 1-2-3 superconductor.

Physical Properties of Superconductors

To appreciate the phenomenon of superconductivity, some of its physical and chemical characteristics will be described in more detail, along with some applications.

Loss of Resistance

A material's passage into the superconducting state is characterized by an abrupt change in its resistivity as it is cooled. As noted in Chapter 7, in ordinary conducting materials the flowing electrons that represent current always encounter some resistance, which can be likened to friction. The source of the resistance is the scattering of electrons from the atoms that make up the conducting material. Scattering arises from the defects (Chapter 6) and from the vibrations of the atoms in the solid. Ohm's law relates the current, I, to the voltage, V (the difference in electrical potential between points in a circuit that causes electrons to flow), and resistance, R, thus: $V = IR$.

Figure 9.14 graphs the resistivity of a superconducting material (the resistance is proportional to resistivity) as a function of temperature. As the sample is cooled, there is initially a smooth decline in resistivity with temperature; with less thermal energy available, the atoms vibrate less, and less scattering results. In a normal metal (Chapter 7), this decline ceases at very low temperatures when scattering becomes limited by fixed defects; a limiting or residual resistivity exists. What characterizes the superconducting state, however, is a sudden drop to zero resistivity at the critical temperature, T_c. Below T_c, a direct current can flow indefinitely in the material so far as anyone has been able to determine.

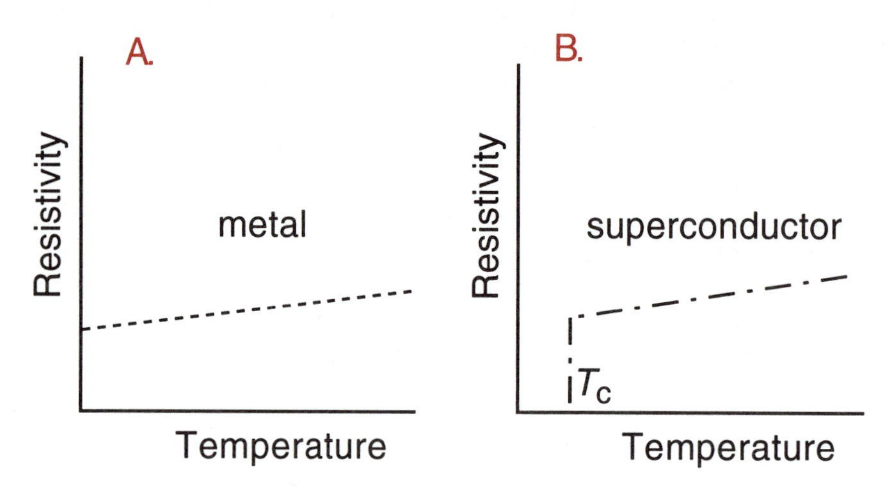

Figure 9.14. Resistivity as a function of temperature for (A) a metal and (B) a superconductor.

The transition from a normal metal to a superconductor can be regarded as a so-called "second-order" phase change. Unlike the phase change associated with the conversion of a liquid to a gas, the enthalpy of the phase change is zero, but there is a change in the specific heat of the material.

The Meissner Effect and Levitation

Besides resistanceless current flow, the other characteristic of a superconducting material is that it is perfectly diamagnetic. The expulsion of magnetic field lines from the interior of a material as it passes from the normal to the superconducting state at T_c is illustrated in Figure 9.15.

This description applies to so-called "Type I superconductors," the class to which many of the early superconductors belong. For "Type II superconductors," which include the ceramic, high-temperature superconductors like 1-2-3, it is only partially correct: At low magnetic field strengths the magnetic field is completely expelled from the superconductor's interior; however, as the magnetic field strength increases, more and more magnetic field lines enter the solid (they do so in a pattern of what are called fluxoids or vortices) until, at sufficiently high magnetic field strengths, the material reverts to being a nonsuperconducting solid.

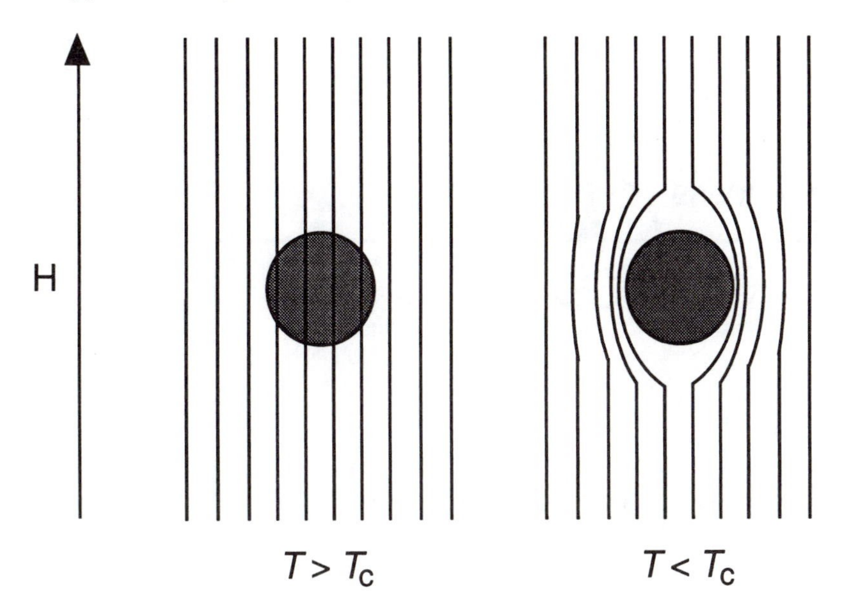

Figure 9.15. The Meissner effect. A superconductor (shaded circle) expels magnetic field lines from its interior.

In order to appreciate how the Meissner effect operates to produce levitation, some of the principles of electromagnetism will be reviewed. As recently as the beginning of the 19th century, electricity and magnetism were believed to be two separate phenomena. In 1820, however, the Danish physicist H. C. Oersted observed a connection between them. As illustrated in Figure 9.16, Oersted found that when he passed an electric current through a wire, a deflection was observed in the magnetic pointer of a compass; in other words, the current had induced a magnetic field. Electromagnets are based on this principle: A magnetic field is induced by passing current through a coil of wire. The very strong magnetic fields produced by passage of a large current through superconducting coils of wire will persist indefinitely in a closed circuit if the wires remain below T_c, because there is no resistance to the flow of current. This effect can be noted in superconducting magnets such as those in modern NMR spectrometers and magnetic resonance imaging (MRI) instruments. Once initiated, the current in a superconducting magnet will continue for years. However, loss of cooling will eventually produce "quenching" of the magnet; if a large current is flowing when superconductivity is lost, resistive heating will quickly boil off the coolant.

Following Oersted's observations, two other scientists, Michael Faraday and Joseph Henry, investigated the related question of whether a magnetic field could induce a current in an electrical conductor. They found that a magnet could induce a current if a magnet in a coil of wire was constantly in motion (or more fundamentally, if the magnetic flux linking the coil is continuously changing), as shown in Figure 9.17. Our present large-scale electrical energy needs are now met by generators that work on this principle.

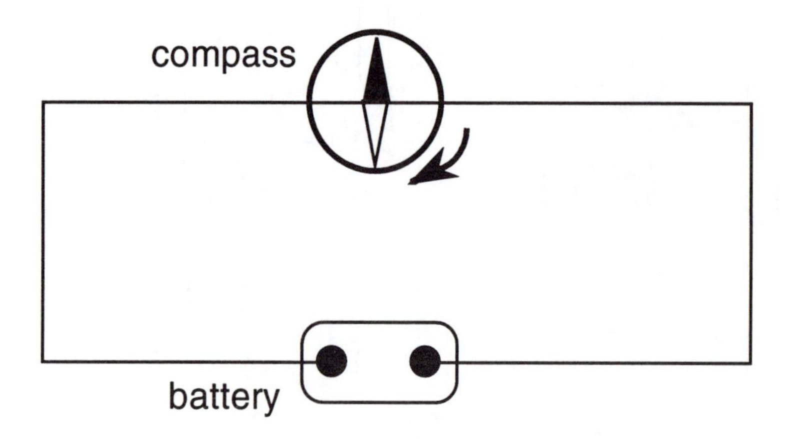

Figure 9.16. Oersted's experiment. The magnetic needle of a compass placed beneath a conductor carrying current will be deflected by the induced magnetic field.

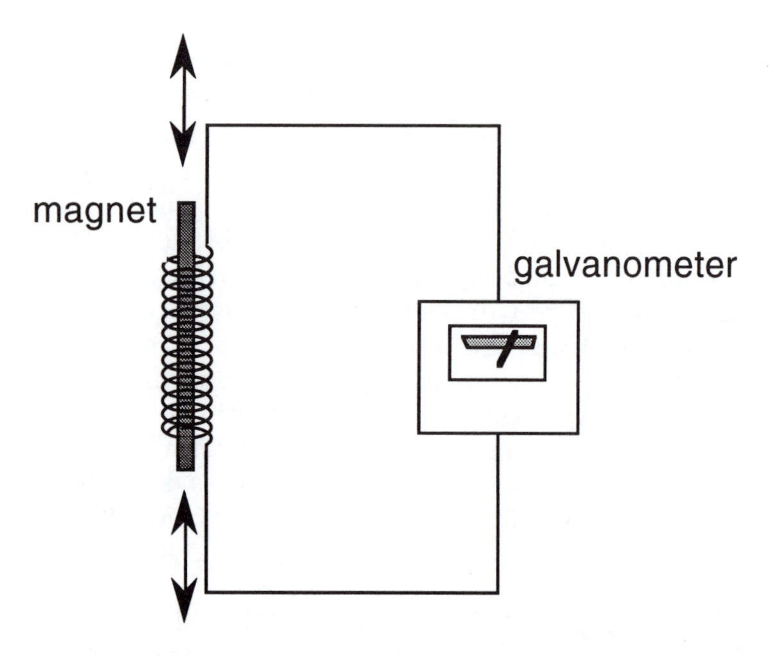

Figure 9.17. Moving a magnet through a coiled wire induces a current (more accurately, a voltage is induced, which drives the current), detected with a galvanometer.

The coupling of electrical and magnetic phenomena is elegantly demonstrated in a levitation experiment. As will be described, if a superconducting pellet is cooled below its critical temperature and a light, strong magnet is placed above the pellet, it hovers over the pellet (*see* Figure 9.18). The magnet retains this position so long as the pellet remains superconducting; once the pellet warms to its normal state, the magnet will no longer remain suspended in air.

Figure 9.18. Levitation of a magnet above a liquid-nitrogen-cooled sample of $YBa_2Cu_3O_{7-x}$. The pellet is sitting atop an inverted Styrofoam cup.

Why does levitation occur? As the magnet approaches the pellet, it will induce a resistanceless supercurrent in the surface of the superconductor. For Type I superconductors, the supercurrent induces its own magnetic field that opposes the external magnetic field. This induced magnetic field leads to cancellation of the magnetic field inside the superconductor, accounting for the perfect diamagnetism of Type I superconductors. However, Type I superconductors do not provide stable levitation: The opposing magnetic fields from the superconductor and magnet can lift the magnet, but there is no force to prevent the lateral movement of the magnet.

In contrast, Type II superconductors provide a mechanism for stable levitation. The magnetic field lines that penetrate the sample resist motion, and thus keep the magnet in a fixed position. The height at which the magnet is levitated reflects the tendency to minimize the total energy of the system: the internal energy of the superconductor will increase as the magnet gets closer to the surface and more magnetic field lines enter the solid; and the gravitational potential energy will increase as the magnet gets further away from the surface. The levitation height reflects the minimum total energy based on the sum of these two contributions.

Demonstration 9.7. Levitation

Materials

Superconducting pellet (*see* Supplier Information)
Styrofoam cup
Strong, small magnet
Liquid nitrogen
Nonmagnetic tweezers

Procedure

- Cool a pellet of the 1-2-3 superconductor with liquid nitrogen using a Styrofoam stand (an inverted coffee cup, for example). **CAUTION: Liquid nitrogen is extremely cold. Do not allow it to come into contact with skin or clothing, as severe frostbite may result. Wear gloves when transferring and using liquid nitrogen.**
- Place a magnet on top of the superconducting pellet using the nonmagnetic tweezers. A standard refrigerator magnet, if small enough, can become airborne. However, the lift is more dramatic with strong, rare-earth-based magnetic compounds like $SmCo_5$ and $Fe_{14}Nd_2B$.
- Spin the levitated magnet. Marking one side with white correction fluid makes the spinning more visible.

Variations

- The experiment can be demonstrated before a large audience by placing an overhead projector on its back and taping, to the lens of the projector, the base of a plastic frame that holds a rotatable mirror. A schematic diagram of the mirror attachment and a picture illustrating its use appears as Appendix 9.2 to this chapter.
- If a small cylindrical magnet is available, a small fragment of another magnet can be attached to its corner. The cylindrical magnet can then be spun after levitation, and its rotation will be evident on the screen from the movement of the magnet fragment that is attached to it.
- The Meissner effect may also be demonstrated by using an enclosed balance in a manner similar to that shown in Figure 2.10. In this case, however, a sample of superconducting material in a cup containing liquid nitrogen is placed on top of the balance lid. The superconducting material repels a magnet supported on a stack of Styrofoam cups on the balance pan and induces a measurable weight change *(30)*.

The Mechanism of Superconductivity

Superconductivity baffled theorists for many years after its discovery. It was not until the late 1950s, some 40-odd years after Kamerlingh Onnes' experiments, that a satisfactory theory was found. Called the BCS theory after its authors, J. Bardeen, L. Cooper, and J. R. Schrieffer, it satisfactorily accounts for virtually all of the properties of the traditional, low-T_c superconductors *(31)*. The cornerstone of the theory is a notion that at first seems counterintuitive: electrons are attracted to each other to form what are called Cooper pairs.

That two electrons, both negatively charged, would feel an attraction seems to violate electrostatic principles. The key to the attraction, though, is that it is mediated by the crystal. Positive ions, whose positions are fixed in the crystal, result in any metal because the valence electrons of atoms composing the metal have been removed from their individual atoms and move about freely (Chapter 7). Figure 9.19 shows that, as these electrons responsible for conductivity travel past the positive ions, the ions are drawn toward the path of the electron by electrostatic attraction. And because the ions are much more massive than the electrons and move more sluggishly, this "wake" of displaced positive charge persists long enough to attract a second electron, and thus forms the Cooper pair. The pair is thus bound by the mutual attraction of each electron for the positive ions in the crystal. Important characteristics of the Cooper-pair electrons are that their spins are paired and the combined momentum of the pairs is not affected by electron scattering. Consequently, scattering does not provide the energy transfer between the electrons and the atoms in the crystal that produces resistance in a normal conductor. In the superconducting state most of the conduction electrons are bound in Cooper pairs.

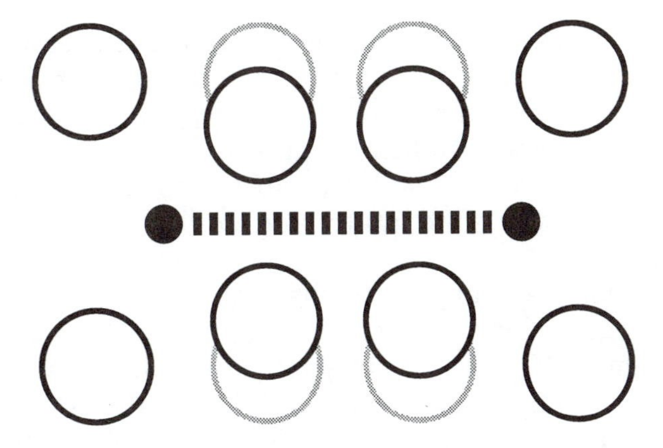

Figure 9.19. The coupling (dashed line) of two electrons (dark circles) into Cooper pairs is mediated by the crystal atoms (open circles), whose positions are affected by electronic motion. The atomic positions shift toward the electrons as they pass by; the original positions are indicated with the lighter outlines. Electrons in Cooper pairs exchange partners frequently.

A crude way to think of these pairs is as diatomic molecules. Like such molecules, they can be dissociated if sufficient energy is present to disrupt the bonding. The value of T_c is a reflection of this energy: as a superconductor is warmed, the number of Cooper pairs drops as T_c is approached.

Temperature is not the only experimental parameter that can affect the formation of Cooper pairs. As already noted, when magnetic fields are applied to a superconductor, supercurrents will be induced in the surface of the solid. If these fields and corresponding currents are sufficiently large, they can impart so much energy to the superconductor that the Cooper pairs will be dissociated and the superconductivity destroyed. Many potential applications of superconductors are precluded by the critical field H_c (and corresponding critical current J_c) at which this conversion from superconducting to normal behavior occurs.

Chemical Properties of a High-T_c Superconductor

Synthesis and Oxidation States

Making pellets of $YBa_2Cu_3O_{7-x}$ ($x \lesssim 0.1$) is straightforward, although several steps are involved *(32, 33)*. In a typical procedure, Y_2O_3, $BaCO_3$, and CuO are ground together and heated to about 950 °C. After cooling, pellets can be pressed and then sintered at 950 °C. Sintering involves heating just below the melting point; this process promotes bonding between the grains composing the pellet, thereby increasing the density and strength of the pellet (*see* Chapter 10). Once the pellet has been sintered, it is heated in oxygen at 500–600 °C and slowly cooled to room temperature.

The oxygen content of 1-2-3 has been estimated by standard chemical methods, which indicate it to be a compound having variable stoichiometry: Instead of having an integral number of oxygen atoms, the best superconducting materials appear to have values of x of about 0.1 or less. As has been discussed in Chapters 3 and 6, compounds with variable stoichiometry are common in the solid state and reflect chemistry associated with crystal defects.

Formal oxidation states can be assigned to the elements composing 1-2-3. Based upon the normal oxidation states of –2 for the oxide oxygen atom, +3 for the yttrium atom, and +2 for each barium atom, the resulting average oxidation state for each copper atom is 7/3 (for $x = 0$). The nonintegral oxidation state can be interpreted to mean that, on average, two-thirds of the Cu is present in Cu^{2+} sites in the crystal and one-third of the Cu is in Cu^{3+} sites.

Given the large number of elements that form oxides and the difficulty in predicting which of these combinations will form new phases, it is perhaps not surprising that 1-2-3 had not been prepared before. Since its discovery, however, thousands of related solids have been prepared. The kind of chemical tuning discussed in connection with solid solutions and periodic properties (Chapters 3 and 7) has been attempted through substitutions for Y, Ba, and Cu. A variety of rare earth elements have been substituted for Y without greatly disturbing the superconducting characteristics of the material. However, substitution for Ba or Cu adversely affects superconductivity *(34)*.

Structure

X-ray diffraction studies (Chapter 4) have established that the 1-2-3 compound is structurally similar to the perovskite family, which is discussed in Chapter 5. Perovskites characteristically have a ratio of two metal atoms for each three oxygen atoms. Representative compounds with the perovskite structure are $CaTiO_3$, $NaNbO_3$, and $LaAlO_3$; in these simple perovskites the metal oxidation states must sum to +6 in order to yield an electrically neutral compound. Figure 5.28 gives the solid-state structure for the mineral perovskite, $CaTiO_3$, which has the larger Ca^{2+} cation in the center of the cubic unit cell, the smaller Ti^{4+} ions at each corner of the cube, and the O^{2-} ions bisecting the edges of the cube.

If 1-2-3 had an idealized perovskite structure, it would possess nine oxygen atoms in its formula (metal-to-oxygen atom ratio of 2:3) and would have the triple-decker structure sketched in Figure 5.29B. The resulting idealized unit cell, consisting of three stacked cubic unit cells, is now tetragonal rather than cubic, having a square base but rectangular sides. The top and bottom compartments of the tetragonal unit cell contain Ba^{2+} ions, and the middle compartment contains a Y^{3+} ion. The Cu ions occupy the corners of the cubes composing the unit cell, and the oxide ions again bisect the cube edges.

Why is the ideal structure not adopted by the 1-2-3 oxide? The answer may lie in the excessively high oxidation state required of Cu. A formula of $YBa_2Cu_3O_9$ leads to an average copper oxidation state of 11/3, which implies contributions from both Cu^{3+} and Cu^{4+} oxidation states. The fact that tetravalent Cu compounds are extremely rare strongly suggests that the average Cu oxidation state must be +3 or less. One avenue for achieving a lower oxidation state is to expel oxygen atoms from the crystal, giving the formula $YBa_2Cu_3O_7$ and a formal oxidation state of 7/3 for copper.

How the structure is tuned by varying the oxygen content is illustrated in Figure 5.29C. Although it is difficult (because of crystal disorder problems) to locate all of the oxygen atoms in the structure, the general geometrical pattern appears to be that 4 of the 12 oxide ions surrounding the Y^{3+} ion have been lost as well as four other oxide ions, two from the top face of the unit cell and two from the bottom.

These oxygen vacancies are defects in the crystal that are governed by chemical equilibria (Chapters 6 and 8). The solid can be regarded as a solvent wherein vacancies act as solutes that are subject to the same kinds of equilibrium expressions used to describe acid–base chemistry (Chapter 8). In this case, the concentration of oxygen vacancies is coupled to the oxidation states of the copper ions composing the solid: As the oxygen content declines, so too does the average copper oxidation state. Specifically, because the solid must remain electrically neutral, each oxide ion that is removed requires compensatory removal of two units of positive charge; this compensatory removal can be accomplished if, for example, two Cu^{3+} centers are converted to Cu^{2+} centers. An equation describing the equilibrium that couples oxygen vacancies, V, with Cu oxidation states is,

$$2\ Cu^{3+} + O^{2-} \rightleftarrows 2\ Cu^{2+} + 1/2\ O_2\ (g) + V$$

The oxygen vacancies in 1-2-3 create sheets and chains of copper atoms (*see* Figure 5.30), linked through the remaining oxygen atoms, that may play a critical role in the superconductivity of the solid. One consequence of this bonding arrangement is that the physical properties of the material, including superconductivity, will show anisotropy, that is, they are dependent on the crystal direction along which they are measured.

Other Superconductors

Efforts to find new families of superconductors continue. After the discovery of 1-2-3, superconductors with critical temperatures as high as ~120 K were found with compositions of $Bi_2Ca_2Sr_2Cu_3O_{10}$ (T_c of 110 K) and $Tl_2Ca_2Ba_2Cu_3O_{10}$ (T_c of ~125 K). Table 9.1 lists a variety of ceramic superconductors and their critical temperatures.

Table 9.1. Examples of Oxide- and Fullerene-Derived Superconductors

Compound	T_c (K)	Reference
$YBa_2Cu_3O_{7-x}$	95	29
$Bi_2Sr_2CuO_6$	9	35
$Bi_2(Sr,Ca)_2CuO_6$[a]	80	36
$Bi_2Sr_{2.7}Ln_{0.3}Cu_2O_8$[b]	80	37
$Bi_2Ca_3Sr_2Cu_4O_{10}$	90	38
$Bi_2Ca_2Sr_2Cu_3O_{10}$	110	39
$Tl_2Ba_2Ca_2Cu_3O_{10}$	125	40
K_3C_{60}	18	41
K_2RbC_{60}	22	42
Rb_2KC_{60}	24	43
Rb_3C_{60}	29	43

[a] Sr and Ca form a solid solution.
[b] Ln = Y, Pr, Nd, Sm, Eu, Gd, Tb, Dy, Ho, Er, or Tm.

More recently, C_{60} was found to react with some of the alkali metals to form solids of composition M_3C_{60} (M = K, Rb, or Cs) that are superconducting with critical temperatures on the order of 30 K. In this structure, the C_{60} molecules form a fcc structure and the cations fill all of the tetrahedral and octahedral holes (*recall* from Chapter 5 that there are two tetrahedral holes and one octahedral hole for each close-packed sphere, which in this case is a C_{60} molecule), and this behavior leads to the indicated stoichiometry. Some of these superconducting solids are also included in Table 9.1.

Applications

Applications of superconductors, at least in principle, increase with critical temperature. This situation results from the fact that liquid helium can be obtained at relatively few sites, whereas liquid nitrogen is readily available; the ease and efficiency of coolant transfer and storage are also greater for liquid nitrogen. Taken one step further, the discovery of a room-temperature superconductor has the potential to bring superconductive devices into every household. Whether these applications come to fruition will depend on whether some formidable problems in the realm of materials science can be overcome.

Focusing on 1-2-3 and related oxides, their brittle nature, characteristic of ceramics, represents a significant challenge if useful shapes like wires, ribbons, and films that are needed for electromagnetic applications are to be fabricated. For example, thin films have potential for use in SQUIDS (superconducting quantum interference devices, which are used for the measurement of very weak magnetic fields) and infrared sensors. Films with critical current densities as high as nearly 10^8 A/cm^2 have been prepared; such films can be grown on the surface of $LaAlO_3$ crystals using synthetic methods that include sputtering, molecular beam epitaxy, and chemical vapor deposition (Chapter 10), followed by treatments in controlled oxygen atmosphere. Films as thin as a few tens of angstroms have been prepared this way.

Superconducting wires and tapes have the mechanical and electrical properties necessary for use in electromagnets and superconducting motors. They are made by a process in which silver tubes are filled with fine powders of $Bi_2Sr_2Ca_2Cu_3O_{10}$ and minor amounts of other constituents that promote crystal growth at elevated temperatures. The filled tubes are drawn into wires, heat-treated, redrawn, and finally treated with oxygen (which diffuses through the silver to produce the correct oxygen stoichiometry in the superconductor). It is remarkable that superconductors that form crystals that are as brittle as glass have been incorporated into flexible wires as long as 100 m!

Commercially successful superconducting wires or tapes must be able to carry a current density of at least 10^4 A/cm^2, and, in some cases, they must retain this capacity when exposed to a strong magnetic field. For example, power transmission cables operate in a low magnetic field, but wires in generators or in magnets are exposed to fields of several teslas (1 tesla is 10,000 G; a refrigerator magnet's strength is on the order of 0.1 G). Some experimental samples of the new superconducting wires exhibit critical current densities of 50,000 A/cm^2 at 20 K compared to 10,000 A/cm^2 for Nb_3Sn wire, the wire used in present commercial applications. Given the substantial progress that has already been made in overcoming fabrication problems, applications of the high-temperature superconductors may await in such diverse areas as the transmission of electrical power, electronics, and magnet-intensive technologies.

The ability of a superconductor to support a resistanceless direct current can be exploited in the lossless transmission of electrical power. At present, a significant fraction of generated electricity is lost as heat through the resistance associated with traditional conductors. Whether the conventional transmission of electricity will be affected by the superconducting ceramic is difficult to assess. Some loss of energy occurs when alternating current, the form of current provided by utilities, is being transferred. A large-scale shift to superconducting technology will also hinge on whether wires can be prepared from the ceramics that retain their superconductivity at 77 K, while supporting large current densities.

Other potential applications of the new superconductors appear in the field of electronics. For example, miniaturization and increased speed of computer chips are limited by the generation of heat and the charging time of capacitors arising from the resistance of the interconnecting metal films. The use of the new ceramics may result in more densely packed chips that could transmit information orders of magnitude more rapidly.

Superconducting electromagnets are an enabling component of several technologies. NMR spectroscopy, one of the most powerful tools for structural characterization in synthetic chemistry, and magnetic resonance imaging (MRI), which is playing an increasingly prominent role in medicine, both rely for their sensitivity on the intense magnetic fields provided by superconducting electromagnets. Levitated vehicles have been designed that employ superconducting magnets (Protoypes like the "mag-lev" train in Japan operate by inducing currents in metal train tracks (eddy currents) and not by the Meissner effect). Similarly, the particle accelerators that serve the high-energy physics community are dependent on high-field superconducting magnets. A controversy that surrounded the Superconducting Super Collider (SSC) also illustrates the political ramifications of new technologies. At issue was whether the multi-billion-dollar project should be constructed using the established liquid-helium-based superconducting technology, or whether construction should be postponed to embrace the developing liquid-nitrogen-based technology. Eventually the former view prevailed.

Ionically Conducting Copper Mercury Iodide

Unlike the 1-2-3 superconductor, which conducts electricity by electronic motion, the salt Cu_2HgI_4 (and the related solid, Ag_2HgI_4) is a good ionic conductor of electricity at temperatures a little above room temperature, where it undergoes an order–disorder phase change. The compound is easily prepared, and its phase change is observable both by increased conductivity and a color change. Other ionic conductors are described in Chapter 8.

Structure and Phase Change

The low-temperature ordered structure of Cu_2HgI_4 is shown in Figure 9.20C. Although the overall unit cell is tetragonal, with square bases and rectangular sides, it can be viewed as two fcc cells of iodide ions, with one cube atop the other. The iodide ions are in the fcc unit cells in the same positions as the sulfur atoms in ZnS (sphalerite form, *see* Chapter 5). All of the Cu^+ and Hg^{2+} cations are in tetrahedral holes of the structure that are formed by the iodide ions (there are two such holes for each iodide (Chapter 5), hence three-eighths of these holes will be occupied by the cations), but with a particular ordering (*see* Figure 9.20) *(42)*. In this low-temperature phase, the solid is brick red.

At a temperature of about 67 °C, disorder sets in, and the cations are randomly distributed about all of the tetrahedral holes in the structure. The phase change is accompanied by a color change to red–brown and a marked increase in electrical conductivity. The color change is due to a small decrease in the band gap (2.1 to 1.9 eV) with the change in structure *(43)*. In this high-temperature phase, the unit cell is a cube, because the X-ray diffraction experiment (Chapter 4) measures the average occupation of the tetrahedral sites, and the disorder makes it appear that, on average, each tetrahedral site contains one-fourth of a copper ion and one-eighth of a mercury ion.

Conductivity Mechanism

Above the transition temperature, Cu_2HgI_4 exhibits ionic conductivity (with some electronic conductivity also). Five-eighths of the tetrahedral holes and all of the octahedral holes formed by the iodide ions are vacant, and these open sites provide possible pathways for the small copper cations to move through the crystal, carrying charge. It is easiest for a copper cation to jump between tetrahedral holes by moving to an octahedral hole and then to the new tetrahedral hole, rather than jumping directly between tetrahedral holes.

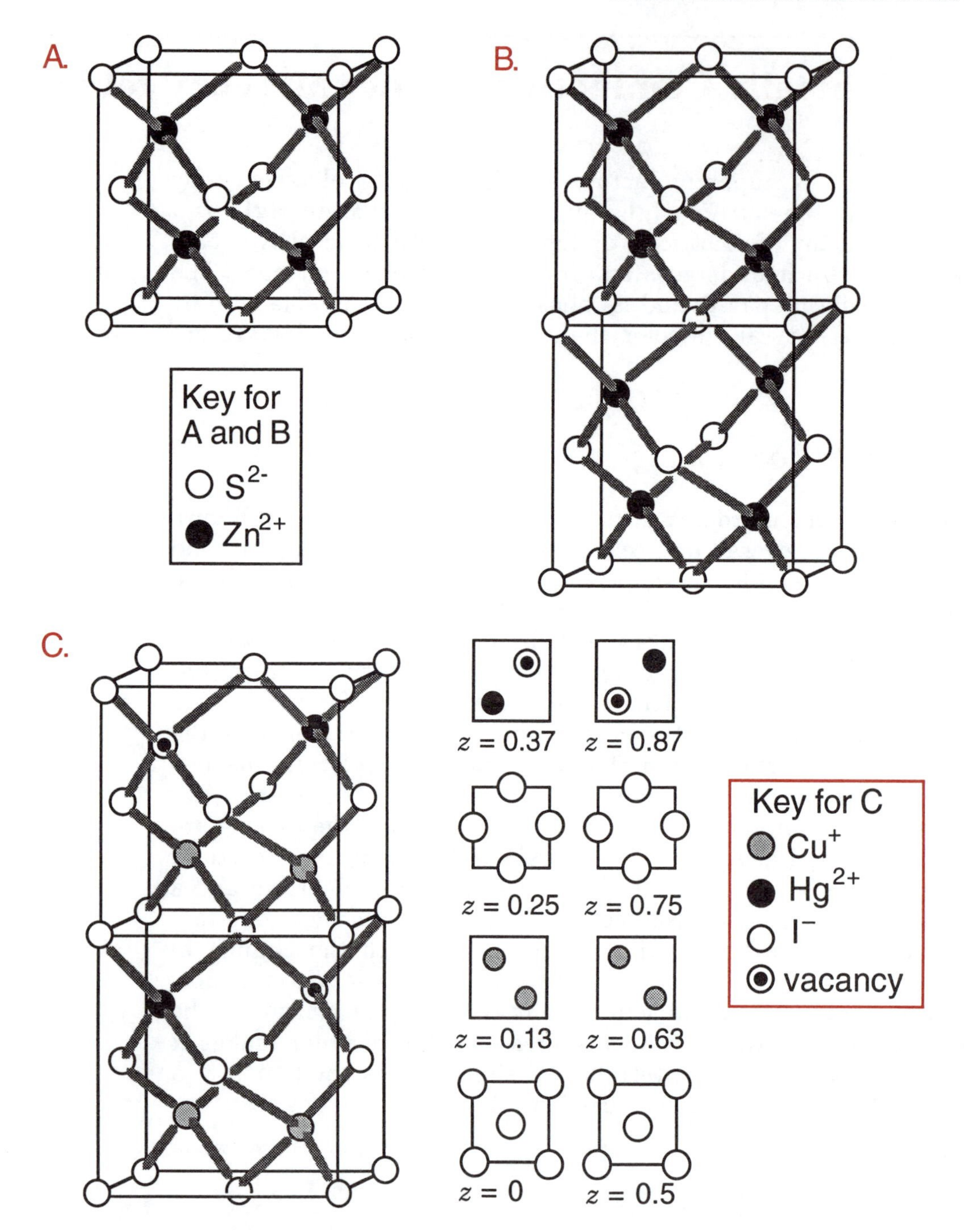

Figure 9.20A–C. The structure of Cu_2HgI_4 is related to the structures of ZnS (sphalerite) and of CaF_2 (fluorite). A: A sphalerite unit cell. B: Two stacked sphalerite unit cells. C: The ordered (low-temperature) structure of Cu_2HgI_4, with the *z* layer sequence. Three-eighths of the tetrahedral holes are occupied (2 of the 10 vacancies are shown; *see* the fluorite-like structure in D).

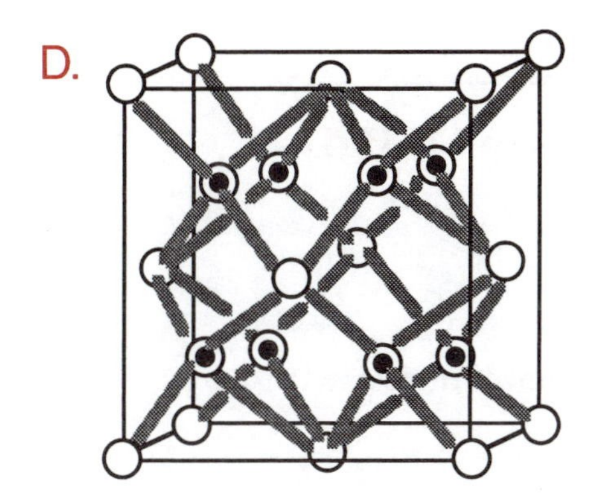

Figure 9.20D. The disordered (high-temperature) structure of Cu_2HgI_4. The cations are randomly distributed throughout all of the tetrahedral holes.

Demonstration 9.8. The Order–Disorder Phase Change in Copper(I) Mercury(II) Iodide

Materials

Cu_2HgI_4 (Samples are readily prepared from CuI, $Hg(NO_3)_2$, and KI; *see* Experiment 12 for the procedure)
Two glass plates or watch glasses (the watch glasses should be of the same size and curvature)
Spatula
Filter paper
Glue or tape
Heat gun
Insulated gloves

Procedure

- Using a spatula, rub some of the red Cu_2HgI_4 powder into a piece of filter paper. Sandwich the paper between two glass plates or watch glasses and glue or tape them together.
- Pick up the glass-encased sample while wearing thermal gloves and blow hot air from a heat gun onto the glass. The color change to red–brown will be readily apparent and, as the sample cools, its red color will return.

Caution: Mercury compounds are toxic. Avoid skin contact and inhaling dust from the compound. Place the waste from this demonstration in an appropriately labeled container and dispose of it according to state and local regulations.

Demonstration 9.9. Conductivity Changes in Cu_2HgI_4

Materials

Cu_2HgI_4
4-cm length of 4-mm o.d. glass tubing
Two brass L-shaped electrodes of about 2.5-mm diameter with wires attached
Rubber band
Ohmmeter and leads
Heat gun or matches

Procedure

- Assemble the apparatus for demonstrating the temperature-dependent conductivity of Cu_2HgI_4, as shown in Figure 9.21.
- Insert one electrode almost half way into the glass tube. Introduce a small portion of Cu_2HgI_4 into the other end, ensuring that there is enough sample in the glass tube so that the ends of the brass Ls will not touch (touching would short the circuit). When packed, the distance between the electrodes should be between 2 and 5 mm.
- Insert the second electrode. Hold the electrodes in place by stretching a rubber band around them.
- Connect the assembled conductivity apparatus in series with an ohmmeter, as shown in Figure 9.21. A heat gun or a match can be used to heat the Cu_2HgI_4.
- As the sample is heated, monitor the resistance. It will decrease, a result indicating that the conductivity is increasing. Upon cooling, the resistance will return to its original value, an effect reflecting the reduced conductivity of Cu_2HgI_4 at room temperature. **Caution: Mercury compounds are toxic. Avoid skin contact and inhaling dust from the compound. Place the waste from this demonstration in an appropriately labeled container and dispose of it according to state and local regulations.**

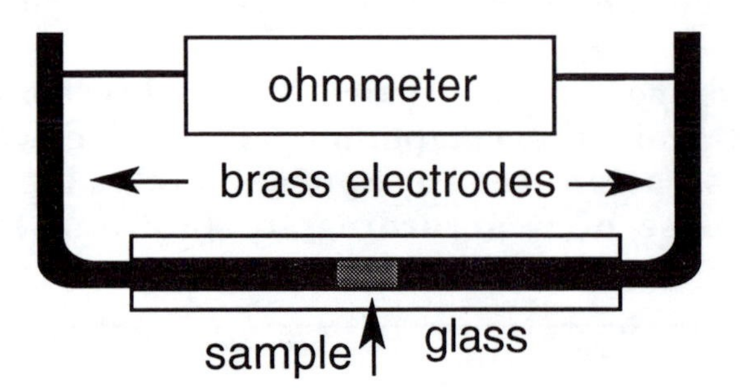

Figure 9.21. An apparatus to measure the conductivity in Cu_2HgI_4.

Laboratory. The synthesis and properties of Cu_2HgI_4 are presented in Experiment 12.

Appendix 9.1. Prediction of Shifts in Equilibria

As noted in the chapter, the phase of a pure material with the lowest free energy (G) at a given pressure (P) and temperature (T) will be the most stable phase. From thermodynamics,

$$dG = V\,dP - S\,dT$$

where V is molar volume and S is entropy. The predictions of Le Chatelier's principle are based on temperature- and pressure-induced changes in the free energies of a pure material.
The slope of a plot of molar free energy versus temperature at a given pressure is the negative of the molar entropy of the phase.

$$\text{at constant } P\text{, slope } \frac{dG}{dT} = -S$$

Figure 9.22A illustrates a typical free energy versus temperature plot. Although the solid is more stable at low temperature, the difference in slopes means that as the temperature increases, the liquid phase approaches the solid phase in stability. At the melting (crossover) point the two phases have equal free energies, and above this temperature the liquid phase has the lower free energy and is more stable. (The slight curvature of the lines occurs because the entropy is not constant but increases with temperature).

The free energy of a pure phase increases with an increase in pressure, and the slope of a plot of molar free energy versus pressure at a given temperature is the molar volume of the phase.

$$\text{at constant } T\text{, slope } \frac{dG}{dP} = +V$$

Figure 9.22B shows a typical plot of free energy versus pressure. Although the liquid is more stable at low pressure, the difference in slopes means that as the pressure increases, the solid phase approaches the liquid phase in stability. At the melting (crossover) point the two phases have equal free energies, and above this pressure the solid phase has the lower free energy and is more stable. (The slight curvature of the lines occurs because the molar volume is not constant but decreases with increasing pressure).

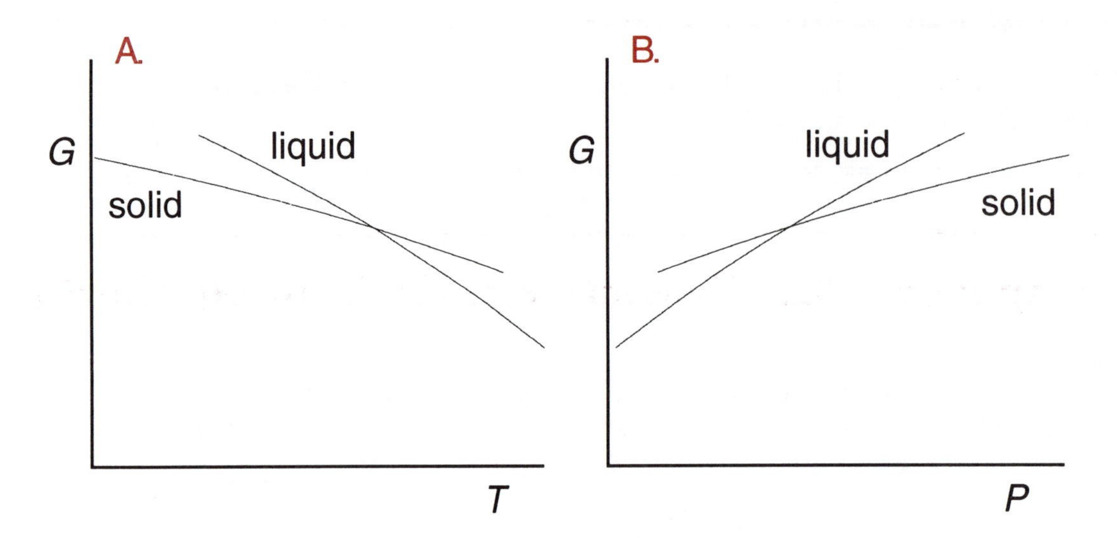

Figure 9.22. Plots of free energy versus temperature (A) and free energy versus pressure (B) for a typical material.

Ice and water are familiar materials that behave differently. The plots of G versus T and G versus P for H_2O are shown in Figure 9.23. In the plot of free energy versus temperature (Figure 9.23A), liquid water has the steeper slope, reflecting its higher entropy. The free energy versus pressure plot for liquid water and ice (Figure 9.23B) shows the anomalous behavior of this substance. Because ice has the greater molar volume, its curve has the steeper slope, meaning that with increasing pressure the denser, liquid phase becomes more stable. This situation is the reverse of the more typical situation shown in Figure 9.22B.

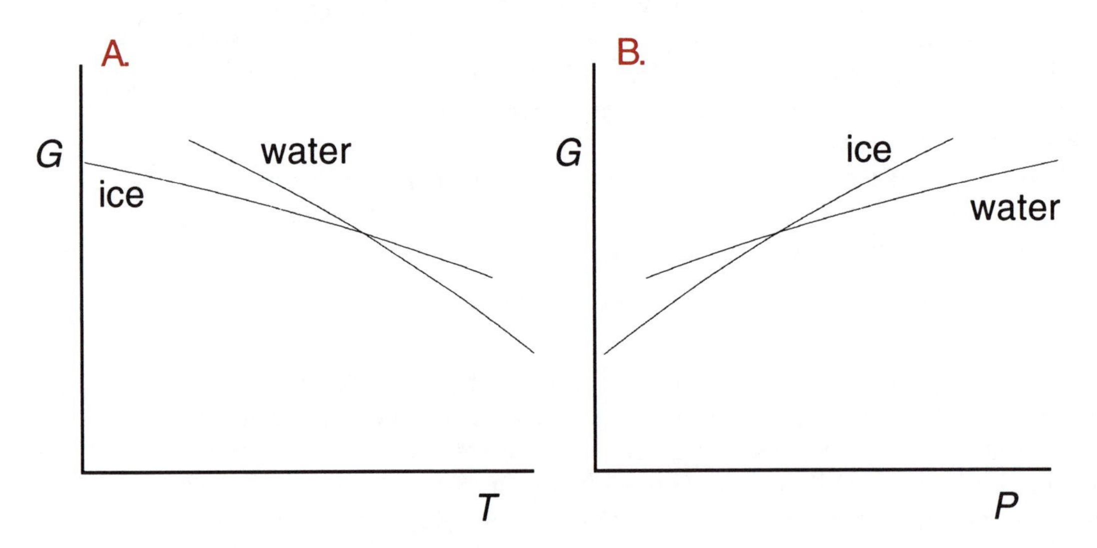

Figure 9.23. Plots of free energy versus temperature (A) and free energy versus pressure (B) for H_2O.

An alternative way to illustrate the effects of temperature and pressure on free energy is to superimpose the pressure effect on Figure 9.22A. This approach is illustrated for the general case in Figure 9.24A. The thicker lines in this figure correspond to the increased free energy for each phase caused by an increase in pressure. The increase is larger for the phase with the larger molar volume (the liquid), and shows that pressure applied to the solid can increase its melting point: the crossover point for the solid and liquid curves moves to higher temperature. Figure 9.24B shows this same plot for H_2O. In this case, the increase in free energy is larger for the solid (ice) and shows that the crossover point moves to lower temperature for this substance, corresponding to a lower melting point with increased pressure.

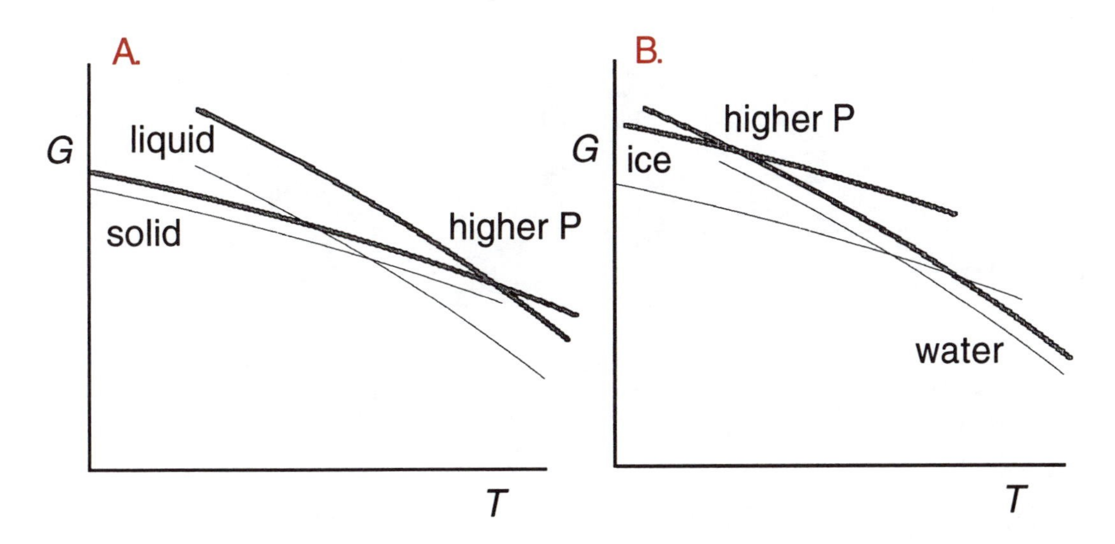

Figure 9.24. Free energy versus temperature at two different pressures for a typical material (A) and for H_2O (B). The thicker curves correspond to higher pressure for each phase.

Appendix 9.2. Construction of the Projector Mirror

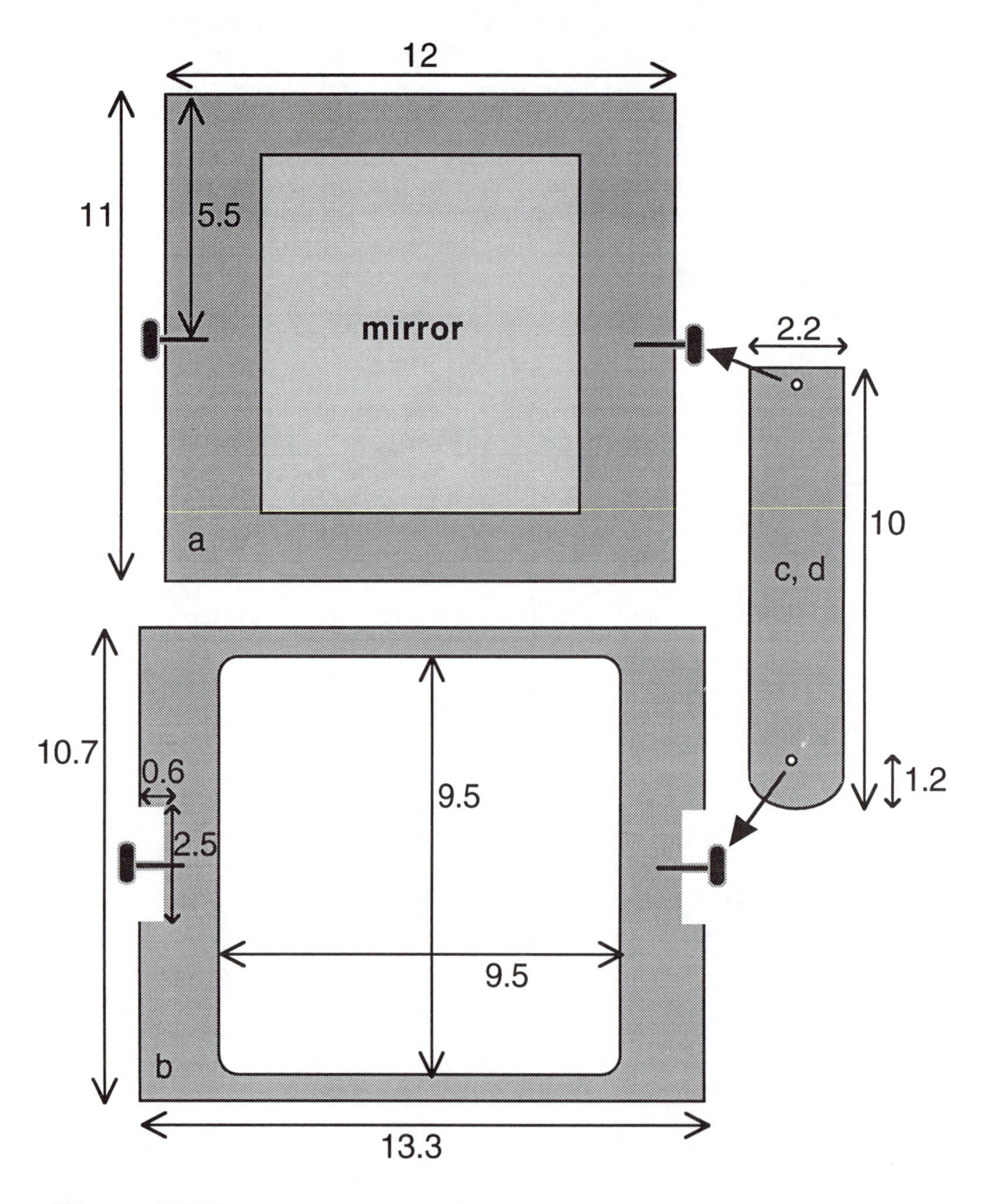

Figure 9.25A. A schematic diagram for a mirror attachment that can be used to display the superconductor levitation experiment with the overhead projector. Pieces a and b are made out of 9-mm thick Plexiglas, and pieces c and d are made out of 5-mm thick Plexiglas. All dimensions given in the diagram are in centimeters. The mirror shown in A is inlaid into the Plexiglas. Four plastic-knobbed thumbscrews are used to connect the pieces. Tapped holes are required for the screws.

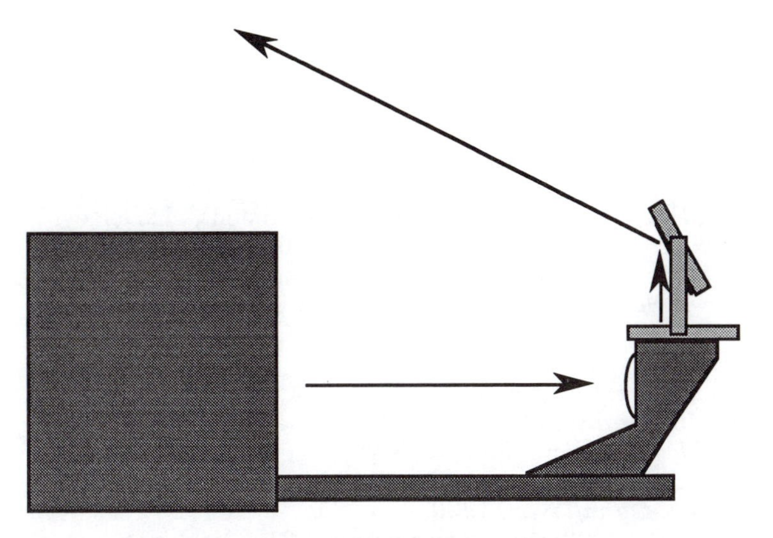

Figure 9.25B. A schematic diagram showing how the mirror attachment and an overhead projector are placed so as to demonstrate levitation. The overhead projector is placed on its back, and the mirror attachment is taped to the head of the projector. The angle of the mirror is adjusted until the image is on the screen. The superconductor is placed on a stand or set of books so that it sits in roughly the middle of the projector glass.

References

1. Shakhashiri, B. Z. *Chemical Demonstrations: A Sourcebook for Teachers*; University of Wisconsin Press: Madison, WI, Vol. 3, 1989, p 310.
2. Louks, L. F. *J. Chem. Educ.* **1986**, *63*, 115–116.
3. Silberman, R. *J. Chem. Educ.* **1988**, *65*, 186.
4. *Mellor's Modern Inorganic Chemistry*; Parkes, G. D., Ed.; Wiley: New York, 1967, p 786.
5. Machlin, E. S. *An Introduction to Aspects of Thermodynamics and Kinetics Relevant to Materials Science*; Giro Press: Croton-on-Hudson, NY, 1991, p 25.
6. Shakashiri, B. Z. *Chemical Demonstrations: A Sourcebook for Teachers*; University of Wisconsin Press: Madison, WI, 1983, Vol. 1, pp 27–33.
7. Based on an in-press paper in *J. Chem. Educ.* by Gisser, K. R. C.; Geselbracht, M. J.; Cappellari, A.; Hunsberger, L.; Ellis, A. B.; Perepezko, J.; Lisensky, G. C.
8. Collen, K. R.; Ellis, A. B.; Perepezko, J. H.; Moberly, W.; Busch, J. D. *Chemistry of Advanced Materials*; Rao, C. D. N., Ed.; Blackwell: London, 1992; p 197.
9. Wayman, C. M. *MRS Bull.* **1993**, *18 (4)*, 49–56.
10. Wang, F. E.; Buehler, W. T.; Pickart, S. J. *J. Appl. Phys.* **1965**, *36*, 3232.
11. Michal, G. M.; Sinclair, R. *Acta Cryst. B* **1981**, *37*, 18.
12. Kudoh, Y; et al. *Acta Metall.* **1985**, *33*, 2049.
13. Shimizu, K.; and Tadaki, T. In *Shape-Memory Alloys*; Funakubo, H., Ed.; Gordon and Breach: New York, 1984; p 7.

14. Geselbracht, M. J.; Penn, R. L.; Lisensky, G. C.; Stone, D. S.; and Ellis, A. B. "Mechanical Properties of Metals," *J. Chem. Educ.*, in press.
15. Murray, J. L. In *Phase Diagrams of Binary Titanium Alloys*; Murray, J. L., Ed.; ASM International: Metal Park, OH, 1987; p 203.
16. Bhatia, A. B. *Ultrasonic Absorption*; Oxford University Press: Oxford, England, 1987.
17. Chalmers, B. et al. In *Progress in Materials Science*; Warlimont, H.; Delaey, L., Eds.; Pergamon: Oxford, England; 1974; Vol. 18; p 104.
18. Collen, K. R.; Ellis, A. B.; Perepezko, J. H.; Moberly, W.; Busch, J. D. In *Chemistry of Advanced Materials*; Rao, C. D. N., Ed.; Blackwell: London, 1992; p 197.
19. Lindquist, P. G.; Wayman, C. M. In *Engineering Aspects of Shape Memory Alloys*; Duerig, T. W., Ed.; Butterworth: London, 1990; p 58.
20. Yi, H. C.; Moore, J. J. *J. Metals* **1990**, Aug., 31.
21. Honma, T. In *Shape-Memory Alloys*; Funakubo, H., Ed.; Gordon and Breach: New York, 1984; p 81.
22. Liu, X.; Stice, J. *J Appl. Manuf. Syst.* **1990**, Jan., 65.
23. Stoeckel, D.; Tinschert, F. *SAE Technical Paper Series 910805*; Society of Automotive Engineers: Warrendale, PA, 1991; p 145.
24. Deschamps, O. *J. Metals* **1991**, *43*, 64.
25. Ellis, A. B. *J. Chem. Educ.* **1987**, *64*, 836–841.
26. Ashcroft, N. W.; Mermin, N. D. *Solid State Physics*; Saunders: Philadelphia, PA, 1976; Chapter 34.
27. Bednorz, J. G.; Müller, K. A. *Z. Phys. B.* **1986**, *64*, 189.
28. Wu, M. K., Jr.; Ashburn, J. R.; Torng, C. J.; Hor, P. H.; Meng, R. L.; Gao, L.; Huang, Z. J.; Wang, Y. Q.; Chu, C. W. *Phys. Rev. Lett.* **1987**, *58*, 908.
29. Cava, R. J.; Batlogg, B.; van Dover, R. B.; Murphy, D. W.; Sunshine, S.; Siegrist, T.; Remeika, J. P.; Rietman, E. A.; Espinosa, G. P. *Phys. Rev. Lett.* **1987**, *47*, 1676.
30. McHale, J.; Schaeffer, R.; Salomon, R. E. *J. Chem. Educ.* **1992**, *69*, 1031–1032.
31. Rose-Innes, A. C.; Rhoderick, E. H. *Introduction to Superconductivity*, 2nd ed.; Pergamon: Oxford, England, 1978.
32. Juergens, F. H.; Ellis, A. B.; Dieckmann, G. H.; Perkins, T. R. I. *J. Chem. Educ.* **1987**, *64*, 851.
33. Grant, P. M. *New Sci.* **1987,** July 30, 36.
34. Dagani, R. *Chem. Eng. News* **1987**, *65* (May 11), 7.
35. Torardi, C. C.; Subramanian, M. A.; Calabrese, J. C.; Gopalakrishnan, J.; McCarron, E. M.; Morrissey, K. J.; Askew, T. R.; Flippen, R. B.; Chowdhry, U.; Sleight, A. W. *Phys. Rev. B* **1988**, *38*, 225.
36. Rao, C. N. R.; Ganapathi, L.; Vijayaraghavan, R.; Ranag Rao, G.; Murthy, K.; Mohan Ram, R. A. *Physica C* **1988**, *156*, 827.
37. Den ,T.; Akinitsu, J. *Jpn. J. Appl. Phys.* **1989**, *28*, L193.
38. Sleight, A. W.; Subramananian, M. A.; Torardi, C. C. *Mat. Res. Soc. Bull.* **1989**, *14* (1), 45.
39. Tarascon, J. M.; McKinnon, W. R.; Barboux, P.; Hwang, D. M.; Bagley, B.B.; Greene, L. H.; Hull, G. W.; Le Page, Y. *Phys. Rev. B* **1988**, *38*, 8885.
40. Torardi, C. C.; Subramanian, M. A.; Calabrese, J. C.; Gopalakrishnan, J.; Morrissey, K. J., Askew, T. R.; Flippen, R. B.; Chowdhry, U.; Sleight, A. W. *Science,* **1988**, *240*, 631.
41. Haddon, R. C. *Acc. Chem. Res.*, **1992**, *25*, 127–130, and references therein.
42. Hahn, H.; Frank, G.; Klingler, W. *Z. Anorg. Allg. Chem.* **1955**, *279*, 279–280.
43. Jaw, H-R. C.; Mooney, M. A.; Novinson, T.; Kaska, W. C.; Zink, J. I. *Inorg. Chem.* **1987**, *26*, 1387–1391.

Additional Reading

- *Chemistry of Superconductor Materials;* Vanderah, T. A., Ed.; Noyes Publications: Park Ridge, NJ, 1992.
- West, A. R. *Solid State Chemistry and Its Applications*; Wiley: New York, 1984, Chapter 13 (ionic conductors).

Memory Metal

- Kauffman, G. B.; Mayo, I. *Invention and Technology.* **1993**, *9* (Fall), 18–23.
- Wayman, C. M. *MRS Bull.* **1993**, *XVIII*, 49–56.

Acknowledgments

This chapter was written with the helpful assistance of Norman Craig, Oberlin College, Department of Chemistry; Frank Weinhold, University of Wisconsin—Madison, Department of Chemistry; and Don Murphy, AT&T Bell Laboratories.

We thank Olivier Deschamps; Atelier D'Art Public; 36 rue Serpente; 75006 Paris, France, for permission to use the photographs in Figure 9.12.

Exercises

1. Choose the correct answer. In the high-temperature phase of NiTi (the unit cell is shown in Figure 9.2), the coordination numbers of the Ni and Ti are
   a. 6 for Ni and 6 for Ti
   b. 6 for Ni and 8 for Ti
   c. 8 for Ni and 6 for Ti
   d. 8 for Ni and 8 for Ti
2. Choose the correct answer. In the phase change during which a material becomes a superconductor
   a. the solid becomes a liquid
   b. the material becomes more electrically resistive
   c. a magnetic field becomes concentrated inside the solid
   d. electrons in the solid pair up
3. What technique lets us determine the atomic positions in NiTi memory metal both before and after the solid has undergone its phase change?
   a. spectroscopy with visible light
   b. measurement of specific heat
   c. electrical resistivity
   d. X-ray diffraction

4. Suppose that you wish to make 12 g of the 1-2-3 superconductor.
   a. Write a balanced equation for the reaction.
   b. How many grams of each starting material (Y_2O_3, $BaCO_3$, CuO, O_2) are needed?

5. Design a device based on a compound such as Cu_2HgI_4 that is insulating at low temperatures but conducting above some threshold temperature.

6. Design a device based on a compound such as Cu_2HgI_4 that is one color at low temperatures but changes to a different color above some threshold temperature. (Such behavior is called thermochromism.)

7. Confirm the formula of Cu_2HgI_4 by examining the unit cell shown in Figure 9.20.

8. Shown in the table are thermodynamic data for white and gray tin; recall that these two forms of tin have very different structures and densities. Consider the reaction, white tin $\rightleftarrows$ gray tin.
   a. For 1 mol of tin, what is the free energy change for the reaction at 298 K, ΔG^o, and therefore in which direction does this reaction go spontaneously at 298 K?

Allotrope	ΔH_f^o (kJ/mol)	ΔG_f^o (kJ/mol)	S^o J/K-mol
White tin	0	0	51.55
Gray tin	–2.09	0.13	44.14

   b. It is often the case that the standard enthalpy and entropy of a reaction, ΔH^o and ΔS^o, respectively, are roughly constant over a broad temperature range. Making this assumption, at what temperature would you predict that the two forms of tin will be in equilibrium with one another (at what temperature will ΔG for the reaction equal zero)?

9. Why do purely solid-state phase changes like that exhibited by NiTi memory metal often involve enthalpies of only a few kilojoules of energy per mole compared to values like 40 kJ to vaporize a mole of water?

10. If a metal can exist in both the hcp and ccp structures, can pressure be used to convert one to the other? Why or why not? [This type of question can be asked of conversions involving any combination of sc, bcc, hcp, and ccp (= fcc) structures for metals; see the packing efficiencies in Table 5.1 to predict the results of an increase in pressure at a given temperature.]

11. Pellets of superconducting 1-2-3 can be used to levitate a magnet. Under what conditions can a magnet be used to levitate pellets of 1-2-3?

12. Figure 2.10 shows a change in measured weight resulting from the magnetic interaction of a paramagnetic or ferromagnetic solid with a magnet. How could a similar experiment be conducted with a pellet of superconducting 1-2-3 and what effect, if any, would this have on the measured weight?

13. How might a "double-decker" levitation experiment be conducted with superconductors and magnets, wherein one object is levitated above a second object, which is simultaneously levitated above a third object? Is a "triple-decker" levitation experiment feasible?

14. Silver mercury iodide, Ag_2HgI_4, undergoes a similar phase transition and corresponding enhancement in ionic conductivity as Cu_2HgI_4, albeit at a slightly lower temperature of about 47 °C. The high-temperature phases of the two solids have the same disordered zinc-blende-like structure. The low

temperature phase of the silver compound, shown as follows, differs from that of the copper compound only in the relative positions of the monovalent cations and vacancies. Show that the unit cell for the Ag_2HgI_4 solid corresponds to this stoichiometry. How do the positions of monovalent cations and vacancies differ in the two structures?

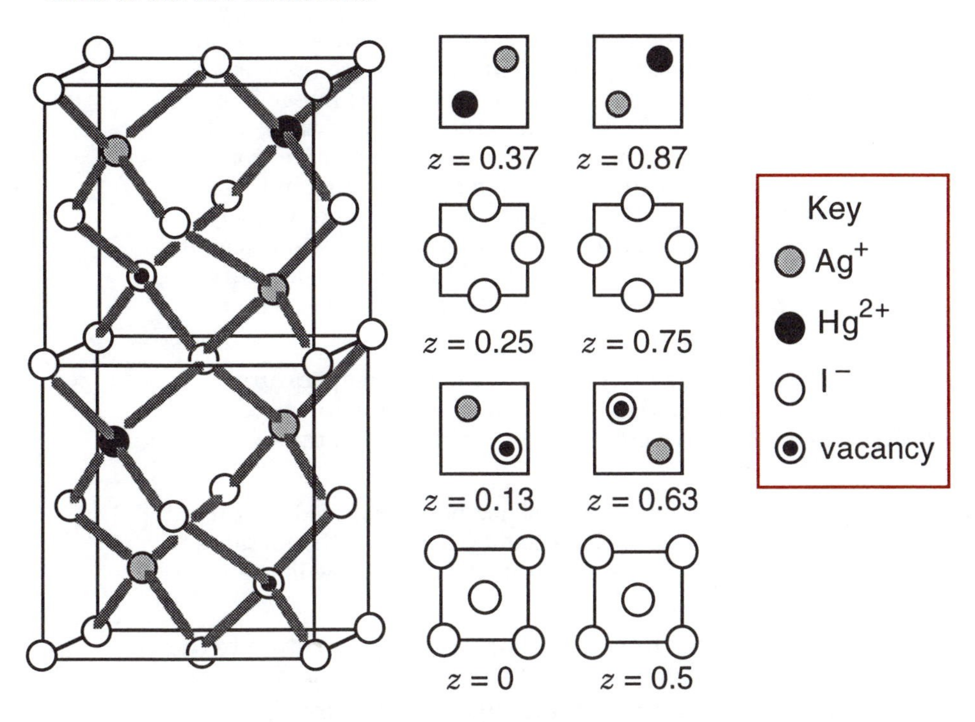

15. The colors of the two phases of Cu2HgI4 reflect their band structures. In both cases, the color is believed to arise from the network of tetrahedral HgI_4^{2-} units composing the solid. The lower energy filled band is derived primarily from filled iodide p orbitals: Which p orbitals are these? (What is the principal quantum number for these p orbitals?) The higher energy unfilled band is derived primarily from mercury s orbitals: Which orbitals are these?

16. Ag_2HgI_4 changes from yellow below the phase transition to orange above it. Interpret this change using Figure 7.18. The shift in the band gap has been interpreted as being due primarily to changes in the widths of the bands. If this is the case, how is the band width changing as the solid goes from the low- to the high-temperature phase? Similarly, Cu_2HgI_4 changes from red at room temperature to dark red–brown above the transition temperature. Interpret these changes in the manner used to describe the silver compound's phase change.

17. Analyze whether solid solutions of Ag_2HgI_4 and Cu_2HgI_4 can be expected to form in the high-temperature form. (Ionic radii are 0.74 Å for Cu^+ and 1.14 Å for Ag^+ in tetrahedral environments.) If so, how would you write the formula for the solid?

18. Determine the formal oxidation state for copper in the other superconductors of Table 9.2, assuming that the oxidation state for lanthanide elements, thallium and bismuth is +3; for oxygen, –2; and for alkaline earth elements, +2.

19. Design a sculpture made of memory metal that will change its shape using electricity. Why can this be done?

20. Using Figure 9.10, what compositions of Ni_xTi_{1-x} would you choose so as to have two samples, one of which is in the low-temperature phase at 0 °C, and the other of which is in the high temperature phase at this same temperature. How could you tell them apart without chemical analysis?

21. Samples of nickel–titanium can also be trained to have "two-way memory," wherein they reversibly adopt one shape in their low-temperature phase and a second shape in their high-temperature phase. This step is accomplished by holding the sample in the desired shape (holding a wire that is linear in its high temperature form in a V-shape, for example) at low temperature, and maintaining this shape at the higher temperature where the sample would normally go back to its high-temperature shape (to a straight line, in this case). If the sample is cycled through this process repeatedly, the two-way memory develops: in this example, the sample will be V-shaped when cool and linear when warm. What happens to the variants of the low-temperature phase during this processing treatment that permits the two-way memory? (See reference 9.)

22. When a thin straight wire of NiTi memory metal in the low-temperature phase is bent and placed in a concentrated solution of bromine in wet methanol, it straightens out as it dissolves. Why? Speculate on the reaction products.

23. Because NiTi memory metal is biocompatible, it has been proposed that it could be inserted into arteries to help unclog them. Given that the metal is to be coiled like a spring in one phase and straight in the other, how is the experiment carried out?

24. Potassium chloride can be converted from its normal rock salt structure to the cesium chloride structure under high pressure at 298 K. The reaction can be written KCl (rock salt structure) $\rightleftarrows$ KCl (cesium chloride structure). Which structure has the greater density? The greater molar volume?

Chapter 10

Synthesis of Materials

The materials tetrahedron of Chapter 1 and the examples of materials in other chapters underscore the special synthetic opportunities and challenges posed by solids: The desired properties of a material or of an interface derived therefrom can be dependent not only on stoichiometric considerations, but on features like atomic-scale defects and microstructure, which are influenced by processing. For example, tangled dislocations cause copper to be more mechanically robust (work hardening, Chapter 6); parts per million of impurity atoms drastically alter the electrical conductivity of semiconductors (doping, Chapter 8); and the concentration of oxygen vacancies controls electromagnetic properties (perovskite superconductivity, Chapter 9).

Nowhere is the vitality of materials chemistry more evident than in our emerging ability to control the arrangement of atoms from the nanoscale to the macroscale. This chapter is intended to provide some context for the synthetic revolution that is the foundation for the Age of Materials.

The first part of the chapter provides the rationale for, and examples of, the traditional, high-temperature, brute force approach to making solids react with one another. Precursor strategies that can be used to coax reactants to form the desired products under milder conditions are presented as well. The second and third parts of the chapter focus on solvent- and gas-phase-based synthetic strategies. These versatile techniques are being used to prepare a broad range of technologically important materials that include diamond films, glass optical fibers, and the multilayered structures of semiconductor diode lasers. In some instances atomic-scale synthetic control can be achieved. Finally, techniques are described for carrying out chemistry inside certain host crystal structures and for preparing large, high-purity crystals of technological importance.

2725–X/93/0329$07.00/1

Solid-Phase Techniques

Diffusion and "Heat-and-Beat" Methods

Many solid-state syntheses are based simply on heating a mixture of solids together with the intent of producing a pure, homogeneous sample of desired stoichiometry, grain size, and physical–chemical properties. These reactions depend on the ability of atoms or ions to diffuse within or between solid particles, Figure 10.1. Diffusion is typically many orders of magnitude slower in solids than in the liquid or gas phases; therefore, high temperatures are required for the reaction to proceed on a reasonable time scale.

> **Laboratory.** Experiment 13 investigates the relative rates of diffusion of gases, liquids, and solids.

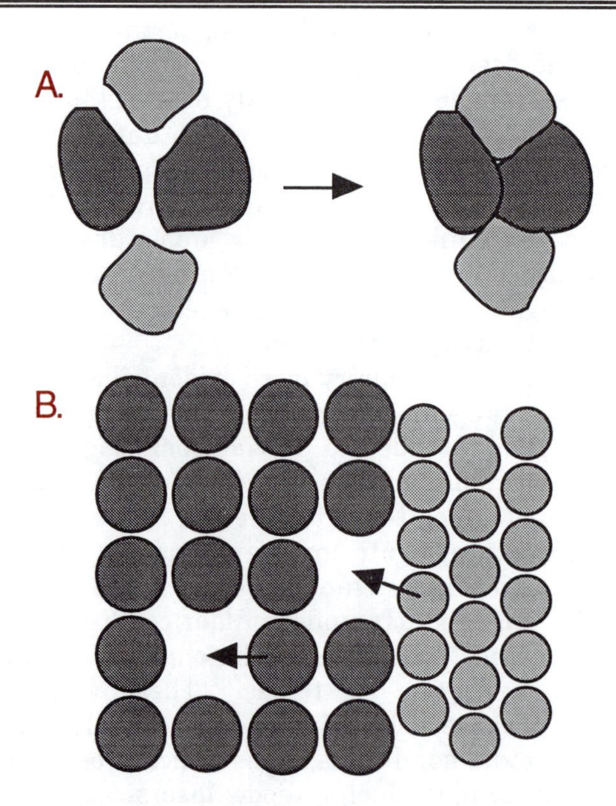

Figure 10.1. What the "heat and beat" methods accomplish. A. Precursor particles are brought into close contact. B. Atoms intermingle by diffusion through vacancies or interstitial positions (*see* Chapter 6).

A dramatic illustration of these principles is provided by the reaction of either nickel or cobalt with aluminum to produce the intermetallic compounds CoAl or NiAl, which have the CsCl structure (Chapter 5; the high-temperature phase of NiTi "memory metal," Chapter 9, also has this structure). Bars of these metal mixtures are easily made, using a die to compact the metal powders and a standard press. The bar requires ignition by sustained exposure of one of its ends to a very hot propane torch. Once ignited, the reaction is sufficiently exothermic (the temperature likely exceeds 1500 °C and may cause localized melting) that it is self-sustaining: the bright orange reaction front travels along the bar with the visual impact of a controlled thermite reaction! Figure 10.2 helps appreciate the extensive shuffling of atoms that has to accompany the conversion of the starting materials, which have the fcc crystal structure, to the CsCl structure of the products. A more quantitative perspective on this kind of atomic shuffling is given in the discussion of the synthesis of the superconducting 1-2-3 compound later in the next section.

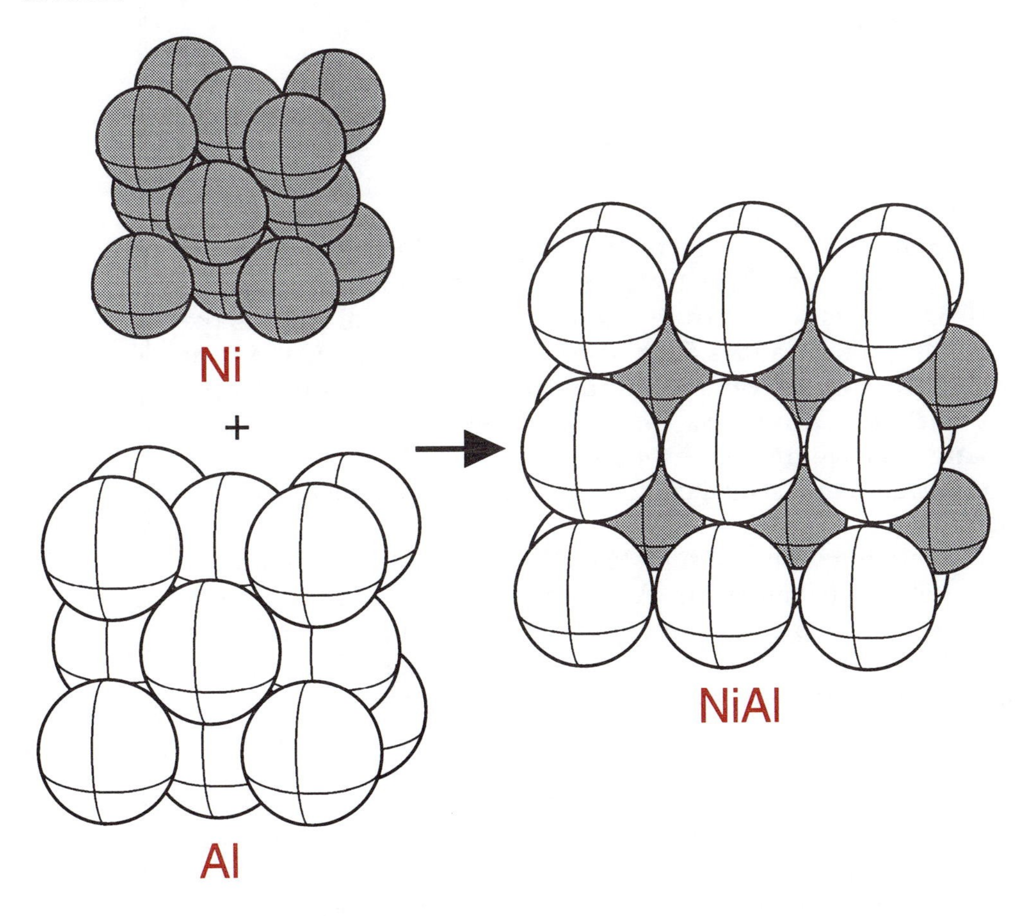

Figure 10.2. Elemental Ni and Al have the face-centered cubic structure, and NiAl has the CsCl structure.

Demonstration 10.1 Synthesis of NiAl and CoAl

Caution: The reaction is vigorously exothermic. Keep a safe distance after initiating the reaction and make sure viewers are at a safe distance of at least several meters. Make sure that all flammable materials have been removed from the demonstration area. Conduct this demonstration with the level of caution used for thermite reactions *(1)*.

Materials

325-mesh elemental powders (aluminum and nickel or cobalt)
A die in which to compact the powders (*see* text and Figure 10.3)
A press (a press used for making KBr pellets for infrared spectra, such as a Carver laboratory press, is suitable)
Lithium stearate
Fire brick (or other surface able to withstand temperatures on the order of 2000 °C)
Propane torch
Fire extinguisher

Procedure

- Weigh out correct molar ratios of aluminum and cobalt, or aluminum and nickel, in order to form CoAl or NiAl. A total mass of 8 g will be sufficient for a mold the size of the one described in Figure 10.3. On this scale, use 5.49 g of Co and 2.51 g of Al to make CoAl; or 5.48 g of Ni and 2.52 g of Al to make NiAl. **Caution: Avoid inhalation of airborne metal particles and direct contact with the powders. Wear a mask and gloves when dispensing and weighing the powders.**
- Place the powders in a sealed container of glass or plastic and shake vigorously.
- Coat a suitable die with lithium stearate to prevent the aluminum from cold-welding to the steel. A die can be manufactured from a stainless steel bar and bolts (*see* Figure 10.3). Lithium stearate can be synthesized by mixing 1:1 molar ratios of a solution of stearic acid in ethanol with an aqueous solution of lithium hydroxide.
- Place the powder evenly in the die and press at 20,000 lbs for a few minutes. Carefully remove the fragile bar from the die.
- Place the pressed bar on a fire brick and heat one end of the bar with a propane torch, using the hottest part of the flame (just above the blue cone). Once the edge begins to glow orange (this may take a minute or two), remove the torch. If the reaction does not proceed, try the other end of the bar.
- The product is much less magnetic than the original mixture.

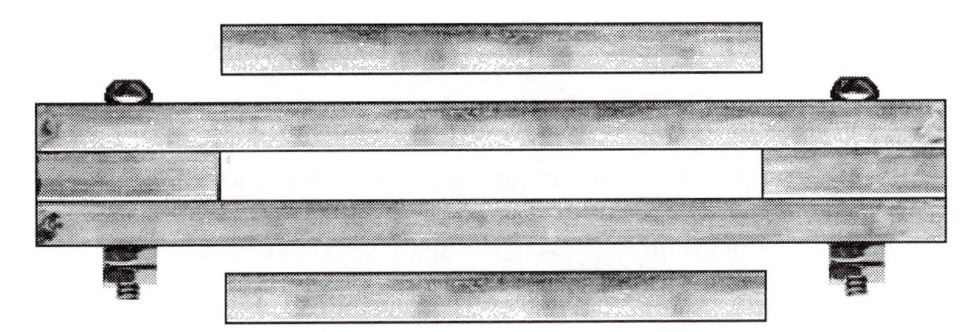

Figure 10.3. Diagram of the center bars and body of the die (actual size) described in Demonstration 10.1. All components are made out of 1/4-inch square stainless steel rods. (The center bars are of the correct length to just fit in the gap in the center of the body of the die.) Holes are drilled through the six bars forming the body, and small nuts and bolts are used to secure the pieces as shown. The body of the die is placed on a piece of cardboard that has a hole cut in it that is the shape of one of the center bars. A center bar is pushed through the hole in the die body and into the hole in the cardboard. In this way, the body of the die is held above the center bar. The cardboard and center bar will rest on the counter (or, on the base of the press). The powder may then be placed on the center bar. The second center bar is then placed on top of the powder. When the press is used, the powder is squeezed between the two center bars. After releasing the pressure, the pressed sample may be removed from the die by carefully removing the top center bar and loosening the bolts.

The reaction of aluminum with nickel or cobalt typifies many aspects of reactions between solids: It is often necessary to heat finely ground, intimate mixtures of solids to temperatures of the order of 1000–1500 °C in order to enhance the rates of diffusion between the reactants. In many syntheses, several cycles of heating for many hours, each followed by regrinding at room temperature, and a final heating may be needed to form a homogeneous sample. With the elevated temperatures and extended times used in such reactions, the products are usually the thermodynamically stable phases (or mixtures of phases); compounds that are stable only at relatively low temperatures may be inaccessible by this synthetic approach.

The preparation of the 1-2-3 superconductor $YBa_2Cu_3O_{7-x}$, described in Experiment 11, is another example of high-temperature synthesis. Samples of $YBa_2Cu_3O_{7-x}$ can be prepared by heating a finely ground, well-mixed powder, consisting of a stoichiometric mixture of Y_2O_3, BaO_2 (or $BaCO_3$), and CuO, at about 950 °C. After heating, the sample is reground and pressed into a pellet and reheated (sintered; described later in this section) at 950 °C. Following sintering, the pellet is heated in an atmosphere of pure oxygen at 500–600 °C, then slowly cooled to room temperature. Each of the steps in this process addresses problems that can arise during a solid-state synthesis.

The necessity for heating finely ground and well-mixed precursor compounds in this synthesis is attributable to the structures of the starting materials and products. In $YBa_2Cu_3O_{7-x}$ all three metal ions are mixed on an atomic scale; particles of the starting materials each contain only one type of ion. Thus, yttrium, barium, and copper ions must all diffuse together in order for the 1-2-3 product to form.

Extensive grinding in a mortar and pestle produces particles with a radius of about 10^{-3} cm. Copper ions are separated by about 4×10^{-8} cm in the particles of CuO; thus about 25,000 copper ions are along the line between the ions in the center of these particles and the surface of the particle. If, for example, yttrium and barium ions are to diffuse into the center of the particle, these ions must push their way past thousands of other positive ions that lie between the surface and the interior of the particle. This process is slow and requires activation by heating.

Regrinding the sample exposes unreacted portions of the reactants. Pressing a pellet increases the contact between the surfaces of the particles that compose the sample. Thus regrinding and pelletizing helps to ensure complete reaction. Subsequent sintering (heating to temperatures a little below the melting point) promotes bonding between the grains that compose the pellet, thereby increasing the density and mechanical strength of the pellet. Sintering is somewhat like forming a snowball in that it is easier to get the snow to stick together if it is near the melting point of ice. For a snowball, squeezing the snow in one's hands helps to increase the snowball's density by fusing grains of snow; this is like pressing a pellet of a solid-state sample before heating.

The oxidation states in many solids that contain ions that can exhibit two or more oxidation states can be controlled by varying the oxidizing or reducing character of the atmosphere above the substance. As discussed in Chapter 9, annealing $YBa_2Cu_3O_{7-x}$ in a pure oxygen atmosphere increases the percentage of ions that are formally Cu(III) in the sample.

Precursor Techniques

In the synthetic technique just described, the reactants are finely divided and mixed manually to produce as intimate a mixture as possible. It is sometimes possible to improve mixing by preparation of a precursor (generally by precipitation from solution) that will contain the reactants in a mixture with atomic scale mixing. These precursors are reactive at lower temperatures and require shorter heating times than mechanically mixed samples of solids: Because the reactant atoms or ions are mixed at the atomic, or near atomic level, in the precursor, the distances that they need to diffuse during reaction are much smaller than between solid particles.

In many cases the precursor is simply a finely divided, coprecipitated, solid mixture or a gelatinous mass called a gel. For example, yttrium iron garnet, $Y_3Fe_5O_{12}$, can be prepared by the reaction of a stoichiometric mixture of finely divided, well-mixed Y_2O_3 and Fe_2O_3 at 1100 °C.

$$3\ Y_2O_3 + 5\ Fe_2O_3 \rightarrow 2\ Y_3Fe_5O_{12} \quad (1)$$

However, the reaction proceeds much more rapidly and at about 900 °C if a precursor consisting of a stoichiometric solid mixture of $Y(OH)_3$ and $Fe(OH)_3$ is used. This mixture is prepared by coprecipitation of the hydroxides from an aqueous solution of stoichiometric amounts of soluble salts of Y^{3+} and Fe^{3+}.

$$3\ Y^{3+}(aq) + 5\ Fe^{3+}(aq) + 24\ OH^-(aq) \rightarrow 3\ Y(OH)_3 / 5\ Fe(OH)_3\ (s, \text{mixture}) \quad (2)$$

$$3\ Y(OH)_3 / 5\ Fe(OH)_3\ (s, \text{mixture}) \xrightarrow{\text{heat}} Y_3Fe_5O_{12}\ (s) + 12\ H_2O \quad (3)$$

Laboratory. Experiment 14 describes the synthesis and magnetic studies of yttrium iron garnets prepared from a co-precipitated precursor.

Stoichiometric control can sometimes be improved if a single compound serves as a precursor. For example, the perovskite (Chapter 5) $LaCoO_3$ is readily prepared from compounds that contain the cobalt and lanthanum ions mixed on an atomic scale in a 1:1 ratio. One such compound is $La[Co(CN)_6]$, which forms as a hydrated precipitate from the reaction between a solution of a lanthanum salt and a solution of $K_3[Co(CN)_6]$. This compound can be converted to $LaCoO_3$ by heating in air for a short time.

$$LaCl_3(aq) + K_3[Co(CN)_6]\ (aq) \rightarrow La[Co(CN)_6]{\cdot}xH_2O(s) + 3\ KCl(aq) \quad (4)$$

$$2\ La[Co(CN)_6]{\cdot}xH_2O(s) + 15\ O_2\ (g) \xrightarrow{\text{heat}} 2\ LaCoO_3(s) + 12\ CO_2(g) + 6\ N_2(g) + 2x\ H_2O(g) \quad (5)$$

A variety of other solid mixtures and stoichiometric compounds also can be used as precursors. For example, mixtures of iron(III) and zinc oxalates readily decompose to form $ZnFe_2O_4$, a compound with the so-called spinel structure.[1] A zinc chromium spinel, $ZnCr_2O_4$, which is used as a magnetic recording medium in cassette recorders, results from the decomposition of the stoichiometric compound ammonium zinc chromate, $(NH_4)_2Zn(CrO_4)_2$.

$$(NH_4)_2Zn(CrO_4)_2\ (s) \xrightarrow{\text{heat}} ZnCr_2O_4(s) + N_2\ (g) + 4\ H_2O\ (g) \quad (6)$$

[1]In the spinel ($MgAl_2O_4$) structure, the oxide ions are in a ccp array with the Mg^{2+} ions in tetrahedral holes and Al^{3+} in octahedral holes.

In principle, coprecipitation and other precursor methods should be applicable to a wide variety of syntheses. However, the reaction conditions usually must be worked out on a case-by-case basis.

Reactions in Liquids

Solvents

Significant advantages are gained by conducting a reaction in a liquid medium: Diffusion rates are much faster in liquids, and the liquid phase can be more readily stirred to enhance mixing. Techniques involving liquids in the synthesis of solids include dissolving reactants in water or other solvents at relatively low temperatures, using molten salts or molten metals as solvents (called fluxes) at moderately high temperatures, and using reactants or products as the molten medium. The formation of lead iodide from lead nitrate and potassium iodide or sodium iodide illustrates the ability of a solvent to enhance reactivity.

Demonstration 10.2. Solvent Influence on Reactivity

Materials

1 g of lead(II) nitrate
1 g of potassium iodide or sodium iodide
Dry test tube and stopper
Water

Procedure

- Place about 1 g of lead(II) nitrate and 1 g of potassium iodide or sodium iodide in a ***dry*** test tube and shake. At best, only a faint tinge of yellow–orange color will be observed on the surface of the powders.
- Add several milliliters of water, and a brightly colored yellow–orange precipitate of lead iodide will form immediately.

Caution: Lead compounds are toxic. Avoid skin contact and inhaling dust from the compounds. Place the waste from this demonstration in an appropriately labeled container and dispose of it according to state and local regulations.

Some solids are prepared from aqueous solution. For example, crystals of alum and Cr-doped alums (Experiment 3) can be grown by direct precipitation from water. Preparation from solution does not require that the reactants be extensively soluble in the solvent. For example, as may be observed in Experiment 12, the ionic conductor Cu_2HgI_4 forms when

HgI_2 and CuI, both of which are only slightly soluble, are heated together in water.

$$HgI_2 + 2\,CuI \rightarrow Cu_2HgI_4\,(s) \tag{7}$$

An important application of synthesis from an aqueous solution is the preparation of molecular sieve zeolites. Zeolites (Chapter 5) are aluminosilicates that have a framework of tetrahedral silicate and aluminate units arranged to form large tunnels and cavities. The tunnels are hydrated and contain additional cations necessary for the electroneutrality of the solid. Dehydrated zeolites can selectively absorb small molecules into the tunnels, and many are catalysts with extensive industrial applications. Most zeolites decompose above about 400 °C, so their synthesis must be accomplished at lower temperatures.

The starting materials for zeolite synthesis are solutions of salts of silicate and aluminate anions in aqueous alkali. Upon mixing, a gel forms from the coprecipitation of the silicate and aluminate anions. Subsequent heat treatment (sometimes under high pressure at temperatures above the normal boiling temperature of water—a so-called hydrothermal treatment) increases the solubility of the gel, and crystals of zeolite are produced from the resulting solution. The process may be represented as follows *(2)*:

$$x\mathrm{NaAl(OH)_4(aq)} + y\mathrm{Na_2SiO_3(aq)} + \mathrm{NaOH(aq)} \xrightarrow{25\ °\mathrm{C}} \mathrm{gel} \tag{8}$$

$$\mathrm{gel} \xrightarrow{25-175°\mathrm{C}} [\mathrm{Na}_x(\mathrm{AlO_2})_x(\mathrm{SiO_2})_y\cdot m\mathrm{H_2O}]\ (\text{zeolite crystal}) \tag{9}$$

Nonaqueous solvents can also be used to advantage in the synthesis of solid-state materials. Organometallic precursors that are soluble in organic solvents can be used to produce extended solids. For example, a zerovalent nickel compound, L_2Ni, reacts with an alkylphosphine and alkylphosphine telluride in boiling toluene over a period of several hours to produce a polycrystalline precipitate of NiTe in quantitative yield *(3)*. In contrast, the direct reaction of the elements requires heating at 400–600 °C for days to weeks in order to produce NiTe.

$$R_3P + L_2Ni + R_3PTe \rightarrow NiTe + 2R_3P + 2L \tag{10}$$

An additional advantage to this route is that by varying the reaction stoichiometry, intermediate metal clusters can be isolated.[2] Clusters may be viewed as model compounds for extended solids, and studying them can provide insight into the mechanism by which the bulk solids form from the reaction of soluble organometallic compounds. Furthermore,

[2] A metal cluster is a compound that has more than two metal atoms in close proximity, but fewer than the number of atoms required to give the properties of a bulk metal. In this particular case, clusters of the general formula $Ni_9Te_6L_8$ and $Ni_{20}Te_{18}L_{12}$ can be isolated.

clusters have different properties than the related extended solids and are interesting materials in their own right.

Melts (Fluxes)

Chemists typically recrystallize molecular compounds by dissolving them in a hot solvent, then allowing them to reprecipitate upon cooling. Solid-state chemists do likewise, except they usually use molten salts rather than lower boiling solvents. For example, crystals of yttrium iron garnet, $Y_3Fe_5O_{12}$, can be produced by dissolving a sample of the garnet in a molten mixture of PbO and PbF_2, followed by slow cooling. The solidified melt is removed from the crystals by dissolving it in acetic acid, in which the garnet is insoluble.

Melts can also be used as solvents for synthesis. Liquid tin (m.p. 232 °C) is used as a solvent for the reaction of ruthenium and phosphorus, producing RuP_2; and molten copper (m.p. 1083 °C) is used as a solvent for the production of MnSi from manganese and silicon.

An interesting technique that can be used for either synthesis or crystal growth is to have the reactants serve as both solvent and container. Many oxides can be prepared in this fashion. For example, $LaTiO_3$ can be prepared from a mixture of molten La_2O_3 and Ti_2O_3. However, this reaction requires extremely high temperatures (La_2O_3 melts at 2315 °C; Ti_2O_3 at 2130 °C) and exotic containers that neither react with the melt nor themselves melt at extremely high temperatures. One way to handle the problem is to create a solid shell (or skull) of the reactants that acts as a sample container. This step can be accomplished by placing a mixture of the reactants in a copper block that is cooled by water passing through tubes in the block, and heating the sample with an electric arc (similar to that of an arc welder) or with an extremely powerful microwave source. The portion of the sample in contact with the cool block does not melt and forms a solid shell that contains the remainder of the sample, which is molten. The molten reactants form product and, after cooling, the crystalline product is recovered by cutting away the skull.

Gas-Phase Synthesis

Chemical Vapor Transport

Arguably, no synthetic technique in the past 20 years has had a greater impact on materials science than chemical vapor methods. Several common examples are presented in this section.

The growth of large, pure single crystals is facilitated by a method called chemical vapor transport. This technique generally involves an equilibrium between solid and gaseous reactants and their gaseous reaction products. The growth of large single crystals of ZnS from microcrystalline ZnS is an example of the process.

At elevated temperatures, solid ZnS reacts with gaseous iodine to establish an equilibrium with gaseous diatomic sulfur and gaseous zinc iodide.

$$ZnS(s) + I_2(g) \rightleftarrows ZnI_2(g) + 1/2\ S_2(g) \quad (11)$$

This reaction is endothermic. If equilibrium is established at 900 °C, cooling the mixture to 800 °C will shift the equilibrium to the left, producing crystalline ZnS.

The reaction is generally run in a sealed tube (Figure 10.4) with microcrystalline ZnS at one end at 900 °C and the empty end of the tube at 800 °C. As ZnS reacts, the gases, which are at equilibrium at 900 °C, diffuse down the tube to the cooler end, where the equilibrium conditions are different, and ZnS reforms, generally as single crystals. Equilibrium is maintained at either end of the tube because the reaction of ZnS and its deposition are faster than the rate of diffusion of gases through the tube. Generally, transport rates are low and only a few milligrams per hour are deposited. However, as is typically the case when crystals are grown slowly, the crystals that form are usually quite large and of very high quality.

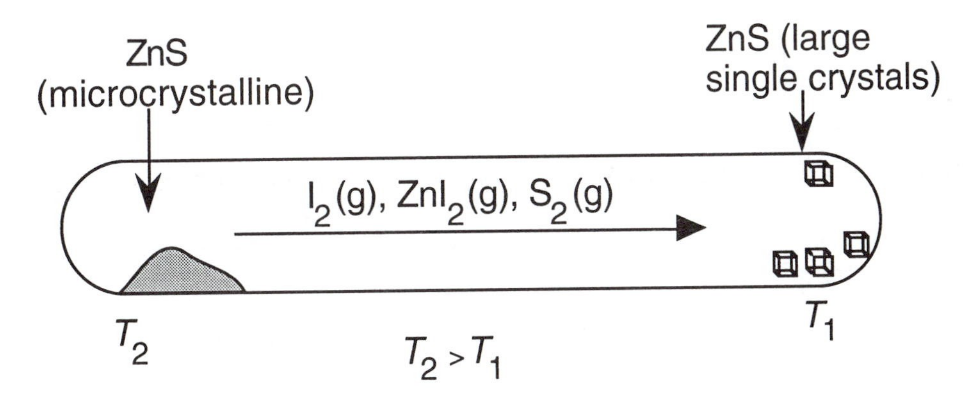

Figure 10.4. The arrangement of reactants, products, and temperatures for the chemical vapor transport of zinc sulfide. (Adapted with permission from reference 2. Copyright 1984 John Wiley and Sons.)

Chemical vapor transport can also be used for synthesis. For example, Cu_3TaSe_4 can be prepared by the reaction of Cu, Ta, and Se at 800 °C in a transport tube containing a little iodine: crystals of Cu_3TaSe_4 form at 750 °C in the cooler end of the tube.

Preparation of Thin Films

Metals can be evaporated from a target in vacuum by ion bombardment, electron bombardment, or heating (Figure 10.5). The evaporated metals condense on a substrate and its surroundings as a film. The thickness of the film depends on the rate of evaporation of the metal and the length of the exposure to the vapor. With experience, films only several atoms thick can be applied, and sandwich structures, consisting of alternating layers of different metals, can be deposited onto the substrate.

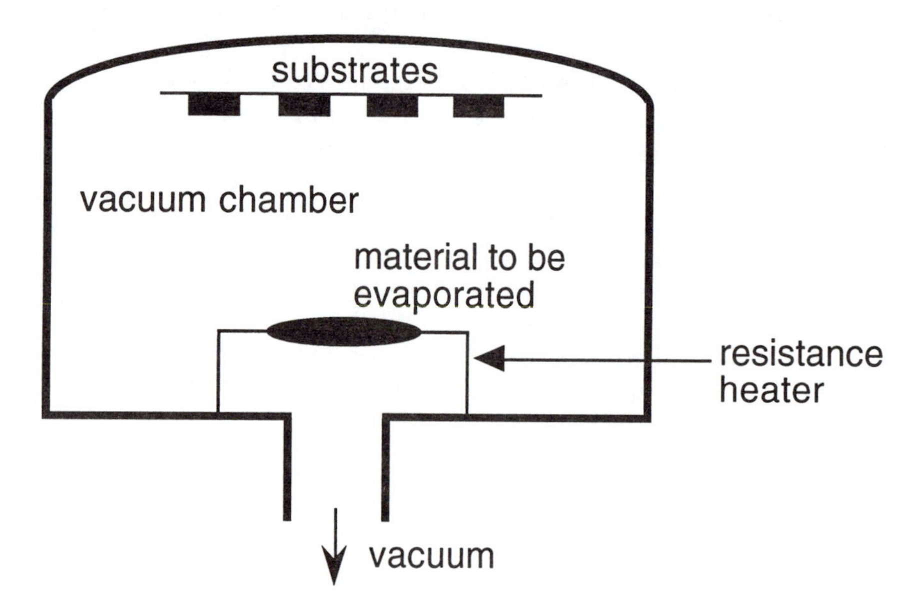

Figure 10.5. Vacuum evaporation equipment for thin film deposition. Atoms evaporate from the source as it is heated and deposit on the substrates. (Adapted with permission from reference 2. Copyright 1984 John Wiley and Sons.)

Evaporation of films followed by heat treatment is one way that silicon is doped during the production of n-type or p-type semiconductors (Chapter 8). Following evaporation of a p-type or an n-type dopant onto a pure silicon substrate, the sample is heated to allow the dopant to diffuse into the silicon.

Chemical Vapor Deposition (CVD) and Molecular Beam Epitaxy (MBE)

One of the most popular techniques for achieving atomic-scale synthetic control is chemical vapor deposition (CVD). In this method, precursor gas molecules are thermally decomposed to yield solids of interest. With appropriate control, solids can be deposited virtually an atomic layer at a time onto a substrate.

Diamond provides an important illustration of the CVD methodology, which can be used to produce thermodynamically metastable products. As noted in Chapters 1, 7, and 9, diamond has many unusual and desirable physical and chemical properties but, until recently, had resisted all but brute force synthetic techniques that relied on extremely high temperatures and pressures. In the early 1980s, however, CVD methods were successfully brought to bear on the problem. In particular, gaseous mixtures of hydrogen containing a small amount of hydrocarbon gas (on the order of 1% methane, for example) were found to yield diamond films under conditions that led to the cracking of the hydrocarbon. Atomic hydrogen appears to play a crucial role in favoring diamond formation. In these low-pressure processes, atomic hydrogen is generated by various techniques, including microwave irradiation, flames, and use of a hot filament *(4)*.

The growth of semiconductors is routinely carried out by CVD methods. Commonly, reactive metal alkyls and nonmetal hydrides are co-reacted by thermally decomposing them onto a heated substrate. For example, equation 12 shows that trimethylgallium and arsine react to produce GaAs and methane:

$$(CH_3)_3Ga(g) + AsH_3(g) \rightarrow GaAs(s) + 3\ CH_4(g) \qquad (12)$$

To make a solid solution of $Al_xGa_{1-x}As$, a third precursor like $(CH_3)_3Al$ would be present as well:

$$x\ (CH_3)_3Al(g) + (1-x)\ (CH_3)_3Ga(g) + AsH_3(g) \rightarrow Al_xGa_{1-x}As(s) + 3\ CH_4(g) \qquad (13)$$

This technique permits the growth of atomically abrupt interfaces and thus of countless tailored solid structures. A solid formed, for example, by growing in repeated alternation 10 atomic layers of GaAs followed by 10 atomic layers of $Al_{0.1}Ga_{0.9}As$, has physically distinct properties from one grown by alternating five layers of each of these compositions. When the dimensions of the tailored features are sufficiently small, quantum size effects can be observed.

The nature of the interfaces formed between dissimilar materials is also important. Because of the chemical similarity of Al and Ga, the unit cell sizes of GaAs and $Al_xGa_{1-x}As$ (both have the zinc blende structure, Chapters 5 and 7, and cubic unit cell lengths of 5.66 Å) are almost identical, meaning that the deposited atomic layers are in near-perfect registry with one another, even when the composition switches from GaAs to $Al_xGa_{1-x}As$ or vice-versa. This kind of growth, wherein the atoms of the deposited layer are in registry with those beneath on the substrate, is called epitaxial growth. Mismatches that occur when there is more structural disparity between the deposited material and the substrate can lead to large concentrations of defects that may degrade the performance of devices constructed from these materials.

The p–n junctions described in Chapter 8 benefit greatly from CVD technology. Samples of GaAs can be grown by CVD in the presence of additional precursor gas molecules that serve as dopant sources. For instance, n-GaAs would result if hydrogen selenide were present during the growth. After a desired thickness of n-GaAs had been deposited, the dopant source could be switched to dimethylzinc to produce the p-type layer atop the n-type substrate and thus a sharp p–n junction (Chapter 8).

A second technologically important method of semiconductor growth that can afford even better control is molecular beam epitaxy, MBE. In this ultrahigh-vacuum deposition process, collimated beams of atoms or molecules impinge on a heated substrate. For example, Ga, As, and Al are heated in separate furnaces that are equipped with shutters, and the heated beams strike a suitable substrate. Like CVD, the MBE process permits the construction of layered structures with atomic-scale control.

Optical Fibers

Optical fibers are prepared using gas-phase techniques and represent an enabling technology that is being used to an ever greater extent to transmit information. In many cases these fibers are replacing metal wire, owing to the much greater density of information that can be carried: light can be modulated at higher frequencies than electrical current. The synthesis and processing of these materials blends several themes of importance to materials chemistry.

In its simplest form, an optical fiber consists of a core of an amorphous material (usually silica, SiO_2, glass) surrounded by another material called the cladding. The amorphous glass is a structure without the long-range, periodically repeating arrangements of atoms that characterize the crystalline solids discussed throughout much of this book. The critical property of the materials composing the optical fiber are the speed of light in the two media, which is c/n, where c is the speed of light in vacuum (3×10^{10} cm/s) and n is the refractive index of the material. When the core of the optical fiber has a larger refractive index than the surrounding cladding, total internal reflection results, meaning that the light is largely confined to the core and can be transmitted along the fiber with relatively little "leakage," as shown in Figure 10.6.

Fibers may have either an abrupt or a gradual change in refractive index. Those in which the change is abrupt are called single-mode fibers and are designed to carry a single wavelength of light, and those with a graded index are called multimode and may carry more than one wavelength. A standard graded index optical fiber has a core diameter of 50–62.5 μm and a cladding diameter of 125 μm. Single-mode fibers have core diameters that are typically 8–9 μm with cladding diameters of 125 μm.

Laboratory. Experiment 15 describes a synthesis of silica.

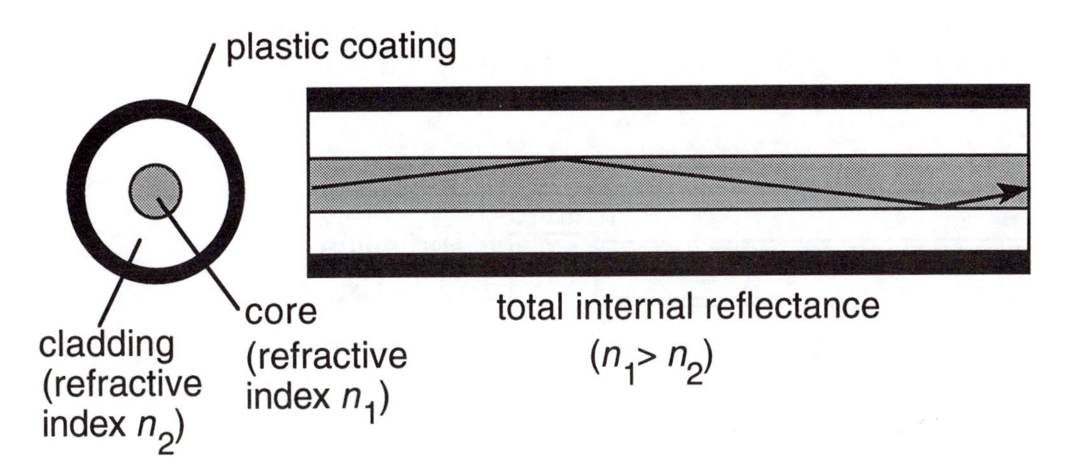

Figure 10.6. Cross-sectional (left) and longitudinal (right) views of an optical fiber. When the refractive index of the cladding is less than the refractive index of the core, light that is sent down the fiber core will tend to stay within the core.

Two materials-related limitations affect the efficiency of light transmission attainable in an optical fiber. First, scattering occurs because of defects in the glass. This so-called Rayleigh scattering increases as the wavelength of transmitted light decreases (it is proportional to λ^{-4}, where λ is the wavelength) so that relatively long wavelengths in the near-infrared are used. Wavelengths of 1300 and 1550 nm correspond to absorption windows (negligible absorption) in the silica core material and have been the focus of development efforts; these are also wavelengths that are accessible to the output of diode lasers corresponding to $In_xGa_{1-x}As_yP_{1-y}$ semiconductor solid solutions (Chapters 3 and 7). A second transmission loss mechanism involves transition metal impurities in the core material, which give rise to metal–oxygen vibrations that absorb over a broad spectral range. Thus, the glass must be very pure. In addition, the glass must be dry, because O–H bonds give rise to overtone bands that appear in the spectral region of interest.

Two general synthetic methods that minimize these problems have been developed for optical fibers. AT&T Bell Laboratories has developed a technique called the MCVD or modified chemical vapor deposition process. In MCVD, high purity $SiCl_4$ and O_2 ,which react to form SiO_2 and Cl_2, are passed through a fused silica (SiO_2) tube. In the first step, a high-purity cladding is deposited on the inside of the tube by passing a mixture of $SiCl_4$, O_2, and a fluorine-containing gas such as CCl_2F_2 or SiF_4 into the tube; the fluorinated gases supply fluorine, which lowers the refractive index of the silica. In the second step, the core material is passed into the tube. Small amounts of $GeCl_4$ (which reacts with oxygen to produce GeO_2) can be added to the $SiCl_4$ as a dopant to increase the refractive index of the core. By carefully monitoring and varying the concentration of reactant gases, the refractive index can be changed abruptly or gradually from the center to the outside of the fiber. The tube is simultaneously heated to high temperatures to fuse the deposited material into a uniform glass. Then the rod is drawn into a fiber and coated with a plastic to increase durability. These processes are shown in Figure 10.7A.

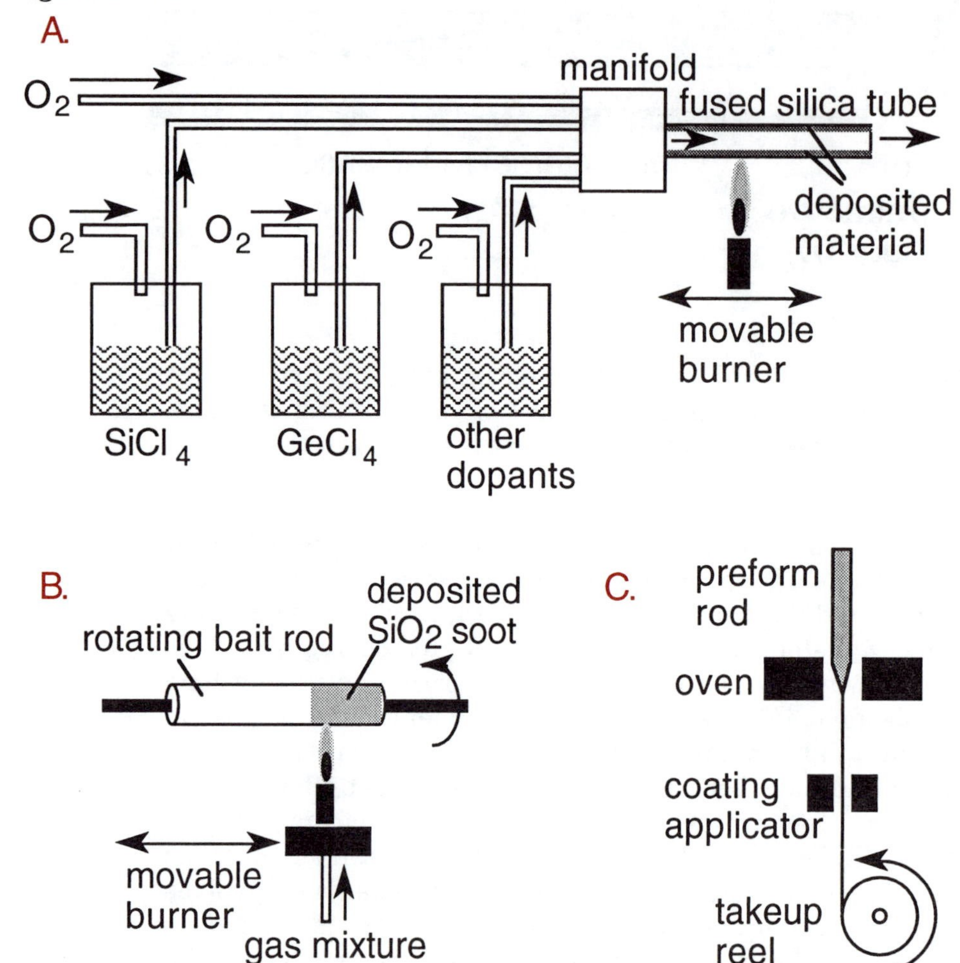

Figure 10.7. A: A schematic diagram of the MCVD process for preparing optical fibers. In this case material is deposited on the inside of a fused silica tube. (Adapted with permission from reference 5.) B: A schematic diagram of the OVD process. In this method, material is deposited on the outside of a bait rod, which is removed prior to pulling the fiber. C: A simplified diagram for the generation of a glass fiber from a preform rod generated by either method of preparation. (Adapted with permission from reference 6. Copyright 1983 McGraw-Hill, Inc.)

Corning Glassworks has developed an alternative process called OVD, or outside vapor deposition, which involves depositing SiO_2 or Ge-doped SiO_2 on the outside of an alumina rod. In this case, $SiCl_4$ and other dopant gases such as $GeCl_4$ are fed into a flame to produce a soot of SiO_2 on a rotating alumina rod called a bait rod. By varying the composition of the gases in the flame, the refractive index of the deposited glass can be varied to create the core and then the cladding. After deposition, the bait rod is removed and the so-called soot blank, the material deposited on the alumina rod, is dried and heated to about 1600 °C to consolidate it into clear glass. Finally, it is drawn into a thin fiber and coated. This process is shown in Figure 10.7B.

Information is transmitted through the fiber by light from a laser or light-emitting diode (LED). The intensity of this light can be changed (turned on and off to correspond to logical 1s and 0s) to carry the digitized information. Many thousands of phone conversations can be carried on one optical fiber simultaneously through time-division multiplexing. In this technique, one second is divided into many segments, and each segment can be considered to be a channel for a separate conversation. If the transmitter and receiver are clocked and coupled, the conversations occur in separate time slots.

Chemie Douce or Soft Chemistry

Soft chemistry (or, in French, *chemie douce*) involves modifying an existing compound under relatively mild conditions to produce a closely related material. In an intercalation reaction, the basic structure of the host is not altered when a guest molecule is inserted. With reference to Figure 10.8, these can include the following reactions of guest molecules, $\bigcirc$ and $\boxplus$, with host solids, $\boxed{\text{Host}}$.

redox

$$n\bigcirc + \boxed{\text{Host}} \rightarrow \boxed{\oplus_n(\text{Host}^{n-})} \tag{14}$$

(where $\bigcirc$ is H^+ (and a reductant), Li, or Na)

ion exchange

$$n\boxplus + \boxed{\oplus_n(\text{Host}^{n-})} \rightarrow n\oplus + \boxed{\boxplus_n(\text{Host}^{n-})} \tag{15}$$

(where $\boxplus$ is H^+ or alkali and transition metal ions)

acid–base

$$n\bigcirc + \boxed{\text{H-Host}} \rightarrow \boxed{\text{H}\oplus_n(\text{Host}^{n-})} \tag{16}$$

(where $\bigcirc$ is an amine and H-Host has acidic protons available)

Because the reactions are generally run at relatively low temperatures, compounds and/or phases that are either metastable or not stable at elevated temperatures can be prepared.

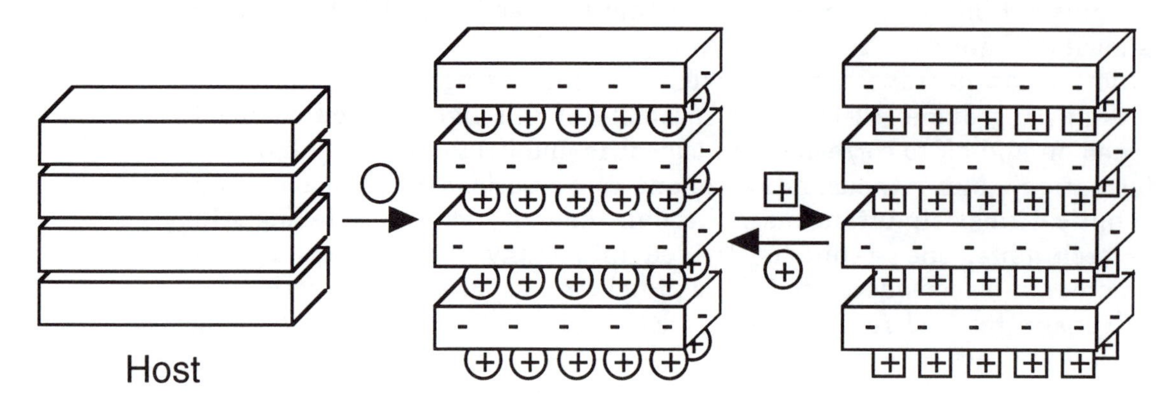

Figure 10.8. Chemical reactions of a layered material such as clays, graphite, MS_2 (M is a transition metal; MoS_2, for example), MOCl (FeOCl, for example), MPX_3 (M is a transition metal and X is S or Se; $CdPS_3$, for example), MO_x (MoO_3, for example), and $M(HPO_4)_2 \cdot xH_2O$ (M is Ti, Zr, Hf, Ge, Sn, or Pb). The large blocks represent layers of the host material which have negative charge when positively charged guests are present.

Compounds such as β-alumina (approximately $Na_2O \cdot 8Al_2O_3$) that have mobile ions can undergo ion exchange. The sodium ion in β-alumina can be replaced by a variety of monovalent cations such as Li^+, K^+, Ag^+, Cu^+, Tl^+, and NH_4^+ by immersing the compound in an appropriate molten salt at about 300 °C.

$$Na_2O \cdot 8Al_2O_3(s) + 2\ AgNO_3(l) \rightarrow Ag_2O \cdot 8Al_2O_3(s) + 2\ NaNO_3(l) \quad (17)$$

Zeolites readily undergo ion exchange of the compensating cation in the channel (for example, the Na^+ ion in $[Na_x(AlO_2)_x(SiO_2)_y]$). This ion exchange is used to soften water (to replace Mg^{2+}, Ca^{2+}, and Fe^{3+} in water with Na^+ by ion exchange) and to prepare catalytically active transition metal and lanthanide derivatives of zeolites.

Oxidation or reduction may accompany loss or gain of an ion. In Experiment 8, the reaction of HCl with zinc metal generates hydrogen atoms, which can intercalate into the cavities in the structure of WO_3. The product of this reaction is H_xWO_3, which is highly colored and conducting due to the donation of electrons from the hydrogen atoms into the conduction band of the WO_3. H_xWO_3 is used in sunglasses and automobile rear-view mirrors.

$$x\ HCl(aq) + x/2\ Zn(s) + WO_3(s) \rightarrow H_xWO_3(s) + x/2\ ZnCl_2(aq) \quad (18)$$

Host solids with acidic protons, such as $Zr(O_3P\text{-}OH)_2 \cdot H_2O$, have been shown to undergo reactions with bases (Figure 10.9).

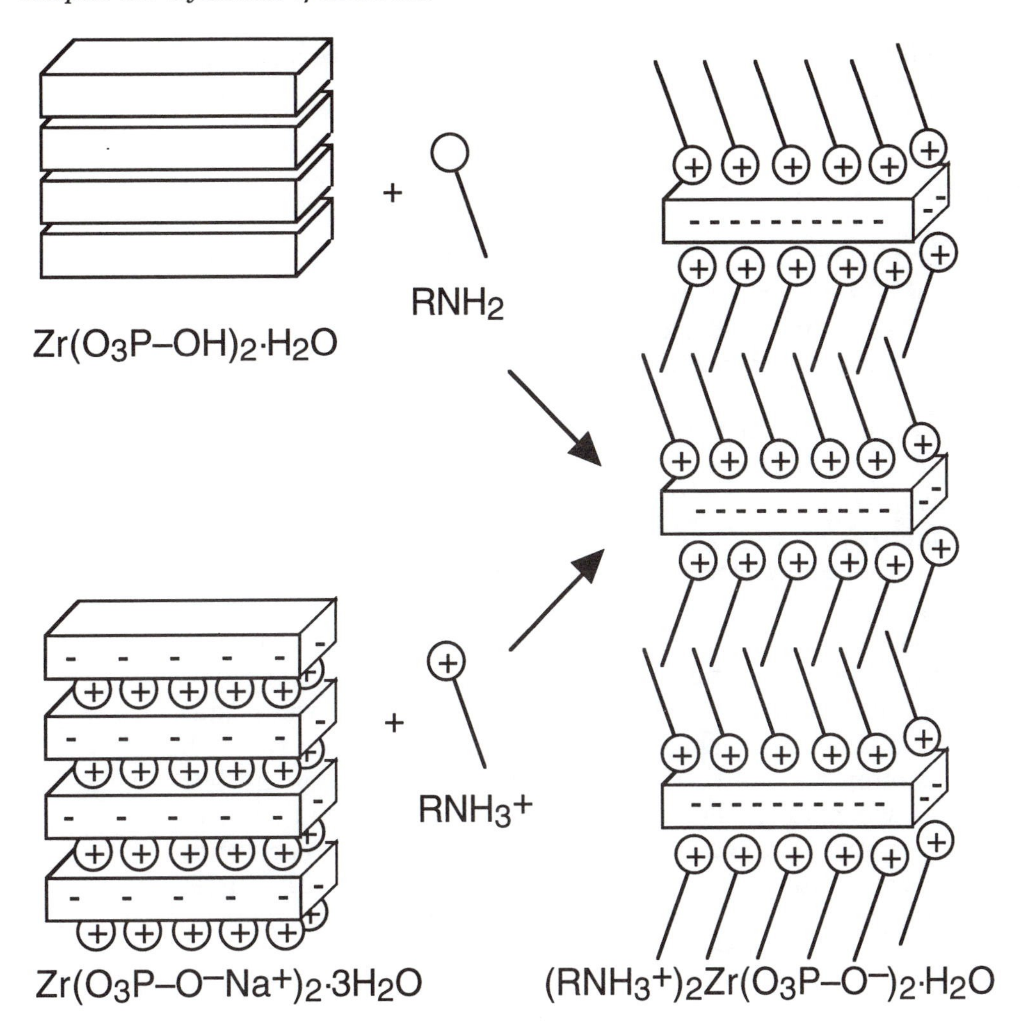

Figure 10.9. Two ways to produce an alkyl ammonium-intercalated zirconium phosphate layered solid. The top reaction shows zirconium hydrogen phosphate reacting with an alkyl amine, RNH_2, in an acid–base reaction. The bottom reaction shows zirconium sodium phosphate reacting with an alkyl ammonium ion, RNH_3^+, in an ion-exchange reaction that releases sodium ions to the solution.

Preparation of Large, Pure Crystals

Workers in processing areas at electronics companies often dress in white coveralls, gloves, cloth boots, caps, and face masks. The protective clothing is not needed to protect the workers: it is used to keep the components clean. As noted in Chapter 8, control of chemical composition at the parts-per-million level (and below) is critical for many electronic applications of solid materials. The electronic applications of microcircuits would be impossible were it not possible to prepare pure samples of silicon as large single crystals.

Zone Refining

Zone refining is a technique for purifying a solid that takes advantage of the fact that as a solution freezes, crystals of the pure solvent are formed. When part of a sample of ocean water freezes, the ice that forms does not contain appreciable concentrations of salt; the salt remains in the liquid water.

Zone refining occurs when a long sample is slowly drawn through a temperature gradient (a change in temperature that occurs over a particular distance) that is sufficiently narrow that only a small section of the sample is melted at any point. Figure 10.10 illustrates one arrangement of sample and furnace. As the sample enters the furnace from the left, the sample at the right end of the container melts first, and impurity phases dissolve in the melt. As the container continues to move through the furnace, the right end of the melt leaves the hot zone and freezes. Because the impurities remain in the melt, the crystal that results is purer than the initial sample. As the sample is drawn through the furnace, the molten zone with its impurities moves toward the left end of the sample. Ultimately, the impurities have been concentrated in the left side of the sample and can be cut off and discarded. Very pure samples can be produced by repeating the process several times, provided the molten sample does not react and pick up impurities from the container or from the atmosphere.

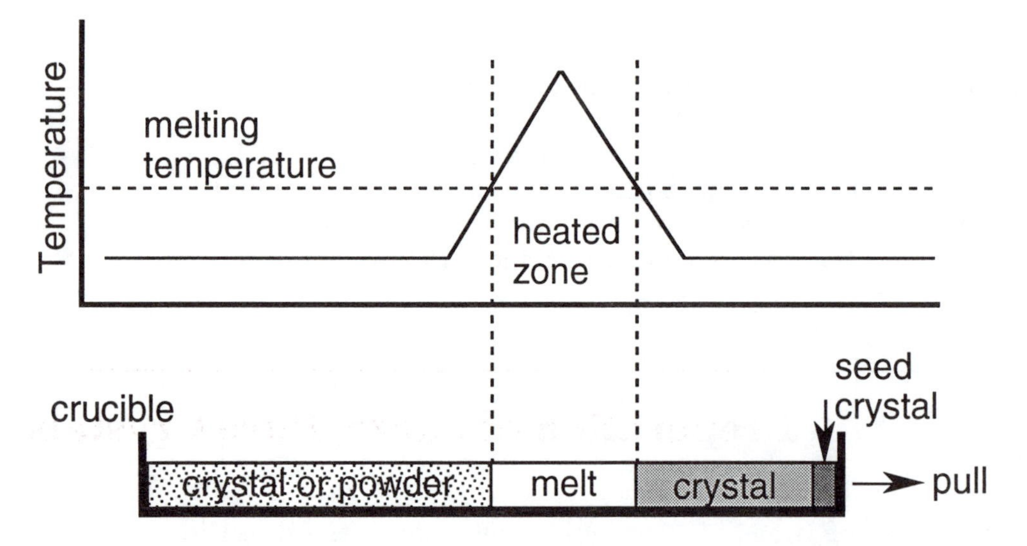

Figure 10.10. A schematic of the zone-refining process. The melting temperature of the pure sample is shown by the horizontal dashed line. (Adapted with permission from reference 2. Copyright 1984 John Wiley and Sons.)

Czochralski Crystal Growth

Large cylinders of single-crystal silicon and other single crystals are produced by the Czochralski process (Figure 10.11). In some ways this process is reminiscent of zone refinement: A crystal forms as a melt passes through a temperature gradient from a hot zone to a cooler zone, and impurities remain in the melt. In this case, however, the sample starts out molten.

To start crystal growth, a seed crystal is placed in contact with the surface of a melt in a furnace with a temperature only slightly above the melting point of the sample. As the seed is pulled from the melt, the melt adheres to the crystal and is pulled into the cooler region of the furnace. It solidifies on the seed and the solidified portion continues to move liquid from the melt into the cooler portion of the furnace. The melt and growing crystal are usually rotated in opposite directions during pulling to maintain a constant temperature and uniform melt. A very pure inert gas is used to provide a nonreactive atmosphere for the process.

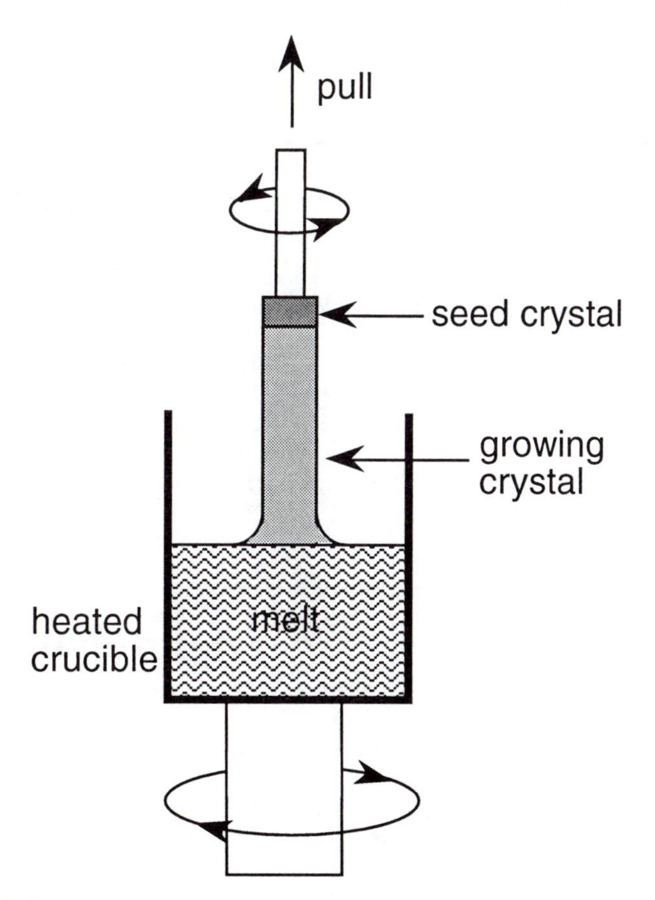

Figure 10.11. Czochralski method for crystal growth. (Adapted with permission from reference 2. Copyright 1984 John Wiley and Sons.)

This method is widely used for growing large single crystals of Si, Ge, GaAs, and other semiconductors. Thin silicon disks (wafers) used to make microelectronic chips are sawed from silicon crystals that can be several inches in diameter and several feet long.

References

1. Shakhashiri, B. Z. *Chemical Demonstrations: A Source Book for Teachers*; University of Wisconsin Press: Madison, WI; Vol. 1, 1983; pp 85–89.
2. West, A. R. *Solid State Chemistry and Its Applications*; John Wiley and Sons: New York, 1984; pp 19–36.
3. Brennan, J. G.; Siegrist, T.; Stuczynski, S. M.; Steigerwald, M. L. *J. Am. Chem. Soc.* **1989**, *111*, 9240–9241.
4. Wold, A.; Dwight, K. *Solid State Chemistry*; Chapman & Hall: New York, 1993; pp 201–202.
5. *Optical Fiber Communications; Volume 1. Fiber Fabrication*; Li, T., Ed.; Academic Press: Orlando, FL, 1985; p 3.
6. Keiser, G. *Optical Fiber Communications*; McGraw-Hill: New York, 1983; p 277.

Additional Reading

- Cheetham, A. K.; Day, P. *Solid State Chemistry: Techniques*; Clarendon Press: Oxford, England, 1987, Chapter 1.
- West, A. R. *Solid State Chemistry and Its Applications*; John Wiley and Sons: New York, 1984; Chapter 2.
- *Intercalation Chemistry*; Whittingham, M. S.; Jacobson, A. J., Eds.; Academic Press: New York, 1982.
- Wold, A.; Dwight, K. *Solid State Chemistry*; Chapman & Hall: New York, 1993; Chapter 6.

Acknowledgments

We thank Scott Shenk, Paul Lemaire, and Jim Flemming of AT&T Bell Laboratories; Scott Slavin of Rohm and Haas; John Geisz, University of Wisconsin—Madison, Department of Chemical Engineering; John Perepezko, and Jim Foley, University of Wisconsin—Madison, Department of Materials Science and Engineering; Mark Thompson, Princeton University, Department of Chemistry; and John Smith, University of Wisconsin—Madison, Communications Service Manager for their help.

Exercises

1. Explain why adding solid lead nitrate to solid potassium iodide at room temperature gives a very low yield of lead iodide, but adding a solution of lead nitrate to a solution of potassium iodide gives essentially a 100% yield of lead iodide.
2. How could the yield of lead iodide from the reaction of solid lead nitrate and solid potassium iodide be increased under anhydrous conditions (without adding water)?
3. Explain why heating Co with Co_3O_4 in air cannot be used to prepare CoO.
4. Write balanced equations for the following syntheses.
   a. $TiVO_4$ from Ti_2O_3 and V_2O_5
   b. $PbNbO_3$ from Pb, PbO, and Nb_2O_5
   c. TiS_2 from Ti and S
   d. $CrCl_2$ from reduction of solid $CrCl_3$ with H_2
   e. MnO from solutions of $MnCl_2$ and NaOH (two steps)
   f. The product resulting from adding $SiCl_4$ to water followed by heating the solid produced in air.
5. Suggest an explanation for why LiI reacts with V_2O_5 to produce $Li_xV_2O_5$ but LiCl will not react.
6. Suggest an explanation for why LiI reacts with V_2O_5 to produce $Li_xV_2O_5$, but CsI will not react.
7. $LaFeO_3$ can be prepared by heating (in air) a mixture of $La_2(CO_3)_3$ and Fe_2O_3 at 1200 °C for 24 hours, grinding the product, and repeating the heating for an additional 24 hours. Alternatively, $LaFeO_3$ can be prepared by heating (in air) the La^{3+} salt of the $Fe(C_2O_4)_3^{3-}$ ion for 8 hours at 1000 °C. Explain why one requires more vigorous reaction conditions.
8. Consider three solids: an equimolar physical mixture of AlAs and GaAs, a $Al_{0.5}Ga_{0.5}As$ solid solution; and a solid consisting of equal total numbers of AlAs and GaAs layers where, using CVD methods, 10 layers of each composition had been deposited at a time. Qualitatively compare these three solids with regard to elemental analyses and their X-ray diffraction patterns.
9. Sketch five different structures in cross section that could be grown by CVD that would analyze as having 1:1 A:Z composition (chemical formula AZ). How would you physically distinguish them?
10. In the spinel ($MgAl_2O_4$) structure mentioned in this chapter, it was noted that the oxide ions were in a ccp arrangement, the Mg^{2+} ions were in tetrahedral holes, and the Al^{3+} ions were in octahedral holes. What fraction of the tetrahedral holes are occupied? What fraction of the octahedral holes are occupied?
11. A structure containing alternating 1000-Å-thick layers of GaAs (E_g = 1.4 eV) and $Al_{0.3}Ga_{0.7}As$ (E_g = 1.9 eV) has the idealized electronic structure:

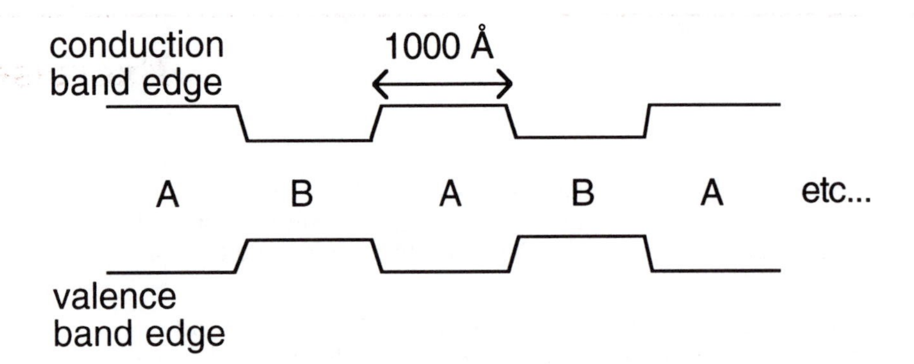

a. Which of the two compositions corresponds to the regions labeled A and which to those labeled B?
b. To which region (A or B) will conduction band electrons go to reach lower energy? To which region (A or B) will valence bond holes go to reach lower energy?
c. In which regions (A, B, both or neither) will the following photon energies be absorbed?
   i. 1.6 eV
   ii. 1.1 eV
   iii. 2.5 eV

12. Sketch part of the structure of the amorphous silica used in optical fibers showing how germanium atoms might be incorporated into the structure as a dopant. How might fluorine atoms be incorporated into the structure?

13. The value of the heat of formation, ΔH_f^o, for $Ni_{0.5}Al_{0.5}$ has been estimated to be –59,000 J/g.
    a. To what reaction does this correspond?
    b. If the specific heat capacity of this alloy is about 25 J/g-°C, estimate the temperature reached during this reaction.

14. Use unit-cell lengths in Appendix 5.6 to suggest combinations of metals that could be grown epitaxially on one another.

15. Explain why ZnSe can be grown epitaxially on GaAs. What other semiconductor combinations might be expected to be amenable to epitaxial growth?

Experiment 1

Heat Capacities of Materials

Arthur B. Ellis, Ann Cappellari,
Lynn Hunsberger, and Brian J. Johnson

Notes for Instructors

Purpose

To determine the specific heats of some materials and to relate them to atomic composition.

Method

Solid samples of known mass are heated in boiling water at a known temperature and transferred to an equal mass of water at a known temperature. The relative temperature changes give the relative specific heat, which is converted to gram specific heat by using the known heat capacity of water; and to the molar specific heat by using the formula weight. Students discover that heat capacity is more strongly related to the number of atoms of matter present than to the mass of matter. For simple solids, the molar specific heat value will be $3R \times p$, where R is the gas constant, 8.3 J/°mol, and p is the number of atoms in the formula, as discussed in Chapter 2.

In this lab, students must decide what experiments should be performed, design the experiments, and pool the results at the end, in addition to gathering data.

2725–X/93/0353$06.00/1

Materials

Solid samples such as aluminum, copper, lead, zinc, brass (a Cu–Zn alloy), pyrite (FeS_2), galena (PbS), quartz (SiO_2), fluorite (CaF_2), dolomite or chalk ($CaCO_3$), and polymers. Samples should be insoluble in water, nonporous, and denser than water. A maximum dimension of 1 inch will allow the samples to fit into a typical Styrofoam cup and still be covered by the mass of water that corresponds to the mass of the sample. The mass of the samples should ideally lie between ~40 and 100 g. One or two of the metals (preferably a relatively dense element such as copper) should either be cut into pieces of about 20 g each so that stacks of 40, 60, 80 and 100 g can be made; or else cut into pieces of about 20, 40, 60, 80, and 100 g.

String

Thermometers

Two Styrofoam cups in a 250-mL beaker *or* 250-, 400-, and 600-mL beakers that are nested.

Large beaker for boiling water

Heat Capacities of Materials

Purpose

To determine the specific heats of some materials and to relate them to atomic composition.

Introduction

Heat is a form of energy, often called thermal energy. Energy can be transformed from one form to another (electrical energy is converted to heat and light in an electric light bulb, for example) but it cannot be created or destroyed; rather, energy is conserved. The higher the temperature of a material, the more thermal energy it possesses. In addition, at a given temperature, the more of a given substance there is, the more total thermal energy the material possesses.

On an atomic level, absorbed heat causes the atoms of a solid to vibrate, much as if they were bonded to one another through springs. A rough sketch of this motion is shown in Figure 1 for a simple cubic structure.

The atoms of the solid can vibrate collectively to progressively greater extents as temperature is increased. This means that the ability of a metal to exchange heat will change as the temperature is changed. Above a certain temperature, however, all vibrations are fully activated and the maximum capacity of the solid to absorb and release heat per degree change in temperature is achieved. In the plot of heat capacity versus temperature shown in Figure 2, the flat region on the right side of the graph corresponds to this situation. For most elements that are metals, this maximum heat capacity is attained by the time room temperature is reached. (Room temperature is approximately 25 °C or 298 K).

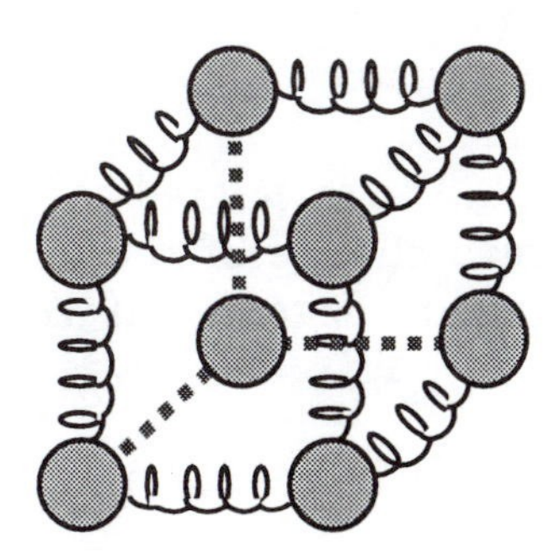

Figure 1. Atoms vibrate as though they were bonded by springs. The shaded circles represent atoms at the corners of a cubic unit cell.

NOTE: This experiment was written by Arthur B. Ellis, Ann Cappellari, and Lynn Hunsberger, Department of Chemistry, University of Wisconsin—Madison, Madison, WI 53706; and Brian J. Johnson, Department of Chemistry, College of St. Benedict and St. John's University, St. Joseph, MN 56374.

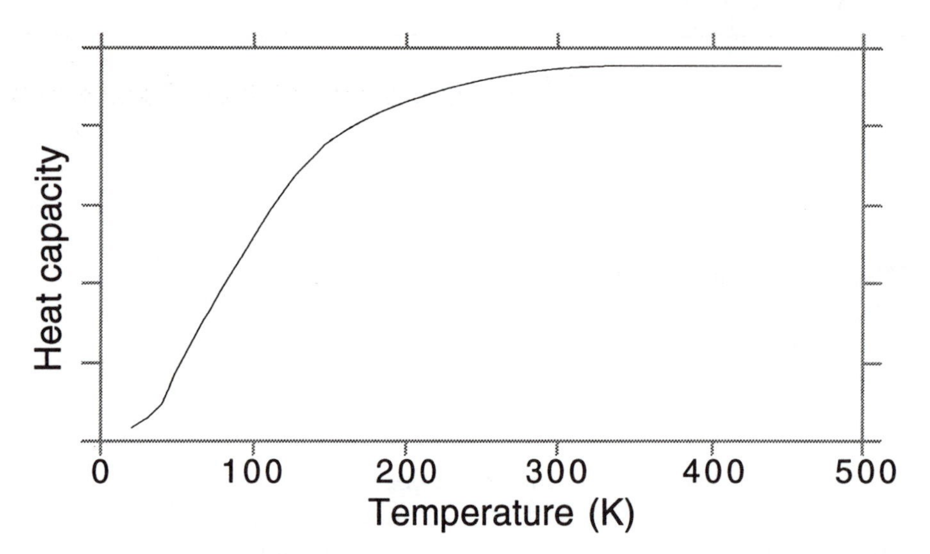

Figure 2. Plot of heat capacity versus temperature for a typical metal.

The heat capacity of a material is often quantified by its "specific heat," which is the amount of heat needed to change the heat content of exactly 1 g of the material by exactly 1 °C. In comparing two materials of the same mass, the material that requires the transfer of more heat to change its temperature by a given amount has the higher specific heat, because more heat will be transferred per gram.

Relative specific heat values can be determined in the following way: When two materials, each initially at a different temperature, are placed in contact with one another, heat always flows from the warmer material to the colder material until the two materials are at the same temperature. At this point, the two materials are in thermal equilibrium, meaning they are at the same temperature and there is no additional net flow of heat.

From the law conservation of energy, the heat lost by the initially warmer object must equal the heat gained by the initially cooler object:

$$(\text{heat lost})_{\text{hot object}} = (\text{heat gained})_{\text{cold object}} \quad (1)$$

The heat quantities in this equation would be calculated as (specific heat) x (mass of the object) x (the temperature change of the object), because specific heat has units of energy per (mass x degrees). If ΔT is the change in temperature, equation 1 becomes

$$(\text{sp. heat} \times \text{mass} \times \Delta T)_{\text{hot object}} = (\text{specific heat} \times \text{mass} \times \Delta T)_{\text{cold object}} \quad (2)$$

If the two materials that are exchanging heat have the same mass, a simpler expression results:

$$(\text{specific heat} \times \Delta T)_{\text{hot object}} = (\text{specific heat} \times \Delta T)_{\text{cold object}} \quad (3)$$

That is, the product (specific heat x ΔT) is the same for the two materials: The larger the specific heat of a material, the smaller its corresponding temperature change and vice versa; the two quantities are inversely related. Another way to write equation 3 is as a ratio:

$$\frac{(\text{specific heat}_{\text{hot object}})}{(\text{specific heat}_{\text{cold object}})} = \frac{(\Delta T_{\text{cold object}})}{(\Delta T_{\text{hot object}})} \tag{4}$$

In the experiment you will perform, you will heat various materials *of known mass* in a hot-water bath *of known temperature* by suspending them on a string; the materials will come into thermal equilibrium with the hot water. You will then transfer each material in turn to a beaker containing water *at a known, lower temperature having the same mass as your sample.* The highest temperature reached after you have transferred the material into the water can be taken as the temperature at which thermal equilibrium has been achieved. You are going to investigate how the temperature change depends on the choice of material being used to transfer heat; on the amount of that material; and on the other experimental conditions you employ. Your group will pool its data to draw conclusions about heat capacity and the atomic composition of the material.

Warm-up Exercises

1. Consider metal chunks (A, B, and C) of three different elements at 100 °C, all of the same mass. Each is then dunked into its own container of water at 20 °C, with the mass of the water being the same as that of the metal chunk. If A, B, and C produce final temperatures of 60, 40, and 30 °C , calculate their specific heats relative to the specific heat of water. That is, if the specific heat of water is considered to be "1", what are the specific heats of A, B, and C?
2. Now suppose that the same hot metals are dunked into a liquid other than water that has a smaller specific heat than water. How will the final temperatures at thermal equilibrium compare to those found in water? Again, the mass of the liquid is the same as that of each of the metals.
3. You can regard equation 3 as being analogous to balancing two people on the two sides of a see-saw. Sketch this situation for metals A and C and indicate what quantities would correspond to the masses of the people, to their distance from the center (the fulcrum) of the see-saw, and what the balancing represents.

General Procedure

Wear eye protection.

Obtain a thermometer.

Partially fill a large beaker with water and heat it to boiling. Note its temperature. Obtain several samples of materials, weigh them, and tie strings securely around each of them. Suspend the metals in the boiling water. (The samples should be suspended, because the bottom of the container being heated may be warmer than the boiling water.)

Either place two Styrofoam cups inside a 250-mL beaker or place a 250-mL beaker inside a 400-mL beaker and place them both inside a 600-mL beaker. Add a volume of room-temperature water of the same mass as your sample to the inner container and note its temperature; the density of water is 1.00 g/mL, so the number of grams of water needed to match the mass of the sample you will immerse in the water is simply the same number in milliliters.

Transfer a sample from the hot-water bath to the room-temperature water in the Styrofoam cup (or 250-mL beaker), stir, and note the maximum temperature reached, which is used to calculate ΔT for the water and for the sample.

Follow the same procedure using other samples of material.

Questions

Answer the following questions collectively, devising and conducting experiments to obtain data where appropriate. Everyone should obtain a copy of all the data collected by the group and use it as the basis for individual written answers. *You should decide as a group what experiments need to be done and who should do what in order to answer the questions.*

1. Divide your efforts so that all of the available materials can be evaluated; ideally, several teams should evaluate each material so results can be averaged. Determine the specific heats of all materials relative to water and describe any trends.
2. If you use different quantities of a particular material, how does the change in temperature caused by transferring the material from the hot to the room temperature water vary (a) if the amount of water stays constant and (b) if the weight of water continues to match the weight of the sample? Support your answers with data.
3. What is likely to happen to the temperature change you will measure if you do not transfer the sample fast enough from the hot water to the cool water? If you don't give the sample time to come to thermal equilibrium with the hot water? How can you test this? Support your answer with data and indicate what effect these errors have on the specific heat you calculate.

4. Speculate on why the cool water is placed in nested containers rather than just one container. Besides being transferred to the water, where else can the heat from your sample go? Does this make your relative specific heat values appear to be larger or smaller? How might you correct for this?

5. Your experiment has provided relative measurements of specific heats of different materials. If you know the actual specific heat of water, you can compare your data with values that have been reported in the scientific literature. Water has a specific heat of 4.18 J/°g, meaning that it takes 4.18 joules of energy to raise the temperature of 1 g of water by 1 °C. Multiply your relative values by the value for water to put units on your specific heat values. Convert your gram specific heats (joules per degree-gram) to molar specific heats (joules per degree-mole), if it makes sense to do so.

6. Does the heat capacity correlate better with the mass of the sample, the number of atoms in the sample, or the density of the sample? (You may use literature values for the density or determine them yourself.) Tabulate your results. What trends do you see? What causes those trends, if any?

Experiment 2

Solid-State Structure and Properties

George C. Lisensky and Ludwig A. Mayer

Notes for Instructors

Purpose

To construct portions of extended three-dimensional solids; on the basis of the structure, to determine the coordination number and geometry for each atom and the empirical formula of the material; to relate the structure to physical properties.

Method

See Chapters 3, 5, and the Experiment 2 laboratory introduction.

Materials

Spheres in radius-ratio sizes, 1.000:0.732:0.414:0.225. The Solid-State Model Kit, available from the Institute for Chemical Education, is convenient and comes with directions for assembling the structures in this experiment.

Copper wire. *See* Demonstration 6.4

Graphite pencil (a normal pencil), diamond-tipped pencil (Aldrich), and glass microscope slides. *See* Demonstration 7.8

Rock salt (sold for water softeners) and flat spatulas to use in cleaving crystals

2725–X/93/0361$06.00/1

Cellophane or Scotch tape

A sample of molybdenum sulfide (rock or gem shop or Wards Natural Science; *see* Supplier Information). *See* Demonstration 5.5

Solid-State Structure and Properties

Purpose

To construct portions of extended three-dimensional solids; on the basis of the structure, to determine the coordination number and geometry for each atom and the empirical formula of the material; to relate the structure to physical properties.

Introduction

Crystalline materials in the solid state, including metals, semiconductors, and ionic compounds, have a patterned arrangement of atoms that in principle can extend infinitely in all three dimensions. This patterned arrangement of objects forms an extended structure that can be described by layers of stacked spheres.

More than two-thirds of the naturally occurring elements are metals. Most of the metallic elements have three-dimensional structures that can be described in terms of close-packing of spherical atoms. Many of the other metallic elements can be described in terms of a different type of extended structure that is not as efficient at space-filling, called body-centered cubic. The packing arrangements exhibited by metals will be explored in this activity.

In any scheme involving the packing of spheres there will be unoccupied spaces between the spheres. This void space gives rise to the so-called interstitial sites. A very useful way to describe the extended structure of many substances, particularly ionic compounds, is to assume that ions, which may be of different sizes, are spherical. The overall structure then is based on some type of sphere-packing scheme exhibited by the larger ion, with the smaller ion occupying the unused space (interstitial sites). Salts exhibiting these packing arrangements will be explored in this lab activity.

The coordination number and geometry for sphere packing schemes are shown in Figure 1.

Another useful and efficient way to describe the basic pattern of an extended structure is to conceive of a three-dimensional, six-sided figure having parallel faces that encloses only a portion of the interior of an extended structure. A cube is one example, but the more general case does not have 90° angles and is called a parallelepiped. If the parallelepiped is chosen so that when replicated and moved along its edges, by a distance equal to the length of that edge, it generates the entire structure of the crystal, it is a unit cell. The unit cell is a pattern for the atoms as well as for the void spaces among the atoms. The contents of the unit cell give the

NOTE: This experiment was written by George C. Lisensky, Department of Chemistry, Beloit College, Beloit, WI 53511; Ludwig A. Mayer, Department of Chemistry, San Jose State University, San Jose, CA 95192.

chemical formula for the solid. Figure 2 corresponds to one large ion for each small ion with a formula of (large ions)$_4$(small ions)$_4$.

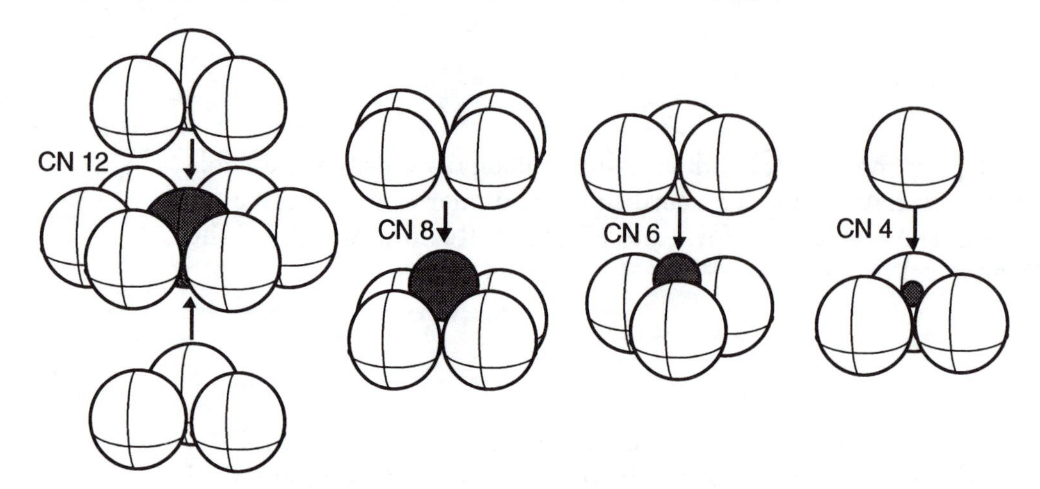

Figure 1. Close-packing of spheres gives a coordination number (CN) of 12 and leaves interstitial sites capable of coordination numbers 6 or 4. Square-packing of spheres leaves an interstitial site capable of coordination number 8.

Large ions	**Small ions**
8 corners $\times \frac{1}{8}$	12 edges $\times \frac{1}{4}$
6 faces $\times \frac{1}{2}$	1 center $\times$ 1
4 large ions	4 small ions

Figure 2. Counting ions in a unit cell.

Warm-up Exercises

1. A cube has ____ corners, ____ edges, and ____ faces.

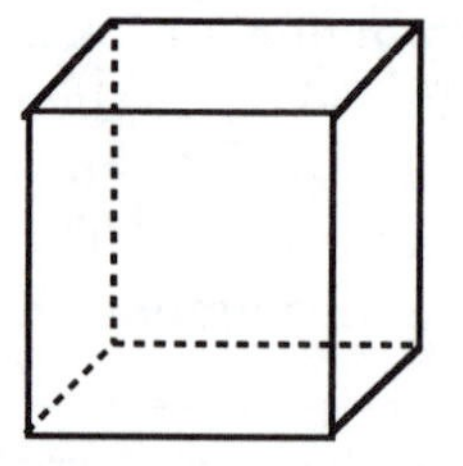

2. In two dimensions, the unit cell is a parallelogram. If the unit cell pattern is moved repeatedly in the plane of the paper in the same directions as its sides and for distances equal to the length of its sides, it must replicate the entire structure. See Structure A as an example, where the original unit cell is shown as a dark parallelogram. The light parallelograms show the start of the process. The replicated structure must include the circles as well as all the spaces between the circles. In structures B, C, and D, draw the outline of their two-dimensional unit cells.

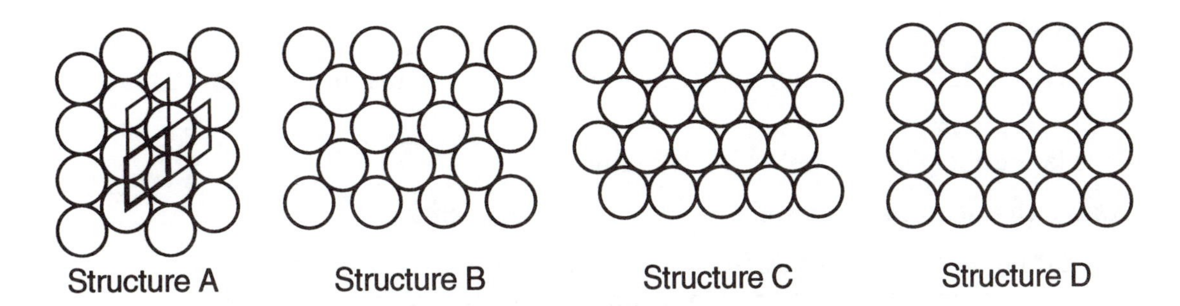

Can a structure have more than one choice of unit cell ? _______

3. If the circle segments enclosed inside each of the bold-faced parallelograms shown here were cut out and taped together, how many whole circles could be constructed for each one of the patterns?

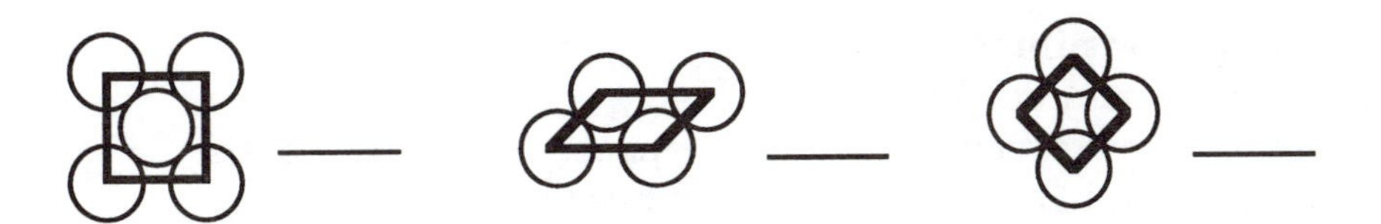

4. Shown at the right is a three-dimensional unit cell pattern for a structure of packed spheres. The center of each of eight spheres is at a corner of the cube, and the part of each that lies within the boundaries of the cube is shown. If all of the sphere segments enclosed inside the unit cell boundaries could be glued together, how many whole spheres could be constructed ? ____

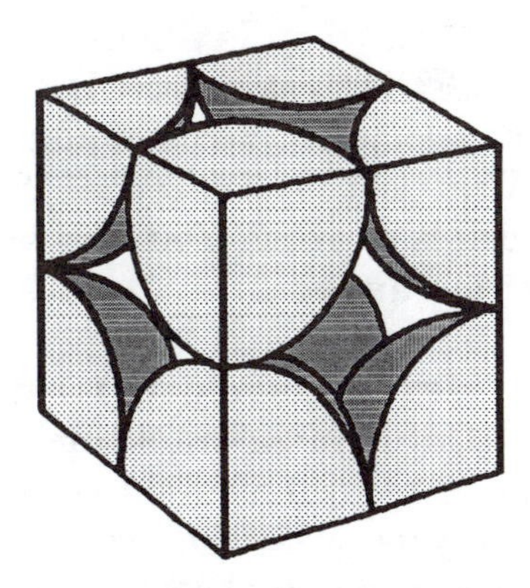

5. Consult your textbook and find examples of metallic elements that adopt the following kinds of structures:

Hexagonal closest packed (hcp): ______________________________
Cubic closest packed (ccp): ______________________________
Body-centered cubic (bcc): ______________________________

Procedure

This investigation involves two teams working together. Team A should build one structure while Team B builds another. Both teams then compare and contrast their structures and together answer the questions. The Work Sheet can be used to examine the structures in more detail. (Some Work Sheet entries will be blank for some structures such as questions about small spheres when only one size sphere is used in the structure; or questions about face-centered spheres when there are none).

For each pair of structures,

1. Identify a unit cell. How many atoms of each type are inside the unit cell?
2. What is the chemical formula for the material?
3. What is the coordination number and geometry of each type of atom?
4. How do the two structures differ? Answer any specific questions asked about each pair. Which of the team A and team B structures are the same?
5. Optional: What is the packing efficiency for each unit cell? Divide the volume occupied (based on the number of spheres in the unit cell and the volume of one sphere) by the volume of the unit cell (measured with a ruler). How does this relate to the density of the material?

Team A: Primitive or Simple cubic
Team B: Body-centered cubic

Which structure is more efficient at filling space? How would the density of the same element in either of these packing structures compare?

Team A: CsCl (first layer Cs)
Team B: CsCl (first layer Cl)

What is the geometric shape of just the large spheres? Of just the small spheres? What is the difference, if any, between the two structures? Which unit cell is correct? You may wish to build more than just one unit cell to check your answer.

Team A: Hexagonal close-packing
Team B: Cubic close-packing

Most of the elemental metals have close packed structures. What is the difference, if any, between the two structures? It may be useful to compare the templates on which the structures are built. How do the packing efficiencies compare qualitatively to those of the simple cubic and body-centered cubic structures?

Bend a piece of copper wire. Does it break? Explain by referring to its structure. (Copper has the cubic close-packed structure.) If you are using the Solid-State Model Kit, lift one of the bottom corner spheres of the cubic close-packing model. What happens? Does it matter which corner sphere is lifted? How do these observations of the model relate to your observations of copper?

Team A: Rock Salt (NaCl)
Team B: Rock Salt (NaCl body diagonal)

What is the structure of just the large spheres? Of just the small spheres? Show how the stoichiometry is related to the cubic unit-cell contents. Which color sphere represents Na^+ and which represents Cl^-?

Take a piece of rock salt, align a metal spatula parallel to one of its faces, as shown in Figure 3, and tap the spatula sharply with something heavier than a pencil and lighter than a hammer. How does the crystal break? The planes of cleavage in a crystal are the planes where the forces between the atoms are weakest. Which plane in the model do you think corresponds to the cleavage plane?

Figure 3. Breaking a crystal to reveal cleavage planes.

Team A: Diamond
Team B: Graphite

Both of these substances are composed exclusively of carbon atoms. Use a regular (graphite) pencil and a diamond-tipped scribe to write on a glass microscope slide. Why is diamond the hardest common substance while graphite is a soft, solid lubricant?

Which form of carbon do you think has the largest density? Why?

Team A: Molybdenum Sulfide
Team B: Gallium Selenide

Press a piece of cellophane tape onto a sample of molybdenum sulfide already stuck on a piece of tape. Peel apart the two pieces of tape. What happens to the molybdenum sulfide? Make a prediction about the properties of gallium selenide. Explain your reasoning.

Team A: Fluorite (CaF_2)
Team B: Zinc Blende (ZnS)

Which color sphere represents Ca and which represents F? Which color sphere represents Zn and which represents S? What is the stoichiometry of calcium fluoride? Of zinc blende? What is the structure of just the large spheres? Of just the small spheres?

Team A: Face-centered cubic
Team B: Face-centered cubic (body diagonal)

Are these structures the same as any structures you have already built?

Work Sheet (Data and Observations)

Unit Cells and Stoichiometry

How many large spheres lie with their centers
at the CORNERS of the unit cell?
on the FACES of the unit cell?
on the EDGES of the unit cell?
completely INSIDE of the unit cell?
How many small spheres lie with their centers
at the CORNERS of the unit cell?
on the FACES of the unit cell?
on the EDGES of the unit cell?
completely INSIDE of the unit cell?
With how many unit cells is a
CORNER-centered sphere shared?
FACE-centered sphere shared?
EDGE-centered sphere shared?
Inside the unit cell is how much of each
CORNER-centered sphere?
FACE-centered sphere?
EDGE-centered sphere?
How many total LARGE spheres are inside the unit cell?
How many total SMALL spheres are inside the unit cell?
What is the formula representation for the unit cell?

Coordination Number

If the structure has more than one size of spheres, nearest neighbor touching spheres will be of unequal size.
Choose a LARGE sphere. What COLOR is it?
How many nearest neighbor touching spheres are
in the layer BELOW?
in the SAME layer?
in the layer ABOVE?
What is the coordination number for large spheres?
Are the large spheres close-packed (CN 12)?
Choose a SMALL sphere. What COLOR is it?
How many nearest neighbor touching spheres are
in the layer BELOW?
in the SAME layer?
in the layer ABOVE?
What is the coordination number for small spheres?

Close-Packed Structures

Are the packing layers ABA or ABCA?

Experiment 3

Solid Solutions with the Alum Structure

William R. Robinson and Brian J. Johnson

Notes for Instructors

Purpose

To grow crystals belonging to a solid solution series, with the variable composition $KAl_xCr_{1-x}(SO_4)_2 \cdot 12H_2O$.

Method

One technique that chemists commonly use to grow single crystals is that of slow solvent diffusion. In this process, the compound of interest is dissolved in a solvent to make a solution. A second liquid (one that is less dense than the first solution, which is miscible with the first solution, and in which the solute compound is insoluble) is carefully layered on top of the solution. As the two liquids mix, the concentration of the second liquid in the solution increases. This increased concentration decreases the solubility of the solute compound, and eventually a solid begins to form. Because the liquids diffuse together slowly, crystallization usually begins at a limited number of sites and crystal growth continues at these sites. Generally speaking, the slower the crystal growth process, the larger and more defect-free the crystals. *See* Chapters 6 and 10.

In this experiment, large crystals of alum–chrome alum solid solutions are produced.

2725–X/93/00369$06.00/1

Materials

Alum, $KAl(SO_4)_2 \cdot 12H_2O$

Chrome alum, $KCr(SO_4)_2 \cdot 12H_2O$

95% ethanol

Beakers

Stirring rods

Filter paper and funnels

13-mm × 150-mm test tubes

Ring stands and test-tube clamps

Disposable pipettes

For optional preparation of alum from an aluminum beverage can,

Aluminum can

1.4 M KOH

6 M H_2SO_4

Open-ended experiments can employ other shapes of containers (to vary the rate of diffusion), other metal salts (such as copper sulfate), potassium sulfate, other chromium salts (such as chromium nitrate or chromium chloride), or other alcohols (such as methanol, propanol, or 2-propanol).

Other Information

If students have difficulty forming two layers without generating a great deal of precipitate, have them try chilling the alcohol or the alum solution, diluting the alum solution slightly, or adding a small layer of water to the test tube (to act as a buffer region) before adding the alcohol.

Generally, the formation of a small amount of precipitate during the layering procedure is unavoidable and will not greatly affect results. The precipitate formed will often sink into the water layer and redissolve.

Likely Results of Open-Ended Experiments

The numbers refer to suggested ideas in the student directions.

1. Generally, the larger the interface between the diffusing solvents, the faster they will mix. Faster mixing will cause the formation of smaller crystals.

2. Unless the ion has the same charge and about the same size as Al^{3+}, a solid solution will not form. Instead, cocrystallization will result and

there will be a mixture of two different types of crystals present. For example, this will happen when copper sulfate is added to alum.

3. A powder will result because of cocrystallization of alum and K_2SO_4.
4. In our hands, addition of $Cr(NO_3)_3$ and $Cr_2(SO_4)_3$ to alum solutions formed crystals that looked like those of Cr-doped alum. On the other hand, addition of $CrCl_3$ formed something that was obviously different.
5. This procedure can work, though care is required.
6. This "seeding" tends to produce larger crystals than if the solution is layered, as there are fewer but better sites for crystal growth to begin.
7. Any alcohol other than ethanol seems to make it very difficult to form two layers without generating a fine precipitate of alum.
8. A dirty beaker may produce smaller crystals, as there will be many sites for crystal growth to begin. The results when tap water is used will depend on the ions present in it.
9. Small crystals will result because of seeding if any solid alum is present.
10. This approach will probably produce large, well-formed crystals that may well be of better quality than those obtained from layering.

Solid Solutions with the Alum Structure

Purpose

To grow crystals belonging to a solid solution series, with the variable composition $KAl_xCr_{1-x}(SO_4)_2 \cdot 12H_2O$.

Introduction

Chemists today use a variety of instrumental techniques to study solids. X-ray diffraction, for example, yields information about the position of atoms and the bond angles in a molecule and is perhaps the most powerful tool for determining the structure of a compound. X-ray studies, electrical conductivity measurements, and many other types of analysis are often done with single crystal samples. In this case, not only must the sample be uncontaminated, but the crystal cannot be a collection of small crystals or crystalline regions (polycrystalline) and it cannot be split or cracked.

Most chemical reactions that result in solid products give powders or polycrystalline materials, rather than single crystals of sufficiently large size. Powders and polycrystalline substances can be converted to single crystals by a variety of techniques. All of these are based on forming or providing a very small number of sites where crystallization begins (nucleation sites), followed by controlled growth at these sites.

One technique that chemists commonly use to grow single crystals is that of slow solvent diffusion. In this process, the compound of interest is dissolved in a solvent to make a solution. A second liquid (a miscible liquid that is less dense than the first solution and in which the solute compound is insoluble) is carefully layered on top of the solution. As the two liquids mix, the concentration of the second liquid in the solution increases. This increased concentration decreases the solubility of the solute compound, and eventually a solid begins to form. Because the liquids diffuse together slowly, crystallization usually begins at a limited number of sites and crystal growth continues at these sites. Generally speaking, the slower the crystal growth process, the larger and more defect-free the crystals.

In this laboratory, you will perform crystallization experiments, as described, and then have a chance to design your own. All of them will be based on growing crystals of alum, $KAl(SO_4)_2 \cdot 12H_2O$. This solid contains potassium ions (K^+), aluminum ions (Al^{3+}), and sulfate ions (SO_4^{2-}). The water molecules are arranged so that six of them are located around each of the cations. Crystals of this compound have long been admired for their beautiful octahedral shape.

NOTE: This experiment was written by William R. Robinson, Department of Chemistry, Purdue University, West Lafayette, IN 47907; and Brian J. Johnson, Department of Chemistry, The College of St. Benedict and St. John's University, St. Joseph, MN 56374

A related compound is chrome alum, $KCr(SO_4)_2 \cdot 12H_2O$. It has the same formula as alum, except that Al^{3+} has been replaced by Cr^{3+}. Furthermore, the cations, anions, and water molecules are arranged in the same way in crystals of chrome alum as they are in alum. This feature is primarily due to the like charge and similarity in size of the aluminum and chromium ions.

When solutions of alum and chrome alum are mixed, the resulting crystals contain chromium and aluminum ions randomly distributed in the M^{3+} ion positions throughout the alum structure. This means that there is a continuous variation in composition possible for such crystals. By varying the amounts of alum and chrome alum present, the composition can be controlled or continuously tuned and for values of x ranging between 0 and 1, is denoted $KAl_xCr_{1-x}(SO_4)_2 \cdot 12H_2O$. Such a family of related solids is called a *solid solution*. Unlike a physical mixture of alum and chrome alum crystals (in which microscopic examination of the solid would reveal only chunks of pure alum and of pure chrome alum), all solid solution crystals with a given composition (for example, $KAl_{0.3}Cr_{0.7}(SO_4)_2 \cdot 12H_2O$) are identical.

An analogy to the solid solution would be a liquid solution of water and alcohol (ethanol). These liquids can be mixed in any proportion to generate homogeneous solutions. No part of a water–ethanol solution appears to be any different than another part, and properties such as density would be continuously variable from that of pure ethanol to that of pure water, depending on the mole ratio of the two compounds. On the other hand, oil and water do not mix, so a homogeneous solution of the two is not possible.

Procedure

Preparation of Crystals of Alum, $KAl(SO_4)_2 \cdot 12H_2O$

Dissolve 3 g of $KAl(SO_4)_2 \cdot 12H_2O$ in 30 mL of deionized or distilled water in a clean beaker. Stir well and make sure that all of the solid has dissolved. If some material remains undissolved, filter the solution. Transfer enough of the solution to a 13-mm × 150-mm test tube so that it is about one-third full. Clamp the test tube to a ring stand so that it will remain steady. Then tilt the test tube at about a 10–15° angle from the vertical. Using a disposable pipette, carefully add ethanol to the test tube by slowly allowing the alcohol to run down the inside of the tube. If this is done with care, the alcohol will not mix with the alum solution but will form a layer on top of it. Continue to add ethanol to the test tube until it is within an inch or so of being full. Stopper the test tube to prevent the ethanol from evaporating, remove it from the clamp, and set it carefully in a location where it can remain undisturbed for several weeks.

Preparation of Crystals of Chrome Alum, $KCr(SO_4)_2{\cdot}12H_2O$

Make a solution of $KCr(SO_4)_2{\cdot}12H_2O$ by dissolving 4 g of solid in 25 mL of deionized or distilled water in a clean beaker. Mix completely, filter if necessary, and layer a sample with ethanol. Set up a crystal-growth apparatus as described for growing alum crystals. **CAUTION: Chrome alum is an irritant. Avoid breathing dust and avoid skin contact. Dispose of this compound and its solutions in an appropriately labeled waste container.**

Preparation of a Solid Solution, $KAl_xCr_{1-x}(SO_4)_2{\cdot}12H_2O$

Using your remaining solutions, measure out a volume of alum solution (6–9 mL) and mix it with the appropriate volume of chrome alum solution so that the total volume is about 10 mL. Layer this sample as described. Repeat this procedure using 6–9 mL of chrome alum solution and a smaller amount of alum solution. Set up a crystal-growth apparatus as described for growing alum crystals.

Harvesting Crystals

When you are ready to collect your crystals, decant the liquid from the test tube into a beaker. If the solution contains chromium it should be placed in a waste container provided by your instructor. If it contains only alum it may be flushed down the sink. Rinse the crystals by pouring a small sample of ethanol into the test tube and gently dislodging the crystals from the walls of the beaker. Swirl the mixture gently and decant the ethanol. Repeat the ethanol wash two more times. The ethanol may be flushed down the sink. Allow the crystals to dry in air on a watch glass. What do the crystals look like?

Optional: Recycling Aluminum

In this procedure, alum is prepared from an aluminum beverage can. **CAUTION: Hydrogen gas is released; use a fume hood. CAUTION: KOH is a strong base. Avoid skin contact.**

With scissors, cut into small portions a 3-cm × 5-cm piece (approximately 1.0 g) of the side of an aluminum can. Place the freshly cut aluminum in a flask and add 50 mL of 1.4 M KOH.

$$2\ Al + 2\ KOH + 6H_2O \rightarrow 2\ KAl(OH)_4 + 3H_2$$

Stir and heat the flask gently until all the aluminum dissolves. Filter off any scraps of paint and impurities that remain. Slowly add 30 mL of 6 M H_2SO_4 with continuous stirring. **CAUTION: H_2SO_4 is corrosive. Avoid skin contact.** A white solid should precipitate and then redissolve.

$$2\ KAl(OH)_4 + H_2SO_4 \rightarrow 2\ Al(OH)_3(s) + K_2SO_4 + 2\ H_2O$$

Heat and stir the mixture until all the white solid disappears. (If solid still remains after 10 minutes, remove the flask from the heat and filter.) Place the solution in an ice bath (ice with a little water) for about 20 minutes. Alum crystals should appear.

$$2\ Al(OH)_3(s) + 3\ H_2SO_4 + K_2SO_4 + 18\ H_2O \rightarrow 2\ KAl(SO_4)_2 \cdot 12H_2O(s)$$

Open-Ended Experiments

Design and perform experiments to answer one question of your own choosing. Be prepared to describe what you are trying to do and how you plan to do it. Sometimes there are unknown or uncontrolled variables in an experiment. (This includes but is not limited to such things as breakage or spillage by the experimenter, contamination of glassware, contamination of solutions, etc.) These effects can be minimized by performing several identical experiments and using well-designed "control" experiments that let you evaluate the effect of changing a single variable at one time. In general, you will need duplicate trials, and you will need to repeat portions of the preceding experiment as a control.

1. What happens to the size of crystals and the rate at which they grow if you use containers with diameters that are different than that of a 13-mm × 150-mm test tube?
2. What happens if you add solutions of metal ions other than chrome alum to alum?
3. Alum contains K^+ ions and SO_4^{2-} ions. What happens to crystal growth if you add a solution of potassium sulfate to the solution of alum?
4. What happens if you use chromium salts other than chrome alum to try to form solid solutions with alum?
5. Can you grow layered crystals by placing a crystal of alum in a solution of chrome alum for a while, then removing it and placing it in an alum solution, and so on?
6. What happens if you grow a crystal of alum and then place it in a fresh solution of alum?
7. Ethanol is an alcohol that has two carbons. What happens to the crystals if methanol (a one-carbon alcohol) is used? What about propanol (a three-carbon alcohol)? 2-Propanol (a three-carbon alcohol that differs in the arrangement of the atoms)?
8. What happens to crystal growth if you use a dirty beaker and tap water to prepare your solutions?
9. What happens if there is some undissolved alum powder present when you layer your solution?

10. What is the effect on crystallization if no ethanol is added and the water is allowed to evaporate slowly?

11. Is there another experiment other than those described here that you would like to try? Check with your lab instructor before attempting it.

Experiment 4

Optical Diffraction Experiments

George C. Lisensky

Notes for Instructors

Purpose

To discover how a diffraction pattern is related to a repeating dot array; to use the diffraction pattern to measure the dimensions of the repeating dot array.

Method

A visible laser beam is directed through a 35-mm slide containing repeating arrays to give a diffraction pattern. *See* Chapter 4.

Materials

Point source of white light, such as a Mini Maglite with the lens removed

Visible laser, such as a solid-state laser pointer with a wavelength of 670 nm

White card (such as an index card)

Rulers

35-mm slides with repeating patterns, available from the Institute for Chemical Education. Alternatively, the patterns in Figure 4 can be drawn on a Macintosh computer with a paint program (HyperCard, MacPaint, SuperPaint, etc.), either by using the rectangle tool with a

2725–X/93/0377$06.00/1

fill pattern, or by using zoom or fat bits to draw the pattern pixel by pixel. A 10.5 × 16.5-inch set of patterns is printed on 11 × 17-inch paper using a laser printer; this set of patterns is illuminated by two 300-W reflector floodlights, and photographed with a 1-second, f-8 exposure onto Kodak Precision Line LPD4 black-and-white, 35-mm slide film. The film is developed using Kodak D11 or D19 developer for 3 minutes, Kodak stop bath for 30 seconds, and Kodak rapid fixer for 5 minutes. The photographic procedure reduces the original 1/72-inch pixels on the paper to (30 mm) /(16.5 inch × 72 $inch^{-1}$) = 0.025-mm pixels on the slide. The smallest pattern (Figure 4F or 4b with alternating black and white pixels) has a repeat distance of approximately 0.05 mm.

Hand lens (the Radio Shack pocket microscope is convenient.)

Optical Diffraction Experiments

Purpose

To discover how a diffraction pattern is related to a repeating dot array; to use the diffraction pattern to measure the dimensions of the repeating dot array.

Introduction

Diffraction of a wave by a periodic array is due to phase differences that result in constructive and destructive interference (illustrated in Figure 1). Diffraction can occur when waves pass through a periodic array if the repeat distance of the array is similar to the wavelength of the waves. Observation of diffraction patterns when beams of electrons, neutrons, or X-rays pass through crystalline solids thus serves as evidence both for the wave nature of those beams and for the periodic nature of the crystalline solids. However, X-rays are hazardous and they require special detectors.

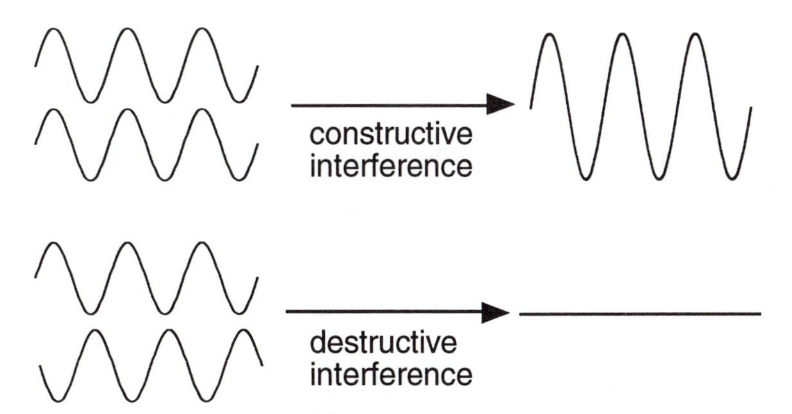

Figure 1. When waves line up (the oscillations are in phase), they add to give a bigger wave. When the peak of one wave is aligned with the trough of another, the waves annihilate each other.

Atoms, with spacings of about 10^{-10} m, require X-rays to create diffraction patterns. In this experiment, you make a change of scale. By using dots with spacings of about 10^{-4} m, visible light can be used instead of X-rays to create diffraction patterns. You will shine red laser light (670 nm wavelength) through a slide containing repeating arrays of dots, and observe Fraunhofer diffraction (see Figure 2).

NOTE: This experiment was written by George C. Lisensky, Department of Chemistry, Beloit College, Beloit, WI 53511.

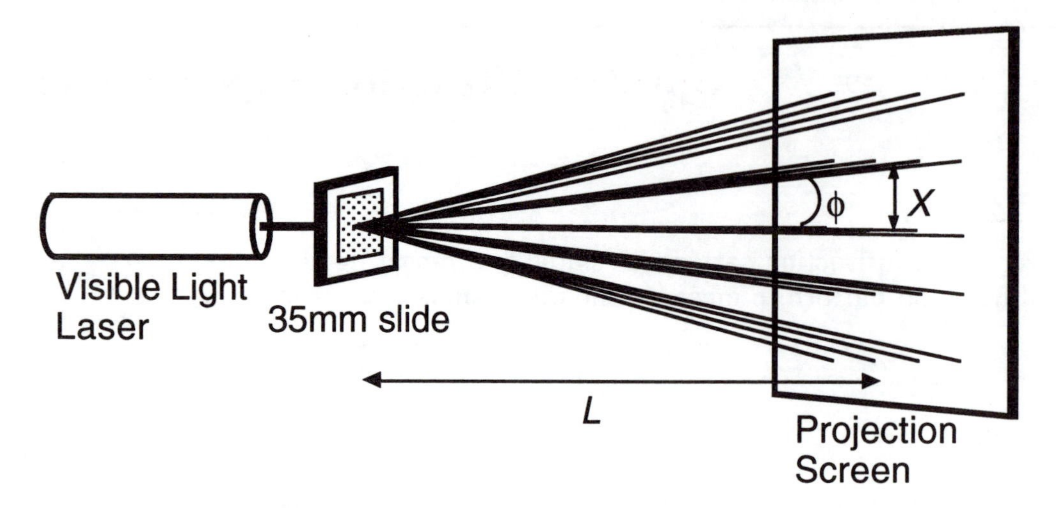

Figure 2. The Fraunhofer diffraction experiment.

Mathematically, the equations for Fraunhofer and Bragg diffraction (the basis of X-ray diffraction) are similar and embody the same functional dependence on the dot spacing (d), wavelength (λ), and scattering angle (ϕ or θ), (*see* Figure 3).

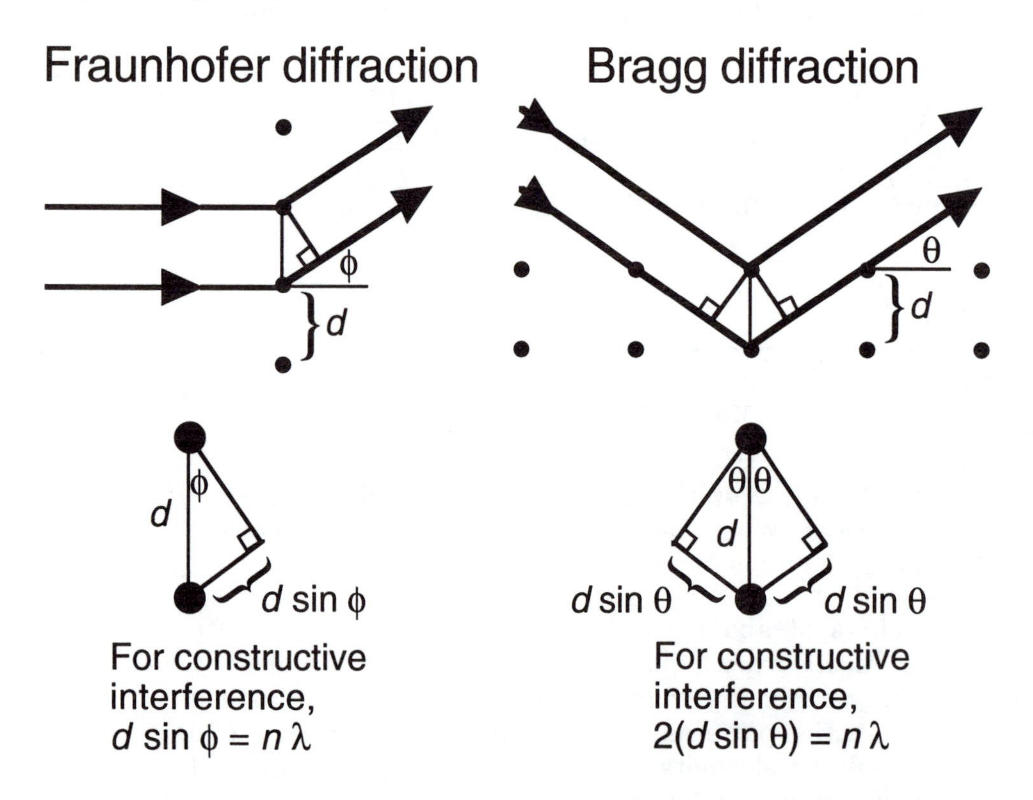

Figure 3. A comparison of Fraunhofer diffraction with Bragg diffraction. When waves are scattered by a periodic array, the path difference between any two waves must be a whole number of wavelengths, n, if the waves are to remain in phase and give constructive interference.

In this experiment, you will first check how the size and orientation of the diffraction pattern is related to the periodic array that produced it, and then you will measure distances in diffraction pattern spacings in order to calculate the repeat distance for the array in the slide. By measuring the distances X and L, shown in Figure 2, and using the trigonometry definition that $\tan \phi = X/L$, you can solve for ϕ. Use of the Fraunhofer equation, $d \sin \phi = n\lambda$, then gives d when λ is known.

General Procedure

Obtain slides containing greatly reduced versions of arrays like those in Figure 4. Each photographic slide contains eight patches. Each patch has a different periodic array.

Look through a slide at a point source of white light. What do you see? Is the slide a diffraction grating? Why?

Shine a diode laser (670-nm wavelength) at a white piece of paper several meters away. Fasten the laser in place. **CAUTION: The laser is potentially dangerous. Do not look directly into a laser beam or shine a laser toward other people, as damage to the eye can occur.** Look through a slide at the laser dot on the paper. What do you see? Why?

Put the slide in the laser beam and watch the paper. What happens? Light travels from the laser to the paper and then to your eye. Are the same results obtained if the beam goes through the slide before it hits the paper as after it hits the paper?

Pretend that you want to sketch the shape of the diffraction pattern from an array. What light source should you use (the white bulb or the laser)? Should you shine the beam through the slide or look at the spot through the slide? Pretend that you want to measure the spacing between the spots. What method is easiest?

Use a hand lens to examine the arrays on the slide. For ease of interpretation, you may prefer to keep a consistent orientation for the slide.

Questions

Answer the following questions, devising and conducting experiments to obtain data where appropriate. "Data" for this experiment will sometimes consist of a sketch of an array and a sketch of the resulting diffraction pattern, and may include measurements of distances.

1. What is the diffraction pattern of a horizontal array of lines? Of a vertical array of lines?
2. What is the diffraction pattern of a square array of dots? Of a rectangular array of dots? Of a parallelogram array of dots where the angle is not 90°? Of a hexagonal array of dots? How do the orientations of the diffraction patterns relate to the orientation of the array of dots?

3. Find two similar arrays that differ only in size. Does an array with a smaller repeat distance give a diffraction pattern with a smaller repeat distance? Does an array with a larger repeat distance give a diffraction pattern with a larger repeat distance?
4. Choose two arrays, carefully measure some distances in the diffraction pattern, and calculate the size of the unit cell for that array. What is the size in millimeters of the spacing between dots in the array on the slides?
5. What happens if you put an additional dot in the array in the center of the unit cell? Does it matter if the dot placed in the center is the same size as those in the original array?

Figure 4. Arrays of dots that can be used to generate diffraction patterns with a laser. The actual patterns on the slide are much smaller.

Experiment 5

X-ray Analysis of a Solid

M. Stanley Whittingham

Notes for Instructors

Purpose

To measure the spacing between crystallographic planes of a layered solid.

Method

X-ray diffraction and the Bragg equation, $n\lambda = 2d \sin \theta$. *See* Chapter 4. When crystalline materials are used in a powder diffractometer, only the crystallographic planes parallel to the sample stage are aligned for diffraction and a set of peaks equally spaced in 2θ are observed. By choosing samples with a natural orientation such as layered solids or crystals with cleavage planes parallel to a cubic unit cell the set of crystallographic planes whose spacing is being measured can be easily identified. By choosing samples with a spacing d large compared with the wavelength, many diffraction orders can be observed.

Materials

X-ray diffractometer

Samples of layered minerals:

Molybdenite (Ward's Natural Science; layer spacing should be 12.295 Å.)

2725–X/93/00385$06.00/1

Vermiculite (type of mica used as packing material; vermiculite often expands like popcorn when heat-treated; select a sample that still looks like mica. The layer spacing is expected to be about 20 Å.)

Biotite (layer spacing should be 10.16 Å × sin 99.05°)

Muscovite (layer spacing should be 20.05 Å × sin 95.77°)

Chlorite (layer spacing should be 14.29 Å × sin 97.13°)

Cubic crystals with cleavage planes:

KCl (layer spacing should be 6.2929 Å)

KBr (layer spacing should be 6.5982 Å)

Other Information

It is possible to compare *d* spacings with literature values of unit cell parameters. When the stacking of the planes is not perpendicular ($\beta \neq 90°$), the measured distance between the planes will be $c \sin \beta$ rather than c; this distinction is small as long as β is close to 90°.

Some symmetries result in systematically absent reflections with the result that only the even diffraction orders are seen in this experiment; labeling orders 2, 4, 6... as 1, 2, 3... will result in an answer that is half the actual spacing.

Sample Data and Calculations

Figures 1–6 show sample data and calculations. Values with * are low intensity and are expected to have a larger error.

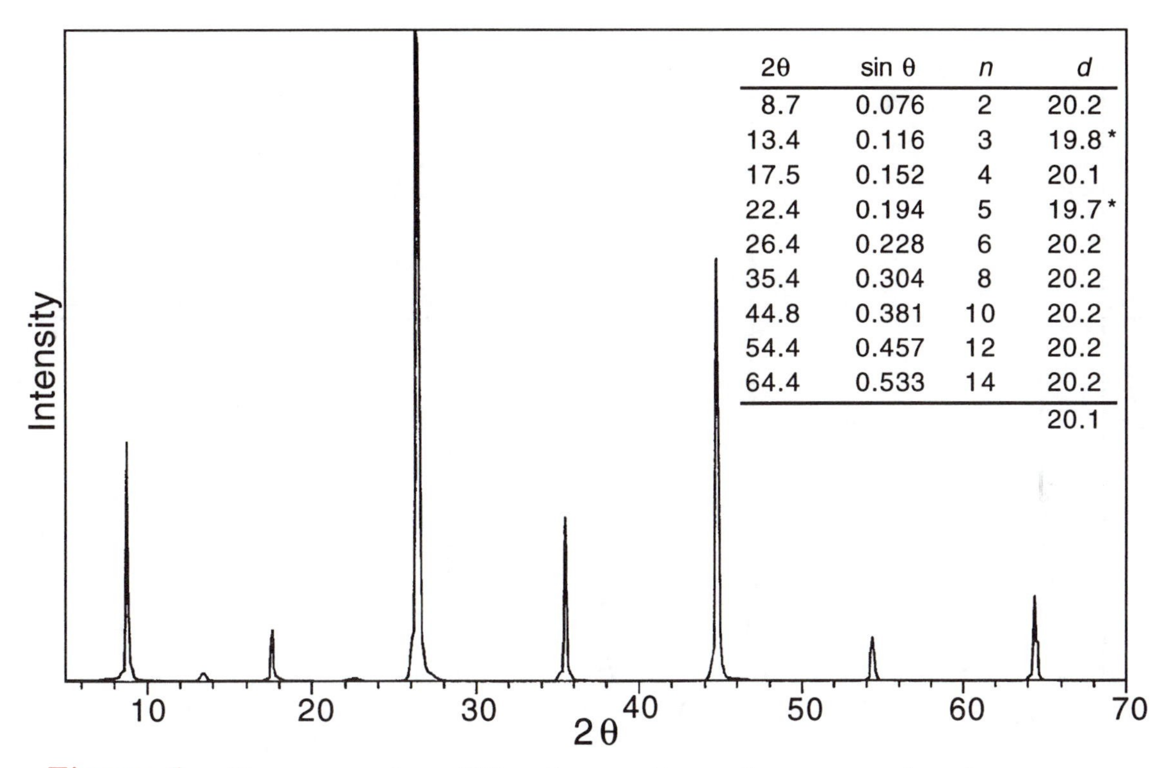

2θ	sin θ	n	d
8.7	0.076	2	20.2
13.4	0.116	3	19.8*
17.5	0.152	4	20.1
22.4	0.194	5	19.7*
26.4	0.228	6	20.2
35.4	0.304	8	20.2
44.8	0.381	10	20.2
54.4	0.457	12	20.2
64.4	0.533	14	20.2
			20.1

Figure 1. X-ray powder diffraction pattern for a sample of vermiculite packing material (Aldrich).

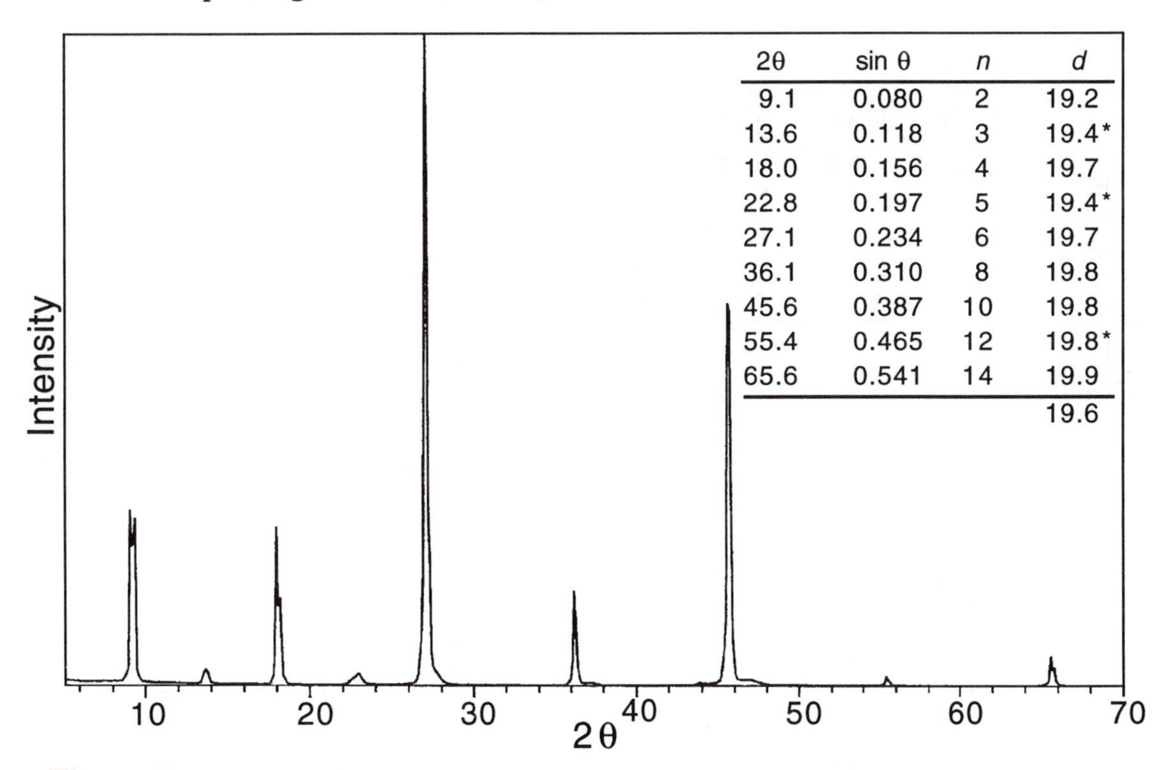

2θ	sin θ	n	d
9.1	0.080	2	19.2
13.6	0.118	3	19.4*
18.0	0.156	4	19.7
22.8	0.197	5	19.4*
27.1	0.234	6	19.7
36.1	0.310	8	19.8
45.6	0.387	10	19.8
55.4	0.465	12	19.8*
65.6	0.541	14	19.9
			19.6

Figure 2. X-ray powder diffraction pattern for a sample of muscovite from Delaware County, PA.

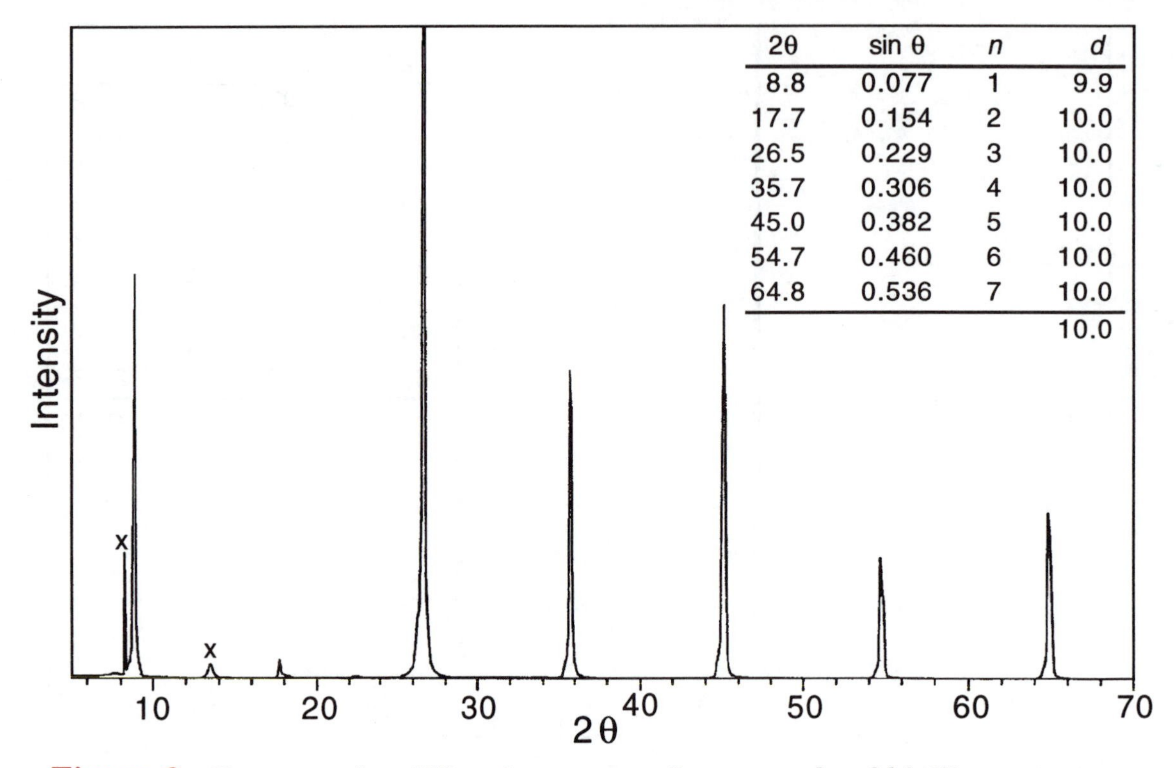

2θ	sin θ	*n*	*d*
8.8	0.077	1	9.9
17.7	0.154	2	10.0
26.5	0.229	3	10.0
35.7	0.306	4	10.0
45.0	0.382	5	10.0
54.7	0.460	6	10.0
64.8	0.536	7	10.0
			10.0

Figure 3. X-ray powder diffraction pattern for a sample of biotite from West Philadelphia, PA. Peaks marked x are from an impurity.

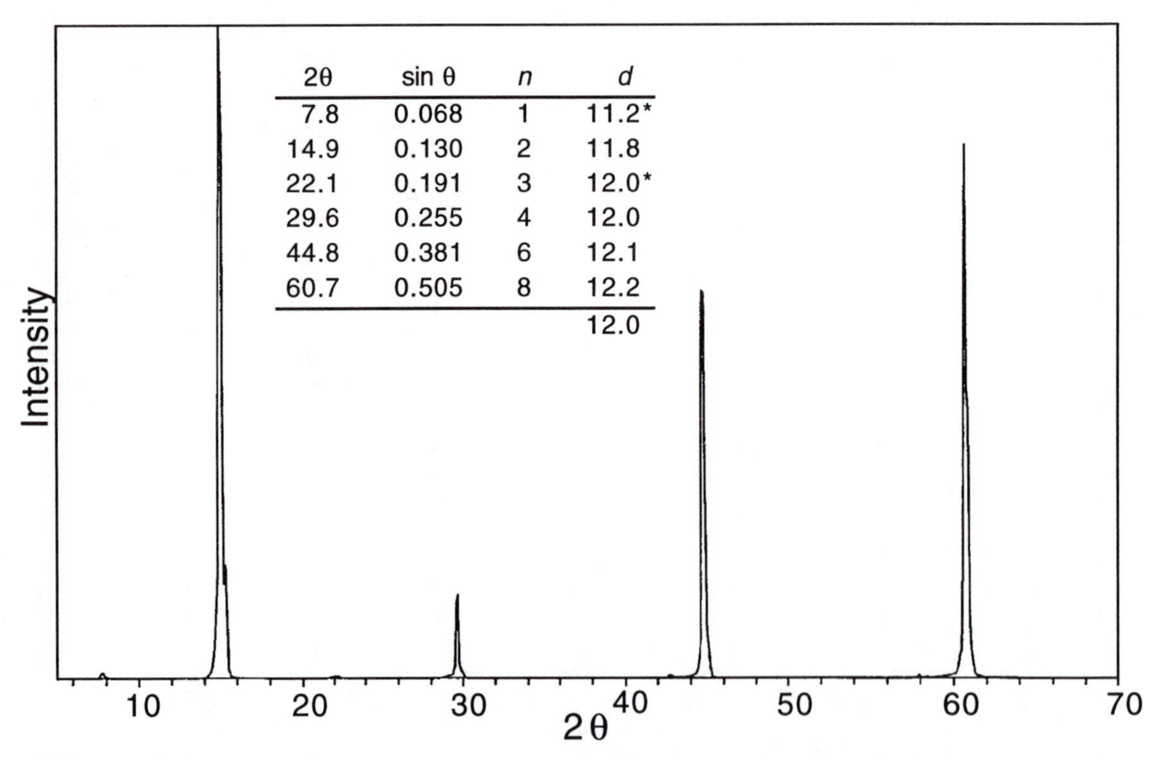

2θ	sin θ	*n*	*d*
7.8	0.068	1	11.2*
14.9	0.130	2	11.8
22.1	0.191	3	12.0*
29.6	0.255	4	12.0
44.8	0.381	6	12.1
60.7	0.505	8	12.2
			12.0

Figure 4. X-ray powder diffraction pattern for a sample of MoS_2 from Ontario.

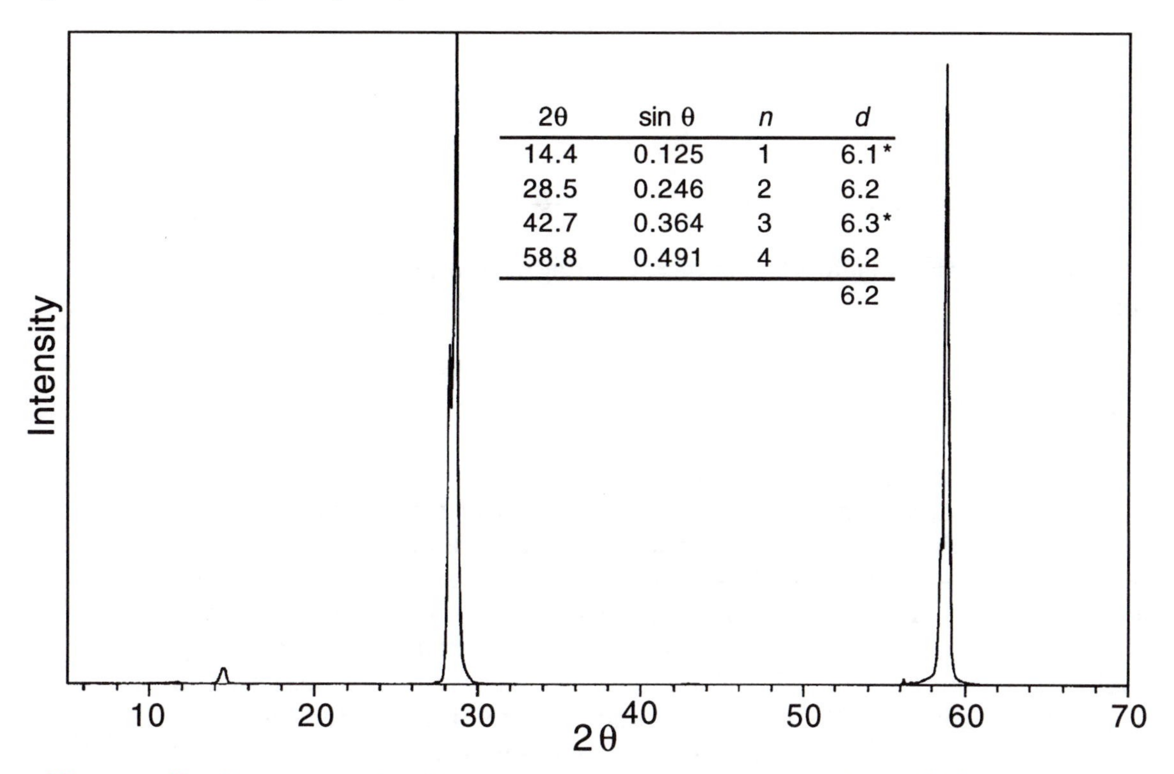

2θ	sin θ	*n*	*d*
14.4	0.125	1	6.1*
28.5	0.246	2	6.2
42.7	0.364	3	6.3*
58.8	0.491	4	6.2
			6.2

Figure 5. X-ray powder diffraction pattern for a sample of KCl (oriented on a cleaved face).

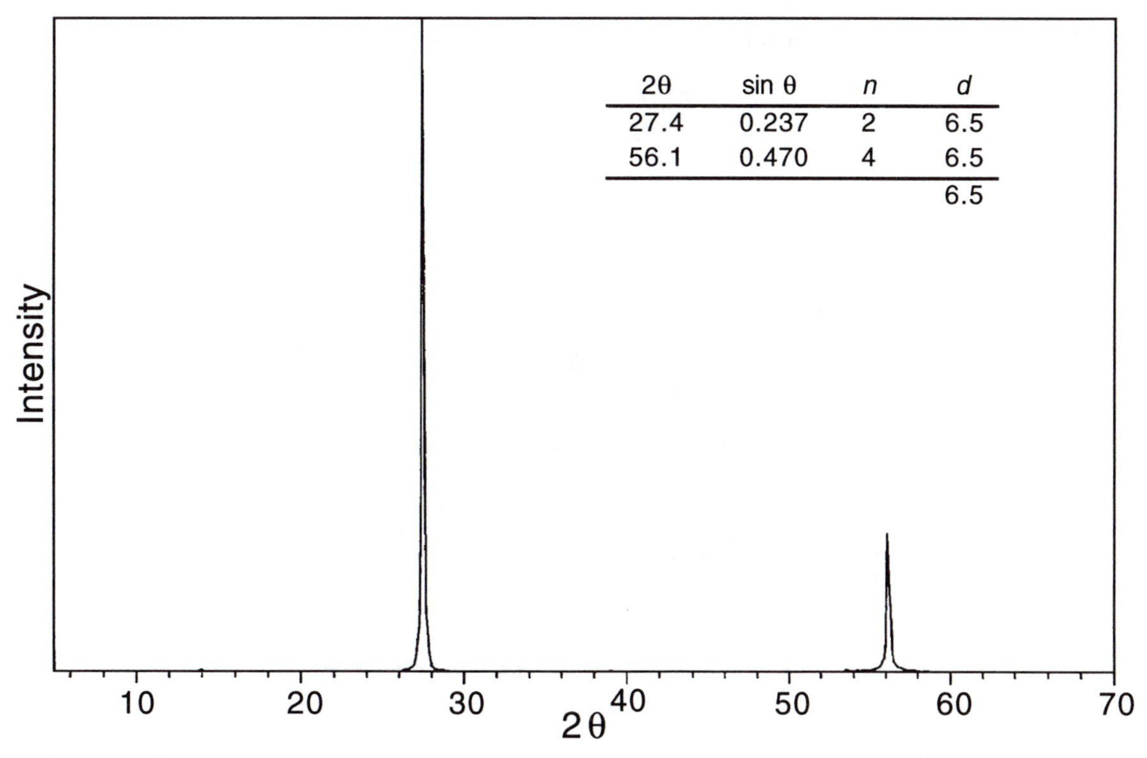

2θ	sin θ	*n*	*d*
27.4	0.237	2	6.5
56.1	0.470	4	6.5
			6.5

Figure 6. X-ray powder diffraction pattern for a sample of KBr (oriented on a cleaved face).

X-ray Analysis of a Solid

Purpose

To measure the spacing between crystallographic planes of a layered solid.

Introduction

X-rays are electromagnetic radiation with a wavelength of about 1 Å (10^{-10} m), the approximate size of an atom. X-ray diffraction has been used in two main ways to probe crystalline structure at the atomic level. First, each crystalline solid has a unique characteristic X-ray powder pattern, which may be used as a "fingerprint" for its identification if the pattern for the known material has been recorded. Second, X-ray crystallography may be used to determine the structure, that is, how the atoms pack together in the crystalline state and the interatomic distances and angles. X-ray diffraction is one of the most important characterization tools used in materials science.

We can determine the size and the shape of the unit cell for any compound most easily by using the diffraction of X-rays.

In Figure 1, the path difference between two waves is:

$$2x = 2d \sin \theta$$

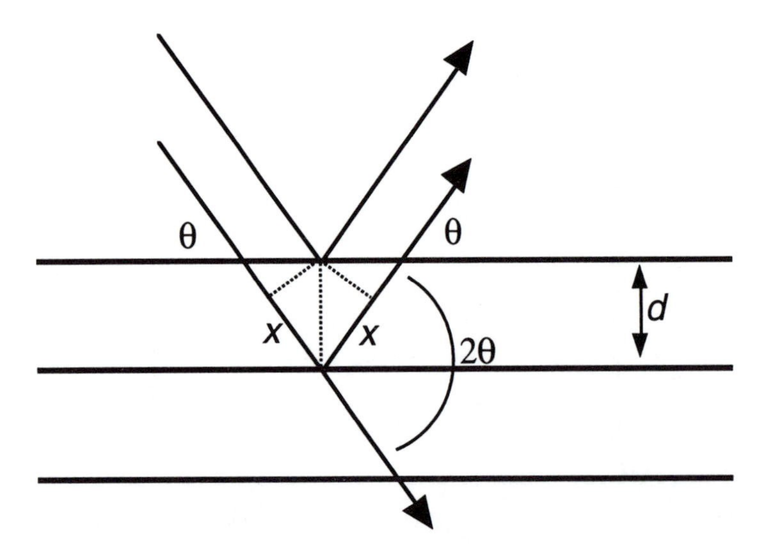

Figure 1. Reflection of X-rays from two planes of atoms in a solid.

NOTE: This experiment was written by M. Stanley Whittingham, Department of Chemsitry, State University of New York at Binghamton, NY 19302.

For constructive interference between these waves, the path difference must be an integral number of wavelengths:

$$n\lambda = 2x$$

This leads to the Bragg equation:

$$n\lambda = 2d \sin \theta$$

For example, when the diffraction pattern of copper metal was measured with X-ray radiation of wavelength λ = 0.711 Å, the first-order (n = 1) Bragg diffraction peak was found at an angle 2θ of 11.3°. To calculate the spacing between the diffracting planes in the copper metal, rearrange the Bragg equation for the unknown spacing d:

$$d = n\lambda/2\sin \theta$$

where θ = 5.64°, n = 1, and λ = 0.711 Å, and therefore $d = 1 \times 0.711/(2 \times 0.4266) = 3.62$ Å.

Figure 2 shows the X-ray diffraction pattern from a layered silicate sample. Strong intensities can be seen for a number of values of n; from each of these lines we can calculate the value of d, the interplanar spacing between the atoms in the crystal.

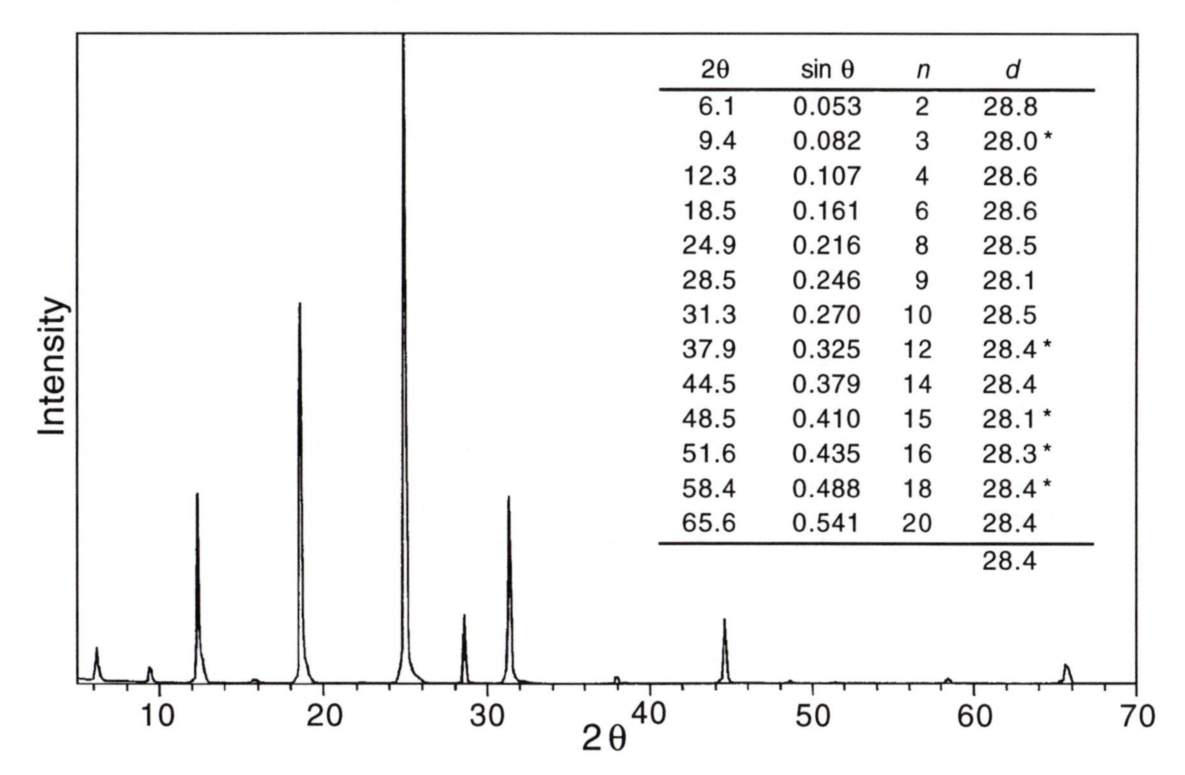

2θ	sin θ	*n*	*d*
6.1	0.053	2	28.8
9.4	0.082	3	28.0 *
12.3	0.107	4	28.6
18.5	0.161	6	28.6
24.9	0.216	8	28.5
28.5	0.246	9	28.1
31.3	0.270	10	28.5
37.9	0.325	12	28.4 *
44.5	0.379	14	28.4
48.5	0.410	15	28.1 *
51.6	0.435	16	28.3 *
58.4	0.488	18	28.4 *
65.6	0.541	20	28.4
			28.4

Figure 2. X-ray powder diffraction pattern of a sample of chlorite from Westchester, PA. The values followed by * are from weak reflections and are expected to be less precise.

The X-ray diffraction experiment requires an X-ray source, a sample and a detector to measure the intensity of the diffracted X-rays. Figure 3 is a schematic diagram of a powder X-ray diffractometer.

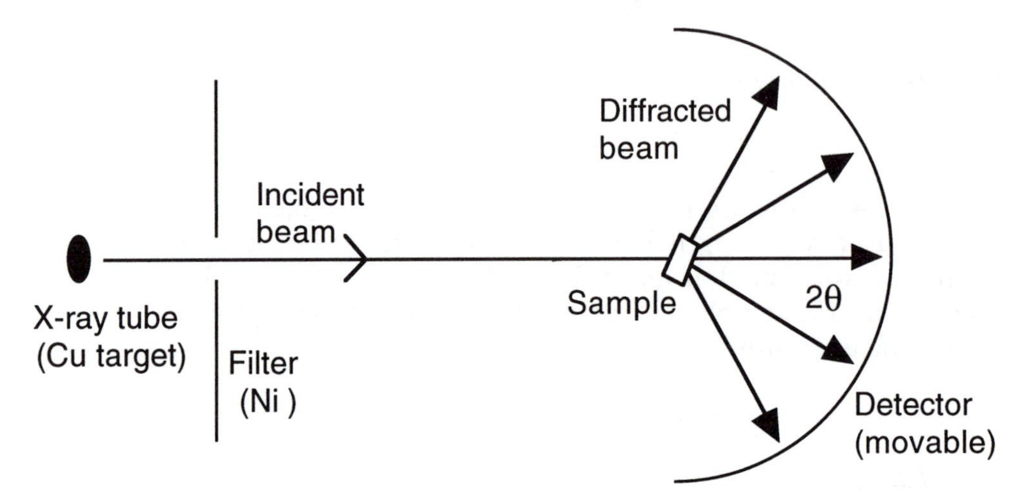

Figure 3. Schematic of an X-ray powder diffractometer.

The X-ray radiation most commonly used is that emitted by copper, whose characteristic wavelength for the so-called K_{α} radiation is $\lambda = 1.5418$ Å. When the incident beam strikes a powder sample, diffraction occurs in every possible orientation of 2θ. The diffracted beam may be detected by using a moveable detector, which is connected to a chart recorder. In normal use, the counter is set to scan over a range of 2θ values at a constant angular velocity. Routinely, a 2θ range of 5° to 70° is sufficient to cover the most useful part of the powder pattern. The scanning speed of the counter 2θ is usually 2° min^{-1} and therefore, about 30 minutes are needed to obtain a trace.

General Procedure

Place your sample onto double-sided tape that is then placed on a sample holder; if you are preparing a powder sample, use a spatula to spread the powder onto the double-side tape.

Slide the sample holder into the diffractometer. **WARNING: X-rays are hazardous to your health. Do not open the chamber while the red alarm light is on.**

Record the scan between $2\theta = 70°$ and $2\theta = 5°$.

Determine the 2θ values for each peak from the chart recorder output. Perform the calculations indicated in Table 1 and then calculate the average repeat distance.

Table 1. Laboratory Calculations

2θ	θ	$\sin\theta$	n	$d = n\lambda/(2\sin\theta)$
			1	
			2	
			3	
			4	
			5	
			6	
			7	
			8	

NOTE: λ = 1.5418 Å for Cu K_α radiation; λ = 0.711 Å for Mo K_α radiation.

Experiment 6

Radon Testing

Craig Gabler, Laura E. Parmentier, and George C. Lisensky

Notes for Instructors

Purpose

To measure the integrated radon concentration over a 30-day period at a location of your choice; to pool class data and construct a map of estimated radon concentrations.

Method

Radon is measured in this experiment by the alpha-track method. The detector is a small piece of plastic. Air being tested diffuses through a filter to the detector. When alpha particles from radon and its decay products strike the detector, they cause damage tracks or defects. At the end of the test, the sample is returned to the laboratory. The plastic is etched in strong base. Etching occurs preferentially at defect sites and thus enlarges the damage tracks. The damage tracks over a given area are counted using a microscope. The number of tracks per area is used to calculate the radon concentration at the site tested.

The detector can be distributed and collected by mail. The method yields an integrated average radon concentration over the 30-day period, and could be used to measure even longer term average concentrations to correct for seasonal variations. The accuracy of the method depends on the area density of tracks counted.

Concentrations greater than 4 pCi/L indicate the need for a follow-up test. If parents or students are concerned about the results, they should be

2725–X/93/0395$06.00/1

referred to the local health department where they can learn how to obtain an approved radon monitoring test. We recommend that copies of *A Citizen's Guide to Radon* (ISBN 0-16-036222-9, U.S. Government Printing Office, Superintendent of Documents, Mail Stop: SSOP, Washington, DC 20402-9328) be available for reference. This document also contains the telephone numbers of places to call for information in each state.

Materials

Samples of CR-39 polymer with a polyethylene film covering both sides. *See* Supplier Information.

Use a small piece of double stick tape to fasten the plastic with its protective covering to an index card. Tape a tissue to one end of the card.

It is possible that no radon will be found, so it is useful to have some CR-39 samples that have been exposed to alpha radiation to give positive results. Exposed and etched samples are also helpful for practice in counting. Direct exposure of the plastic for 24 hours to a lantern mantle containing thorium will result in a sufficient number of alpha tracks after etching. ^{232}Th is a weak alpha emitter with a half-life of 14 billion years. (A nonradioactive lantern mantle based on yttrium rather than thorium recently was introduced, so be sure to read the label when purchasing. When used as a lantern mantle, thorium oxide or yttrium oxide provides a catalytic surface with a high melting point for the combustion reaction. For more information, see Breedlove, C. H. *J. Chem. Educ.* **1992**, *69*, 621.)

Sample holder (small key ring) and paper clip

6 M NaOH (240 g NaOH or 312 mL 50% NaOH, diluted to 1 L).

Test tube (16 × 125 mm)

Boiling water

Microscope with 10× objective (100× total magnification)

Sample Results

See the experiment for discussion of the calculations. Counting the tracks in 10 different fields of view gave 3, 4, 4, 5, 9, 7, 4, 3, 8, and 6 for an average of 5.3 tracks.

Area of view at 10×: $A = \pi\,(\text{diameter}/2)^2 = \pi\,(0.17\text{ cm}/2)^2 = 0.023\text{ cm}^2$

Days exposed: 24.19 days

Tracks/cm²/day: 5.3 tracks/0.023 cm²/24.19 days = 9.5 tracks/cm²/day

Activity: $$\frac{370\text{ pCi/L}}{2373\text{ tracks/cm}^2\text{/day}} = \frac{x\text{ pCi/L}}{9.5\text{ tracks/cm}^2\text{/day}};\ x = 1.5\text{ pCi/L}$$

Radon Testing

Indoor radon gas is a national health problem. Radon causes thousands of deaths each year. Millions of homes have elevated radon levels. Homes should be tested for radon. When elevated levels are confirmed, the problem should be corrected.

Surgeon General Health Advisory, 1987

Purpose

To measure the integrated radon concentration over a 30-day period at a location of your choice; to pool class data and construct a map of estimated radon concentrations.

Introduction

Radon-222 is a naturally occurring, radioactive gas, with a half-life of 3.824 days. Radon comes from the natural radioactive breakdown of uranium-238, which has a half-life of 4.47 billion years; uranium is found in nearly all soils (20–30 ppb average earth abundance). Radon enters homes through cracks or other holes in the foundation.

The small plastic disk that you will use to detect radon is a high-clarity polymer that is often used in eyeglasses, poly[ethylene glycol bis(allyl carbonate)], called CR-39. The chemical formula of the monomer is $C_{12}H_{18}O_7$. Solid CR-39 is sensitive to penetration by alpha particles, but is insensitive to beta and gamma radiation. Alpha particles cause damage in the plastic, probably due to disruption of the polymer linkage along the path of penetration. Although the damage is not visible to the eye, chemical etching of the sample by NaOH occurs preferentially in the damaged regions. The etched samples then exhibit damage "tracks" when viewed under a microscope. The number of tracks in a given area of the plastic disk can be used to estimate the radon level at the sampled location.

Procedure

Exposing the Detector

The detector should remain undisturbed at the sampling location for at least 3 weeks. Take or mail the detector to your home or other location. Include with the detector these instructions:

NOTE: This experiment was written by Craig Gabler, Centralia High School, Centralia, WA 98531; and Laura E. Parmentier and George C. Lisensky, Beloit College, Beloit, WI 53511.

Radon Detection

1. A small, plastic disk that is sensitive to the presence of radon gas is taped to the middle of the enclosed index card. Please peel off the thin plastic film covering the disk (there really is a film there!) and very loosely tuck the tissue over the card and disk. The disk needs to be protected from dust and particles, but air needs to be able to flow freely over the disk.
2. Place the card with the disk in a location where it can remain undisturbed for 3 weeks. The best location to test for radon is a basement or any room or garage that is close to the ground.
3. Please record the following information about this experiment:

   Date detector was put in place ______ Time detector was put in place __________

   Location of detector ________________ Floor level of location __________________

   Date detector was removed ________ Time detector was removed ____________

4. At the end of the 3-week period, please fold this paper over the index card and disk (to protect the disk) and mail back. We will analyze the disk on ___________. **Please mail this back so that it is received by ___________.**

Etching the Disk

After 3 or more weeks of exposure, remove the disk from the index card and peel off the polyethylene film from the back. Slip the ring of an "etch clamp" over the top of the disk and fasten onto a paper clip that has been bent so there is a hook at each end (see Figure 1A). Hook the disk inside a test tube, and add enough 6 M NaOH solution to cover the disk in the tube. **CAUTION: Hot 6 M NaOH is extremely corrosive. Wear goggles at all times. Rubber gloves are strongly recommended.**

Heat the test tube containing the disk and NaOH solution in a boiling water bath for 40 minutes. At the end of the 40 minutes, remove the disk and rinse thoroughly with lots of water.

Counting the Alpha Tracks

Determine the area (in square centimeters) of the field of view of your microscope at low power (10×) by looking through the microscope at a ruler.

Practice looking at alpha tracks on a previously exposed and etched disk. (This disk was exposed to a camping lantern mantle which contains thorium!) Your disk will likely have fewer tracks. Note the various shapes of the tracks, which depend on how the alpha particles entered the solid (see Figure 1B): The circular-shaped tracks are due to alpha particles that entered straight in (perpendicular to the disk), and the more tear-drop shapes are due to alpha particles that entered at an angle.

Place your disk on a microscope slide and count the number of tracks *in 10 different fields of view.*

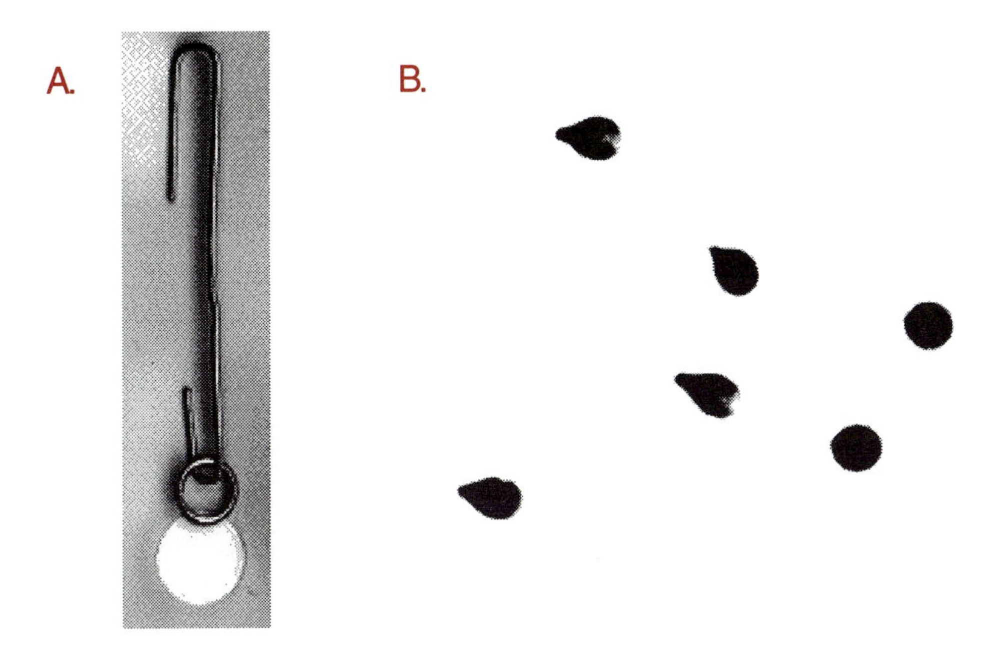

Figure 1. A. Disk ready for etching. The disk is pinched between the rings of the "etch clamp." B. Alpha tracks after etching, viewed through microscope.

Calculations

Experiments have shown that tracks due to background alpha radiation are negligible. Determine the average number of tracks in a field of view for your disk, and then determine the average number of tracks per square centimeter per day (including any fraction of a day). Control plastic disks that were sent to a radon facility in which the radon level was known to be 370 picocuries per liter were found to exhibit 2370 tracks/cm^2/day by this etching and counting technique. Given this relationship between picocuries per liter and tracks per square centimeter per day, calculate the radon level in picocuries per liter of air for your sample. How does your sample compare with the 4-pCi/L guidelines set by the Environmental Protection Agency? Concentrations greater than 4 pCi/L indicate the need for a follow-up test using an EPA-approved radon monitoring service.

The precision of this experiment is not good because of day-to-day and seasonal variations in radon levels (longer exposure times give better average results), and because of lack of operator experience in counting alpha tracks. Commercially available alpha-track detectors give results that vary by about 25% for a 30-day exposure at a 4-pCi/L radon level.

Experiment 7

Periodic Properties and Light-Emitting Diodes

Arthur B. Ellis, Lynn Hunsberger, and George C. Lisensky

Notes for Instructors

Purpose

To relate observed properties such as the color, wavelength and energy of light, and excitation voltage for a series of compound semiconductors to composition and periodic trends.

To observe how electrical resistance of a metal changes with temperature and how electrical resistance of a semiconductor changes under illumination.

Method

A pair of circuits for light-emitting diodes (LEDs) are constructed and used to study a series of four LEDs from a family of solid-solution semiconductors that have the zinc blende structure. The light emitted by the LEDs and the voltage drop across the devices is determined; in a sense the latter is the opposite of the photoelectric effect, since here there is a threshold electrical energy needed to produce light. Both measurements are correlated with chemical composition and bonding features, including trends in internuclear distance and electronegativity differences. Liquid nitrogen is used to contract the structure, and the effect on the band gap is observed. In addition, the electrical resistance of a CdS photocell and of a copper wire are measured with a multimeter. *See* Chapters 5 and 7.

2725–X/93/0401$06.00/1

Materials

Water-clear red LED—Mouser 592-SLH34-VT3. Label these LEDs with the formula $GaP_{0.40}As_{0.60}$

Water-clear orange LED—Mouser 592-SLH34-DT3. Label these LEDs with the formula $GaP_{0.65}As_{0.35}$

Water-clear yellow LED—Mouser 592-SLH34-YT3. Label these LEDs with the formula $GaP_{0.85}As_{0.15}$

Water-clear green LED—Mouser 592-SLH34-MT3. Label these LEDs with the formula $GaP_{1.00}As_{0.00}$

$Al_xGa_{1-x}As$ LEDs (optional, see Supplier Information)

Note: If the clear LEDs that all look identical when unlit are not marked or labeled in some manner, the LEDs are likely to become mixed up.

CdS photocells (low light resistance) —Mouser 338-76C348 or Radio Shack

150 mH RF choke (long thin copper wire in a coil)—Mouser 43LJ415

1-kΩ resistors—Mouser 29SJ250-1K

1-MΩ resistors—Mouser 29SJ250-1M

6-inch molded black battery snap—Mouser 12BC011

Break-away IC sockets for LEDs—Mouser 151-5530

Soldering pencils—Mouser 381-FS030 or Radio Shack

Solder—Mouser 533-24-6040-39 or Radio Shack

Wooden spring clothespins

Diffraction gratings—Flinn holographic grating AP1714 (8 × 10-inch sheet) or Edmund card-mounted diffraction viewer C39,502

Multimeters (with alligator clips) —Mouser 585-DM231

LED reference strips (made in-house; see Figure 7.22)

9-V batteries

Foam cups

Liquid nitrogen

Diamond and zinc blende structural models

Answers to Warm-up Exercises

1. P has a smaller atomic radius than As. Shorter bond distances often correlate with better orbital overlap and a larger band gap. (See Chapter 7 for some qualifications to this statement.) GaP would be expected to have a larger band gap than GaAs, on this basis.

2. Al is less electronegative than Ga. The increased electronegativity difference between Al and As relative to Ga and As suggests a greater ionic contribution to the bond and hence a larger band gap. Thus, AlAs would be expected to exhibit a larger band gap then GaAs.

3. In order of increasing band gap energy:

   $GaP_{0.40}As_{0.60} < GaP_{0.65}As_{0.35} < GaP_{0.85}As_{0.15} < GaP_{1.00}As_{0.00}$

4. In order of increasing band gap energy:

   $Al_{0.05}Ga_{0.95}As < Al_{0.15}Ga_{0.85}As; < Al_{0.25}Ga_{0.75}As < Al_{0.35}Ga_{0.65}As$

5. Bond distances will contract as the material is cooled, and the band gap energy is expected to increase, as described in answer 1.

Results

Typical results expected are shown in Figure 1 and Table 1.

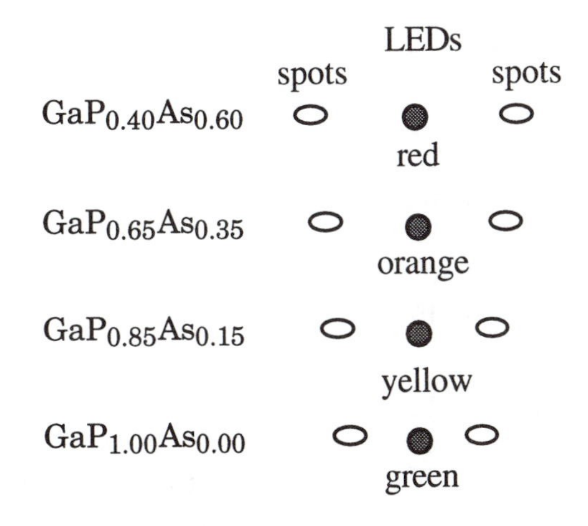

Figure 1. Color, composition, and diffraction spacing.

Table 1. LED Measurements

Composition	Color	RT λ, nm (eV)[a]	LN_2 λ, nm (eV)[b]	Volts[c]
$GaP_{0.40}As_{0.60}$	Red	637 (1.95)	621 (2.00)	1.84 (1.40)
$GaP_{0.65}As_{0.35}$	Orange	593 (2.09)	579 (2.14)	2.09 (1.51)
$GaP_{0.85}As_{0.15}$	Yellow	560 (2.21)	—	2.20 (1.58)
$GaP_{1.00}As_{0.00}$	Green	547 (2.27)	—	2.28 (1.62)
$Al_{0.35}Ga_{0.65}As$	Red	625 (1.98)	605 (2.05)	
$Al_{0.25}Ga_{0.75}As$	—	666 (1.86)	642 (1.93)	
$Al_{0.15}Ga_{0.85}As$	—	729 (1.70)	703 (1.76)	
$Al_{0.10}Ga_{0.90}As$	—	773 (1.60)	741 (1.67)	
$Al_{0.05}Ga_{0.95}As$	—	835 (1.48)	803 (1.54)	

[a]Experimental values of LED band maxima at room temperature using a spectrometer.
[b]Experimental values of LED band maxima at liquid nitrogen temperature using a spectrometer.
[c]Experimental values at liquid nitrogen temperature using a voltmeter and the 1-MΩ circuit. Values in parentheses obtained at room temperature.

Resistance

As the temperature of a metal decreases, its resistance also decreases. The small resistance to electrical current in a metal can be attributed to the vibrations of the metal atoms which scatter the mobile electrons. As the temperature is lowered, the vibrational motion of the atoms decreases, and conduction increases (i.e., resistance decreases). A resistance of 150 Ω at room temperature drops to 15 Ω at liquid nitrogen temperature.

When light of sufficient energy strikes the semiconductor, electrons are excited from their localized positions. Resistance decreases when the CdS semiconductor is exposed to visible light. Typically, a resistance of 1 kΩ in room light decreases to 15 Ω when illuminated in direct sunlight or a projector beam.

Models

How many nearest neighbors
does each carbon in the diamond structure have? 4
does the Zn in ZnS have? 4
does the S in ZnS have? 4
What is the name for the shape defined by the nearest neighbors
to each C in diamond? tetrahedral
to each Zn in ZnS? tetrahedral
to each S in ZnS? tetrahedral
How many atoms of carbon are in the diamond unit cell?

$8 \times 1/8 = 1$ corner atom
$6 \times 1/2 = 3$ face-centered atoms
4 interior atoms
8 total atoms

How many atoms of Zn are in the ZnS unit cell?
$(8 \times 1/8) + (6 \times 1/2) = 4$ atoms of zinc
How many atoms of S are in the ZnS unit cell? 4 interior atoms

Draw the z layer sequence showing the positions of atoms for diamond and zinc blende.

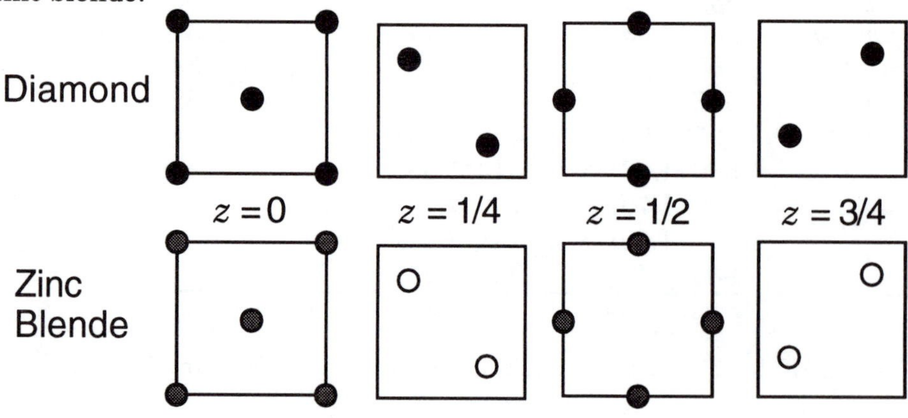

Estimate the packing efficiency for the diamond structure. Show your work. How does it compare to a closest packed structure?

$$\text{Packing Efficiency} = \frac{\text{volume occupied by atoms}}{\text{total volume of unit cell}} \times 100\%$$

Diamond built with the SSMK has eight 1-inch spheres (2.54 cm diameter) in the unit cell. The length of one side of the unit cell in the SSMK (measured with a ruler) is 5.9 cm.

$$V_{sphere} = \frac{4}{3}\pi r^3 = 1.33\pi(1.27\text{ cm})^3 = 8.56\text{ cm}^3\text{ (per sphere)}$$

$$V_{sphere}(\text{total}) = 8 \times 8.56\text{ cm}^3 = 68.5\text{ cm}^3$$

$$V_{cell} = (\text{length of cube edge})^3 = (5.9\text{ cm})^3 = 205.4\text{ cm}^3$$

$$\text{Packing Efficiency} = \frac{68.5\text{ cm}^3}{205.4\text{ cm}^3} \times 100 = 33\%$$

A rigorous geometric calculation gives 37%. The packing efficiency for a closest packed structure is ~74%. The packing efficiency for the diamond structure is much lower than this value.

Periodic Properties and Light-Emitting Diodes

Purpose

To relate observed properties such as the color, wavelength and energy of light, and excitation voltage for a series of compound semiconductors to composition and periodic trends; to observe how electrical resistance of a metal changes with temperature and how electrical resistance of a semiconductor changes under illumination.

Introduction

The terms insulator, semiconductor, and metal are used to classify solids based on the amount of energy, called the *band gap energy*, needed to excite electrons from localized bonds to a higher energy state where the electrons are free to move about the solid. In metals the very small amount of energy required (very small band gap) leads to a large number of delocalized electrons and the high electrical and thermal conductivity of metals. Insulators require a relatively large amount of energy (large band gap) to produce mobile electrons. Semiconductors are the intermediate case corresponding to a moderate band gap.

The energy needed to remove an electron from a localized bond in the solid can be supplied by adding energy in the form of heat, electricity, or light. When an excited electron returns to the localized bond, the energy is released, sometimes in the form of light.

Periodic Trends

The distance between the nuclei (the sum of the atomic radii of the bonded atoms) is one factor that determines the strength with which an electron is held in a localized bond. In the solid elements C, Si, Ge, and α-Sn that exhibit the diamond structure, Figure 1, the atomic radii increase and the band gaps decrease rapidly going down their column of the periodic table. Diamond is an electrical insulator, Si and Ge are semiconductors, and α-Sn is a metal.

NOTE: This experiment was written by Arthur B. Ellis and Lynn Hunsberger, Department of Chemistry, University of Wisconsin—Madison, Madison, WI 53706; and George C. Lisensky, Department of Chemistry, Beloit College, Beloit, WI 53511.

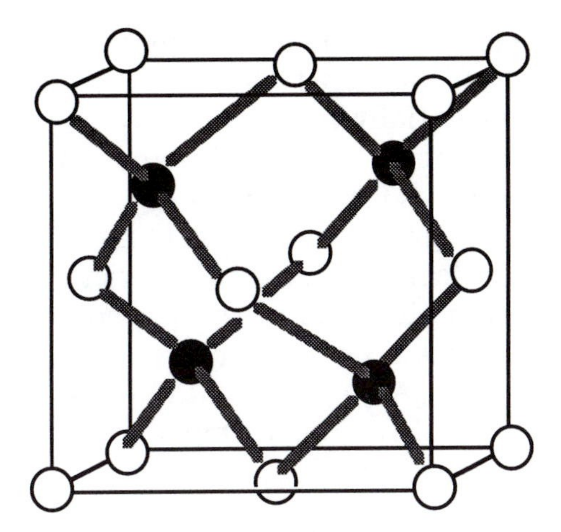

Figure 1. Drawing of a unit cell where all of the atoms are bonded to four other atoms. When all of the atoms are the same element, this is the diamond crystal structure. When the lighter colored spheres are different elements than the darker colored spheres, this structure has AZ stoichiometry and is called zinc blende.

One can imagine making a solid with the same total number of electrons as C, Si, Ge, or Sn using AZ stoichiometry. For example, if two carbon atoms are replaced with one boron atom and one nitrogen atom, the total number of valence electrons in the solid is conserved. Other combinations are indicated in Figure 2. These compounds have the zinc blende structure (Figure 1).

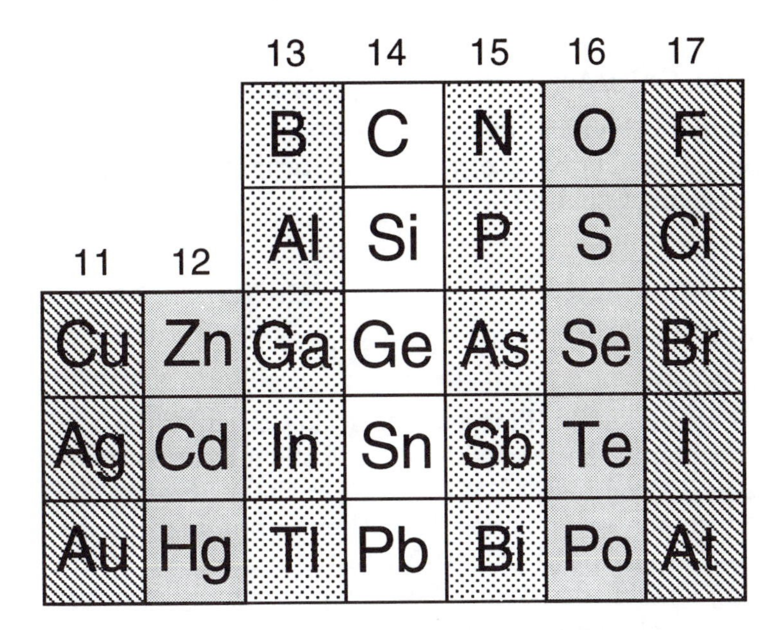

Figure 2. A portion of the periodic table emphasizing the formation of AZ solids that are isoelectronic with the Group 14 solids. Isoelectronic pairs are indicated with similar shading; for example Ge, GaAs, ZnSe, and CuBr.

The strength by which an electron is held in a localized bond also depends on the difference in electronegativity between the atoms. Ge, GaAs, ZnSe, and CuBr all have essentially the same size unit cell, but the increasing difference in electronegativity adds an ionic bonding contribution and the band gap increases for this series.

The ability to form solid solutions provides a chemical means for tuning band-gap energies that would be lacking were we restricted to the AZ stoichiometries obtainable with the elements in the periodic table. The GaP_xAs_{1-x} ($0 \le x \le 1$) solid solutions will be used in this experiment, since $GaP_{0.40}As_{0.60}$, $GaP_{0.65}As_{0.35}$, $GaP_{0.85}As_{0.15}$, and $GaP_{1.00}As_{0.00}$ are sold commercially with wires attached as LEDs (Light Emitting Diodes).

Warm-up Exercises

1. Does P or As have a larger atomic radius? Considering only atomic radii, would GaP or GaAs have a larger band gap energy? Explain.
2. Is Ga or Al more electronegative? Considering only the electronegativity difference, would GaAs or AlAs have a larger band gap energy? Explain.
3. Considering only atomic radii, rank the following in order of increasing band gap energy:

   $GaP_{0.40}As_{0.60}$, $GaP_{0.65}As_{0.35}$, $GaP_{0.85}As_{0.15}$, $GaP_{1.00}As_{0.00}$
4. Considering only the electronegativity, rank the following in order of increasing band gap energy:

   $Al_{0.35}Ga_{0.65}As$, $Al_{0.25}Ga_{0.75}As$; $Al_{0.15}Ga_{0.85}As$; $Al_{0.05}Ga_{0.95}As$
5. What usually happens to the bond distances of a material when it is cooled? Considering only bond distance, would a material's band gap be larger when warm or cold? Explain.

Procedure

Wear eye protection. Work in pairs.

One member of the team should build the single socket circuit shown in Figure 3. The second member of the team should build the identical circuit in which a 1-MΩ resistor is used instead of the 1-kΩ resistor. Solder one end of the battery snap to the resistor and the other to one lead of the socket. Complete the socket circuit by soldering the remaining end of the resistor to the remaining lead of the socket. Soldering is usually easier if you melt some solder onto each end individually before trying to join them together. A wooden spring clothespin is a convenient way to hold the socket while soldering.

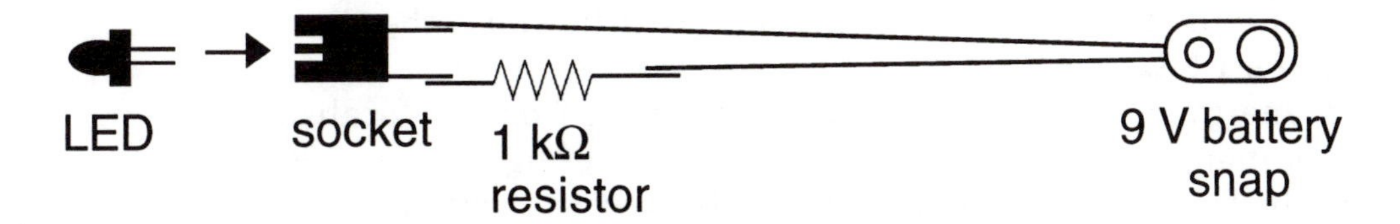

Figure 3. The circuit for the LED socket.

Color and Composition

Obtain samples of the four different GaP_xAs_{1-x} compositions. The LEDs all look identical so do not mix them up.

Using the circuit containing the 1-kΩ resistor, connect the battery snap to a 9-V battery. The circuit is "turned on" by inserting the LED into the socket. Does it matter which way the LED is inserted?

Plug each kind of LED into the socket of the circuit containing the 1-kΩ resistor. For each composition of LED, record the color of light emitted.

Wavelength and Composition

Obtain a reference strip of LEDs. View this strip through a diffraction grating oriented so that the light from the LEDs is diffracted away from the reference strip (Figure 4). For each LED, the central spot due to the undiffracted source will be the brightest. Compare the spacing length between the first diffraction spots to either side of the central spot.

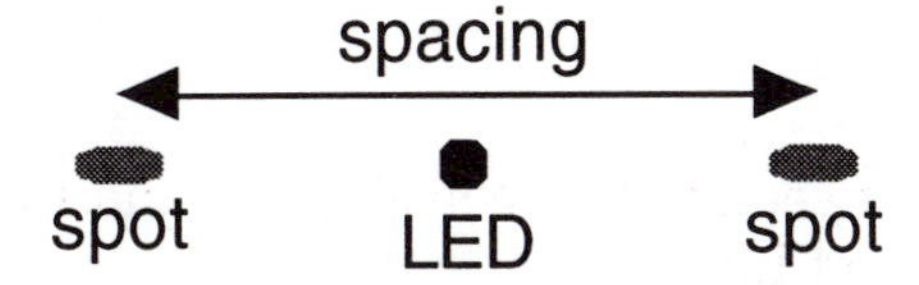

Figure 4. An LED viewed through a diffraction grating.

List the LED compositions in order of increasing diffraction spacing. The light separation by the diffraction grating is proportional to the wavelength of light, so a list of increasing spacing lengths is also a list of increasing wavelengths.

Band-Gap Energy and Composition

When a voltage is applied across the LED, nothing will happen unless the energy (voltage is proportional to energy) is sufficient to excite an electron from its localized bond. When the energy is increased beyond the band-gap energy, current can begin to flow. Measurement of the voltage for a minimum current flow thus provides another estimate of the band-gap energy.

To approximate the conditions of minimum current flow, use the circuit you built with the 1-MΩ resistor.

How much energy is required to excite an electron from its localized bond? Is the minimum voltage required for emission of light by an LED dependent on the wavelength of light emitted? To answer these questions, measure the voltage across the LED. (If the voltage is more than 3 volts, then current is not flowing through the LED and its electrical leads should be reversed.) Thermal excitation of localized bond electrons also contributes to the current; while the data obtained will not represent the actual band gap energies, the data will show the correct trend.

List the LED compositions in order of increasing voltage. Voltage is proportional to energy, so a list of increasing voltages is also a list of increasing band gap energies.

Composition, Color, Wavelength, and Energy Comparisons

You should have three lists based on the LED compositions: predicted relative band gap energies (Warm-up Exercises), observed relative diffraction spacings (wavelengths), and observed relative voltages (band gap energies). Do these lists agree?

Use the information in Table 1 to make a fourth list of composition based on the observed colors. Are your observations consistent?

Summarize your results.

Table 1. Relationship between Color, Wavelength, and Energy of Light.

Color of Light	Approximate Wavelength (nm)	Energy eV (kJ/mol)
Ultraviolet	<400	>3.1 (300)
Violet	410	3.0 (290)
Violet–blue	430	2.9 (280)
Blue	480	2.6 (250)
Blue–green	500	2.5 (240)
Green	530	2.3 (225)
Green–yellow	560	2.2 (215)
Yellow	580	2.1 (205)
Orange	610	2.0 (195)
Red	680	1.8 (175)
Purple–red	720	1.7 (165)
Infrared	>720	<1.7 (165)

Temperature and the Band Gap

For this part, you will need to work with an additional group. Both groups should insert their $GaP_{0.40}As_{0.60}$ LED into their 1-kΩ circuit. Practice holding both LEDs in a single column and viewing through the diffraction grating. LEDs of the same composition should give the same spacing. Check with your instructor if the spacing is not the same.

Dip only one lighted LED into a foam cup of liquid nitrogen for a few seconds. Does the *color* change? **CAUTION: Liquid nitrogen is extremely cold. Do not allow it to come into contact with skin or clothing, as severe frostbite may result.**

View through the diffraction grating the cold LED and, as a reference, a second room temperature LED of the same composition. Did the wavelength of the cold LED shift? Did the wavelength get longer or shorter?

Let your LED warm back to room temperature. Are any observed changes reversible?

Repeat this experiment using a pair of $GaP_{0.65}As_{0.35}$ LEDs. Do you observe the same changes?

Do your observations agree with your predictions in the Warm-up Exercises?

CAUTION: Do not dip the reference strip in liquid nitrogen. The circuit board used to hold the LEDs aligned in position will crack. The $GaP_{0.85}As_{0.15}$, and $GaP_{1.00}As_{0.00}$ LEDs have multiple peaks due to a very small amount of added impurity atoms used to enhance the light intensity, and these multiple peaks shift in wavelength and relative intensity with temperature; interpretation is not as simple.

Electrical Resistance in Metals and Semiconductors

Resistance measures the difficulty with which an electron moves through a material.

Use a digital voltmeter on the resistance setting (ohms or Ω) to record the resistance of a very long (about 150 m) length of thin copper wire wound in a coil, sealed in plastic and sold as an inductor or choke coil. The small resistance to electrical current (flow of electrons) in metals is due to vibrations of the atoms that interfere with the flow of electrons.

Dip the copper coil in liquid nitrogen. Does the resistance change? Why?

Use a digital voltmeter on the resistance setting (ohms or Ω) to record the resistance of a short length (about 6 cm) of CdS semiconductor. Such samples are sold commercially with wires attached as photocells. Does the resistance change with light exposure? Why?

Caution: *Do not dip the photocell in liquid nitrogen; the plastic casing will crack.*

Semiconductor Models

Obtain or build a model of the diamond and zinc blende structures.

1. How many nearest neighbors does each atom have?
2. How are the nearest neighbors arranged about the central atom? (What is the name for the shape they assume?) For ZnS, do this determination for the Zn atoms and the S atoms separately.
3. How many atoms of each type are in the unit cell? *(Hint: Remember some atoms are not entirely within the unit cell.)*
4. Draw the z layer sequences showing the positions of the atoms. Hint– In both structures there are atoms in the plane that forms the bottom of the unit cell cube, in planes with z = 1/4, 1/2, and 3/4 of the way up the unit cell, as well as on the top of the unit cell cube. Draw the atoms as they intersect each of these five planes.
5. For the diamond structure, estimate the packing efficiency by making the appropriate measurements on the model and recalling that the spheres have diameters of 1 inch and $V_{sphere} = \frac{4}{3}\pi r^3$.

   How does the packing efficiency compare to that of a close-packed structure?

Experiment 8

Hydrogen Insertion into WO_3

M. Stanley Whittingham

Notes for Instructors

Purpose

To reductively insert protons into solid tungsten trioxide and observe the effect on color and conductivity.

Method

If yellow–green WO_3 is placed in a beaker and hydrochloric acid is added to it, nothing happens. But if a few chips of zinc are added, hydrogen is produced, causing the solid to turn a very dark blue in a chemical reduction. Its electrical conductivity also increases dramatically, from less than 10^{-6} /Ω-cm to around 10^{-2} /Ω-cm. The color and conductivity changes observed are due to the intercalation of hydrogen atoms into the cavities in the WO_3 structure, and the donation of their electrons to the conduction band of the WO_3 matrix, making H_xWO_3. These electrons make the material behave like a semiconductor, and at higher values of x, a metal. If the H_xWO_3 is left exposed to the air, it will re-oxidize and return to its original yellow–green color.

$$\underset{\text{Yellow–green}}{WO_3} \underset{O_2}{\overset{Zn + HCl}{\rightleftarrows}} \underset{\text{Blue}}{H_xWO_3}$$

2725–X/93/0413$06.00/1

Materials

Tungsten trioxide, WO_3 (Aldrich, 99%+; Johnson–Matthey 99.8%)

3 M HCl

Zinc metal (filings)

Filter paper, Büchner funnel, suction flask, and water aspirator

Glass capillary tubes, 1.5 × 100 mm (melting point size)

18-gauge copper wire

Ohmmeter

Rubber band

Answers to Warm-up Exercises

$$Zn + 2H^+ \rightarrow H_2 + Zn^{2+}$$

$$WO_3 + \frac{x}{2} H_2 \rightarrow H_xWO_3$$

$$4H_xWO_3 + xO_2 \rightarrow 4WO_3 + 2xH_2O$$

Hydrogen Insertion into WO_3

Purpose

To reductively insert protons into solid tungsten trioxide and observe the effect on color and conductivity.

Introduction

Tungsten trioxide has an idealized structure consisting of WO_6 octahedra joined at their corners. It may also be considered as having the perovskite structure of $CaTiO_3$ with all of the calcium sites (in the center of the unit cell shown in Figure 1) vacant. When an atom is inserted into the center of the WO_3 structure, the structure is called a tungsten bronze.

Black spheres = W
White spheres = O
Shaded spheres = M

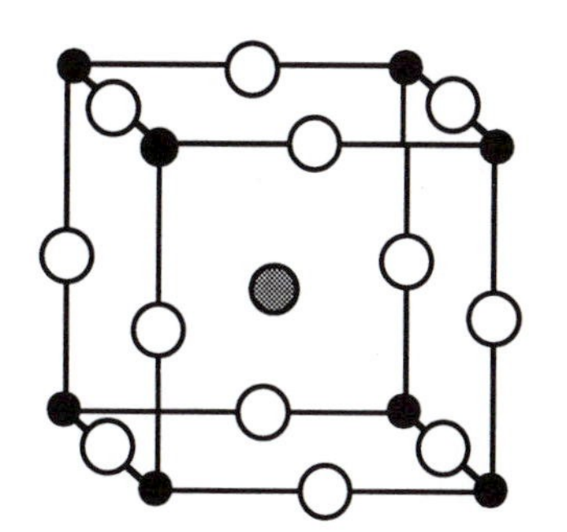

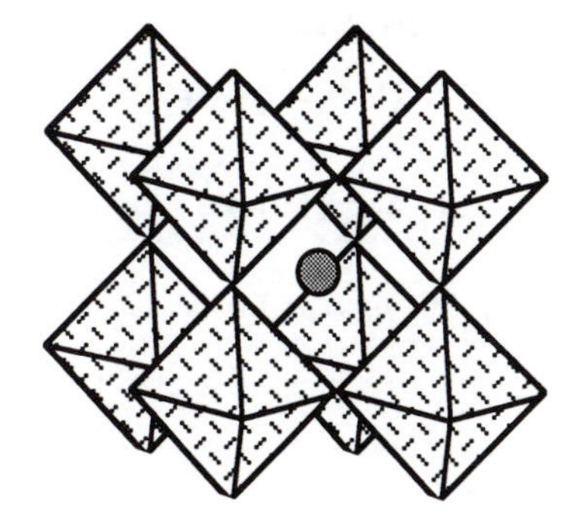

Figure 1. Two views of the idealized M_xWO_3 unit cell.

These compounds have the formula M_xWO_3, where M is usually Na or K and $0 < x < 1$. The color of the compound is controlled by the stoichiometry; thus $Na_{0.9}WO_3$ is yellow, but $Na_{0.3}WO_3$ is blue–black (*see* Figure 2). The intense colors of these solids have led to their use in paint pigments.

M can be an alkali metal, Ca, Sr, Ba, Al, In, Tl, Sn, Pb, Cu, Ag, Cd, the rare earth elements, H^+, or NH_4^+. In this experiment, you are going to make H_xWO_3.

The intercalation of hydrogen atoms into the cavities in the WO_3 structure and the donation of their electrons to the conduction band of the WO_3 matrix will cause color and conductivity changes. These electrons make the material behave like a metal, with both its conductivity and color being derived from free electron behavior. This coloration reaction is now being used in electrochromic displays for sunglasses and rear-view mirrors in cars, and has been proposed for adjusting the transmission of

---

NOTE: This experiment was written by M. Stanley Whittingham, Department of Chemistry, State University of New York at Binghamton, Binghamton, NY, 19302.

light through the glass panes in large buildings. In these applications the redox coloring reaction is carried out electrochemically.

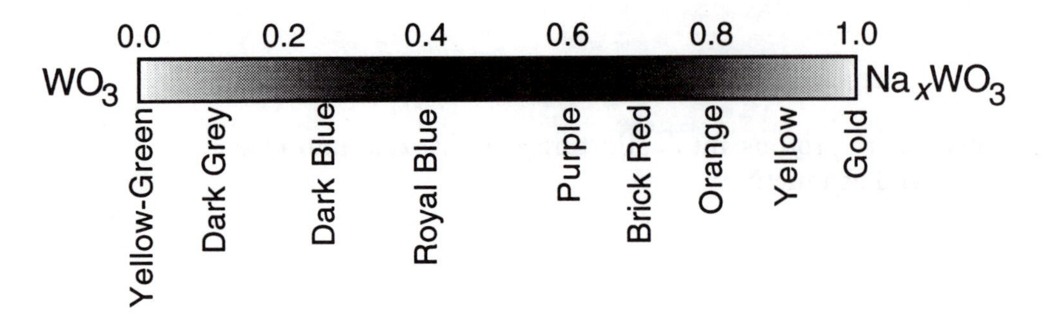

Figure 2. Change in color of Na_xWO_3 as x (shown at the top) varies.

Warm-up Exercises

Balance the equations.

a. $Zn + 2H^+ \rightarrow H_2 +$

b. $WO_3 + \frac{x}{2} H_2 \rightarrow$

c. $H_xWO_3 + xO_2 \rightarrow$

Procedure

Wear eye protection.

Synthesis

Place 0.5 g of WO_3 into a 150-mL beaker. **CAUTION: WO_3 is an irritant. Avoid creating or breathing dust. Avoid eye and skin contact. Wash hands thoroughly after handling.** What is the color of WO_3? Carefully pour about 50 mL of 3 M hydrochloric acid onto the WO_3. Does anything happen? Add less than 1 g of zinc filings to the acid and observe what takes place. **CAUTION: Keep hydrogen away from open flames.**

After all reaction has ceased, note the colors of all the products. Suction-filter off the solid product. Wash twice with water and air-dry for a few minutes.

Change in Conductivity upon Intercalation

Make a tube to measure the conductivity of WO_3 by first inserting a straight piece of copper wire (18-gauge) into a glass capillary tube, as shown in Figure 3. The fit should be tight. Push the open end of the tube into some WO_3 so that some of the material is stuck in the tube. Flip the tube upside down and tap the copper wire end against a hard surface so that the WO_3 falls to the wire. Repeat this until there is about 1 cm of the material in the tube. Finally, insert another length of copper wire into the open end of the glass capillary, so the WO_3 is packed between the two copper electrodes. In order to ensure a tightly packed tube, carefully bend the copper wire electrodes and place a rubber band around the electrodes:

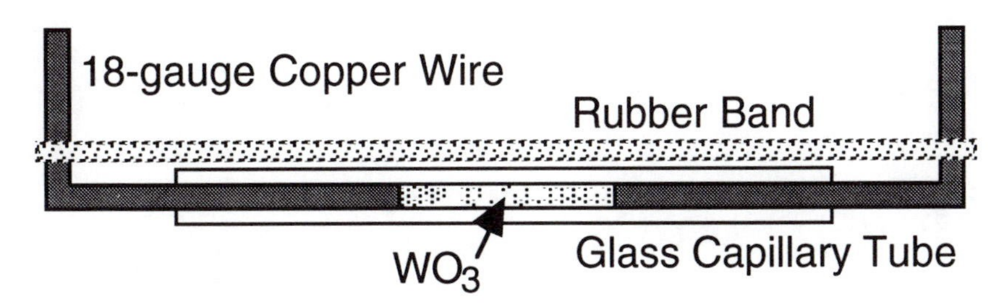

Figure 3. The apparatus used to measure the conductivity of WO_3 and H_xWO_3.

Prepare a similar tube packed with H_xWO_3. Make sure that the length of the sample in the tube is about the same.

Using an ohmmeter (adjust the meter to read in the 10–100-kΩ range), measure the electrical resistance between the electrodes of the WO_3 sample. Repeat the measurement on the H_xWO_3 sample. How much difference is there?

Re-oxidation of H_xWO_3 upon Exposure to Oxygen in the Air

Has the color of your sample changed at all since you made it? Leave your sample in your lab drawer for 1 week, then measure the conductivity again.

What happens when you leave H_xWO_3 exposed to the air? Does the conductivity change? Why?

Cleanup

Place the glass capillaries containing WO_3 and H_xWO_3, as well as any remaining solid, in an appropriately labeled waste container.

Experiment 9

Absorption of Light

Ann M. Cappellari, Frederick H. Juergens,
Arthur B. Ellis and George C. Lisensky

Notes for Instructors

Purpose

To determine the effect of solution path length, concentration, and color on the fraction of light intensity transmitted through a solution.

Method

Current as measured from a solar cell in this experiment depends directly on the amount of light reaching the cell (*see* Chapter 8). The intensity of a monochromatic laser beam is varied by passing the beam through solutions that vary in path length, concentration, and color in this experiment.

The fraction of light intensity transmitted through a solution (see student directions) is the transmittance of the solution and the –log(transmittance) is the absorbance. The –log(transmittance) is directly proportional to path length and to concentration, where the proportionality constant is a function of the wavelength and the color of the solution. This relationship is commonly referred to as Beer's law. *See* Demonstration 8.4 for data using different path lengths and concentrations.

2725–X/93/0419$06.00/1

Materials

Laser (670 nm, 5 mW)

Small portion of a solar cell (Radio Shack), approximately the size of the laser beam, with wires soldered to opposite sides. To trim a solar cell, score once with a diamond pencil (Aldrich) and then snap along the line. *See* Figure 1. Alternatively a larger mounted solar cell (Edmund Scientific) can be masked with black electrical tape to leave only a beam-size portion of the solar cell showing. Packaged solar cells are more robust in student hands and are easier to keep aligned in use.

Digital multimeter (ammeter)

Disposable plastic cuvettes (1 cm × 1 cm)

0.15 M $CuSO_4$ (37.5 g of $CuSO_4 \cdot 5H_2O$/L). Add a few drops of H_2SO_4 if the solution is cloudy.

0.03 M $CuSO_4$ (7.5 g of $CuSO_4 \cdot 5H_2O$/L)

Dark red aqueous solution, prepared by adding red food coloring to water (McCormick red food color does not absorb the 670-nm laser light, and is a mixture of FD&C Red No. 40 and FD&C Red No. 3).

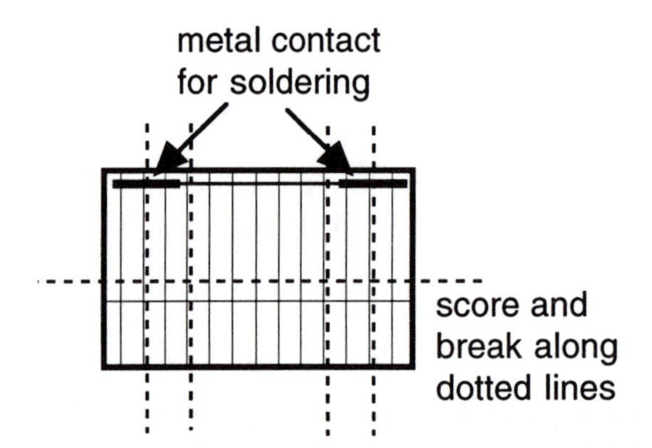

Figure 1. A Radio Shack solar cell can produce four usable pieces (two from the upper left and two from the upper right), because each piece must contain enough of the metal contact to solder a connection to the front. The entire back is metal-coated so the second wire can be soldered anywhere on the back side; the cell is easiest to tape down if the two wires are soldered to opposite ends of the piece.

Absorption of Light

Purpose

To determine the effect of solution path length, concentration, and color on the fraction of light intensity transmitted through a solution.

Procedure

Wear eye protection.

Clamp or fasten a small laser pointer and solar cell with attached wires such that the laser beam strikes the solar cell, as shown in Figure 1. One way to do this is to tape the laser and the wires to the bench top. **CAUTION: Do not look into the laser or aim the laser at another person, as it can damage the eye.**

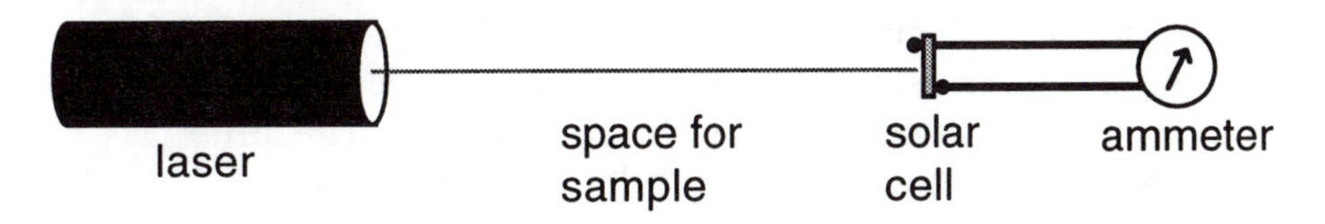

Figure 1. Arrangement of equipment.

The solutions to be studied are put in a 1-cm × 1-cm plastic cuvette, and the cuvette is placed in the beam. Both the sample cuvette and the solar cell should be oriented perpendicular to the laser beam.

Current from a solar cell depends directly on the amount of light reaching the solar cell; to measure light intensity, connect the solar cell to a digital multimeter reading milliamps. To measure the fraction of light intensity transmitted through a sample placed in the beam, you will also need to record the value obtained when your sample is not present (a blank):

$$\text{fraction of light intensity transmitted} = \frac{\text{milliamps for sample}}{\text{milliamps for blank}}$$

The solar cell also responds to changes in room light. Using a small solar cell helps, but try to maintain constant lighting in the room.

NOTE: This experiment was written by Ann M. Cappellari, Frederick H. Juergens, and Arthur B. Ellis, Department of Cheimstry, University of Wisconsin—Madison, Madison, WI, 53706; and George C. Lisensky, Department of Chemistry, Beloit College, Beloit, WI, 53511.

Investigations

1. How does the path length of a 0.03 M copper sulfate solution affect the amount of light reaching the solar cell? Keep the concentration of the solution constant and vary the path length by using more than one cuvette.
2. How does the concentration of the copper(II) solution affect the amount of light reaching the solar cell? Keep the path length of the solution constant by using only one cuvette and vary the concentration by diluting a 0.15 M copper sulfate solution.
3. Some light intensity may also be lost by reflection at the interfaces (where the index of refraction changes) even for clear solutions. The beam encounters interfaces in going from air to plastic, plastic to water, water to plastic, and plastic to air. Keeping the cell perpendicular to the beam minimizes reflections, but a better choice of blank can help: measure the milliamps for the blank when a water-filled cuvette is in the beam. Using a water-filled cuvette as the blank, measure the change in light intensity due to only the presence of copper sulfate in the solution. Measure the change in light intensity due to only the presence of red food coloring in the solution. Is there a difference in how much light passes through the blue solution and through the red solution? Does color matter?

Questions

Plot the current as a function of the tested variables. Also plot log(current) as a function of the tested variable. Which is more linear?

Write an equation to describe the relationship between light intensity (solar cell current) and the variables you tested.

If you measured a 50% reduction in light intensity when the beam passes through a solution of copper sulfate in a single cuvette in your apparatus, what would be the concentration of the copper sulfate?

Would you suggest any changes in the experimental procedure if you wanted to measure concentrations of red food coloring in a solution?

Experiment 10

A Shape Memory Alloy, NiTi

Kathleen R. C. Gisser, Margret J. Geselbracht, Ann Cappellari, Lynn Hunsberger, Arthur B. Ellis, John Perepezko, and George C. Lisensky

Notes for Instructors

Purpose

To explore the physical properties of NiTi (nitinol) and to determine the transition-temperature range of the NiTi solid-state phase change.

Method

The temperature (and phase) of the sample is adjusted by placing a sample in a water bath. NiTi in the low-temperature phase makes a soft "thud" sound when dropped on a surface; in the high-temperature phase it makes a louder "ring" sound. *See* Figure 1. This feature allows the temperature range over which a NiTi sample changes its phase to be determined acoustically. *See* Chapter 9.

Materials

NiTi wire (*see* Supplier Information)

Low-temperature phase NiTi rods (*see* Supplier Information)

High-temperature phase NiTi rods (*see* Supplier Information)

Beakers

Hot plates

2725–X/93/0423$06.00/1

Tongs

Candles and matches

String (thin string or thread)

Foam cups

Liquid nitrogen

Results

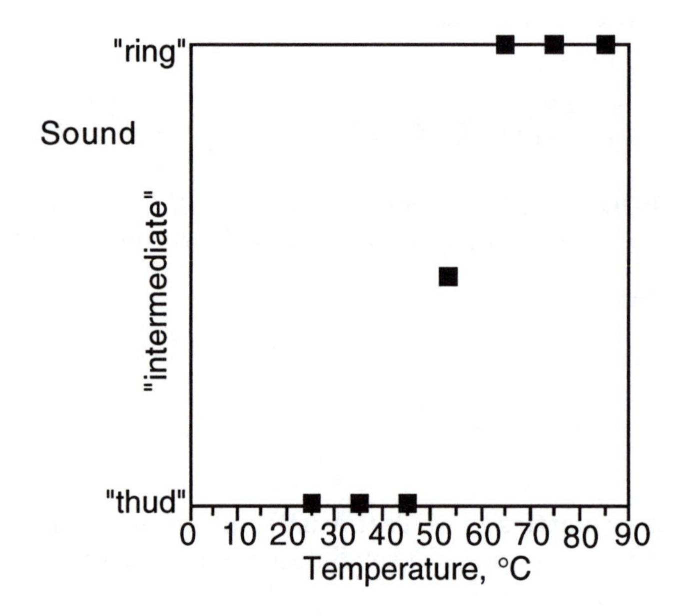

Figure 1. Typical results for the transition temperature study. The phase transition usually occurs over about a 10°C interval.

A Shape Memory Alloy, NiTi

Purpose

To explore the physical properties of NiTi (nitinol) and to determine the transition-temperature range of the NiTi solid-state phase change.

Introduction

Nitinol is an acronym for an alloy of nickel and titanium that was discovered in the early 1960s at the Naval Ordnance Laboratory (Nickel Titanium Naval Ordnance Laboratory). Its relative inertness to many chemical reagents led scientists to study nitinol's value as a rust-resistant alloy for ships. However, it was also discovered that the alloy could be mechanically deformed, and by applying heat, restored to its original shape. Compounds that "remember" their original shape are called shape memory alloys. NiTi is probably the most well-known of the shape memory alloys; other such alloys include gold–cadmium, copper–aluminum, and copper–aluminum–nickel.

The shape to which a piece of NiTi will return (or "remember") is set into the alloy by annealing at temperatures of 500–550 °C. At these temperatures, the nickel and titanium atoms crystallize in the CsCl structure, as shown in Figure 1. Each titanium atom is surrounded by eight nickel atoms, and each nickel atom is surrounded by eight titanium atoms. This high-temperature crystalline phase is called austenite. When the sample is cooled to lower temperatures, the nickel and titanium atoms shift slightly to form a less symmetric crystalline structure called martensite. The transition between these two crystal structures is an example of a solid–solid phase transition. If the alloy is bent while in the martensite phase, the original shape of the metal can be recovered by heating the sample back to the austenite phase.

The temperature range over which the alloy passes from the martensite phase to the austenite phase (or vice versa) will vary, depending upon the composition of the alloy. Although the ratio between nickel and titanium atoms in NiTi is close to 1:1, even a slight deviation from this value can cause a noticeable change in the temperature of the transition (Figure 2).

NOTE: This experiment was written by Kathleen R. C. Gisser, Margret J. Geselbracht, Ann Cappellari, Lynn Hunsberger and Arthur B. Ellis, Department of Chemistry, University of Wisconsin—Madison, Madison, WI 53706; John Perepezko, Department of Materials Science and Engineering, University of Wisconsin—Madison, Madison, WI 53706; and George C. Lisensky, Department of Chemistry, Beloit College, Beloit, WI 53511.

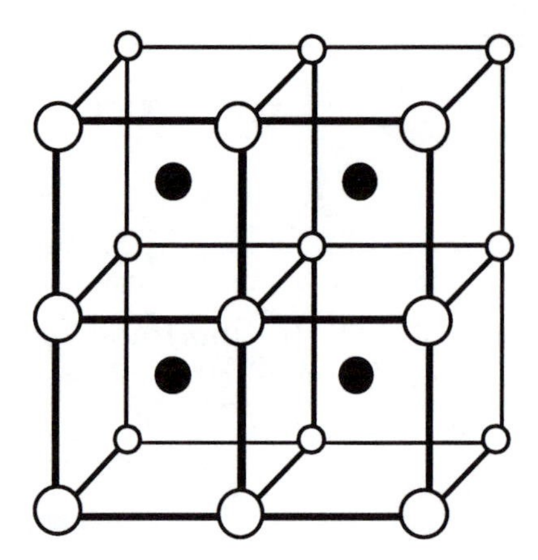

Figure 1. Four unit cells of the crystalline structure of the high-temperature phase of nitinol. Filled circles represent nickel atoms, and open circles represent titanium atoms (or vice versa).

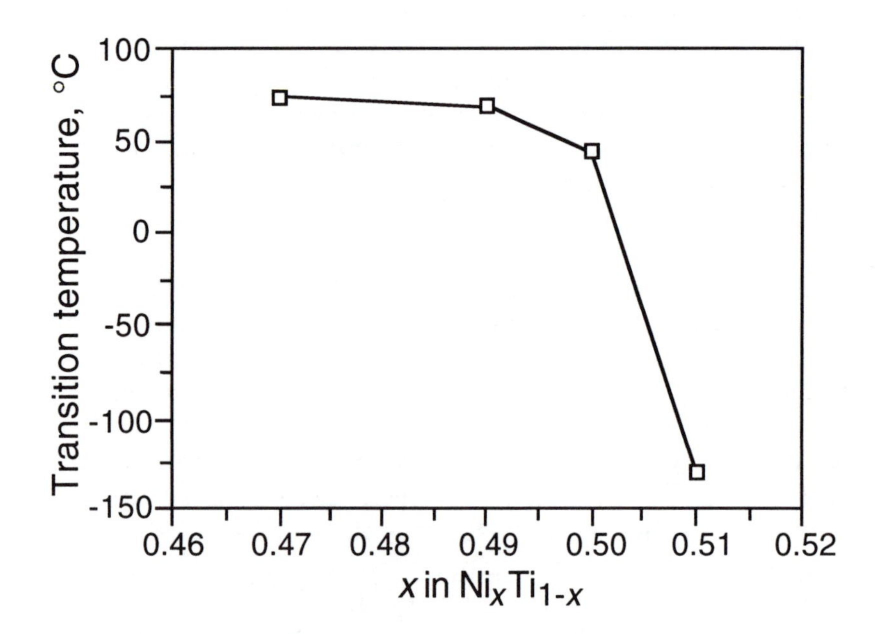

Figure 2. The effect of nickel concentration on the transition temperature. Data from a variety of sources using a variety of methods. (Adapted with permission from Murray, J. L. In *Phase Diagrams of Binary Nickel Alloys*; Nash, P. Ed.; ASM International: Materials Park, OH, 1991; p 345.

In general, the transition temperature decreases as the amount of nickel increases. Therefore, depending upon the exact composition of the alloy, a sample of NiTi could exist in either the austenite or martensite phase at room temperature.

Procedure

Wear eye protection.

Physical Properties of NiTi

Warm some water to 50–60 °C. Obtain a piece of NiTi *wire*, and bend it into a new shape. Dip the wire into the warm water. What happens? Try it again after the wire cools. (What happens if you gently try to bend the wire while it is still hot?)

Grasp both ends of the wire. Place the center of the *wire* in the center of a candle flame, and try to bend the wire into a V-shape with the vertex of the V in the flame. Initially the wire will resist bending. However, as it becomes hot, it will deform into a V-shape. When it reaches this point, remove it from the flame immediately. (Do not hold the wire in the flame long enough for the ends of the wire to get hot.) Cool the wire by waving it in the air or blowing on it. After the wire has returned to room temperature, straighten it into a new shape. Dip the wire into warm water. What happens?

Obtain two NiTi *rods*, one in the low-temperature phase and one in the high-temperature phase. Compare the physical properties of the two phases. How are they similar? How do they differ? Gently try to bend each phase. What happens?

Drop each of the rods on your bench top. What do you notice? Can you identify the phase by the sound?

Tie a string around the low-temperature-phase rod, dip it in warm water (50–60 °C) for a few minutes, then quickly remove the rod and drop it on your bench top. What happens and why?

Tie a string around the high-temperature-phase rod, and dip it in an insulated cup containing liquid nitrogen for a few minutes, then quickly remove the rod and drop it on your bench top. **CAUTION: Liquid nitrogen is extremely cold: 77 K = –196 °C = –321 °F. Do not spill it on your skin or clothing. Severe frostbite or freezing of the flesh can occur. Remove clothing that becomes saturated with liquid nitrogen, because the liquid may be held within the spaces in the fabric and thus freeze the skin underneath.** What happens and why? If you wish to try to bend the rod while cold, use a stack of paper towels as insulation. **CAUTION: Do not touch the cold rod with your bare hands, as frostbite may result.**

Determining the Transition Temperature Range of NiTi

Now that you can recognize the two phases of NiTi by their physical properties, estimate the transition temperature for the solid-state phase change. By dropping the rod, monitor the sound of a low-temperature-phase *rod* as it is slowly warmed in a water bath. Plot your data as relative "ring" or "thud" (y-axis) as a function of temperature (x-axis). Then reverse the process by collecting the same data as the rod is slowly cooled in a water bath and plotting on the same graph as before. What is the transition-temperature range for the change from the low-temperature phase to the high-temperature phase? What is the transition temperature range for the change from the high-temperature phase to the low-temperature phase?

As you are designing and performing experiments, you might consider the following points in order to get good results:

a. A beaker resting on a hot plate may not be at the same temperature as the water inside, so both the thermometer and the rod should be suspended in the water. Should the water be stirred? How long should the rod be left in the water to adjust to the temperature of the water bath? How quickly should the rod be dropped once it is removed from the water?

b. The rod should be dropped from a roughly reproducible initial position that is approximately parallel to the surface it will strike and from a height that ensures an easily recognizable sound.

c. You may find it useful to use your high-temperature-phase rod as a reference.

d. To slowly increase the temperature of water in a beaker, the hot plate will need to be on the lowest setting.

e. Recording data every 5 to 10 °C should be sufficient.

Questions

1. a. Draw a curve through the points on your graph that represent the martensite-to-austenite phase transition. Label the phase-transition-temperature range on your graph.

   b. Draw a curve through the points on the graph that represent the austenite-to-martensite phase transition. Label the phase-transition temperature range on your graph.

   c. How does the curve for the martensite-to-austenite phase transition compare to the graph representing the austenite-to-martensite phase transition? Are the curves the same?

2. Compare your phase-transition temperatures with those obtained by other students in the class. How do these values compare with yours? How might you account for any differences?

3. On the basis of the properties of NiTi, list several (at least three) possibly useful applications for this metal. Be creative!

Experiment 11

A High-Temperature Superconductor, $YBa_2Cu_3O_{7-x}$

M. Stanley Whittingham

Notes for Instructors

Purpose

To prepare $YBa_2Cu_3O_{7-x}$ and observe its properties in liquid nitrogen.

Method

Stoichiometric amounts of Y_2O_3, BaO_2, and CuO are ground together and heated at high temperature. The product is cooled in liquid nitrogen and its conductivity and effect on a small magnet are observed. *See* Chapter 9.

Materials

Yttrium oxide, Y_2O_3

Barium peroxide, BaO_2

Copper(II) oxide, CuO

Furnace (capable of operating up to 1000 °C)

Pellet press

Alundum combustion boats (VWR or Fisher Chemical)

2725–X/93/0429$06.00/1

Liquid nitrogen

Foam cup

Plastic forceps

Small, strong magnet

Four-probe conductance test circuit and voltmeter (*See* Figure 6.)

A High-Temperature Superconductor, $YBa_2Cu_3O_{7-x}$

Purpose

To prepare $YBa_2Cu_3O_{7-x}$ and observe its properties in liquid nitrogen.

Introduction

H. Onnes, a Dutch physicist, discovered in 1911 that mercury loses all resistance to electrical flow when cooled to about 4 K; thus, a current once started will flow continuously. Such a phenomenon is known as superconductivity. At ordinary temperatures, metals have some resistance to the flow of electrons, because atomic vibrations scatter the electrons. As the temperature is lowered, the atoms vibrate less and the resistance declines smoothly, until, if the material can become a superconductor it reaches a so-called critical temperature, T_c. At this point, the resistance drops abruptly to zero (Figure 1). If an electrical current is started in a superconducting ring, it will continue forever.

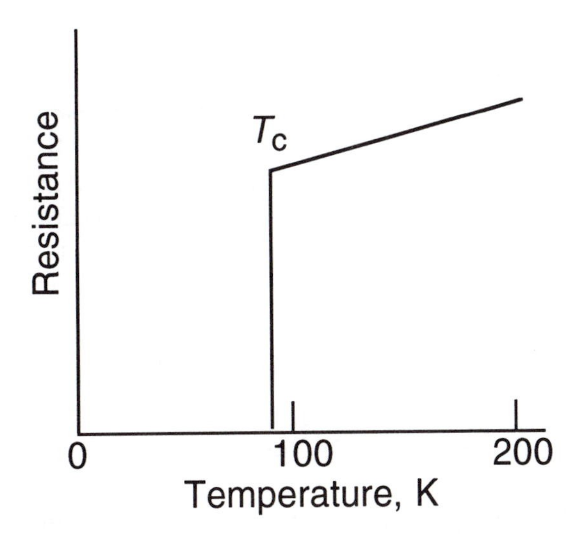

Figure 1. Electrical resistance of a $YBa_2Cu_3O_{7-x}$ superconductor.

Superconductors are also perfectly diamagnetic (i.e., they expel a magnetic field from their interior); this property was discovered in 1933 and is known as the *Meissner effect*. When a magnet approaches a superconductor, it induces a current in the superconductor. Because there is no resistance to the current, it continues to flow and thus induces its own magnetic field which then repels the magnet's field. If the magnet is sufficiently small and strong, the repulsion will be enough to

NOTE: This experiment was written by M. Stanley Whittingham, Department of Chemistry, State University of New York at Binghamton, Binghamton, NY, 19302.

counterbalance the pull of gravity and the magnet will levitate above the surface of the superconductor. In the $YBa_2Cu_3O_{7-x}$ superconductor, there is actually some penetration of the magnetic field that helps to provide a stable levitation position for the magnet above the superconductor pellet.

Even though until recently the highest known critical temperature was only 23 K, observed in the intermetallic compound Nb_3Ge, superconductors found a number of applications. The most common of these is for superconducting magnets for nuclear magnetic resonance instruments used for medical imaging and in the research laboratory; you may find these in many chemistry departments. These instruments require liquid helium as the refrigerant, which is scarce and expensive. In a major breakthrough in 1986 (which resulted in a Nobel Prize in 1987), J. G. Bednorz and K. A. Müller at the IBM research labs in Zurich discovered superconductivity at over 30 K in copper-containing oxides. A massive worldwide research effort culminated in the discovery of a metallic oxide of yttrium, barium, and copper that was superconducting at about 92 K. This meant that liquid nitrogen (b.p. 77 K), which is cheaper and easily handled in a Dewar flask (the laboratory version of a Thermos bottle), could be used as a coolant rather than liquid helium. Related oxides have now been found that superconduct at temperatures up to 130 K.

The superconducting compound, $YBa_2Cu_3O_{7-x}$, is readily prepared by heating an intimate mixture of yttrium oxide, barium peroxide, and cupric oxide at approximately 930 °C for 10–12 hours. In this stage the crystalline structure is formed by the interdiffusion of ions, but it has a deficit of oxygen ($x \approx 0.5$). By cooling the material to 500 °C and annealing at this temperature for 10–12 hours, it reacts further with oxygen from the air; x is reduced to less than 0.1. The overall reaction (with only the metal ions balanced) is given by

$$1/2\ Y_2O_3 + 2\ BaO_2 + 3\ CuO \rightarrow YBa_2Cu_3O_{7-x}$$

This compound is often called the "1-2-3" material from the molar ratios of Y:Ba:Cu. The heating–cooling synthesis sequence is shown graphically in Figure 2.

The unit cell structure of $YBa_2Cu_3O_7$ is shown in Figure 3; although it appears complex, its basic building block is the simple perovskite structure. The main feature of the structure that is thought to be important in the superconductivity is the existence of the CuO two-dimensional sheets, extending through the material. Identify these sheets in Figure 3.

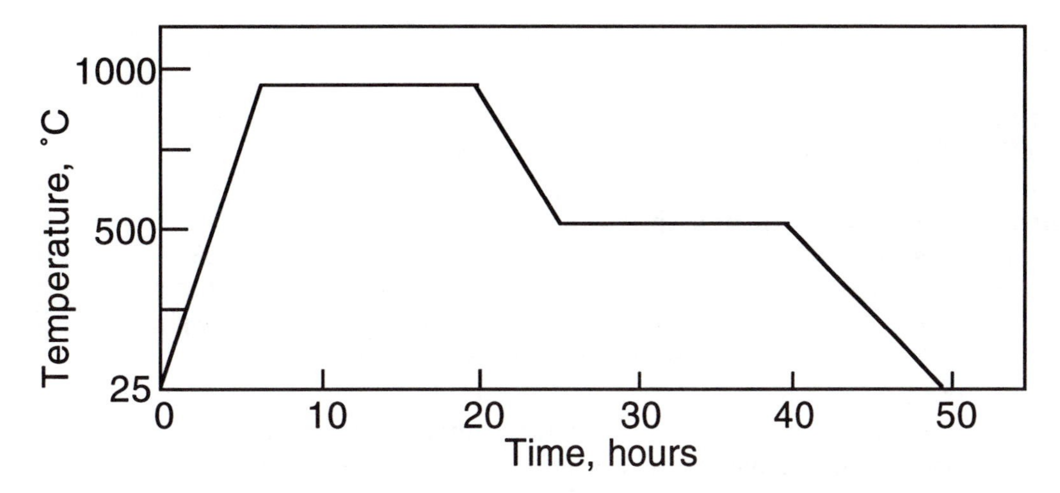

Figure 2. Heating-cooling sequence for synthesis of the 1-2-3 superconductor.

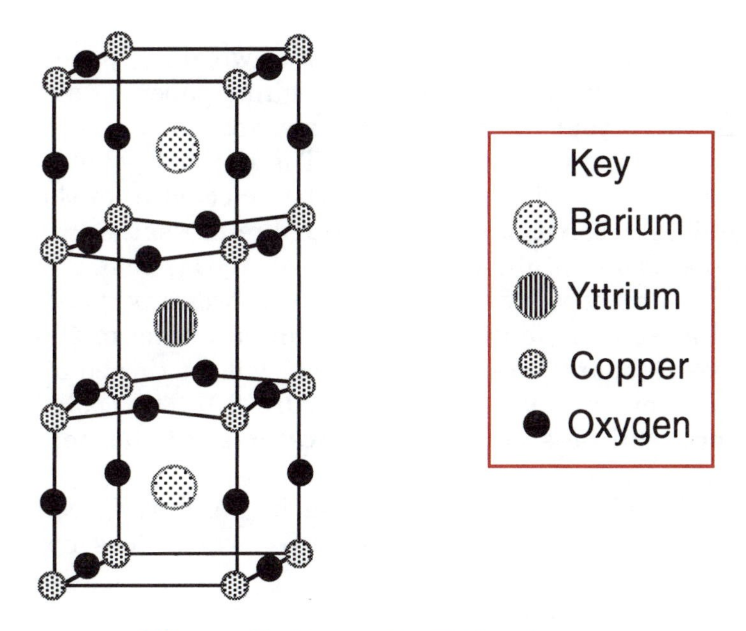

Figure 3. Unit cell of $YBa_2Cu_3O_7$.

Warm-up Exercises

1. Calculate the weights of BaO_2 and CuO required to react *stoichiometrically* (1Y:2Ba:3Cu) with 0.60 g of Y_2O_3 to produce $YBa_2Cu_3O_7$. (Do not balance the amount of oxygen.)
2. What is a superconductor? Discuss two major physical properties usually associated with superconducting solids.

Procedure

Wear eye protection.

CAUTION: The chemicals used in this experiment are toxic. Avoid creating or breathing dust when grinding. Avoid eye and skin contact. Wash your hands thoroughly after handling.

Weigh out onto a piece of weighing paper 0.60 g of yttrium oxide, Y_2O_3, and transfer it to a small *dry* beaker. Weigh out stoichiometrically equivalent amounts (you calculated these as a warm-up exercise) of barium peroxide, BaO_2, and cupric oxide, CuO, transferring each in turn to the same beaker.

These three materials must now be thoroughly mixed to obtain good results. *In a hood* place the powdered materials in a mortar. Mix and grind the material with a pestle for about 10 minutes; use a spatula to scrape the material off the sides of the mortar when it cakes there. Your final powder should be a uniform color with no lumps and no black or white spots or patches visible. What color is your powder? Why must you mix the starting powders?

Scrape the powdered mixture onto a creased piece of weighing paper. Divide the mixture in half using a second piece of weighing paper. Make two pellets with this powder: your instructor will show you how to press each mixture into a pellet using the pellet press. (The pressed pellet is quite fragile and may shear crosswise or crumble when ejected from the die. If it shears or crumbles, crush it in your mortar and re-press it.)

The pellets will be placed in an alundum (a form of Al_2O_3) boat and heated in a furnace. The furnace will be heated to 930 °C over a period of about 8–12 hours, held at 930 °C for 12–16 hours, allowed to cool to 500 °C and held there for 12–16 hours. Finally the furnace will be turned off and allowed to cool to room temperature. (See Figure 2.) The cooled pellets, in their beakers, will be stored in a desiccator until the next laboratory period.

The finished pellet should be dark gray to black. A dark green material is a second phase, of composition Y_2BaCuO_5, which does not superconduct.

Model Building

The properties of a superconductor obviously will depend on the structure and packing of the atoms in its crystalline form. X-ray and neutron diffraction studies were most important in the elucidation of the 1-2-3 oxide structure, which belongs to a structural family known as perovskites. Perovskites, ABO_3, generally have a ratio of two metal atoms, AB, for three oxygen atoms.

Perovskite Structure

The perovskite structure is named after the mineral $CaTiO_3$. This structure is made up by corner-sharing of TiO_6 octahedra with Ca ions in the large cavities at the corners of the unit cell.

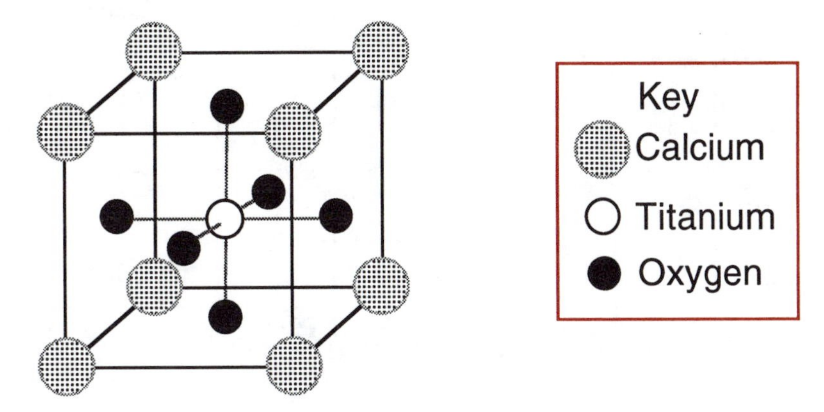

Figure 4. The perovskite structure of $CaTiO_3$.

Study the structure in Figure 4 or use a model (for the SSMK, build alternate perovskite using template C) to answer the following questions.

1. What is the coordination number of:

   Calcium______ Titanium______ Oxygen______

2. Show that the unit cell of this compound corresponds to the formula $CaTiO_3$. Remember that an ion shown as part of a unit cell does not contribute a whole ion to the cell unless it is wholly enclosed within the cell, as the Ti ion is here. (Remember: How many unit cells is each atom in?)

3. Determine the oxidation state of the Ti if Ca and O have their normal oxidation states.

4. How does the structure of WO_3 differ from that of $CaTiO_3$?

Structure of the Superconductor $YBa_2Cu_3O_7$

The 1-2-3 superconductor has a structure similar to perovskite. The resulting unit cell consists of three stacked cubic unit cells with Y and Ba sharing the Ca positions, and Cu taking the Ti positions; it is considered to be orthorhombic rather than cubic, having an almost square base but rectangular sides (a = 3.817 Å, b = 3.882 Å, c = 11.671 Å, where a, b, and c are lengths of the sides of the unit cell.)

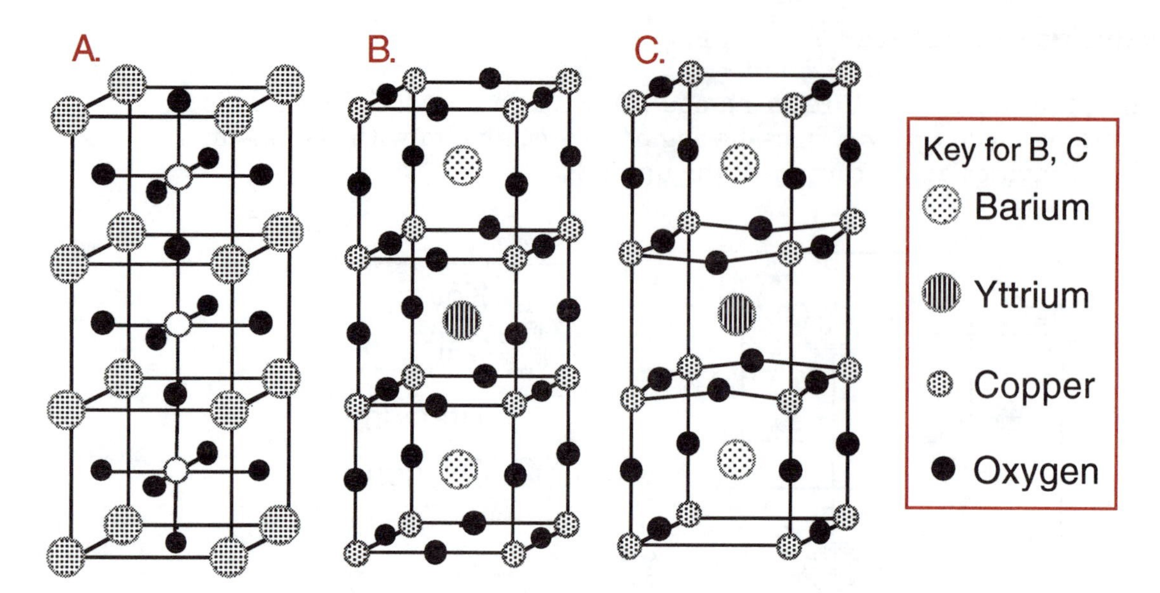

Figure 5. Idealized structure of $YBa_2Cu_3O_7$ obtained from X-ray diffraction data, showing evolution from the perovskite structure shown in Figure 4. A: Stacking of three perovskite units. B: Shift of origin. C: Removal of oxygen to give correct chemical composition. (Reprinted with permission from Whittingham, M. S. *Mater. Res. Soc. Bull.*, p 41, August 1990.)

Study the structures in Figure 5 or use a model (for the SSMK use template C and modify the directions for perovskite) to answer the following questions.

1. Show that the unit cell obtained by X-ray diffraction analysis corresponds to the formula $YBa_2Cu_3O_7$.
2. The copper in $YBa_2Cu_3O_7$ may be considered to be a mixture of +2 and +3 oxidation states. If Y has a +3 oxidation state and Ba and O have their normal oxidation states, what fraction of the copper is in each of the two oxidation states?

Properties of a Superconductor

Meissner Effect

Using the *plastic* forceps, remove the superconductor pellet from the beaker and place it in a cut-off foam cup. (If there is loose material on the pellet scrape it off gently with a spatula. If the pellet has sheared, use the thickest piece, and place it flat side up.) Then place a small magnet on top of the pellet with *plastic* forceps.

Obtain some liquid nitrogen in a second foam cup. **CAUTION: Liquid nitrogen is extremely cold: 77 K = –196 °C = –321 °F. Do not spill it on your skin or clothing. Severe frostbite or freezing of the flesh can occur. Remove clothing that becomes saturated with**

liquid nitrogen, because the liquid may be held within the spaces in the fabric and freeze the skin underneath.

Carefully pour some of the liquid nitrogen into the cut-off cup, covering the pellet. Some of it will boil away as the pellet, magnet, and cup cool; add more as necessary. When the pellet cools below the critical temperature, the magnet should "levitate" above the superconductor. Touch it gently with your forceps and it should spin.

Allow the liquid nitrogen to evaporate and the pellet and magnet to warm to room temperature.

Loss of Electrical Resistance

The loss of resistance below the critical temperature can be measured by measuring the voltage drop across the pellet in a circuit. The four-probe apparatus used is shown schematically in Figure 6. A battery generates a current, I, which passes through the pellet. The voltage drop, $V = IR$, along the pellet due to the pellet's resistance, R, at room temperature is measured with the voltmeter. When the material becomes superconducting, $R \to 0$, so $V \to 0$, and no voltage drop should be measured. (The resistor in the circuit prevents the battery from being shorted when the superconductor loses its resistance. A four-probe technique eliminates contact resistances.)

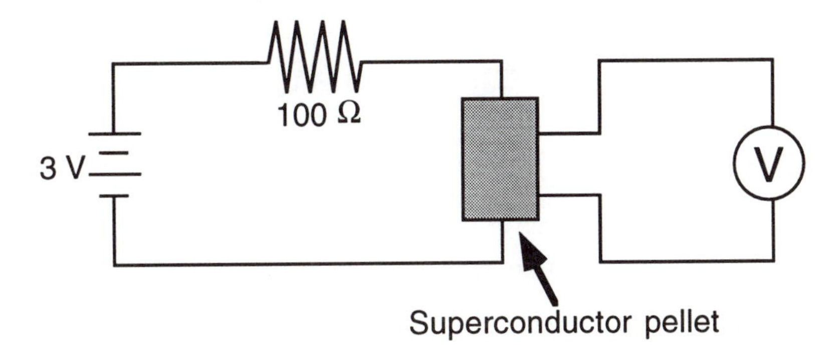

Figure 6. The four-probe conductance test circuit.

Carefully insert the pellet into the four-pronged holder. Connect the outer leads to the battery and the inner leads to the voltmeter.

After placing the holder in the foam cup, pour liquid nitrogen slowly and carefully into the cup as in the Meissner experiment. **CAUTION: The resistor in the test circuit becomes quite hot. Do not touch the resistor or spill liquid nitrogen on it. Thermal shock could cause the resistor to break, and you could burn your fingers.** When the pellet cools below T_c, it should become superconducting. What happens to the voltmeter reading at this point?

Disassemble the experiment as follows: (1) Disconnect the battery to prevent overheating the resistor, (2) let the nitrogen evaporate, (3) let the sample and holder return to room temperature, and (4) remove the sample from the holder.

Questions

1. Discuss your observations. Does your product exhibit the characteristics of superconductivity? Does it appear to be uniform? If not, what do you hypothesize went wrong? How would you identify the presence of additional phases?
2. Why did you compress your reactants into pellets for your solid-state reaction? Hint: What is the difference between reactions in solution and in the solid state?
3. The dream of the materials chemist is to discover a material that would be superconducting above room temperature, so that no refrigerant would be required. Suggest how the properties of such superconductors could be valuable in each of the following fields: computers, electronics, power generation, power transmission, and rail transportation.

Experiment 12

A Solid Electrolyte, Cu_2HgI_4

William R. Robinson

Notes for Instructors

Purpose

To synthesize copper(I) tetraiodomercurate(II), Cu_2HgI_4, and study its changes in color and conductivity as the compound undergoes a phase transition upon heating.

Method

Bright red, thermochromic Cu_2HgI_4 is synthesized from $CuSO_4$, KI, and $Hg(NO_3)_2$; when heated, the compound changes to a dark brown color, but returns to its original red color when it is allowed to cool. The resistance of Cu_2HgI_4 dramatically decreases when it is heated as a result of enhanced ion mobility in the high-temperature structure.

Materials

0.5 M Copper (II) sulfate solution

1.0 M Potassium iodide solution

6 M Acetic acid solution

Sodium sulfite, Na_2SO_3

0.05 M Mercury(II) nitrate solution

2725–X/93/0439$06.00/1

Stirring hot plate

Filter paper, Büchner funnel, suction flask, and water aspirator

Distilled water

Glass capillary tubes, 1.5 × 100 mm (one end sealed), optional

Glass capillary tubes, 1.5 × 100 mm (open both ends)

18-Gauge copper wire

Ohmmeter

Rubber band

Solid-State Model Kit

A Solid Electrolyte, Cu_2HgI_4

Purpose

To synthesize copper(I) tetraiodomercurate(II), Cu_2HgI_4, and study its changes in color and conductivity as the compound undergoes a phase transition upon heating.

Introduction

Cu_2HgI_4 is thermochromic; it reversibly changes color with temperature. At low temperatures this compound is bright red, and at high temperatures it is dark brown. In the low-temperature form, the Cu^+ and Hg^{2+} ions are arranged in separate alternating layers, packed between layers of I^- ions (Figure 1). In the high-temperature form, the I^- ions occupy the same positions as before, but the metal ions now randomly occupy all the tetrahedral holes in the iodide array. The transition from one crystal form to the other takes place fairly sharply at a distinct temperature.

In the high-temperature form, the number of metal ions is smaller than the number of positions among which these ions are distributed. Thus, it is easy for the metal ions to move through the crystal by simply moving into unoccupied positions. In the high temperature form the metal ions diffuse through the crystal much like they would diffuse through an aqueous solution. With suitable equipment we can easily demonstrate that the electrical conductivity of the high-temperature form is much larger than that of the low-temperature form.

Copper(I) tetraiodomercurate(II) is prepared by the reaction of copper(II) sulfate with potassium iodide, followed by the addition of mercury iodide. The copper(II) ion is reduced by the iodide ion, forming solid CuI (*see* the Appendix).

$$2Cu^{2+}(aq) + 4I^-(aq) \rightleftarrows 2CuI(s) + I_2(aq)$$

In the presence of excess iodide ion, the iodine undergoes further reaction to form the triiodide ion.

$$I_2(aq) + I^-(aq) \rightleftarrows I_3^-(aq)$$

The net ionic equation therefore becomes,

$$2Cu^{2+}(aq) + 5I^-(aq) \rightleftarrows 2CuI(s) + I_3^-(aq)$$

NOTE: This experiment was written by William R. Robinson, Department of Chemistry, Purdue University, West Lafayette, IN 47907.

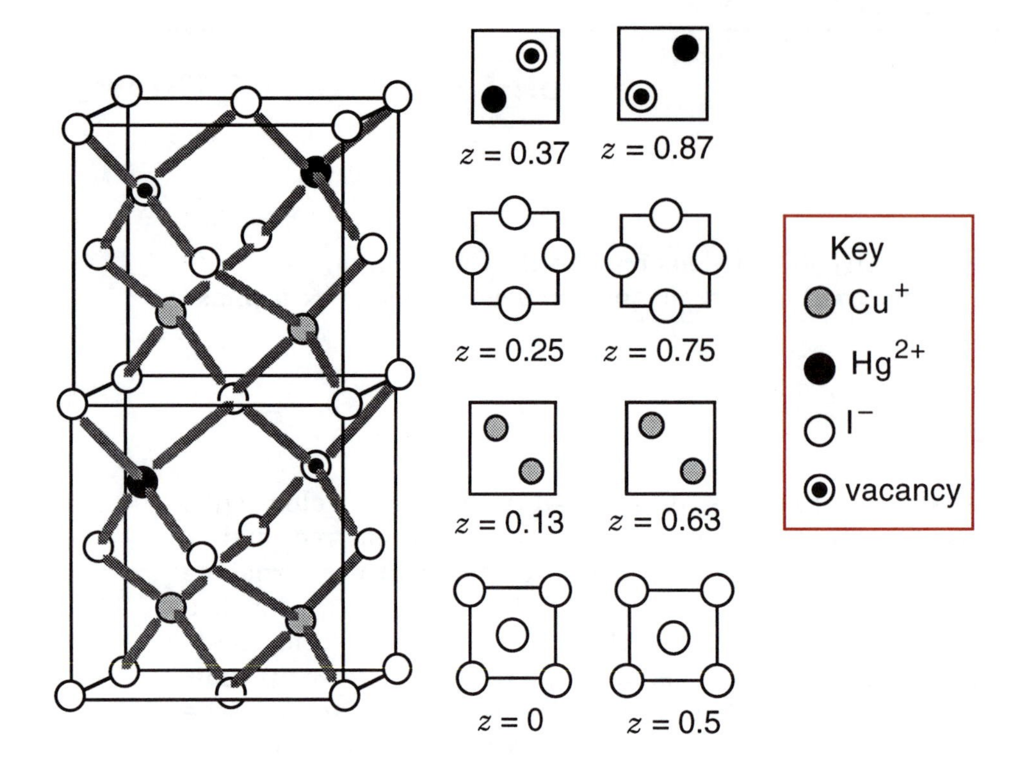

Figure 1. The structure of the low-temperature form of Cu_2HgI_4.

Sodium sulfite is added to this mixture to reduce the triiodide ion back to iodide,

$$I_3^- (aq) + SO_3^{2-} (aq) + 3H_2O \rightarrow 3I^- (aq) + SO_4^{2-} (aq) + 2H_3O^+$$

The solid copper(I) iodide can be separated from the reaction mixture by carefully pouring off the excess solution of supernatant liquid.

A mercury(II) iodide precipitate is synthesized by the reaction of mercuric nitrate with potassium iodide,

$$Hg^{2+} (aq) + 2I^- (aq) \rightleftarrows HgI_2 (s)$$

Finally, Cu_2HgI_4 is prepared by adding the solid copper(I) iodide to the mixture containing the mercury(II) iodide precipitate.

Procedure

Wear eye protection.

Synthesis of Cu_2HgI_4

CAUTION: The chemicals used in this experiment, particularly mercury-containing compounds, are toxic. Avoid

creating or breathing dust. Avoid eye and skin contact. Wash your hands thoroughly after handling.

Add 5 mL of 0.5 M $CuSO_4$, 6 mL of 1 M KI, and several drops of 6 M acetic acid to 50 mL of deionized water in a 150-mL beaker. A precipitate will form. Weigh out about 0.2 g of Na_2SO_3. Dissolve this Na_2SO_3 in about 10 mL of water and add the solution, stirring continuously, to the beaker containing the CuI precipitate. Allow the precipitate of copper(I) iodide to stand for several minutes. Pour off as much of the supernatant solution as possible without losing more than a few percent of the precipitate.

Add 25 mL of 0.05 M $Hg(NO_3)_2$, 3 mL of 1 M KI, and 100 mL of deionized water to a 250-mL beaker. Add the suspension of copper(I) iodide just prepared, using a stream of deionized water from a wash bottle to wash all of the CuI from the beaker.

Heat the mixed suspension of CuI and HgI_2 almost to the boiling point for about 20 minutes on a hot plate while stirring with a magnetic stirrer. During this digestion period, a single dark-brown solid should form. Suction filter this solution while it is still hot. Wash the precipitate with portions of acetone. **CAUTION: Acetone is quite volatile and flammable. No open flames should be present.**

Let the solid air-dry for 10 minutes.

Determination of the Transition Temperature

The transition temperature of Cu_2HgI_4 lies between 40 and 90 °C, a temperature that can be conveniently studied using a water bath. You should be able to determine the transition temperature by finding the temperature at which a color change takes place as a sample of Cu_2HgI_4 cools. A small amount of Cu_2HgI_4 can be rubbed onto the surface of a piece of filter paper, and the filter paper can then be placed into a beaker of water of known temperature. It is convenient to start with a beaker of hot water and to approach the transition temperature with the stepwise addition of cold water. Record the measured transition temperature. An alternative approach to the determination of the transition temperature involves packing a small amount of dry Cu_2HgI_4 precipitate into a piece of capillary tubing that has one end sealed. The capillary tube is used as a small test tube and immersed in a hot-water bath. The color is observed as the temperature of the water bath is decreased by the addition of cold water.

Comparison of Electrical Conductivities

Insert a straight piece of copper wire (18-gauge) into a glass capillary tube (Figure 2). The fit should be tight. Push the open end of the tube into some Cu_2HgI_4 so that some of the material is stuck in the tube. Flip the tube upside down and tap the copper wire end against a hard surface so that the Cu_2HgI_4 falls to the wire. Repeat this tapping until there is about 1 cm of the material in the tube. Finally, insert another length of copper wire into the open end of the glass capillary, so the Cu_2HgI_4 is packed between the two copper electrodes. In order to ensure a tightly packed tube, carefully bend the copper wire electrodes and place a rubber band around the electrodes:

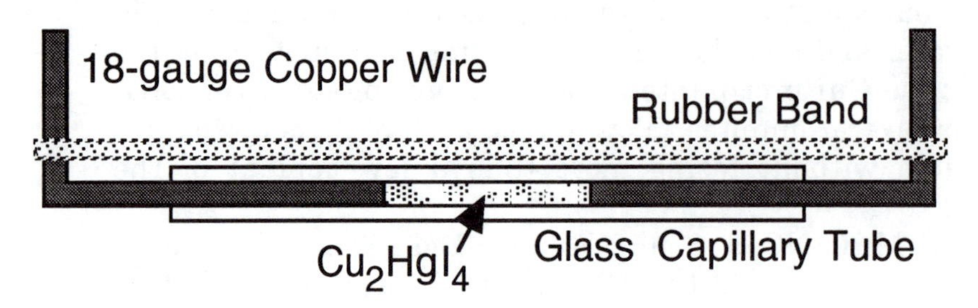

Figure 2. Apparatus for measuring conductivity.

Using an ohmmeter (adjust the meter to read in the 1–10-MΩ range), measure the electrical resistance between the electrodes of the Cu_2HgI_4 sample. Heat the tube with a heat gun (you may use a match instead, but try not to blacken the tube) and measure the electrical resistance between the electrodes. How does the conductivity change as the color changes?

Dispose of the Cu_2HgI_4 in the waste container provided.

The Crystal Structures of Cu_2HgI_4

At room temperature, Cu_2HgI_4 crystallizes in a tetragonal unit cell with unit cell lengths $a = b = 6.09$ Å and $c = 12.24$ Å. All of the angles in a tetragonal cell are 90°. Note that the c-axis length is almost double the a and b axis lengths (the unit cell is almost made from two cubes) as shown in Figure 1.

Study Figure 1 or build a model of the structure of the low-temperature form of Cu_2HgI_4. (For the SSMK, use template D, and modify the directions for expanded face-centered cubic zinc blende.) What is the coordination number and the geometry about the Cu(I) and Hg(II) ions? Even if the iodide ions did form a perfect face-centered cubic array, the cell would not be cubic. Why not?

When Cu_2HgI_4 is heated above the transition temperature, the iodide ions move into the exact positions of a face-centered cubic array, and the Cu^+ and Hg^{2+} ions diffuse through the solid by hopping from tetrahedral site to tetrahedral site. On average throughout the crystal, each tetrahedral site contains one-fourth of a Cu^+ ion and one-eighth of a Hg^{2+} ion. An X-ray measurement of a crystal measures the average of a great many unit cells, so the crystal appears cubic with $a = b = c = 6.10$ Å. Why is the height of the unit cell only half as large as for the low-temperature form?

Build a model of the high-temperature form of Cu_2HgI_4. Use one color of spheres to represent the position of the iodide ions and second color spheres to indicate the location of the tetrahedral sites, which are either empty, occupied by a Cu^+ ion or occupied by a Hg^{2+} ion. (For the SSMK, use template D, and modify the directions for expanded face-centered cubic zinc blende.)

As copper or mercury ions diffuse through the crystal, they can move to an adjacent, empty tetrahedral hole by one of two paths. A direct jump to an adjacent tetrahedral hole is possible, or a jump to an octahedral hole followed by a jump to the second tetrahedral hole is possible. Consider the position of an ion when it is halfway between two tetrahedral holes and when it is halfway between a tetrahedral hole and an octahedral hole. Which of these two paths would require the smallest distortion of the iodide lattice? Remember that the iodide ions are touching.

Appendix. The Stability of Cu(I) Compounds

The synthesis of Cu_2HgI_4 contains some interesting chemistry. Beginning with an aqueous copper(II) sulfate solution that is treated with potassium iodide, a precipitate is formed, but not the one that might first be expected. Rather than obtaining a precipitate of copper(II) iodide, CuI_2, the precipitate analyzes as copper(I) iodide, CuI. Copper(II) is reduced to Cu(I), and some of the iodide (I^-) is oxidized to I_2.

When an element can exist in more than one oxidation state in aqueous solution, each oxidation state will have a different thermodynamic stability. The relative stability of two oxidation states in aqueous solution is most conveniently expressed in terms of the electrochemical potential for the reaction.

$$M^{a+} + (a - b)\,e^- \rightleftarrows M^{b+} \quad \text{where } b < a$$

The potential for a solution containing the ions M^{a+} and M^{b+} is given by the Nernst equation,

$$E = E^\circ - \frac{RT}{nF} \ln \frac{[M^{b+}]}{[M^{a+}]}$$

where n is the number of electrons per ion transferred at the electrode; F is the Faraday constant, 96,480 C/mol; E is the potential of the solution; E° is the standard potential; $[M^{a+}]$ is the concentration of M^{a+} ions in the solution; $[M^{b+}]$ is the concentration of M^{b+} ions in the solution; R is the gas constant; and T is the absolute temperature.

Therefore, any species added to the solution that decreases the concentration of either M^{a+} or M^{b+} and so alters the ratio $[M^{b+}]/[M^{a+}]$ will cause an observable change in the potential and in the relative stability of M^{a+} and M^{b+}. If $[M^{a+}]$ is decreased, then the observable potential will become less positive, that is, the higher oxidation state will become more stable. Alternatively, if $[M^{b+}]$ is decreased, the observed potential will become more positive and the lower oxidation state will become more stable.

Copper(I) iodide is a very insoluble salt ($K_{sp} = 5.1 \times 10^{-12}$). Thus, the addition of iodide ion to a solution of copper(I) will decrease the Cu^+ concentration and increase the ease of reduction of Cu^{2+} to Cu^+. The concentration of Cu^+ in a solution containing iodide ions is so low that even the weakly reducing iodide ion is strong enough to reduce Cu^{2+} to Cu^+.

Experiment 13

Diffusion in Solids, Liquids, and Gases

David Johnson

Notes for Instructors

Purpose

To estimate the relative diffusion rates for gases, liquids, and solids.

Method

Measure the time that it takes for diffusion to occur over a certain distance for vanilla (vapor), propylene glycol in water (liquid), and Cu^{2+} in glass (solid).

Materials

Stopwatch or timer

Vanilla extract

Red food coloring (McCormick Red Food Color contains water, propylene glycol, FD&C red No. 40 and FD&C red No. 3)

Petri dish

Short pieces of 1/4-inch o.d. glass rod (a stirring rod)

Copper(II) chloride, $CuCl_2{\cdot}2H_2O$ (m.p. 435 °C)

Test tube

2725–X/93/0447$06.00/1

Bunsen burner

Crucible tongs

Results

Estimates of the diffusion coefficient, D, in units of cm^2/s for a gas, liquid, and solid are 10^{-1}, 10^{-3}, and $< 10^{-7}$ (at 500 °C), respectively. The estimate for the liquid state can be quite large and probably indicates that some mechanism of mixing in addition to diffusion is contributing to the molecular motion observed.

Diffusion in Solids, Liquids, and Gases

Purpose

To estimate the relative diffusion rates for gases, liquids, and solids.

Introduction

Diffusion is an important chemical and physical phenomenon that has a bearing on our daily lives. Diffusion of gases and other molecules across membranes in our bodies allows many biological processes to occur and thus keeps us alive by providing oxygen and energy sources to our cells. Paint dries because water or a solvent can diffuse to the surface and evaporate. Diffusion is one of the ways in which molecules move across large distances. Molecules can also move in other ways such as when a mechanical fan is used to draw fresh air into a room.

Perhaps one of the most interesting uses of diffusion is the enhancement of the color of gemstones such as rubies and sapphires. These two gemstones, which are important for manufacturing lasers, microcircuits, and even the glass plates above supermarket checkout scanners, are aluminum oxide with various metal oxides as impurities and can be synthetically prepared. Miners of sapphires and rubies have long known that the color of poor-quality stones can be enhanced by packing them in titanium and iron oxides (for blue color) or chromium oxide (for red color) and baking them at temperatures near 2050 °C, the melting point of the aluminum oxide crystal that makes up most of the stone. This procedure turns low-grade gems into colorful, high-quality stones. Similar procedures are used in making some types of stained glass.

The process of diffusion can depend on many factors. Some of these factors are temperature, pressure, the size and mass of the diffusing species, and the state of the material (whether it is a gas, liquid, or solid). All of these factors can affect the speed at which the two materials are intermixed, as well as the distance that one material can diffuse into another. Increasing the temperature increases the kinetic energy of atoms and molecules and can dramatically increase the rate of diffusion. Within a few hundred degrees of room temperature, the rate of diffusion of ions in solution doubles for every increase of 10 °C.

The velocity of a typical molecule near room temperature is very high - 100 to 1000 m/second. Yet it can take minutes, hours, or even years for molecules to diffuse a few centimeters. How can this be?

In a gas, collisions between molecules occur about every 10^{-7} m and cause the molecules to change directions. Therefore, a long and tortuous

---

NOTE: This experiment was written by David Johnson, Department of Chemistry, University of Oregon, Eugene, OR 97403. We acknowledge helpful comments from Mark Ediger, Department of Chemistry, University of Wisconsin—Madison.

path must be followed in order to go 1 m in any particular direction.

Diffusion in a liquid is similar to diffusion in a gas except that the molecules are much closer together. They may travel only 10^{-10} m before a collision causes them to change direction. In addition, "cages" formed by surrounding molecules may trap the molecule in one place for many collisions. Thus, diffusion occurs at a slower rate. If some ink were dropped in a container of water, the molecules of the ink would undergo billions of collisions before they reached the edges of the container.

The atoms in the solid for the most part simply vibrate in a fixed position. This being the case, how can diffusion occur? The crystal structure of a solid has vacancies in it, whose concentration is governed by chemical equilibrium. (The concentration generally increases rapidly with increasing temperature.) The vacancies allow some atoms to diffuse through the solid, as shown in Figure 1.

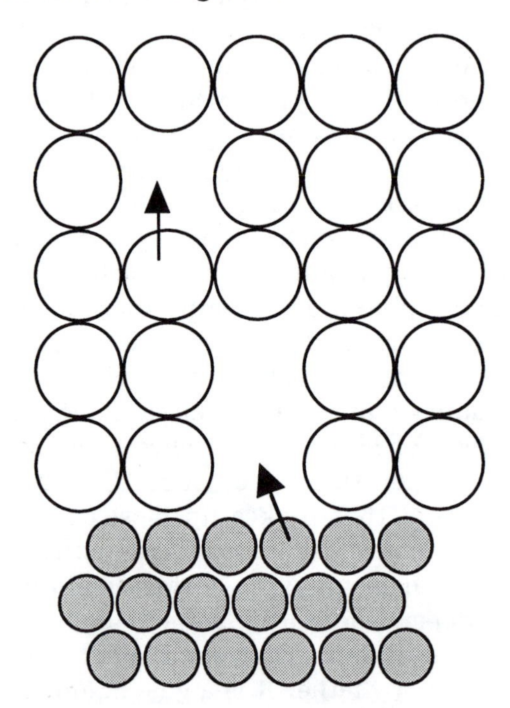

Figure 1. Solid diffusion. The atoms of one solid (open circles) can migrate into vacancies, allowing the atoms of the second solid (shaded circles) to diffuse into the first solid. If the atoms are small enough, they may be able to fill vacancies by migrating between the atoms of the first solid.

In this lab you will measure the time required for molecules, or atoms, to move over a certain distance in the gas, liquid, and solid phases. From this information you will be able to estimate the diffusion coefficient in each of these states. Mechanical mixing may occur during some measurements, and can lead to overestimates of the diffusion constant.

Warm-up Exercise

If you were to try to enhance the color of a sapphire by heat treatment (that is, by packing a piece of sapphire in titanium oxides and iron oxides and heating it to about 2000 °C), you would find that it takes 600 hours at this temperature for the color to diffuse 0.4 mm. Calculate the value of the diffusion coefficient, D, in units of cm^2/s.

Procedure

Wear eye protection.

Diffusion Between Two Gases

You will need to be in a group of four for this part of the experiment. Find an area of the lab where no one is working and where there are few or no drafts. Have each person stand in a line with 1 m between each person, as shown in Figure 2. The first person will open the bottle with vanilla and pour a few drops on a paper towel on the table or bench top.

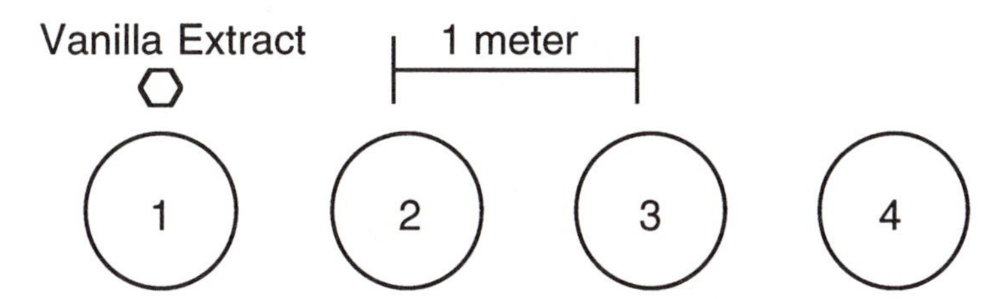

Figure 2. Arrangement of people (1–4) for the gaseous diffusion experiment.

Record the time required for each person (corresponding to a different distance) to smell the vanilla (Table 1). Discard the paper towel in a plastic bag when finished with a trial. Repeat the experiment twice more (after the odor has dissipated).

Table 1. Laboratory Data for Diffusion between Two Gases

Distance (m)	Trial 1 Time (seconds)	Trial 2 Time (seconds)	Trial 3 Time (seconds)
1			
2			
3			

Diffusion Between Two Liquids

You will need a Petri dish, a stopwatch, and the bulls-eye diagram in Figure 3. Fill the Petri dish with room temperature water and center it on the diagram. Wait at least several minutes for the water to stop moving. Put some food coloring (dissolved in an aqueous solution of propylene glycol) in a dropper. Holding the dropper upright, lower the end of the dropper under the surface of the water and inject a couple of drops of coloring in the center circle of the diagram. Start the stopwatch when the colored ring crosses the first circle. Record the time that the color crosses each of the next rings until you can no longer see the ring clearly (use Table 2, try to get four or five data points). The rings are 0.5 cm apart.

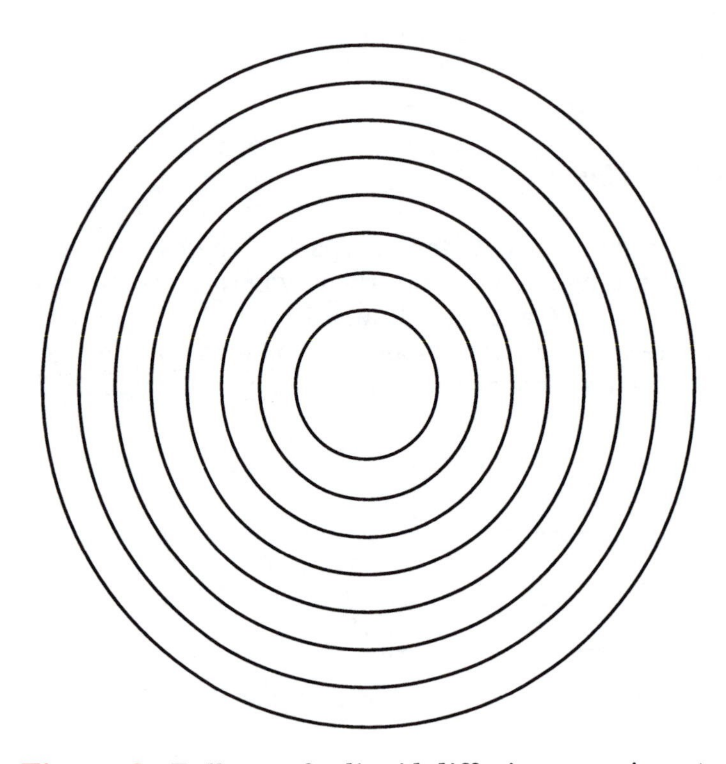

Figure 3. Bulls-eye for liquid diffusion experiment.

Table 2. Laboratory Data for Diffusion between Two Liquids

Distance (cm)	Trial 1 Time (seconds)	Trial 2 Time (seconds)	Trial 3 Time (seconds)
0.5			
1.0			
1.5			
2.0			

Rinse out the Petri dish and repeat the experiment twice.

Calculations

Calculate the average time for each of the distances in the gas and liquid diffusion experiments.

The diffusion coefficient, *D*, is proportional to velocity times distance or to $(\text{distance})^2$ per time. *D* has the units square centimeters per second. Estimate *D* for both the gas and the liquid diffusion.

Diffusion in Solids

Partly fill a test tube with $CuCl_2 \cdot 2H_2O$. **CAUTION: Hydrogen chloride gas is produced when $CuCl_2 \cdot 2H_2O$ is heated. Do this experiment under a fume hood.**

Gently heat the test tube with a Bunsen burner. Adjust the heat to melt the copper chloride without spattering. You may be able to hear the $CuCl_2$ crackle as it starts to melt (m.p. 435 °C).

Obtain a glass rod that is a little longer than the height of the test tube. Put one end of the glass rod into the liquid copper chloride for at least 10 minutes. Record the actual time the glass is in contact with the melt. **CAUTION: Be careful of the molten copper chloride and the test tube. They are VERY HOT!**

Turn off the Bunsen burner. Remove the glass rod from the test tube and place on a flame-resistant surface for a few minutes to cool.

Rinse the glass rod in water and observe the color. If you wish to make the color more defined, heating the glass in the yellow (reducing) part of a Bunsen burner flame will turn the glass red. After the glass has cooled, carefully cut the glass so that you can see a cross-section of the rod. Try to observe how far into the glass the Cu^{2+} ions were able to travel. Estimate the value of *D* as before.

Experiment 14

Magnetic Garnets, $Y_xGd_{3-x}Fe_5O_{12}$

Margret J. Geselbracht, Ann M. Cappellari, Arthur B. Ellis, Maria A. Rzeznik, and Brian J. Johnson

Notes for Instructors

Purpose

To determine the magnetic properties of the $Y_xGd_{3-x}Fe_5O_{12}$ ($0 \le x \le 3$) family of solid solutions.

Method

Each student prepares $Y_xGd_{3-x}Fe_5O_{12}$ for a particular value of $0 \le x \le 3$ (*see* Chapter 3). The sample is prepared from a mixed metal hydroxide precursor by firing in a furnace at 900 °C for 18–24 hours. A visible color change from reddish-brown to olive green indicates that a reaction has taken place. The strength of the attraction of the garnet product to a strong magnet is observed at room temperature, dry-ice temperature, and liquid-nitrogen temperature. Above a characteristic compensation temperature (*see* literature values in Table 1), a sample pellet will be attracted to a strong magnet.

2725–X/93/0455$06.00/1

Table 1. Compensation Temperature, T_{comp}, for $Y_xGd_{3-x}Fe_5O_{12}$

Composition	T_{comp} (K)
$Y_3Fe_5O_{12}$	—[a]
$Y_{2.7}Gd_{0.3}Fe_5O_{12}$	—[b]
$Y_{2.1}Gd_{0.9}Fe_5O_{12}$	—[b]
$Y_{2.0}Gd_{1.0}Fe_5O_{12}$	70
$Y_{1.8}Gd_{1.2}Fe_5O_{12}$	93
$Y_{1.2}Gd_{1.8}Fe_5O_{12}$	158
$Y_{1.0}Gd_{2.0}Fe_5O_{12}$	175
$Y_{0.6}Gd_{2.4}Fe_5O_{12}$	218
$Gd_3Fe_5O_{12}$	289

[a]No compensation temperature was observed down to 5 K.
[b]No compensation temperature was observed down to 73 K.
SOURCE: Data were taken from Harrison, G. R.; Hodges, L. R., Jr. *J. Appl. Phys.* **1962**, *33*, 1375–1376 and Geselbracht, M. J.; Cappellari, A. M.; Ellis, A. B.; Rzeznik, M. A.; Johnson, B. J. *J. Chem. Educ.*, in press.

Materials

1 M aqueous solution of $Y(NO_3)_3 \cdot 5H_2O$ (an average of about 3 mL per student, *see* Table 2)

1 M aqueous solution of $Gd(NO_3)_3 \cdot 6H_2O$ (3 mL per student)

1 M aqueous solution of $FeCl_3 \cdot 6H_2O$ (10 mL per student)

6 M NaOH (10 mL per student)

pH paper

Funnels and filter paper

Drying oven (120 °C)

Firing oven (900 °C)

Strong magnet (cow magnet or rare earth magnet; *see* Supplier Information)

Table 2. Volumes of Reagents to Synthesize $Y_xGd_{3-x}Fe_5O_{12}$

x	mL of Gd^{3+} Solution	mL of Y^{3+} Solution
0	6.0	0
0.6	4.8	1.2
1.0	4.0	2.0
1.2	3.6	2.4
1.5	3.0	3.0
1.75	2.5	3.5
2.0	2.0	4.0
2.25	1.5	4.5
3	0	6.0

X-ray Powder Diffraction

The powder diffraction pattern of the sample with the nominal composition $Y_{1.5}Gd_{1.5}Fe_5O_{12}$ is shown in Figure 1. The major product of the reaction is the garnet phase, $R_3Fe_5O_{12}$ (R is a rare earth element). The peaks indicated with an asterisk in Figure 1 are due to $RFeO_3$, which is a common impurity in samples of $R_3Fe_5O_{12}$. Grinding the olive green solid, repelletizing, and firing a second time at 900 °C (or higher temperatures) for 24 hours led to a noticeable decrease in the intensities of the peaks of the impurity phase in the diffraction pattern. The presence of a small amount of the $RFeO_3$ phase in the samples does not adversely affect the investigation of the magnetic properties in this experiment.

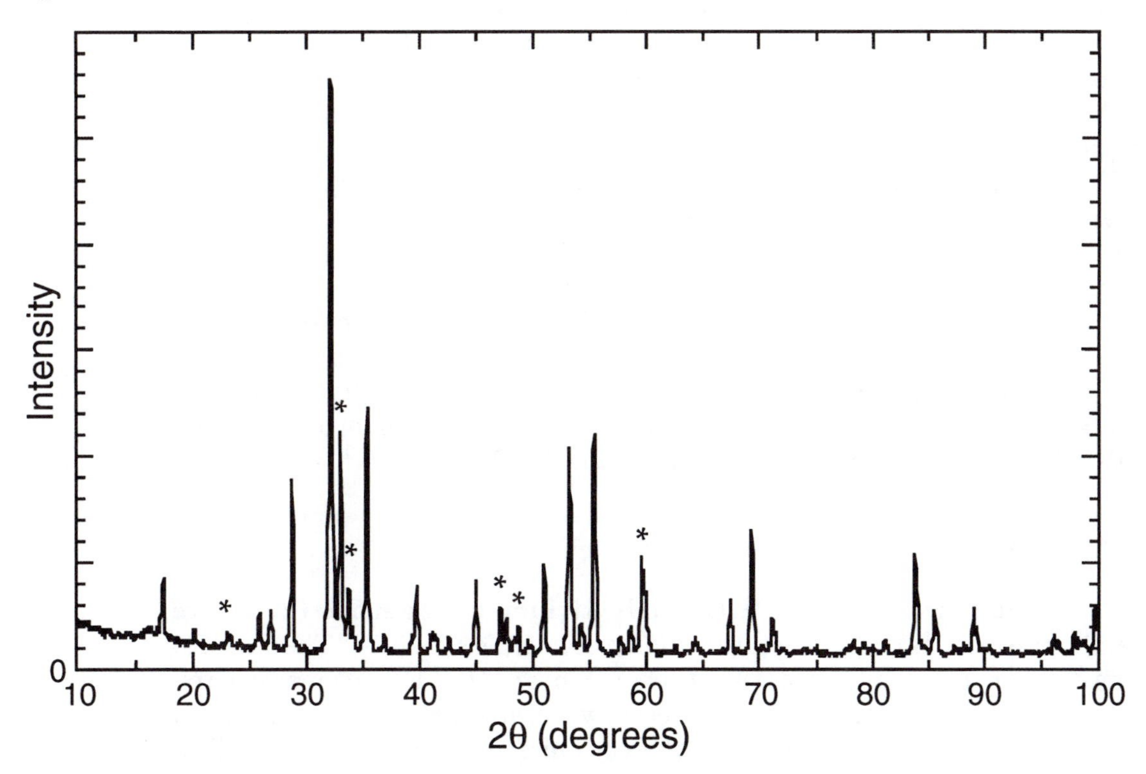

Figure 1. X-ray powder diffraction pattern of a sample of $Y_{1.5}Gd_{1.5}Fe_5O_{12}$ prepared via a mixed metal hydroxide precursor. Diffraction peaks that are marked with an asterisk correspond to the impurity $RFeO_3$ (R is Y or Gd).

Acknowledgement

We thank Professor Francis DiSalvo of Cornell University for the original suggestion of this laboratory experiment.

Magnetic Garnets, $Y_xGd_{3-x}Fe_5O_{12}$

Purpose

To determine the magnetic properties of the $Y_xGd_{3-x}Fe_5O_{12}$ ($0 \leq x \leq 3$) family of solid solutions.

Introduction

In the majority of molecules and solids that we encounter, all of the electrons are paired, and the molecule or solid is said to be diamagnetic. Some solids exhibit interesting magnetic behavior due to the cooperative effect of many electrons in the solid acting in concert. Ferromagnetism is such a property and is exploited in the use of permanent magnets, magnetic recording media, and transformers.

Garnets are a family of solids exhibiting cooperative magnetic behavior. Their magnetic properties depend on composition (the presence of magnetic ions), on the underlying crystal structure (the geometrical arrangement of the ions in three dimensions), and on temperature. The garnets can form solid solutions that permit changing the composition of the solid without disrupting the crystal structure. This condition allows the magnetic properties of the family to be tuned while preserving the crystal structure. Rare earth iron garnets, $R_3Fe_5O_{12}$, can be prepared with all of the rare earth ions, R, except La, Ce, Pr, or Nd.

Structure

Gadolinium iron garnet (GIG) and yttrium iron garnet (YIG) are members of the garnet structural family of complex oxides. The general formula for a garnet is $C_3A_2D_3O_{12}$ where the C cations occupy dodecahedral sites, the A cations occupy octahedral sites, and the D cations occupy tetrahedral sites in the crystal structure. The unit cell of the garnet structure has cubic symmetry and contains eight formula units for a total of 160 atoms. A portion of the unit cell of $Gd_3Fe_5O_{12}$ is drawn in Figure 1 with only the cations and selected oxygen atoms shown for clarity.

In the case of $Gd_3Fe_5O_{12}$, Gd^{3+} occupies dodecahedral sites and Fe^{3+} occupies both octahedral and tetrahedral sites in the structure. However, a wide variety of cations in different valence states can reside in the cation

NOTE: This experiment was written by Margret J. Geselbracht, Ann M. Cappellari, and Arthur B. Ellis, Department of Chemistry, University of Wisconsin—Madison, Madison, WI 53706; Maria A. Rzeznik, Department of Chemistry, University of California Berkeley, Berkeley, CA 94720; and Brian J. Johnson, Department of Chemistry, St. John's University and the College of St. Benedict, St. Joseph, MN 56374.

sites, as the primary consideration for site occupancy is ionic size. This capacity leads to many compositions that form the garnet structure.

Because of the similarity in ionic radii of the rare earth ions, many rare earth iron garnets will form solid solution phases. In the case of yttrium (ionic radius = 0.900 Å) and gadolinium (ionic radius = 0.938 Å), the complete family of solid solutions can be prepared: $Y_xGd_{3-x}Fe_5O_{12}$ ($0 \le x \le 3$). Substitution of yttrium for gadolinium occurs on the dodecahedral site. Gd^{3+} is slightly larger than Y^{3+}; hence, the lattice parameter, a (the length of the side of the cubic unit cell), would be expected to increase with increasing gadolinium content. This trend is clearly seen in the data presented in Table 1.

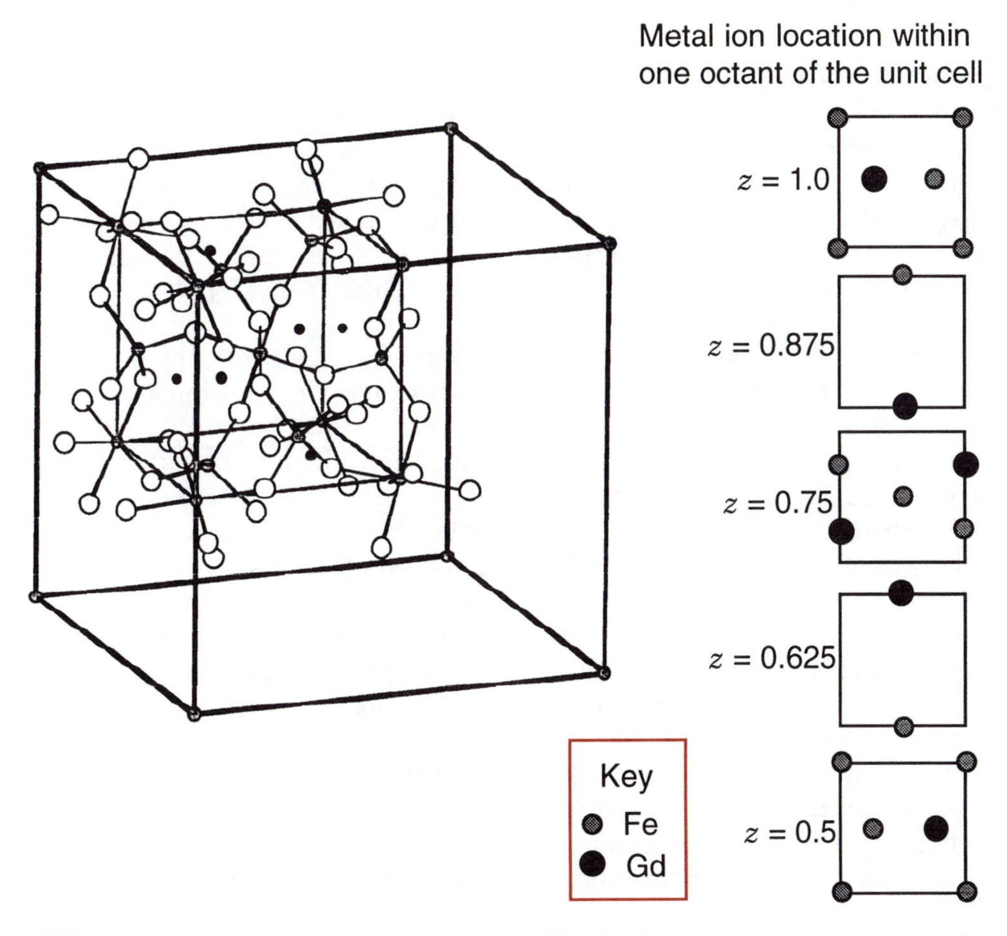

Figure 1. Portion of the unit cell of the garnet structure of $Gd_3Fe_5O_{12}$ is shown at left. All of the metal ions in one-eighth of the unit cell are drawn along with the oxygen ions that complete the octahedral or tetrahedral coordination around the iron atoms. The rest of the atoms have been deleted for clarity. At the right, horizontal sections perpendicular to the z direction are shown to indicate the positions of the iron and gadolinium ions within this octant of the unit cell.

Table 1. Variation in Cubic Unit-Cell Parameter, a, with x for $Y_xGd_{3-x}Fe_5O_{12}$

Composition	Lattice Parameter (Å)
$Y_3Fe_5O_{12}$	12.370
$Y_{2.5}Gd_{0.5}Fe_5O_{12}$	12.382
$Y_2Gd_1Fe_5O_{12}$	12.402
$Y_{1.5}Gd_{1.5}Fe_5O_{12}$	12.423
$Y_1Gd_2Fe_5O_{12}$	12.437
$Y_{0.5}Gd_{2.5}Fe_5O_{12}$	12.450
$Gd_3Fe_5O_{12}$	12.468

Magnetic Properties

For the material $Y_3Fe_5O_{12}$, the Fe^{3+} ions (with five unpaired electrons) in the octahedral holes have their electron spins aligned in the opposite direction from those of the Fe^{3+} ions (also with five unpaired electrons) in the tetrahedral holes. However, because three tetrahedral sites and two octahedral sites are present in the garnet formula, a net magnetic moment of five unpaired electrons per formula unit results. No magnetic contribution comes from the closed-shell yttrium ion. Thus, $Y_3Fe_5O_{12}$ is strongly magnetic at all temperatures.

At the other extreme in the solid solution, $Gd_3Fe_5O_{12}$ has the same net five unpaired electrons from the two kinds of iron sites. But, in addition, each Gd^{3+} ion (there are three in the garnet formula) has seven unpaired electrons. The electrons in the gadolinium sites have their electron spins aligned opposite to those of the net five unpaired electrons of the iron atoms in tetrahedral sites. This feature suggests that the magnetic moment of $Gd_3Fe_5O_{12}$ should be 16 ($=3 \times 7 - 5$) unpaired electrons. This result is true if the magnetic moment is measured at very low temperatures. However, the unpaired electrons associated with Gd^{3+} ions and with Fe^{3+} ions thermally randomize their spins to different extents as a function of temperature. The net result is a compensation temperature, T_{comp}, where the net magnetization is zero. In $Gd_3Fe_5O_{12}$, the compensation temperature is just below room temperature.

The effect of combining both yttrium and gadolinium in the garnet structure in the solid solution compounds is that the compensation temperature becomes a strong function of composition, that is, it can be tuned.

Procedure

Wear eye protection.

CAUTION: The yttrium and gadolinium compounds used in this experiment are oxidizing agents and irritants. Avoid creating or breathing dust. Avoid eye and skin contact. Wash your hands thoroughly after handling.

Synthesis

You will prepare one member of the $Y_xGd_{3-x}Fe_5O_{12}$ family, that is, you will be assigned a particular value of x. A description of the synthesis of $Y_{1.5}Gd_{1.5}Fe_5O_{12}$ via a mixed metal hydroxide precursor follows. Other compositions in the series can be prepared by varying the volumes of $Gd(NO_3)_3$ and $Y(NO_3)_3$ used, such that the amount of $Gd(NO_3)_3$ plus the amount of $Y(NO_3)_3$ totals 6 mL.

Put 10 mL of 1 M $FeCl_3$ in a beaker. Add 3 mL of 1 M $Gd(NO_3)_3$ solution and 3 mL of 1 M $Y(NO_3)_3$ solution. Add 5–10 mL of 6 M NaOH dropwise to the metal ion solution in order to precipitate a reddish-brown solid. Decant the solution, and wash the remaining solid repeatedly with water until the wash is no longer basic (test with pH paper). Filter and dry the solid in a drying oven (120 °C) overnight.

Press the powder into 1/2-inch-diameter pellets using a standard pellet press, and fire in a furnace at 900 °C for 18–24 hours. A visible color change from reddish-brown to olive green after firing indicates that a reaction has taken place.

Magnetic Properties

Determine the strength of the attraction of your pellet of garnet to a strong magnet at room temperature, dry-ice temperature, and liquid-nitrogen temperature. Discuss your procedure with other students in the class and agree on a method that will allow comparison of the magnetic effects as a function of the garnet composition.

Experiment 15

The Sol–Gel Preparation of Silica Gel Sensors

Anne M. Buckley, Martha Greenblatt, Peter G. Allen, and George C. Lisensky

Notes for Instructors

Purpose

To observe how cross-linking affects the physical properties of a silicate polymer containing an acid-base indicator, or other reagent appropriate for chemical sensing.

Method

Acid-catalyzed polymerization of $Si(OCH_2CH_3)_4$ gives linear molecules that are occasionally cross-linked and tangle to form additional branching and gelation. Base-catalyzed polymerization gives more highly branched polymers and discrete structures linked by gelation. Acid-base indicators or reagents for qualitative tests can be trapped in the gel for demonstrating the porosity of the structure.

Materials

Tetraethyl silicate, $Si(OCH_2CH_3)_4$ (Aldrich, tetraethylorthosilicate, 98%)

Absolute ethanol

Hydrochloric acid (conc)

2725–X/93/0463$06.00/1

Aqueous ammonia (conc)

Bromthymol blue, sodium salt (Prepared indicator solutions are less desirable, because they will change the amount of water in the reaction mixture.)

Potassium iodide

Potassium thiocyanate

Solutions of Fe^{3+}, Cu^{2+}, Ag^{+}, or Pb^{2+}

Sonicator (If magnetic stirring is used, the mixing requires 1–2 hours; with a sonicator only 20 minutes is needed.)

Graduated cylinder

Plastic or glass disposable pipettes

pH indicator paper

Drying oven set at 60 °C

Beakers, graduated cylinders, test tubes

Answers to Student Questions

1. Calculation of $Si(OCH_2CH_3)_4 : CH_3CH_2OH : H_2O$ mole ratio

   15.0 mL of $Si(OCH_2CH_3)_4$: 15.5 mL of CH_3CH_2OH : 19.0 mL of H_2O

   15.0 mL of $Si(OCH_2CH_3)_4 \frac{0.936\ g}{mL} \frac{mol}{208.33\ g} = 0.0674$ mol of $Si(OCH_2CH_3)_4$

   15.5 of mL $CH_3CH_2OH \frac{0.789\ g}{mL} \frac{mol}{46.07\ g} = 0.2655$ mol of CH_3CH_2OH

   19.0 mL of $H_2O \frac{1.00\ g}{mL} \frac{mol}{18.0\ g} = 1.056$ mol of H_2O

   or a mole ratio of 1 $Si(OCH_2CH_3)_4$: 4 CH_3CH_2OH : 16 H_2O

2. Balanced equation for complete hydrolysis of $Si(OCH_2CH_3)_4$

   $Si(OCH_2CH_3)_4 + 4H_2O \rightarrow Si(OH)_4 + 4CH_3CH_2OH$

3. The approximate shrinkage between the wet and dried (in oven 7 days) alcogels is to one-fourth their original size. In the wet alcogel the SiO_2 network surrounds pores that are filled with CH_3CH_2OH and H_2O. The gel is left in a drying oven, so the solvent gradually evaporates and the gel ages by additional cross-linking of unreacted –OH and –OR groups. This cross-linking will cause the network to collapse somewhat and result in a reduction in volume. (However, the dried alcogel still contains residual –OH and –OR species as well

as ROH and H_2O. If these are to be removed completely to form a xerogel, the gel must be heated up to about 500 °C in an oxygen atmosphere.)

4. Positive indicator tests for the oven-dried samples illustrate that the gel is still porous. (The pore structure is somewhat collapsed in the dried alcogel, relative to the wet alcogel, so the porosity of the wet alcogel is expected to be much greater. The porosity of the wet alcogel can be retained, before further shrinkage occurs in the oven, by the supercritical drying of the gel to form an aerogel.)

5. The acid-catalyzed xerogels have extremely fine microstructural features, and the low electron-density contrast in the xerogel suggests that the pores are extremely small and evenly spread. The individual silica particles cannot be resolved. The base-catalyzed xerogels are particulate, and the individual silica particles may be resolved (~10 nm) and are less densely packed than in the acid-catalyzed case.

   Base catalysis favors rapid hydrolysis of silicon alkoxides compared to condensation. These conditions favor the formation of highly branched silica clusters that do not interconnect prior to gelation. These clusters form relatively large polymers that grow at the expense of the smaller clusters. On drying by evaporation, the clusters impinge, but the polymers do not readily deform or shrink. Thus, the gel dries as a random packed array of identifiable particles. Consequently, silica xerogels prepared using base-catalyzed conditions are expected to be particulate. However, acid catalysis favors a reduced rate of hydrolysis, causing relatively small polymers. When these are dried, strong surface tension forces are generated by the removal of solvent from the small regions between the polymers. As these weakly cross-linked polymers impinge, they readily deform and produce a dense gel structure. This means that the gels prepared from acid hydrolysis have extremely fine microstructural features and are microporous.

The Sol–Gel Preparation of Silica Gel Sensors

Purpose

To observe how cross-linking affects the physical properties of a silicate polymer containing an acid-base indicator, or other reagent appropriate for chemical sensing.

Introduction

Sols are dispersions of colloidal particles (size 1–100 nm) in a liquid. A *gel* is an interconnected, rigid network with pores of submicrometer dimensions and polymeric chains whose average length is greater than a micrometer.

The sol–gel process is the name given to any one of a number of processes that involve a solution or sol that undergoes a transition to a rigid, porous mass, a gel. One particular example of a sol–gel process is the reaction of $Si(OCH_2CH_3)_4$, ethanol, and water. These three reactants form a one-phase solution that goes through a sol–gel transition to a rigid, two-phase system of solid silica (SiO_2) and solvent-filled pores. In this lab experiment you will carry out a sol–gel preparation of silica gels.

The fundamental reaction of the sol–gel process is hydrolysis and polymerization of a silicon alkoxide. Hydrolysis occurs when $Si(OCH_2CH_3)_4$ and water are mixed in a mutual solvent, generally ethanol:

$$Si(OCH_2CH_3)_4 + H_2O \rightarrow Si(OCH_2CH_3)_3OH + CH_3CH_2OH$$

The intermediates that exist as a result of partial hydrolysis include molecules with Si–OH groups, which are called silanols. Complete hydrolysis of $Si(OCH_2CH_3)_4$ to $Si(OH)_4$ would yield silicic acid, but complete hydrolysis does not occur. Instead, condensation may occur between either two silanols or a silanol and an ethoxy group to form a bridging oxygen or a siloxane group ($-\overset{|}{\underset{|}{Si}}-O-\overset{|}{\underset{|}{Si}}-$), and a water or ethanol molecule is eliminated. An example of the condensation between two silanols with the elimination of water is

$$Si(OCH_2CH_3)_3OH + Si(OCH_2CH_3)_3OH \rightarrow (CH_3CH_2O)_3SiOSi(OCH_2CH_3)_3 + H_2O$$

NOTE: This experiment was written by Anne M. Buckley and Martha Greenblatt, Rutgers, The State University of New Jersey, Piscataway, NJ 08855; and Peter G. Allen and George C. Lisensky, Beloit College, Beloit, WI 53511.

Then hydrolysis of $(CH_3CH_2O)_3SiOSi(OCH_2CH_3)_3$ will produce, for example, $(CH_3CH_2O)_2Si(OH)OSi(OCH_2CH_3)_3$ that can undergo further polymerization.

The hydrolysis and polycondensation reactions are initiated at numerous sites within the $Si(OCH_2CH_3)_4$ and H_2O solution as mixing occurs. When a sufficient number of interconnected Si–O–Si bonds are formed in a region, they interact cooperatively to form colloidal particles or a sol. With time, the colloidal particles and condensed silica species link together to form a three-dimensional network. At gelation, the viscosity increases sharply, and a solid object in the shape of the mold results. The product of this process at the sol–gel transition is called an *alcogel*. Once through the sol–gel transition, the solvent phase can be removed. If it is removed by conventional drying, such as evaporation, *xerogels* result. If it is removed via supercritical (high-temperature) evacuation, *aerogels* result. *See* Figure 1.

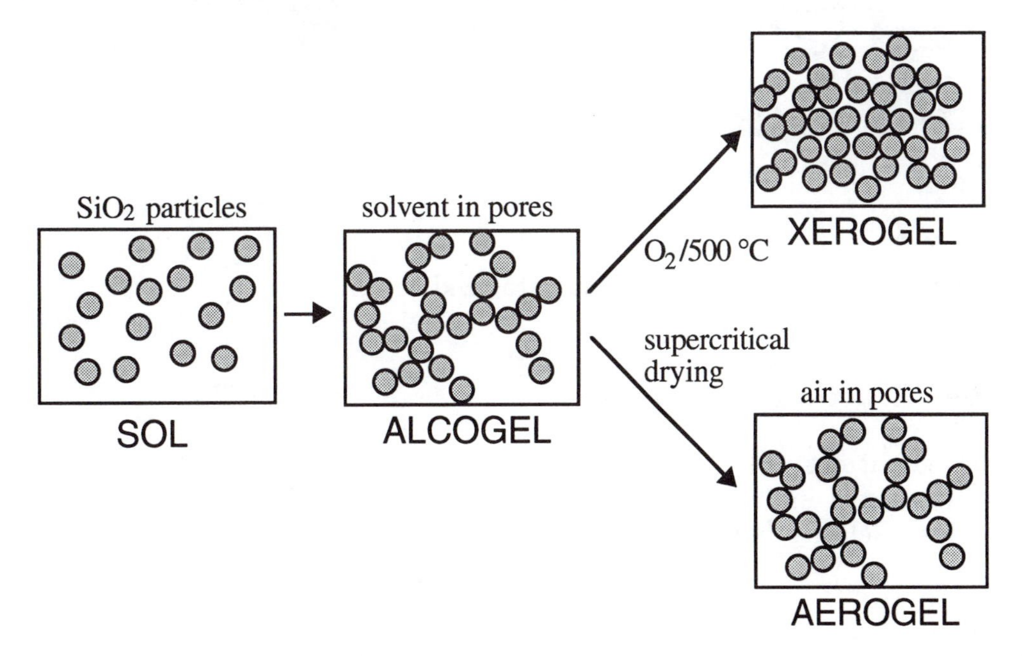

Figure 1. Formation of alcogels, xerogels, and aerogels.

If the solvent phase is removed via supercritical evacuation, the aerogel has a very low density. These aerogels have very good thermal insulating properties when sandwiched between glass plates and evacuated. Xerogels are denser than aerogels, have high surface areas and are often microporous. They can be used as catalyst supports, ionic conductors (when appropriately doped), and precursors for a wide range of glasses, ceramics, coatings, films, and fibers, depending upon the method of preparation. Also, several commercialized processes of sol–gel technology are in use, and an increasing amount of research is being carried out in this field. The hydrolysis and condensation reactions are not, of course, limited to silicon alkoxides, but can be applied to many metal alkoxide systems. Thus, the sol–gel processing of more complex ceramics is now rapidly evolving to include the synthesis of superconductors and coatings on optical memory disks.

Acid-Catalyzed Reaction

At low pH levels, that is, acidic conditions (slow hydrolysis), the silica tends to form linear molecules that are occasionally cross-linked. These entangle and form additional branches, resulting in gelation. *See* Figure 2.

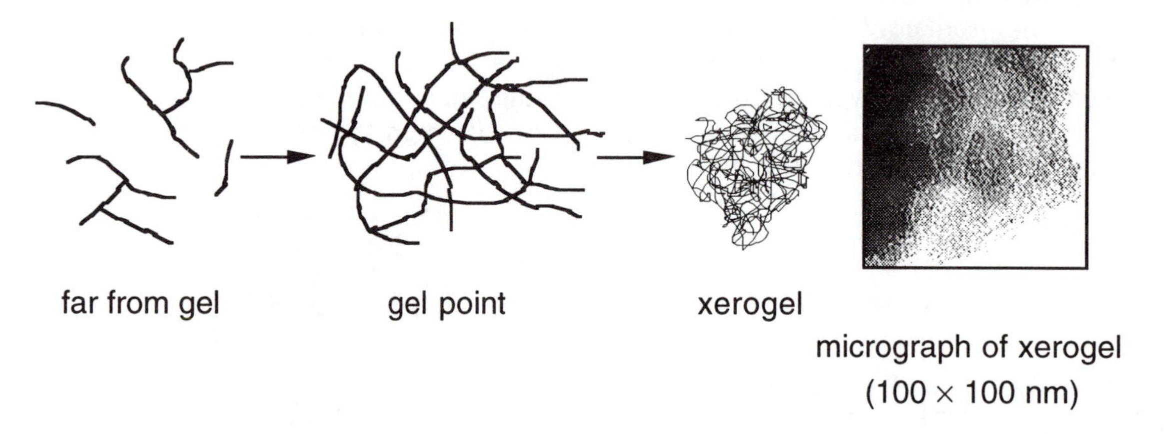

Figure 2. A representation of acid-catalyzed gelation.

Base-Catalyzed Reaction

Under basic conditions (faster hydrolysis), more highly branched clusters form that do not interpenetrate prior to drying and thus behave as discrete species. Gelation occurs by linking of the clusters. *See* Figure 3.

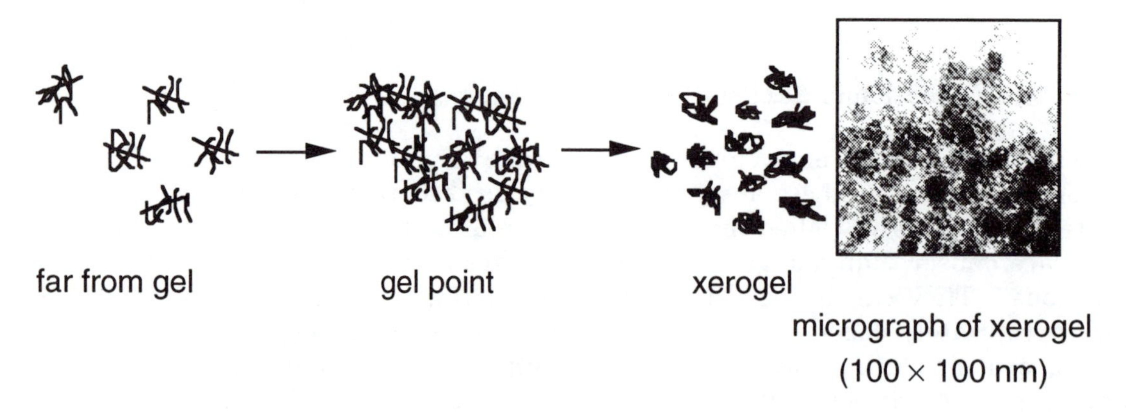

Figure 3. A representation of base-catalyzed gelation.

To understand how acid- and base-catalyzed sol–gel reactions lead to different microstructures in the gels, we now consider what happens when the solvent is removed by evaporation to form xerogels. For polymer gels such as these, removal of solvent is expected to collapse the network of pores, gradually resulting in additional cross-linking as unreacted –OH and –OR groups come into contact. The different structures of the slowly (acid-catalyzed) and rapidly (base-catalyzed) hydrolyzed gels respond differently to the removal of solvent during drying. High-density, low-pore-volume gels are formed in weakly cross-linked systems (acid-catalyzed gels). As the polymers impinge on one another, they deform readily and form a dense gel structure. When hydrolysis is more rapid (base-catalyzed gels), polymers are larger and more highly cross-linked; on impingement the polymers will not deform as readily. The gel dries to a more or less randomly packed array of identifiable particles around which are large voids. Think of the two cases as the ability to further entangle a plate of spaghetti (acid-catalyzed) versus the inability to further entangle a plate of tumbleweeds (base-catalyzed).

Procedure

Wear eye protection.

CAUTION: $Si(OCH_2CH_3)_4$ is irritating to eyes, mucous membranes, and internal organs. Be sure to work in the fume hood as much as possible and wear gloves. Dispose of all waste in an organic waste solvent jug.

Measure 15.5 mL of absolute ethanol into each of two Erlenmeyer flasks. In a fume hood, pour about 30 mL of the liquid $Si(OCH_2CH_3)_4$ (tetraethyl silicate) into a small beaker and immediately recap the stock bottle. Use this $Si(OCH_2CH_3)_4$ to measure out and add exactly 15.0 mL of $Si(OCH_2CH_3)_4$ to each of the ethanol-containing flasks.

Acid-Catalyzed Hydrolysis

Measure out 19.0 mL of distilled water and mix in 2–3 drops of concentrated HCl. Pour this aqueous acid solution into the $Si(OCH_2CH_3)_4$–ethanol solution. Do any layers form? Test the pH of the water by dipping a glass rod into the solution and allowing a small drop to touch a piece of indicator paper. The pH should be about 3. Adjust the pH to this value by adding more drops of acid as necessary.

Base-Catalyzed Hydrolysis

Measure out 19.0 mL of distilled water and mix in 8–10 drops of concentrated aqueous ammonia. Pour this aqueous base solution into the $Si(OCH_2CH_3)_4$–ethanol solution. Do any layers form? Test the pH of the water by dipping a glass rod into the solution and allowing a small drop to touch a piece of indicator paper. The pH should be about 10. Adjust the pH to this value by adding more drops of base as necessary.

Sol–Gel Formation

Place both flasks in a sonicator for 20 minutes (or mix with a magnetic stirrer for 1–2 hours if no sonicator is available). If multiple layers are still present, add a few more drops of concentrated HCl (acid-catalyzed) or concentrated aqueous ammonia (base-catalyzed) and mix them in the sonicator for another 10 minutes or stir for another hour. Continue until the layers are completely mixed.

Remove some of the sol from the *acid-catalyzed* reaction flask, and fill a Petri dish to a level of about 5 mm. Take the remainder of the sol and divide it among three small test tubes. Add a few crystals of the sodium salt of bromthymol blue to one of the tubes. In the second tube, add a few crystals of potassium iodide. Add a few crystals of potassium thiocyanate to the third tube. What is the color of the sols? Put the three labeled tubes in a drying oven at 60 °C. Let the contents of the Petri dish evaporate slowly at room temperature.

Remove some of the sol from the *base-catalyzed* reaction flask, and fill a Petri dish to a level of about 5 mm. Take the remainder of the sol and divide it among three small test tubes. Add a few crystals of the sodium salt of bromthymol blue to one of the tubes. In the second tube, add a few crystals of potassium iodide. Add a few crystals of potassium thiocyanate to the third tube. What is the color of the sols? Let the contents of the Petri dish evaporate slowly. Put the three labeled tubes in a drying oven at 60 °C.

After 1 week, take the tubes from the oven. Did any changes occur? Examine the contents of the Petri dishes. Record your observations.

Indicator Tests

Place some of the *acid-catalyzed* product that contains bromthymol blue in a covered beaker. Also place a small open container of concentrated aqueous ammonia in the beaker (a source of NH_3 gas). What happens? What does this tell you about the porosity of the product?

Place some of the *base-catalyzed* product that contains bromthymol blue in a covered beaker. Also place a small open container of concentrated HCl in the beaker (a source of HCl gas). What happens? What does this tell you about the porosity of the product?

Add some of a KI-containing product to a Ag^+ or Pb^{2+} solution. What happens? Why? Do the acid- and base-catalyzed products do the same thing?

Add some of a KSCN-containing product to a Fe^{3+} or Cu^{2+} solution. What happens? Why? Do the acid- and base-catalyzed products do the same thing?

Questions

1. Calculate the mole ratio of $Si(OCH_2CH_3)_4$ to CH_3CH_2OH to H_2O used in the experiment (the density of $Si(OCH_2CH_3)_4$ is 0.936 g/mL; the density of ethanol is 0.789 g/mL).
2. What is the ratio of $Si(OCH_2CH_3)_4$ to H_2O required for complete hydrolysis to $Si(OH)_4$? (Hint: Write a balanced equation.)
3. Did you observe any changes in volume after mixing? After drying? Why does this happen?
4. What evidence do you have for the porosity of the gels?
5. Study the electron micrographs of the acid- and base-catalyzed xerogels. What are the differences in the microstructures of the two types of gels? Can you suggest a reason for the differences in the microstructures?

Appendix. Hydrolysis and Condensation Mechanisms

Both hydrolysis and condensation may occur by acid- or base-catalyzed bimolecular nucleophilic substitution reactions. The acid-catalyzed mechanisms are preceded by rapid protonation of the OR or OH substituents bonded to Si, whereas under basic conditions hydroxyl or silanolate anions attack Si directly. The mechanisms are outlined in more detail as follows.

Acid-Catalyzed Hydrolysis

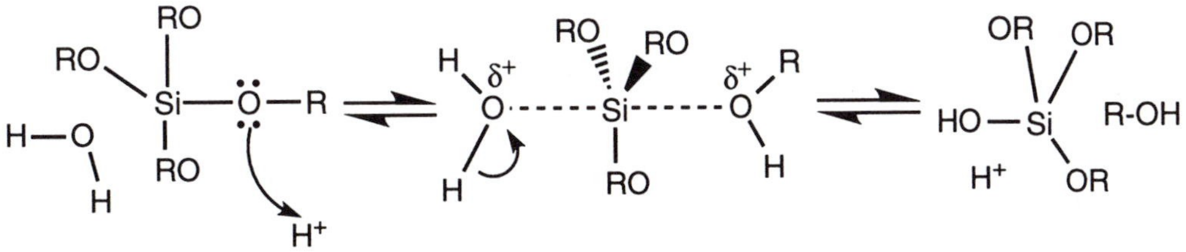

Base-Catalyzed Hydrolysis

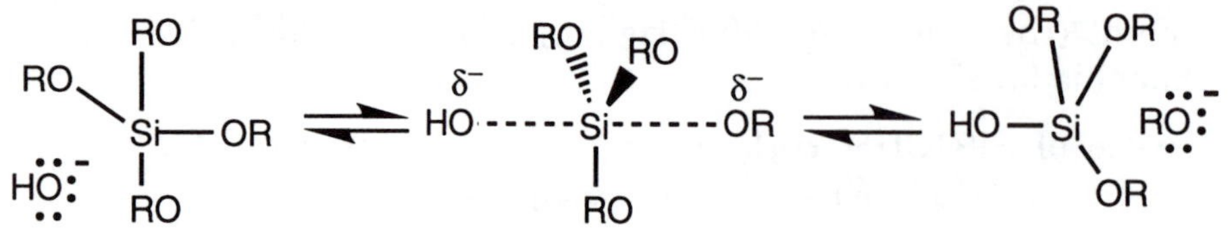

After one or more –OR groups has been replaced by –OH, a condensation reaction can occur.

Acid-Catalyzed Condensation

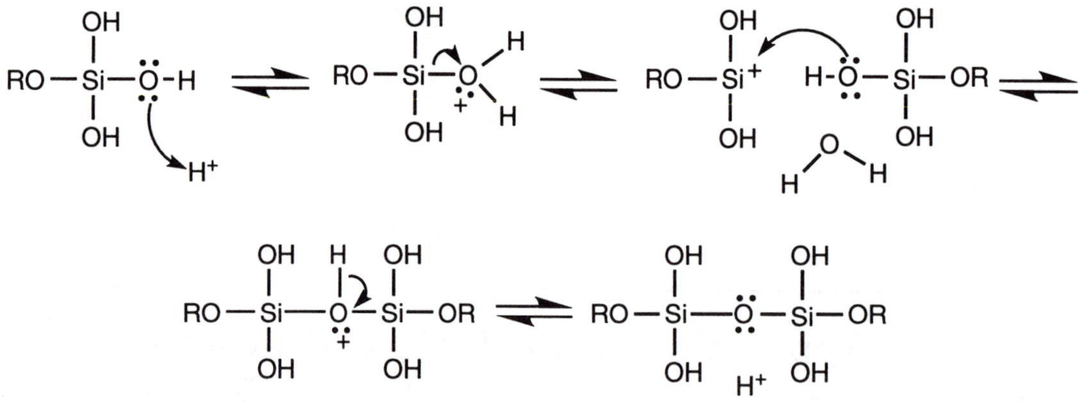

Base-Catalyzed Condensation

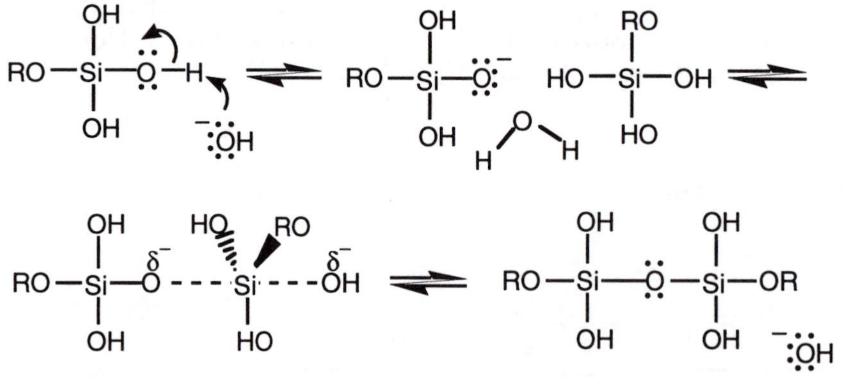

Further hydrolysis of the OR groups can occur, resulting in further condensation and eventual formation of a polymer.

Appendix 1

Glossary

Acceptor	a type of impurity atom in a semiconductor that removes electrons from the valence band and thereby creates mobile holes in the valence band.
Alloy	a solid comprising two or more elements that has metallic properties.
Amphibole	a class of silicate minerals that have double chains of fused silicate [$(SiO_4)^{4-}$] tetrahedra.
Anisotropy	a property that varies with direction.
Annealing	heating a solid at temperatures of generally one-half to two-thirds of the melting point on the kelvin scale, followed by slow cooling. This process allows some atomic rearrangement, removing some defects, and thereby improving the crystallinity of the solid.
Antifluorite structure	an (expanded) face-centered cubic array of anions, with cations in all of the tetrahedral holes (the reverse of the fluorite structure). Li_2O has the antifluorite structure.
Atomic force microscope	an instrument that can image atoms and operates by sensing the force between surface atoms of a sample and a probe tip.

2725–X/93/0473$06.00/1

Austenite — (1) the high-temperature phase of iron (also known as γ-iron), which is stable between ~900 and 1400 °C; (2) the high temperature phase in a two-phase systems that undergoes a martensitic phase transition (NiTi, for example).

Band — a collection of orbitals, each delocalized throughout the solid, that are so closely spaced in energy as to be nearly continuous.

Band gap — the energy separation between the top of the valence band and the bottom of the conduction band.

Berthollide compounds — materials that have variable compositions, such as NbH_x, where x can vary from 0.64 to 1.0.

Biasing — applying a voltage, often done to alter the electrical and optical output of a device such as a light-emitting diode (LED).

Biodegradable — capable of being decomposed in the environment.

Body-centered cubic (bcc) — a type of unit cell that has identical atoms at the corners of a cube and in the center of the cube.

Bohr radius — the distance between the nucleus and the first electronic orbital in the Bohr model of the hydrogen atom (the most probable distance for finding the electron in the hydrogen atom ground state in quantum mechanics). It is a unit of distance equal to ~0.53 Å.

Buckminsterfullerene — an allotropic form of carbon with the formula C_{60}. The molecule has the same shape and symmetry as a soccer ball.

Carrier concentration — the number of charge carriers per unit volume, often expressed in units of carriers (electrons, holes or ions) per cubic centimeter.

Cesium chloride structure — a structure formed by two interpenetrating simple cubic arrays. The unit cell consists of a cube having one kind of atom at the corners of the cube and a second kind of atom in the center.

Chemical vapor deposition (CVD)	a method for growing solids; typically gaseous precursor species are decomposed at a heated substrate, after which they deposit the desired material.
Conductance	the ability to carry current. It has units of ohm^{-1}, also called mhos or siemens, and is the reciprocal of resistance.
Conducting polymers	long-chain molecules, usually possessing extensive conjugation, that are capable of carrying current.
Conduction band	a band that when partially occupied by mobile electrons, permits their net movement in a particular direction, corresponding to the flow of electricity through the solid.
Constructive interference	the condition that occurs when two (or more) waves are in phase and interfere to give a wave with enhanced amplitude and intensity.
Cooper pairs	electron pairs that form in superconducting solids; formation of these pairs is a characteristic of the phase change that occurs at the critical temperature of the solid.
Critical current (J_c)	the highest current density at which the superconducting state may be maintained in a solid.
Critical field (H_c)	the highest magnetic field strength at which the superconducting state may be maintained in a solid.
Critical temperature (T_c)	the highest temperature at which the superconducting state may be maintained in a solid.
Crystalline solid	a solid that is composed of identical patterns (unit cells) of atoms, ions, and/or molecules that repeat continuously until the boundaries of the solid are reached.
Cubic close packed (ccp)	planes of close-packed atoms or ions that are stacked ABCABC....

Cubic hole — the hole formed in the center of eight atoms or ions that sit at the corners of a cube.

Curie temperature — the temperature above which thermal energy is sufficient to destroy the alignment of electron spins in a ferromagnetic material, and thus cause the material to become paramagnetic.

Czochralski crystal growth — a technique for growing ultrapure crystals from a melt of the same composition.

De Broglie wavelength — the wavelength associated with a particle due to its momentum. $\lambda = h/mv$, where λ is the de Broglie wavelength, h is Planck's constant, m is the mass of the particle, and v is its velocity.

Destructive interference — the condition that occurs when two (or more) waves are out of phase and interfere to produce a wave with smaller amplitude and intensity.

Diamagnetism — the magnetic property that results when all electrons in a material are paired, a condition causing the material to be slightly repelled by a magnetic field.

Dielectric constant — a measure of the ease with which a material is polarized by an electric field.

Diffraction — the scattering of light from a regular array, producing constructive and destructive interference. In crystallography, diffraction of X-rays is observed from the regular and repeating arrays of atoms in a crystalline solid.

Diode laser — a type of laser in which a p–n semiconductor junction is used to produce coherent light.

Dislocation — a type of one-dimensional (line) defect in a crystal. Movement of dislocations facilitates the motion of atoms in a crystal.

Domain — a microstructural region within a sample. For example, in a ferromagnetic material, a region in which the electron spins are aligned.

Donor — a type of impurity atom in a semiconductor that adds mobile electrons to the conduction band.

Dopant	an impurity atom that is deliberately added to a semiconductor.
Ductile	the ability of a material to undergo changes in shape (plastic deformation), such as being drawn into wires, rather than breaking; a term often applied to metals.
Elastic deformation	a temporary change in shape in response to a stress; removal of the stress restores the original shape (contrast plastic deformation).
Electrical conductivity	the ability of a material to carry an electric current. The symbol for conductivity is σ, and it has units of ohm^{-1} cm^{-1}. It is also the reciprocal of resistivity.
Epitaxial growth	the growth of one crystalline material on a substrate of another crystalline material in such a way that the deposited atoms are in registry with atoms on the surface of the substrate.
F-center	a type of defect in ionic compounds in which an electron is trapped in an anion (often halide) vacancy in the crystal. It is also sometimes called a color center.
Face-centered cubic (fcc)	a type of unit cell in which there are identical atoms at the corners and in the centers of the faces of a cube. Identical to the cubic close-packed (ccp) arrangement.
Fermi level	the thermodynamic electrochemical potential or the energy in a solid, below which it is more likely that the electronic states are occupied with electrons and above which it is more likely that the electronic states are not occupied; because it is defined statistically, there may or may not be an energy level at this energy.
Ferrite	the low-temperature phase of iron, also known as α-iron; stable below ~900 °C.
Ferrofluid	a suspension of a magnetic solid like Fe_3O_4 (Fe_3O_4 is actually ferrimagnetic, *see* Chapter 2) in a liquid that responds to an external magnetic field.

Ferrimagnetism	a phenomenon in which the internal magnetic moments of multiple spin sets of unpaired electrons within the domain of the solid do not cancel and therefore leave a net spin.
Ferromagnetism	a phenomenon in which the internal magnetic moments of unpaired electrons within a domain of the solid are aligned and act cooperatively.
Fluorite	the common name for the mineral CaF_2.
Fluorite structure	a structure in which there is a face-centered cubic array of cations and a simple cubic array of anions. This is the structure of CaF_2.
Flux	a melted sample of a material, usually a metal or salt, that is used as a solvent for a reaction or crystallization, often at elevated temperature.
Frenkel defect	a type of defect in a solid in which an atom or ion (usually the smaller cation) is found in an interstitial site rather than in its normal site.
Garnet	a family of complex oxides with the general formula $C_3A_2D_3O_{12}$, where the C cations occupy dodecahedral sites, the A cations occupy octahedral sites, and the D cations occupy tetrahedral sites in the crystal structure—an example is $Gd_3Fe_5O_{12}$, where Fe(III) ions occupy both A and D sites.
Grain	a small crystalline region randomly oriented within the bulk crystal.
Grain boundary	the intersection of two grains within a crystal.
Hexagonal close packed (hcp)	a type of crystalline structure in which close-packed layers of atoms or ions are stacked ABABAB....
Hole (h^+)	a fictitious mobile particle that behaves as though it is a positively charged particle; holes are produced in the valence band when electrons from the valence band are promoted to the conduction band or to an acceptor dopant.

Hole	an empty site in a crystalline solid (*see* cubic hole, octahedral hole, or tetrahedral hole).
Hydrodesulfurization (HDS)	removal of sulfur from petroleum feedstocks.
Hysteresis	a phenomenon in which some property of a material depends on the sample's history; for example, the structure and electrical conductivity of a material such as NiTi over the temperature range corresponding to its phase change may depend on whether the sample was heated or cooled to reach its current temperature.
Insulator	a type of material having a lower energy valence band that is nearly completely filled with electrons and a higher energy conduction band that is nearly completely empty of electrons as a result of a large energy gap between the two bands. Such materials are poor conductors of electricity.
Interstitial atom	an atom in a normally unoccupied hole between atoms in a crystal.
Isomorphous substitution	the replacement of one atom with another of similar size such that there is no significant change in the structure.
Isotropic	a property that is the same in any direction.
Law of Dulong and Petit	the observation that many metallic elements have molar heat capacities of 25 J/mol-deg at and above room temperature; used as an early method for determining atomic weights.
Light-emitting diode (LED)	a semiconductor p–n junction that is optimized to release light of approximately the band-gap energy when electrons fall from the conduction band to the valence band under forward bias.
Malleable	able to be shaped by hammering, pressing, or bending; a term often applied to metals.

Martensite — the original name given to the hard material formed during the quenching of steel, it more generally refers to the low-temperature phase in a martensitic phase transformation.

Material life cycle — the extraction of raw materials, synthesis and processing, product preparation, use, and waste disposal considerations that are related to the development of a useful material.

Materials chemistry — the study of the synthesis, processing, and chemical and physical properties of solids.

Meissner effect — the expulsion of a magnetic field (perfect diamagnetism) from a superconductor.

Metal — a material with a partially filled energy band. Metals are generally malleable, ductile, good reflectors of electromagnetic radiation, and good conductors of heat and electricity. Metals are usually identified by having electrical conductivities that decrease with increasing temperature.

Mho — a unit of conductance; the reciprocal of ohm.

Microstructure — the arrangement of atoms on a scale of micrometers (10^{-6} m).

Molecular beam epitaxy (MBE) — a technique for growing solids, atomic layer by atomic layer, based on precursors delivered to a target substrate by collimated beams of atoms or molecules.

Molybdenite — the common name for the mineral MoS_2, whose layered structure allows it to be used as a lubricant.

n-type semiconductor — a semiconductor that has been doped with an electron donor such that the number of mobile conduction band electrons is greater than the number of mobile valence band holes at thermal equilibrium.

Nanostructure — the arrangement of atoms on a scale of nanometers (10^{-9} m).

Nitinol	a "smart material" that is an alloy with the formula NiTi; the solid has shape-memory applications.
Nonstoichiometric compounds	another name for compounds of variable composition; such substances are also called berthollide compounds. In these materials a variable amount of atoms is present in the holes of a host structure.
Octahedral hole	the hole formed in the center of six atoms or ions sitting at the corners of an octahedron.
Optical fiber	a glass or plastic cable capable of transmitting light with high efficiency.
Order-disorder phase change	a phase change that involves reorientation or repositioning of the atoms in a structure from a more orderly arrangement to a more random arrangement or vice versa.
p–n junction	the region of contact on the atomic scale between a p-type and an n-type semiconductor.
p-type semiconductor	a semiconductor that has been doped with an electron acceptor so that the number of mobile valence-band holes exceeds the number of mobile conduction-band electrons at thermal equilibrium.
Paramagnetism	a phenomenon characterized by an attraction of a material to a magnetic field due to the presence of unpaired electrons.
Perovskite	the mineral $CaTiO_3$; also, a class of structures that have the perovskite structure.
Phonons	quantized, collective vibrations of the atoms in a solid.
Piezoelectricity	a phenomenon whereby the application of an electric field to certain solids causes them to change shape. Conversely, their mechanical deformation produces an electrical signal.
Pinning	the process of preventing the movement of atoms or defects within a crystal.

Pixel — the smallest dot that can be displayed on a computer screen or output (picture element).

Plastic deformation — the ability to re-shape a material because of the sliding of atomic planes in a crystal relative to one another (contract elastic deformation).

Point defect — a zero-dimensional defect such as a vacancy (missing atom), substitutional impurity, or interstitial atom.

Poisson ratio — a mechanical property of a material,

$$\mu = \frac{\text{fractional decrease in width}}{\text{fractional increase in length}} = \frac{\Delta w/w}{\Delta l/l}$$

Polymorph — a substance that crystallizes in two or more different forms, such as the wurtzite and sphalerite forms of ZnS or graphite, diamond, and fullerene forms of C.

Precipitation hardening — generation of a precipitate of an impurity phase within a host crystal that serves to roughen slip planes and makes the motion of dislocations more difficult. This process generally results in harder but more brittle materials.

Pyroxene — a class of silicate minerals that have a one-dimensional single chain of linked silicate tetrahedra.

Quenching — the process of quickly immersing a hot object in a cold medium like water or liquid nitrogen to lock in a particular structural arrangement of atoms. Also the rapid heating of a superconducting electromagnet when the superconductivity is lost.

Radius ratio — in ionic compounds, the radius of the cation divided by the radius of the anion.

Re-entrant foam — a polymer foam that has been isotropically compressed at an elevated temperature to produce a solid with a negative Poisson ratio

Reflectivity — the amount of light reflected relative to the total incident light.

Resistance	opposition to the flow of electric current. It has the symbol Ω and units of ohms and is the reciprocal of conductance.
Resistivity	the reciprocal of conductivity, given the symbol ρ. It is related to resistance by the equation $R = \rho L/A$, where R is the resistance, ρ is the resistivity, L is the length of the wire, and A is its cross-sectional area.
Rock salt structure	the common name for NaCl and other salts having the same structure consisting of two interpenetrating arrays of face-centered cubic cations and anions.
Scanning tunneling microscope (STM)	an instrument that is capable of imaging individual atoms through the quantum mechanical tunneling of electrons between an electrically conducting atomic tip and the substrate to be imaged.
Schottky defect	a type of vacancy defect in an ionic compound, consisting of a pair of vacant cation and anion sites.
Semiconductor	a type of material having a lower energy valence band that is nearly completely filled with electrons and a higher energy conduction band that is nearly completely empty of electrons, with a modest energy gap between the two bands. Such materials generally exhibit electrical conductivity that increases with temperature because of an increase in the number of charge carriers. *See also* n-type semiconductor, p-type semiconductor.
Shear	the application of antiparallel forces to opposite sides of an object.
Siemen	a unit of conductance; equivalent to a ohm^{-1} or a mho.
Silicate ion	the SiO_4^{4-} ion.
Simple cubic	a type of unit cell in which there are identical atoms at the corners of a cube.

Sintering	a method for increasing the density of a polycrystalline or powdered material by heating it to near its melting point. This step shrinks or removes pores and increases grain size.
Slip plane	a plane of atoms that is able to slide past an adjacent plane of atoms with relative ease.
Smart material	a material that is capable of sensing changes in its environment and responding to them.
Sodium chloride structure	a structure formed by interpenetrating face-centered cubic arrays of cations and anions (also called rock salt structure).
Solid solution (substitutional)	a homogeneous pure solid in which one type of atom (or ion) has randomly substituted for a similar atom (or ion) in a structure.
Solid solution hardening	improving the hardness of a metal by adding impurity atoms that are either much larger or smaller than the host atoms. This process roughens the slip planes and makes the motion of dislocations more difficult.
Specific heat capacity	the amount of heat it takes to raise 1 g of a substance by 1 °C.
Sphalerite	the name for the mineral ZnS (also called zinc blende) in which there is a face-centered cubic array of anions with cations in half of the tetrahedral holes.
Sputtering	a preparative technique in which gaseous ions bombard a metal cathode, and thereby enable the freed metal atoms to form a metallic coating on a substrate.
SQUID	superconducting quantum interference device. This device, based on junctions of a superconductor with an insulator, provides a sensitive method for detecting magnetic fields.
SSMK	abbreviation used for the Solid-State Model Kit.

Substitutional stoichiometry	substitution of one atom or ion for another without significantly changing the overall structure.
Surfactant	a material such as a soap or detergent that reduces surface tension; the molecule typically has a polar head and a nonpolar tail.
Tetrahedral hole	the hole formed in the center of four atoms or ions sitting at the corners of a tetrahedron.
Thermal conductivity	the transfer of heat via the vibration of atoms.
Thermal spray coating	a process in which a metal, ceramic, or plastic material is heated, melted, and sprayed onto a surface. The particles of the material flatten on impact and bond to the surface.
Thermoluminescence	light energy that is released when a sample is exposed to heat.
Tunneling	the movement of an electron due to its wave nature through a classical barrier.
Unit cell	a repeating unit of an extended structure; a parallelepiped in three dimensions, whose translation along each of the edges of the unit cell by the length of the edge, allows the entire structure to be constructed.
Vacancy	an absent atom or ion in a crystal.
Valence band	the highest energy band that lies at the bottom at the band gap.
Van der Waals forces	weak interatomic forces that dominate interactions between nonpolar molecules or atoms.
Variable stoichiometry	a situation in which varying amounts of an atom are present in the interstices of a host compound.
Variant	a possible direction for shifting or shearing in a crystal.

Wurtzite	a mineral form of ZnS in which the anions form a hexagonal close-packed arrangement and the cations are in half of the tetrahedral holes.
Zeolite	an aluminosilicate that has a framework of tetrahedral aluminate and silicate units that form a series of tunnels and cavities.
Zinc blende	a common name for the mineral ZnS (also called sphalerite) in which the anions form a face-centered cubic arrangement and the cations are in half of the tetrahedral holes.
Zinc blende structure	a structure in which there is a face-centered cubic array of anions with cations in half of the tetrahedral holes.
Zone refining	a process for obtaining pure materials that involves melting a small region of the impure solid and gradually moving the molten zone. The impurities tend to concentrate in the melt, and the recrystallized region is highly purified.

Appendix 2

Supplier Information

Note: Prices and part numbers are subject to change! The list given here is not meant to be exclusive. Other suppliers may be able to provide the materials at competitive prices.

General Sources of Materials

Aldrich Chemical
P. O. Box 2060
Milwaukee, WI 53201
(800) 558-9160
Fax (800) 962-9591

Edmund Scientific
101 E. Gloucester Pike
Barrington, NJ 08007–1380
(609) 573–6250

Flinn Scientific, Inc.
P.O. Box 219
Batavia, IL 60510–0219
(708) 879–6900
Fax (708) 879–6962

NADA Scientific LTD.
P. O. Box 1336
Champlain, NY 12919–1336
(800) 233–5381
Fax (514) 246-4188

Materials Education Council
106 Materials Research Laboratory
University Park, PA 16802
(814) 865-1643
Fax (814) 863-7040
(Educational aids: Journal of Materials Education, videotapes, books)

Arbor Scientific
P.O. Box 2750
Ann Arbor, MI 48106-2750
(800) 367-6695
(Piezoelectric demonstrator P6-2110)

Chapter 1

Re-entrant foam

Foam Aide
Auburn Hills, MI
(800) 221–7388

- Request open cell (10–20 pores per inch) foam. $250 minimum.

Chapter 2

Ferrofluid

Ferrofluidics Corporation
40 Simon Street
Nashua, NH 03061–2000
(603) 883–9800

- EMG 905 Ferrofluid, 400 Gauss, 10 Centipoise. Special academic offer: 15 mL. at a cost of $50.00. Check or money order must accompany order.
- Ferrofluid-filled bottle cell $25 each Minimum order, $100.

Heat capacity samples

NADA Scientific LTD. (*See* General Sources.)

- C15–7650 includes 100 g samples (with hole for string) of Cu, Fe, and Al $54.00
- C15–7651 includes 100 g samples (with hole for string) of Cu, Fe, Pb, and Al $69.00
- C15–7652 includes 100 g samples (with hole for string) of Cu, Fe, Pb, Al, Zn, and brass $99.00
- F-35-1800 1 mole cylinders of Cu, Fe, Al, and C $46.00

Magnets

Magnet Sales and Mfg. Co.
11248 Playa Court
Culver City, CA 90230–6100
(310) 391–7213, Fax (310) 390–4357

- 35NE209620 is a strong Nd–Fe–B magnet (1.5 inch × 0.3 inch × 0.3 inch) that is convenient for the demonstrations described in this book. They sell for $16.24 each in small quantities.

Farm supply store

- Cow magnet (0.5 inch diameter × 3 inch), $3. (Cow magnets are fed to cows and lodge in the upper stomach. Their purpose is to prevent scraps of baling wire from passing through the digestive system.)

NADA Scientific LTD. (*See* General Sources.)

- B10-3300 Neodymium Magnet (14 mm diameter × 5 mm), set of 2, $23.00

Edmund Scientific (*See* General Sources.)
- D38,428 Nd–Fe–B magnet (0.187 inch diameter × 0.0625 inch), $3.25
- D35,104 Nd–Fe–B magnet (0.5 inch diameter × 0.125 inch), $25.25
- D31,101 Cow magnet (0.5 inch diameter × 3 inch), $14.95/pair

Piezoelectricity

Flinn Scientific , Inc. (*See* General Sources.)
- AP1576 lighter $13.75

Aldrich Chemical (*See* General Sources.)
- Z11,523-1 lighter $20.25

Educational Innovations
151 River Road
Cos Cob, CT 06807
- popper $3.50 each (plus $3.95 shipping and handling)

Teaching Scanning Tunneling Microscope

Burleigh Instruments
PO Box E, Burleigh Park
Fishers, NY 14453–0755
(716) 924–9355

Chapter 4

Lasers

Beta Electronics
2209 Cloverdale Court
Columbus, OH 43235
(800) 5–GO–BETA
(614) 792–5595
- B400 or LP670 Laser pointer (670 nm, 5 mW), $94.95 (discounts on 5 or more).

Edmund Scientific (*See* General Suppliers)
- D38,914 Laser pointer (670 nm, 5 mW), $189.00.

Melles Griot
1700 Kettering St.
Irvine, CA 92714
(800) 836–2626

Optical Transforms

Institute for Chemical Education
Department of Chemistry
University of Wisconsin–Madison
1101 University Ave.
Madison, WI 53706
(608) 262–3033

- 90–002 Optical Transform Kit (2nd ed.)—includes background information, transparency masters, directions for a student laboratory, and one copy of each of the three slides described in Chapter 4; $6.
- 93–003S 10 copies of 35-mm Discovery slide, Figure 4.4, $6.
- 90–002S 10 copies of 35-mm Unit Cell slide, Figure 4.6, $6.
- 93–004S 10 copies of 35-mm VSEPR slide, Figure 4.8, $6.
- 93–005S 10 copies of 35-mm Plane Groups slide, $6.

Chapter 5

Computer Programs

Eugene Myers and Carlos Blanco
Department of Computer Science
Richard Hallick and Jerome Jahnke
Departments of Biochemistry and Cellular Biology
University of Arizona
Tucson, AZ 85721

- MacMolecule, Appendix 5.2, can be obtained by anonymous FTP at joplin.biosci.arizona.edu. User-contributed image files will also be available at this address in the /pub/MacMolecule directory. Macintosh.

J. A. Bertrand
School of Chemistry
Georgia Institute of Technology
Atlanta, GA 30332–0400
(404) 894–4050
e-mail: jbertrand@chemistry.courier.gatech.edu

- Three-dimensional modeling of crystal and molecular structures program and tutorial for stereo viewing (red and green) and manipulation of crystal structures. IBM and compatibles.

W. R. Robinson, submitted to *J. Chem. Educ. Software*

- A Window on the Solid State, tutorial and lecture demonstration programs that introduce structure and unit cells of metals. Microsoft Windows.

Minerals

Wards Natural Science Est.
5100 W. Henrietta Rd.
P.O. Box 92912
Rochester, NY 14692
(716) 359–2502

- 49E-1686 10 gram Molybdenite natural crystal $16.50 ea.

Model kits

Carolina Biological Supply Co.
2700 York Road
Burlington, NC 27215
(919) 584–0381
- Kit that employs atom centers with plastic straw connectors

Darling Models
P.O. Box 1818
Stow, OH 44224
(216) 688–2080
- Inexpensive octahedral and tetrahedral pieces

Flex-O-Models
P.O. Box 844
Everett, MA 02149
(617) 884–7643
- Pre-assembled models.

Geoscience Resources
2990 Anthony Road
Burlington, NC 27215
(800) 742–2677
- Kit that employs atom centers with plastic straw connectors

Halsteducate Company
2140 Lincolnwood Dr.
Evanston, IL 60201
(708) 866–7069
- Rock salt model that uses space-filling spheres with connectors

Institute for Chemical Education
Department of Chemistry
University of Wisconsin–Madison
1101 University Ave.
Madison, WI 53706
(608) 262–3033
Fax (608) 262–0381
- Solid State Model Kit described in Chapters 3 and 5 of this book

Klinger Educational Products
112–19 14th Road
College Point, NY 11356–1353
(800) 522–6252
- Kit that employs atom centers with plastic straw connectors and preassembled models

Learning Things Incorporated
P.O. Box 436
Arlington, MA 02174
(617) 646–0093
- Kit that employs atom centers with plastic straw connectors

Molecular Models
Edgerton, WI 53534
(608) 884-9877

- Large Styrofoam Spheres

NADA Scientific LTD. (*See* General Sources.)

- Transparent cubes that contain the unit cells for the bcc, fcc, diamond, and NaCl structures

Chapter 6

Irradiated NaCl for F-center demonstrations

Penn State Breazeale Reactor
University Park, PA 16802
(814) 865–6351

- Make check payable to Penn State University and label envelope "salt demo". For $10, they send enough NaCl for about twenty demonstrations.

Piano Wire

Schaff Piano Supply Co.
Lake Zurich, IL 60047
(708) 438–4556

- $\frac{1}{3}$W–16$\frac{1}{2}$ $\frac{1}{3}$ lb (~90 ft) of 16.5-gauge piano wire $4.45

Salt crystals for cleavage and F-center experiments

International Crystal Laboratory
11 Erie Street
Garfield, NJ 07026
(201) 478–8944

- 0002-4538 Random cuttings, chunks of KCl, $45 / 100 g.
- 0002-4539 Random cuttings, chunks of KBr, $55 / 100 g.
- 0002-4789 Random cuttings, chunks of NaCl, $37 / 100 g.

Chapter 7

Conductivity/Resistance experiments

Mouser Electronics
2401 Highway 287 North
Mansfield, TX 76063–4827
(800) 346–6873

- 43LJ415 150-mhenry RF choke (150 m of coiled copper wire, 150Ω at room temperature, 15Ω at 77 K)
- 338-76C348 CdS photocell, $1.53 each in lots of 10. Lowest resistance when illuminated is desirable (also found in Radio Shack stores)

•585–DM231 3 1/2 digit multimeter to read current, voltage, and resistance (with continuity tester); $50 less academic discount.

Diamond Tip Pencil

Aldrich Chemical (*See* General Suppliers)

•Z22,554-1 Engraving pen, $5.60

Graphite intercalation demonstration

Alfa/Johnson Matthey
P.O. Box 8247
Ward Hill, MA 01835–0747

•10135 Graphite rod (6.3 mm × 60 cm long), 99%, $23 per 4 pieces.

Carbone of America
P.O. Box 747
Bay City, MI 48707
(517) 894–2911

•014145–08 Ultra Carbon F synthetic graphite powder, 200 mesh, 16 oz. $ 7.50 / oz. (minimum order $75).

Chicago Miniature Lamp Works
Chicago, IL

•1493 miniature bulb 6.5 V, 2.75 Amps, D.C., bayonet base. Light bulb having 1–2 Ω resistance with two leads soldered to bottom.

Experiments with light emitting diodes

Mouser Electronics
2401 Highway 287 North
Mansfield, TX 76063–4827
(800) 346–6873

Parts to construct circuits for LEDs

•1236005 one piece molded black battery snaps, $0.15 each in lots of 100.

•29SJ250 1-kΩ Transohm Carbon Film Resistors, $0.018 each in lots of 200.

•29SJ250 1-MΩ Transohm Carbon Film Resistors, $0.018 each in lots of 200.

•151–5530 Breakaway IC socket strips, $1.40 per 32 sockets (enough to mount 16 LEDs).

•381–FS030 Soldering pencils $4.79 each (Radio Shack also carries these).

•533–24–6040–39 Solder $11.52 (one spool).

Rohm super-bright LED lamps with water-clear lenses, $0.15 each in lots of 100:

•592–SLH56–VT3 red $GaP_{0.40}As_{0.60}$
•592–SLH56–DT3 orange $GaP_{0.65}As_{0.35}$:N
•592–SLH56–YT3 yellow $GaP_{0.85}As_{0.15}$:N
•592–SLH56–MT3 green $GaP_{1.00}As_{0.00}$:N

Digi-Key Corp.
701 Brooks Ave. South
Thief River Falls, MN 56701–0677
(800) 344–4539

Panasonic LEDs with clear, flat lenses (bulk discounts available):

•P431	red	$GaP_{0.40}As_{0.60}$	$2.67 per 10
•P434	orange	$GaP_{0.65}As_{0.35}$:N	$3.21 per 10
•P433	yellow	$GaP_{0.85}As_{0.15}$:N	$3.60 per 10
•P432	green	$GaP_{1.00}As_{0.00}$:N	$3.60 per 10

Quantum Devices, Inc.
112 Orbison, P.O. Box 100
Barneveld, WI 53507
(608) 924–3000
Fax (608) 924–3007

$Al_xGa_{1-x}As$ LEDs:

- •LED Sample Kit Model A, 630–730 nm central wavelength $98.95 (3 each of 630 , 660, 680, 700, and 730 nm).
- •LED Sample Kit Model B, 700–840 nm central wavelength $98.95 (3 each of 700, 730, 780, 800, 820, and 840 nm).

Chapter 9

Memory metal rods and wires

Shape Memory Applications, Inc.
1034 West Maude Ave., Suite 603
Sunnyvale, CA 94086–2818
(408) 730–5633
Fax (408) 730–5632

- •NiTi shape memory wire, 0.03-inch diameter × 3 inches long, $0.30 per piece.
- •austenitic NiTi rod ("rings") , 0.1-inch diameter × 2 1/2 inches long, $3 per piece.
- •martensitic NiTi rod ("thuds"), 0.1-inch diameter × 2 1/2 inches long, $3 per piece.

Note: A minimum order may be required.

Mondo-tronics
524 San Anselmo Ave. #107
San Anselmo, CA 94960
(800) 374–5764
Fax (415) 455–9333

- •3–074 Book and sample kit contains 1 meter total of wires with three different diameters, $29.90.
- •wires with diameters of 50, 100, and 150 micrometers may be purchased separately.

NADA Scientific Ltd. (*See* General Sources.)

- •P70-3800D Memory metal spring with a hook, $20.00.
- •P70-3800A Memory metal (0.75 mm diameter × 100 mm), $11.00.

Superconductors

Colorado Superconductor Inc.
Department F4
P.O. Box 8223
Fort Collins, CO 80526
(303) 491–9106
Fax (303) 490–1301

- K1 Levitation kit contains a superconducting pellet, rare earth magnet, nonmagnetic tweezers, and an instruction manual. $29. Colorado Superconductor sells other kits that allow the measurement of the critical temperature, critical current, and critical field. They also sell kits that are based on a Bi-based superconductor.

Flinn Scientific, Inc. (*See* General Sources.)

- AP1489 Superconductivity Demonstration Kit includes superconductor, magnet, nonmetallic forceps, and instruction manual. $30.00.

Chapter 10

Optical Fibers

Edmund Scientific (*See* General Sources.)

- D41,213 Optical fiber sample kit includes 35-feet of unjacketed fibers and 1-foot pieces of jacketed multifiber light guides, $23.95.

Experiment 1

Heat capacity samples (*see* supplies listed under Chapter 2)

Experiment 4

Lasers and optical transform slides (*see* supplies listed under Chapter 4)

Experiment 5

Mineral samples (*see* Wards Natural Science listed under Chapter 5)

Experiment 6

Radon test disks

Alpha Trak
141 Northridge Dr.
Centralia, WA 98531
(206) 736–3884

Fax (206) 736–0897

- Introductory Kit (100 plastic test disks, 100 etch clamps, one exposed but unetched disk, one exposed and etched disk, a 35 mm slide of the alpha tracks as viewed through the microscope, a photocopy of the slide, and a suggested procedure), $75.
- Refill Kit (100 plastic test disks), $60.

Experiment 7

LEDs and related equipment (*see* supplies listed under Chapter 7)

Diffraction gratings

Edmund Scientific (*See* General Sources.)

- C39,502 card-mounted diffraction viewer, $20/ pkg. of 25.

Flinn Scientific, Inc. (*See* General Sources.)

- AP1714 8 × 10 sheet holographic grating, $21 each.

Experiment 9

Encapsulated solar cells

Edmund Scientific (*See* General Sources.)

- G39,809 2 3/16 inches long × 1 3/8 inches wide × 1/4 inch deep, $2.20 each for orders of 3 or more.

Experiment 10

NiTi (*see* supplies listed under Chapter 9)

Appendix 3

Journal of Chemical Education Review (January 1982–July 1993)

Acid–Base Chemistry

"Periodicity in the Acid–Base Behavior of Oxides and Hydroxides," Rich, R. L. **1985**, *62*, 44.
"The Generalized Lewis Acid–Base Theory: Surprising Recent Developments," Brewer, L. **1984**, *61*, 101.
"Clay Minerals as Solid Acids and Their Catalytic Properties. A Demonstration Test with Montmorillonite," Helsen, J. **1982**, *59*, 1063.

Atomic Structure and Periodic Properties

"Periodic Properties in a Family of Common Semiconductors: Experiments with Light Emitting Diodes," Lisensky, G. C.; Penn, R.; Geselbracht, M. J.; Ellis, A. B. **1992**, *69*, 151.
"The Solvation of Halide Ions and Its Chemical Significance," Sharpe, A. G. **1990**, *67*, 309.
"Band Breadth of Electronic Transitions and the Particle in a Box Model," Olsson, L. F. **1986**, *63*, 756.
"The Particle in a Box Revisited," El-Issa, B. D. **1986**, *63*, 761.
"Relativistic Effects at the Freshman Level," Banna, M. S. **1985**, *62*, 197.
"Atomic Volume and Allotropy of the Elements," Singman, C. N. **1984**, *61*, 137.
"On the Occurrence of Metallic Character in the Periodic Table of the Elements," Edwards, P.; Sienko, M. J. **1983**, *60*, 69.
"Metalloids," Goldsmith, R. H. **1982**, *59*, 526.

Bonding

"Conducting Midshipmen—A Classroom Activity Modeling Extended Bonding in Solids," Lomax, J. F. **1992**, *69*, 794.
"A Thermochemical Note on the Bonding in Metallic Crystals," Tykodi, R. J. **1989**, *66*, 306.
"Coulombic Models in Chemical Bonding I. Description and Some Applications of a Coulombic Model," Sacks, L. J. **1986**, *63*, 288.
"An Ionic Model for Metallic Bonding," Rioux, F. **1985**, *62*, 383.
"A Model to Illustrate the Brittleness of Ionic and Metallic Crystals," Birk, J. **1985**, *62*, 667.

Colloids and Surfaces

"Synthesis and Optical Properties of Quantum-Size Metal Sulfide Particles in Aqueous Solution," Nedeljkovic, J. M.; Patel, R. C.; Kaufman, P.; Joyce-Pruden, C.; O'Leary, N. **1993**, *70*, 342.
"Scanning Tunneling Microscopy of Silicon and Carbon," Braun, R. D. **1992**, *69*, A90.
"Instructional Materials for Teaching Surface Analysis," Williams, K. R.; Young, V. Y. **1991**, *68*, 380.
"An Improved Demonstration of Colloidal Flocculation," Crees, O. L.; Senogles, E. **1986**, *63*, 715.
"Colloid Titration—A Rapid Method for the Determination of Charged Colloids," Ueno, K.; Kina, K. **1985**, *62*, 627.
"Analytical Chemistry of Surfaces Part I: General Aspects," Hercules, D. M.; Hercules, S. H. **1984**, *61*, 402.
"Analytical Chemistry of Surfaces. Part II: Electron Spectroscopy," Hercules, D. M.; Hercules, S. H. **1984**, *61*, 483.
"Analytical Chemistry of Surfaces. Part III: Ion Spectroscopy," Hercules, D. M.; Hercules, S. H. **1984**, *61*, 592.
"The Smart Electron (ESCA and Auger)," Adler, I.; Yin, L. I.; Sang, T. T.; Coyle, G. **1984**, *61*, 757.
"Electric Birefringence (Apparatus for Colloids)," Trimm, H. H.; Parslow, K.; Jennings, B. R. **1984**, *61*, 1114.

Density

"Method for Separating or Identifying Plastics," Kolb, K.; Kolb, D. **1991**, *68*, 348.
"Determination of the Density of Crystalline Solids in the Undergraduate Laboratory," Craig, R. **1989**, *66*, 599.
"Density Gradient Columns for Chemical Displays," Guenther, W. B. **1986**, *63*, 148.
"A Quick Method for Determining the Density of Single Crystals," Gutiérrez-Zorrilla, R.; Gutiérrez-Zorrilla, J. M. **1985**, *62*, 167.
"Molarity (Atomic Density) of the Elements as Pure Crystals," Pauling, L.; Herman, Z. S. **1985**, *62*, 1086.

Electrochemistry

"Coloring Titanium and Related Metals by Electrochemical Oxidation," Gaul, E. **1993**, *70*, 176.

"The Aluminum Can as Electrochemical Energy Source," Lehman, T. A.; Renich, P.; Schmidt, N. E. **1993**, *70*, 495.
"An Easy-To-Do Plating Experiment," Herrmann, M. S. **1992**, *69*, 60.
"A Lemon Powered Clock," Letcher, T. M.; Sonemann, A. W. **1992**, *69*, 157.
"The Energy Conversion Go-Around Kit Developed by NEED: Ten Teaching Modules That Use the Products of Chemistry," Kauffman, G. B.; Zafran, R. **1992**, *69*, 366.
"The Construction and Use of Commercial Voltaic Cell Displays in Freshman Chemistry," Shearer, E. C. **1990**, *67*, 158.
"Solid State Electrochemical Measurements of the Phenazine–Iodine Complex," Aronson, S. et al., **1990**, *67*, 432.
"Redox Ion Exchanger Fuel Cell," Rawat, J.; Ansari, A. A. **1990**, *67*, 808.
"Electrochemistry of the Zinc–Silver Oxide System. Part I. Thermodynamic Studies Using Commercial Miniature Cells," Smith, M. J.; Vincent, C. A. **1989**, *66*, 529.
"Electrochemistry of the Zinc–Silver Oxide System. Part II. Practical Measurements of Energy Conversion Using Commercial Miniature Cells," Smith, M. J.; Vincent, C. A. **1989**, *66*, 683.
"A Simple Platinum Electrode," Worley, J. D. **1986**, *63*, 274.
"A Lithium-Powered Heart," *Proc. Natl. Acad. Sci. U.S.A.*, **1986**, *63*, 844.
"Photoelectrochemical Solar Cells," McDevitt, J. T. **1984**, *61*, 217.
"Anodizing Aluminum with Frills," Doeltz, A. E.; Tharaud, S.; Sheehan, W. F. **1983**, *60*, 156.
"Electrosynthesis Technology," Weinberg, N. L. **1983**, *60*, 268.
"Electrochemical Engineering," Alkire, R. C. **1983**, *60*, 274.
"Electrochemistry of the Hall–Heroult Process for Aluminum Smelting," Haupin, W. E. **1983**, *60*, 279.
"Fuel Cells and Electrochemical Energy Storage," Sammells, A. F. **1983**, *60*, 320.
"Photoelectrochemistry: Introductory Concepts," Finklea, H. O. **1983**, *60*, 325.
"Energetics of the Semiconductor Electrolyte Interface," Turner, J. A. **1983**, *60*, 327.
"Dye Photooxidation at Semiconductor Electrodes: A Corollary to Spectral Sensitization in Photography," Spitler, M. T. **1983**, *60*, 330.
"Excited-State Processes of Relevance to Photoelectrochemistry," Ellis, A. B. **1983**, *60*, 332.
"Chemically Derivatized Semiconductor Photoelectrodes," Wrighton, M. S. **1983**, *60*, 335.
"An Overview of the Progress in Photoelectrochemical Energy Conversion," Parkinson, B. **1983**, *60*, 338.
"A Photoelectrochemical Solar Cell. An Undergraduate Experiment," Boudreau, S. M.; Rauh, R. D.; Boudreau, R. A. **1983**, *60*, 498.
"Dyeing of Anodized Aluminum," Grotz, L. C. **1983**, *60*, 763.
"Photoelectrochemistry: Inorganic Photochemistry at Semiconductor Electrodes," Wrighton, M. S. **1983**, *60*, 877.
"Photochemical Conversion and Storage of Solar Energy," Kutal, C. **1983**, *60*, 882.
"Thermodynamics of a Galvanic Cell," McSwiney, H. D. **1982**, *59*, 165.
"Some Tungsten Oxidation–Reduction Chemistry," Pickering, M.; Monts, D. L. **1982**, *59*, 693.

Kinetics and Catalysis

"Experiments Illustrating Heterogeneous Catalysis," Luengo, M.; Sermon, A. **1991**, *68*, 251.
"The Preparation of Pure Zeolite NaY and Its Conversion to High Silica Faujasite," Blatter, F.; Schumacher, E. **1990**, *67*, 519.

"Selective Oxidation and Ammoxidation of Olefins by Heterogeneous Catalysis," Grassel, R. K. **1986**, *63*, 216.
"Preparation and Evaluation of a Synthetic Zeolite Catalyst. An Undergraduate Laboratory Experiment," Copperthwaite, R. Hutchings, G. J.; van der Reit, M. **1986**, *63*, 632.
"Conversion of Methanol to Hydrocarbons Using a Zeolite Catalyst," Bibby, D. M.; Copperthwaite, R.; Hutchings, G. J.; Johnston, P.; Orchard, S. W. **1986**, *63*, 634.
"The Production of Oxygen Gas: A Student Catalysis Experiment (Spinels)," Onuchukwu, A. I.; Mshelia, P. B. **1985**, *62*, 809.
"Treat 'Em to Tchaikovsky," Whitman, M. **1983**, *60*, 229.
"Getting a 'Bang' Out of Chemical Kinetics," Hague, G. R., Jr. **1983**, *60*, 355.
"The Kinetics of Photographic Development. A General Chemistry Experiment," Byrd, J. E.; Perona, M. J. **1982**, *59*, 335.
"Cyclohexanol Dehydration. A Simple Experiment in Heterogeneous Catalysis," Costa, A. **1982**, *59*, 1066.

The Lanthanides

"BCS Primer," Sahu, D.; Langner, A.; George, T. F. **1990**, *67*, 738.
"On the Lanthanide and 'Scandinide' Contractions," Lloyd, D. **1986**, *63*, 502.

Miscellaneous

"Using the Multimeter in the Chemistry Teaching Laboratory: Part 1. Colorimetry and Thermometry Experiments," Andres R. T.; Sevilla , F. III **1993**, *70*, 514.
"Diode Lasers," Baumann, M. G. D.; Wright, J. C.; Ellis, A. B.; Kuech, T.; Lisensky, G. C. **1992**, *69*, 89.
"A Simple Inexpensive Device for Measuring the Critical Temperature of a High-Temperature Superconductor," Green, D. B.; Douphner, D.; Hutchinson, B. **1992**, *69*, 343.
"Demonstrating High-Temperature Superconductivity in the Chemistry Lab through the Meissner Effect," McHale, J.; Schaeffer, R.; Salomon, R. E. **1992**, *69*, 1031.
"The Rocky Road to Chemistry," Walker L.; Lee, **1990**, *67*, 325.
"A Catalog of Reactions for General Chemistry," Tykodi, R. J. **1990**, *67*, 665.
"Apparatus for Two-Probe Conductivity Measurements on Compressed Powders," Wudl, F.; Bryce, M. R. **1990**, *67*, 717.
"Designer Solids and Surfaces," Mallouk, T.; Lee, H. **1990**, *67*, 829.
"The Bismuth-Sodium Nitrate Reaction," Hill, W. D., Jr. **1989**, *66*, 709.
"Two Approaches to the Determination of Lead in Brass: DSC and AAS," Choi, S.; Larrabee, J. A. **1989**, *66*, 864.
"A Reaction Involving Oxygen and Metal Sulfides," Hill, W. D., Jr. **1986**, *63*, 441.
"Lasers: A Valuable Tool for Chemists," Findsen, E. W.; Ondrias, M. R. **1986**, *63*, 479.
"Photochromic Glass," Araujo, R. J. **1985**, *62*, 472.
"Phototropic Glasses," Eiswirth, M.; Schwankner, R. J. **1985**, *62*, 641.
"Geochemistry for Chemists," Hostettler, J. D. **1985**, *62*, 823.
"Multi-Image or Lap-Dissolve Slide Techniques and Visual Images in the Large Lecture Section," Bodner, G. M.; Cutler, A.; Greenbowe, T. J.; Robinson, W. R. **1984**, *61*, 447.
"A Simultaneous Analysis Problem for Advanced General Chemistry Laboratories (Mg_3N_2 and MgO)," Leary, J. J.; Gallaher, T. N. **1983**, *60*, 673.
"Photochemistry and Beer," Vogler, A.; Kunkely, H. **1982**, *59*, 25.
"Infrared Lasers in Chemistry," John, P. **1982**, *59*, 135.

"Lasers—An Introduction," Coleman, W. F. **1982**, *59*, 441.
"Chemical Aspects of Dentistry," Helfman, M. **1982**, *59*, 666.
"Elemental Trivia," Feinstein, H. I. **1982**, *59*, 763.
"Solar Energy: Hydrogen and Oxygen," Farrell, J. J. **1982**, *59*, 925.
"Solid State Chemistry," Boldyreva, E. V. **1993**, *70*, 551.

Nonmetals and Semiconductors

"A Simple Paper Model for Buckminsterfullerene," Vittal, J. J. **1989**, *66*, 282.
"A Facile Hückel MO Solution of Buckminsterfullerene Using Chemical Graph Theory," Dias, J. R. **1989**, *66*, 1012.
"Planck's Constant from a CdS Photoconductivity Cell," Sturm, J. E. **1989**, *66*, 1052.
"Simple Determination of the Energy Gap in a Semiconductor: A Physical Chemistry Experiment," Lehman, T. A. **1986**, *63*, 643.
"Some Chemical and Electronic Considerations of Solid State Semiconductor Crystals," Hinitz, H. J. **1986**, *63*, 956.
"Diamond: Some Interesting Physical Properties," Bickford, F. R. **1984**, *61*, 401.
"Chemistry of Printed Circuit Substrates: Physical and Chemical Requirements," Freeman, J. H. **1984**, *61*, 875.
"The Chemistry of Printed Circuit Substrates: Some of the Latest Developments," Freeman, J. H. **1984**, *61*, 993.
"Preparation of a Phosphor, ZnS: Cu^{2+}," Suib, S.; Tanaka, J. **1984**, *61*, 1099.
"An Inexpensive Zone Refining Apparatus," Needham, G. F.; Boehme, G.; Willett, R.D.; Swank, D. D. **1982**, *59*, 63.

Polymers

"Polymers and the Visible Spectrum: An 'Enlightening' Activity," Levine, E. H. **1992**, *69*, 122.
"A Program to Simulate a Nematic Liquid Crystal in 2-D for Use in Computer-Assisted Instruction," Hemenway, E. C.; Wade, A. **1992**, *69*, 123.
"Polymer Blends: Superior Products from Inferior Materials," Seymour, R. B.; Kauffman, G. B. **1992**, *69*, 646.
"Polyurethanes: A Class of Modern, Versatile Materials," Seymour, R. B.; Kauffman, G. B. **1992**, *69*, 909.
"Classroom Demonstration of Polymer Principles: Part V: Polymer Fabrication," Rodriguez, F. **1992**, *69*, 915.
"Biodegradable Films Based on Partially Hydrolyzed Corn Starch or Potato Starch," Sommerfeld, H.; Blume, R. **1992**, *69*, A151.
"Identification of Polymers in University Class Experiments," Bowence, H. **1990**, *67*, 75.
"A Polymer Viscosity Experiment with No Right Answer," Rosenthal, L. **1990**, *67*, 78.
"Piezoelectric Polymers," Seymour, R.; Kauffman, G. **1990**, *67*, 763.
"Classroom Demos of Polymer Principles: Part IV: Mechanical Strength," Rodriguez, R. **1990**, *67*, 784.
"Polymeric Fast Ion Conductor: A Physical Chemistry Experiment," Guo, W.; Fung, B. M.; Frech, R. E. **1989**, *66*, 783.
"Viscoelasticity of Cheese," Chang, Y. S.; Guo, J. S.; Lee, Y.; Sperling, L. H. **1986**, *63*, 1077.
"Determination of the Metal–Metal Distance in the Mixed Valence Complex $K_2Pt(CN)_4Br_{0.3}{\bullet}3H_2O$," Sedney, D. L.; Tanner, M. **1985**, *62*, 254.
"Nylon 6—A Simple, Safe Synthesis of a Tough Commercial Polymer," Mathias,

L. J.; Vaidya, R. A.; Canterberry, J. B. **1984**, *61*, 805.
"Polydiacetylenes. An Ideal Color System for Teaching Polymer Science," Patel, G. N.; Yang, N. L. **1983**, *60*, 181.
"The Laboratory for Introductory Polymer Courses," Mathias, L. J. (includes Appendix of previous polymer publications in *J. Chem. Educ.*), **1983**, *60*, 990.
"One Dimensional $K_2Pt(CN)_4Br_{0.3} \bullet 3H_2O$ —A Structure Containing Five Different Types of Bonding," Masuo, S. T.; Miller, J. S.; Gebert, E.; Reis, A. H., Jr. **1982**, *59*, 361.
"Some Experiments in Sulfur–Nitrogen Chemistry," Banister, A. J.; Smith, N. R. M. **1982**, *59*, 1058.

Structure

"A Discussion of the Term "Polymorphism", Reinke, H.; Dehne, H.; Hans, M. **1993**, *70*, 101.
"Chemical Principles Revisited: The Importance of Understanding Structure," Galasso, F. **1993**, *70*, 287.
"Demonstrating the Optical Principle of Bragg's Law with Moiré Patterns," Kinnegang, A. J. **1993**, *70*, 451.
"A Simple and Reliable Chemical Preparation of $YBa_2Cu_3O_{7-x}$ Superconductor: An Experiment in High Temperature Superconductivity for an Advanced Undergraduate Laboratory," Djurovich, P. I.; Watts, R. J. **1993**, *70*, 497.
"Close Packing of Identical Spheres," Ostercamp, D. L. **1992**, *69*, 162.
"Hands-on Model for Close Packing," Martin, D. F. **1992**, *69*, 495.
"An Introduction to Fullerene Structures," Boo, W. O. J. **1992**, *69*, 605.
"A Paper-Pattern System for the Construction of Fullerene Molecular Models," Beaton, J. M. **1992**, *69*, 610.
"Some Questions on Buckminsterfullerene, C_{60}, for General and Inorganic Chemistry Students," Hecht, C. E. **1992**, *69*, 645.
"Simple Generation of C_{60} (Buckminsterfullerene): A Process Suitable for Undergraduate Research Laboratories," Iacoe, D. W.; Potter, W. T.; Teeters, D. **1992**, *69*, 663.
"C_{60} and C_{70} Made Simply," Craig, N. C.; Gee, G. C.; Johnson, A. R. **1992**, *69*, 664.
"Fractal Structure for the Overhead Projector," Silverman, L. **1992**, *69*, 928.
"Finding the Face-Centered Cube in the Cubic Closest Packed Structure," Birk, J.; Coffman, P. R. **1992**, *69*, 953.
"Inexpensive Laboratory Experiments on Crystal Growth of Water Soluble Substances in Gel Media," Das, I.; Pushkarma, A.; Sharma, A.; Jaiwai, K.; Chand, S. **1992**, *69*, A47.
"Diffraction of a Laser Light by a Memory Chip," Klier, K.; Taylor, J. A. **1991**, *68*, 155.
"The Optical Transform: Simulating Diffraction Experiments in Introductory Courses," Lisensky, G. C.; Kelly, T. F.; Neu, D. R.; Ellis, A. B. **1991**, *68*, 91.
"Direct Visualization of Bragg Diffraction with a He–Ne Laser and an Ordered Suspension of Charged Microspheres," Zare, R. N.; Spencer, B. H. **1991**, *68*, 97.
"A Simple Bragg Diffraction Experiment with Harmless Visible Light," Segschneider, C.; Versmold, H. **1990**, *67*, 967.
"The Brillouin Zone —An Interface Between Spectroscopy and Crystallography," Kettle , S.; Norrby, L. J. **1990**, *67*, 1022.
"Construction of the Seven Basic Crystallographic Units," Li, T.; Worrell, J. H. **1989**, *66*, 73.
"Ionic Crystals and Electrostatics: The Cluster Model Versus the Standard Model," Recio, J. M.; Luana, V.; Pueyo, L. **1989**, *66*, 307.
"Fourier Analysis and Structure Determination. Part III. X-ray Crystal Structure Analysis," Chesick, J. **1989**, *66*, 413.

"An Experimental Study of the Liesegang Phenomenon and Crystal Growth in Silica Gels," Sharbaugh III, A. H.; Sharbaugh, Jr., A. H. **1989**, *66*, 589.
"A General Approach to Radius Ratios of Simple Ionic Crystals," Sharma, B.D. **1986**, *63*, 504.
"The 17 Two-Dimensional Space-Group Symmetries in Hungarian Needlework," Hargittai, I.; Lengyei, G. **1985**, *62*, 35.
"Crystal Growth in Gels," Suib, S. L. **1985**, *62*, 81.
"Predictions of Crystal Structure Based on Radius Ratio: How Reliable Are They?" Nathan, L. C. **1985**, *62*, 215.
"Crystal Systems," Schomaker, V.; Lingafelter, E. C. **1985**, *62*, 219.
"A General Approach to Radius Ratios of Simple Ionic Crystals," Gül, H. **1985**, *62*, 384.
"Interstitial Solid Solutions: Cooperation of Energy and Geometry," Lindsay, C. G. **1985**, *62*, 675.
"The Synthesis and Characterization of Some Fluoride Perovskites," Langley, R. H. Schmitz, C. K.; Langley, M. B. **1984**, *61*, 643.
"The Seven One-Dimensional Space-Group Symmetries Illustrated by Hungarian Folk Needlework," Hargittai, I.; Lengyei, G. **1984**, *61*, 1033.
"Models as an Aid to Courses in Crystallography and Mineralogy," Brady, K. T. **1983**, *60*, 36.
"Association of Ions and Fractional Crystallization," Scaife, C. W. J.; Dubs, R. L. **1983**, *60*, 418.
"Restrictions Upon Rotation and Inversion Axes in Crystals," Sharma, B. D. **1983**, *60*, 462.
"An Experiment with Manifold Purposes. The Chemical Reactivity of Crystal Defects Upon Crystal Dissolution," Lazzarini, A. F.; Lazzarini, E. **1983**, *60*, 519.
"Liquid Crystals—The Chameleon Chemicals," Brown, G. H. **1983**, *60*, 900.
"Models for the Study of Cubic Crystallographic Point Groups," Gremillion, A. F. **1982**, *59*, 194.
"Easily Constructed Model of Twin Octahedrons Having a Common Plane," Yamana, S.; Kawaguchi, M. **1982**, *59*, 196.
"Crystallographic and Spectroscopic Symmetry Notations," Sharma, B. D. **1982**, *59*, 554.
"An Easily Constructed Icosahedron Model," Yamana, S.; Kawaguchi, M. **1982**, *59*, 578.
"Some Simple AX and AX_2 Structures," Wells, A. F. **1982**, *59*, 630.
"Crystal Systems and General Chemistry," Sharma, B. D. **1982**, *59*, 742.

Thermodynamics

"The Use and Utility of Phase Science," Laughlin, R. G. **1992**, *69*, 26.
"Boiling and Freezing Simultaneously—With a Feeble Vacuum Pump!" Ellison, M. **1992**, *69*, 325.
"Copper/Aluminum Surprise," Bindel, T. H. **1990**, *67*, 165.
"Lattice Enthalpies of Ionic Halides, Hydrides, Oxides, and Sulfides," Holbrook, J. et al., **1990**, *67*, 304.
"Thermal Decomposition of a Natural Manganese Dioxide," Puerta, M. C.; Valerga, P. **1990**, *67*, 344.
"Free Energy Minimization Calculation of Complex Chemical Equilibria: Reduction of SiO_2 with Carbon at High Temperature," Wai, C. M.; Hutchinson, S. G. **1989**, *66*, 546.
"Thermodynamic Analysis of Ionic Compounds: Synthetic Applications," Yoder, C. H. **1986**, *63*, 232.
"Thermodynamic Inefficiency of Conversion of Solar Energy to Work," Adamson, A. W.; Namnath, J.; Shastry, V. J.; Slawson, V. **1984**, *61*, 221.

"Inorganic Thermochromism: A Lecture Demonstration of a Solid State Phase Transition," Willett, R. D. **1983**, *60*, 355.
"Ternary and Quaternary Composition Diagrams," MacCarthy, P. **1983**, *60*, 922.
"Cesium Neonide: Molecule or Thermochemical Exercise?" Blake, P. G.; Clack, D. W. **1982**, *59*, 637.
"Potential-pH Diagrams," Barnum, D. W. **1982**, *59*, 809.
"The Iron–Iron Carbide Phase Diagram," Long, G. J.; Leighly, Jr., H. **1982**, *59*, 948.

The Transition Metals

"Microscale Techniques for Determination of Magnetic Susceptibility," Woolcock J.; Zafar, A. **1992**, *69*, A176.
"Linear Chain Magnetism," Carlin, R. L. **1991**, *68*, 361.
"Thermal Decomposition of a Natural Manganese Dioxide," Puerta, M. C.; Valerga, **1990**, *67*, 344.
"Ferrimagnetism Demo," Knox, K. **1989**, *66*, 337.
"Thermochromic Tetrachlorocuprate(II)," Choi, S.; Larrabee, J. A. **1989**, *66*, 774.
"Production or Recovery of Silver for Laboratory Use," Hill, J. W.; Bellows, L. **1986**, *63*, 357.
"Metal Substitutions in Wartime Coinages," Akers, H. A. **1984**, *61*, 47.
"The Extraction of Gold and Its Simulation with Copper," Bradley, J. D.; Brand, M.; Loulié, J. **1984**, *61*, 634.
"Analysis of 1982 Pennies," Miller, J. M. **1983**, *60*, 142.
"External Weighing with Analytical Balances. Determination of Magnetic Susceptibilities of Inorganic Compounds," Toma, H. E.; Ferreira, A. M. C.; Osorio, V. K. L. **1983**, *60*, 600.
"Titanium Dioxide," Davis, K. A. **1982**, *59*, 158.
"Determination of Pyrite in Coal and Lignite by Thermogravimetric Analysis," Hyman, M.; Rowe, M. W. **1982**, *59*, 424.
"The Corrosion and Preservation of Iron Antiques," Walker, R. **1982**, *59*, 943.

Addendum: August 1993–August 1994

"Rare Earth Iron Garnets: Their Synthesis and Magnetic Properties," Geselbracht, M. J.; Cappellari, A. M.; Ellis, A. B.; Rzeznik, M. A.; Johnson, B. J. **1994**, *71*, 696.
"The Sol-Gel Preparation of Silica Gels," Buckley, A. M.; Greenblatt, M. **1994**, *71*, 599.
"Nickel-Titanium Memory Metal: A "Smart" Material Exhibiting a Solid-State Phase Change and Superelasticity," Gisser, K. R. C.; Geselbracht, M. J.; Cappellari, A. M.; Hunsburger, L.; Ellis, A. B.; Perepezko, J.; Lisensky, G. C. **1994**, *71*, 334.
"Mechanical Properties of Metals: Experiments with Steel, Copper, Tin, Zinc, and Soap Bubbles, " Geselbracht, M. J.; Ellis, A. B.: Penn, R. L.; Lisensky, G. C.; Stone, D. S. **1994**, *71*, 254.
"Really, your lattices are all primitive, Mr. Bravais!," Kettle, S. F. A.; Norrby, L. J. **1993**, *70*, 959.
"Photodegradation of Methylene Blue: Using Solar Light and Semiconductor (TiO_2)," Nogueira, R. F. P.; Jardim, W. F. **1993**, *70*, 861.

Appendix 4

Answers to Selected Problems

David C. Boyd
Department of Chemistry, University of St. Thomas
St. Paul, MN 55105

Todd K. Trout
Department of Chemistry, Mercyhurst College
Erie, PA 16546

Chapter 1

1. a. "Synthesis and Processing" b. "Properties" c. "Properties" and "Performance" d. "Synthesis and Processing"

3. The re-entrant foam has a negative Poisson ratio as described in the text. If the re-entrant foam is compressed, the denominator of the Poisson ratio (the fractional increase in length) will be negative. As a result, the numerator of the Poisson ratio (the fractional decrease in width) must be positive in order to satisfy the condition that the ratio be negative. Thus, compressing the foam (decrease in length) results in a decrease in width. The size of the cross-section will decrease when the foam is compressed.

4. Pulling on re-entrant foam causes the foam to expand in cross-sectional area, and compressing it causes the foam to shrink in cross-sectional area (*see* problem 3). If re-entrant foam is inserted into a hole, compression of the foam during insertion causes its cross-section to shrink, assisting the process. If an object is hung on the foam, the foam resists being pulled out of the hole by its expanding cross-section.

5. In general, any common material can be researched and traced from the raw material stage to the waste disposal stage. However, only certain products are actually recycled or reused to complete the cycle. Other materials might be said to "spiral" rather than cycle in that they cannot be reused for their

2725–X/93/0505$09.75/1

original purpose. (For example, a polystyrene food container cannot be recycled into anything that comes in contact with food, so such material might be made into "speed bumps", plastic lumber, etc.) Waste disposal questions alone can also make for interesting library research projects. We have found that students are able to find a reasonable amount of information about the following:

petroleum	rubber	paper	aluminum
nuclear fuel	nuclear weapons	carbon	batteries
glass	water	antifreeze	cotton
mercury	steel	motor oil	diapers
polyethylene	other plastics		

Good starting points for research include
- *Encyclopedia of Physical Science and Technology*
- *McGraw-Hill Encyclopedia of Science and Technology*
- *Kirk-Othmer Encyclopedia of Science and Technology*
- *Ullmann's Encyclopedia of Industrial Technology*
- *Concise Encyclopedia of Polymer Processing and Application*
- *Handbook of Polymer Science and Technology*

Local experts may also be willing to serve as resource people.

Chapter 2

1. The molar masses of tantalum and niobium are 180.95 g/mol and 92.91 g/mol respectively, a ratio of 1.95:1.00. With nearly identical unit cell sizes, the difference in mass results in solids with densities that differ by a factor of 1.94:1.00.

2. The diameter of a tungsten atom is (2)(137 pm) = 274 pm
$= 2.74 \times 10^{-10}$ m $= 2.74 \times 10^{-8}$ cm = 2.74 Å

3. molar heat capacity = (3)(R)(number of ions per ionic formula unit)
RbCl: molar heat capacity = (3)(8.31 J mol^{-1} K^{-1})(2) = 50.0 J mol^{-1} K^{-1}
MgF_2: molar heat capacity = (3)(8.31 J mol^{-1} K^{-1})(3) = 74.8 J mol^{-1} K^{-1}
AlF_3: molar heat capacity = (3)(8.31 J mol^{-1} K^{-1})(4) = 99.8 J mol^{-1} K^{-1}
Al_2O_3: molar heat capacity = (3)(8.31 J mol^{-1} K^{-1})(5) = 124 J mol^{-1} K^{-1}

4. For example, aluminum molar heat capacity = 24 J mol^{-1} K^{-1}
(24 J $mol^{-1}K^{-1}$)(1 mol Al/26.98 g) = 0.89 J g^{-1} K^{-1}

5. For example, aluminum:
molar heat capacity of Al = 24 J mol^{-1} K^{-1}
gram heat capacity of Al = 0.89 J $g^{-1}K^{-1}$ (*see* problem 4)
gram heat capacity of H_2O = 4.184 J $g^{-1}K^{-1}$
heat absorbed by water = heat released by Al
heat absorbed by water = (gram heat capacity H_2O)(mass of H_2O)(ΔT),
heat released by Al = (gram heat capacity Al)(mass of Al)(ΔT),
Al and H_2O will thermally equilibrate to the same final temperature.
T_f = 310 K or 37 °C.

6. Heat released by metal = (gram heat capacity of metal)(mass of metal)(ΔT),
Heat absorbed by water = (gram heat capacity H_2O)(mass of H_2O)(ΔT),
heat absorbed by water = heat released by metal

∴ the gram heat capacity of metal is calculated to be 0.491 J g^{-1} K^{-1}

The gram heat capacity and molar heat capacity differ by a factor of molar mass. Assuming the molar heat capacity of this metal to be the theoretical value for the molar heat capacity of a metallic element (25 J mol^{-1} K^{-1}, as per Table 2.1), the molar mass of the metal is

$$25 \text{ J mol}^{-1}\text{ K}^{-1} = (0.491 \text{ J g}^{-1}\text{K}^{-1})(\text{molar mass})$$

$$\text{molar mass} = (25 \text{ J mol}^{-1}\text{K}^{-1})/(0.491 \text{ J g}^{-1}\text{ K}^{-1}) = 51 \text{ g mol}^{-1}$$

The metal is predicted to be vanadium.

7. As a result of sodium bromide's solubility in water, the heat capacity of NaBr cannot be measured in water, because the energetics of dissolution and solvation will be observed together with the simple transfer of thermal energy. The measurement can be performed in ethyl alcohol, as sodium bromide will not dissolve in ethyl alcohol. The procedure to be used follows:

   As a first step, measure the heat capacity of ethyl alcohol. This can be accomplished by heating a sample of aluminum of known mass to a known temperature. For example, place the aluminum sample into boiling water and allow the aluminum to achieve thermal equilibrium with the water. Place the hot piece of aluminum into a preweighed sample of ethyl alcohol of known temperature and measure the temperature change to obtain the heat capacity of ethyl alcohol.

   To measure the heat capacity of NaBr, heat a dry, preweighed sample of NaBr in an oven and place the sodium bromide in a sample of ethyl alcohol of known mass and temperature. The rise in temperature together with the heat capacity of ethyl alcohol allows calculation of the heat capacity of NaBr.

9. (400 Ni atoms)(2.48 Å/Ni atom) = 992 Å or 9.92×10^{-8} m

10. $PV = nRT$
    $n = PV/RT$ $\qquad P = (1.0 \times 10^{-10}$ torr)(1 atm/760 torr) = 1.3×10^{-13} atm
    $n = (1.3 \times 10^{-13}$ atm)$(1.0 \times 10^{-3}$ L)/[(0.0821 L-atm/mol-K)(296 K)]
    $n = 5.3 \times 10^{-18}$ mol gas
    $(5.3 \times 10^{-18}$ mol)$(6.02 \times 10^{23}$ gas molcules/mol) = 3.2×10^{6} molcules

11. Because there are only two degrees of freedom in Flatland, a mol of monatomic gas atoms will have an average thermal energy of (2)(1/2 RT) = RT, and a molar heat capacity of R or approximately 8.3 J mol^{-1} K^{-1}.

12. (0.38 J g^{-1} K^{-1})(63.55 g Cu/1 mol Cu) = 24 J mol^{-1} K^{-1}
    (24 J $mol^{-1}K^{-1}$)(1 mol Ag/107.87 g) = 0.22 J g^{-1} K^{-1}

13. Bi: (0.120 J g^{-1} °C^{-1})(208.98 g/1 mol Bi) = 25.1 J mol^{-1} °C^{-1}
    Zn: (0.388 J g^{-1} °C^{-1})(65.37 g/1 mol Zn) = 25.4 J mol^{-1} °C^{-1}
    Fe: (0.460 J g^{-1} °C^{-1})(55.85 g/1 mol Fe) = 25.7 J mol^{-1} °C^{-1}
    S: (0.787 J g^{-1} °C^{-1})(32.06 g/1 mol S) = 25.2 J mol^{-1} °C^{-1}

    Note that these molar heat capacities are comparable and close to the theoretical value of 25 J mol^{-1} °C^{-1} as described in the text.

14. a. In the absence of an external magnetic field, the magnetic domains of iron remain randomly oriented; thus, there is little net magnetization and no attraction to the iron filings.
    b. The magnet aligns or orients the domains with the applied external field, creating a large net magnetization in the nail. The nail then serves as the

source of a magnetic field that orients the domains of the iron filings, attracting the filings.

15. The equation: molar heat capacity = $3Rp$ (where p is the number of atoms in a formula unit) predicts a molar heat capacity for elemental iron (Fe) of $3R$. Calcium oxide, a 1:1 ionic substance composed of Ca^{2+} and O^{2-} ions, has a predicted molar heat capacity of $6R$ as $p = 2$. Since the formula weights of calcium oxide and elemental iron are about the same, the gram heat capacity of CaO should be about twice that of Fe.

16. Vanadium in VO has a formal oxidation state of +2 and an odd number of electrons (3), making it paramagnetic. In contrast, V_2O_5 has a formal oxidation state of +5 and no unpaired electrons, making it diamagnetic. Thus, VO is more strongly attracted to a magnetic field.

17. Stored chemical energy, present in the foods we eat, is used to give us the energy needed to squeeze the trigger to cock a spring-loaded hammer in the lighter. The cocked spring contains potential energy, which is converted to kinetic energy as the hammer accelerates towards the piezoelectric crystal. When the hammer strikes the crystal, the energy is converted to heat and is also used to mechanically induce a momentary net dipole moment that will generate a change in the electric field of the crystal. This can cause sparking in the nearby combustible gas and ignite it. The combustion results from an exothermic reaction between the fuel and oxygen.

18. Exercise machines with a tunable resistance to motion is one possibility.

Chapter 3

1. As examples, we do problems a, c, and l.

a. CsCl

8 Cl^- corner-shared ions: (8)(1/8) = 1 Cl^-
1 Cs^+ ion within the cell: (1)(1) = 1 Cs^+
Thus the unit cell is composed of 1 Cs^+ and 1 Cl^- ion. The empirical formula is CsCl.

c. Cu_2AlMn

8 Mn corner-shared atoms: (8)(1/8) = 1 Mn
6 Mn face-shared atoms: (6)(1/2) = 3 Mn
12 Al edge-shared ions: (12)(1/4) = 3 Al
1 Al atom within the cell: (1)(1) = 1 Al
8 Cu atoms within the cell (8)(1) = 8 Cu
Thus the unit cell is composed of four Mn atoms, four Al atoms, and eight Cu atoms. The empirical formula for the solid is Cu_2AlMn

l. CaC_2

8 Ca^{2+} corner-shared ions: (8)(1/8) = 1 Ca^{2+}
6 Ca^{2+} face-shared ions: (6)(1/2) = 3 Ca^{2+}
8 C^- edge-shared ions: (8)(1/4) = 2 C^-
8 C^- face-shared ions: (8)(1/2) = 4 C^-
2 C^- ions within the cell: (2)(1) = 2 C^-
Thus the unit cell is composed of four Ca^{2+} and eight C^- ions. The empirical formula for the solid is CaC_2.

2. As examples, we do problems a and b.

a. The formula of the compound that crystallizes in a cubic unit cell with a rubidium ion on each corner and a bromide ion in the center of the cell is RbBr:

8 Rb^+ corner-shared ions: $(8)(1/8) = 1\ Rb^+$
1 Br^- ion within the cell: $(1)(1) = 1\ Br^-$

The empirical formula is therefore RbBr.

The percent composition of RbBr is calculated as follows:
Molar mass of RbBr = (85.47 g) + (79.91 g) = 165.38 g
%Rb = (85.47 g/165.38 g)(100) = 51.68 %
%Br = (79.91 g/165.38 g)(100) = 48.32 %

The masses of rubidium cations and bromide anions required to prepare 5.00 g of RbBr are
(5.00 g RbBr)(1 mol/165.38 g) = 3.02×10^{-2} mol RbBr
(3.02×10^{-2} mol RbBr)(1 mol Rb^+/mol RbBr) = 3.02×10^{-2} mol Rb^+.
(3.02×10^{-2} mol Rb^+)(85.47 g Rb^+/1 mol) = 2.58 g Rb^+.
(3.02×10^{-2} mol RbBr)(1 mol Br^-/mol RbBr) = 3.02×10^{-2} mol Br^-.
(3.02×10^{-2} mol Br^-)(79.91 g Br^-/1 mol) = 2.41 g Br^-.

b. 8 Ni^{2+} corner-shared ions: $(8)(1/8) = 1\ Ni^{2+}$
6 Ni^{2+} face-shared ions: $(6)(1/2) = 3\ Ni^{2+}$
12 O^{2-} edge-shared ions: $(12)(1/4) = 3\ O^{2-}$
1 O^{2-} ion within the cell: $(1)(1) = 1\ O^{2-}$

The empirical formula is therefore NiO.

NiO is 78.58 % Ni and 21.42 % O

The masses of nickel cations and oxygen anions required to prepare 5.00 g of NiO are 3.93 g Ni^{2+} and 1.07 g of O^{2-}.

3.

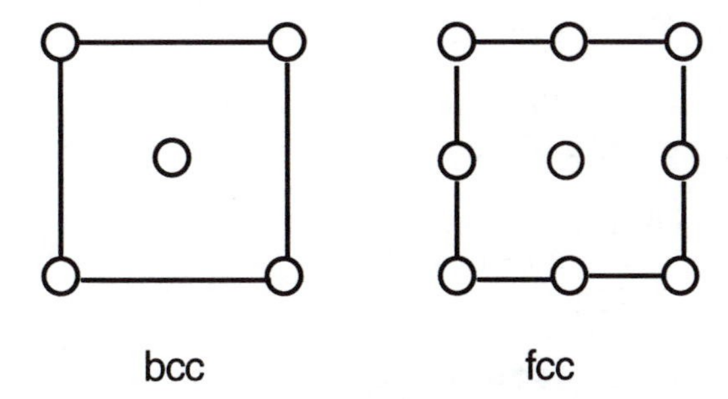

4. As an example, we do the calculation for barium. Barium crystallizes in a cubic unit cell with an edge of length 5.025 Å. There is a barium atom on each of the corners of the unit cell and in the center of the unit cell.
Contents of a unit cell:

(8 corner Ba atoms)(1/8 occupancy) = 1 Ba atom
(1 Ba atom within the cell)(1 occupancy) = 1 Ba atom

There is a total of 2 Ba atoms per unit cell.

Mass of a unit cell:
(2 Ba atoms/unit cell)(1 mol Ba/6.022×10^{23} atoms)(137.33 g/mol Ba)
= 4.561×10^{-22} g per unit cell

Volume of a unit cell:
$(5.025\ \text{Å})^3(1 \times 10^{-8}\ \text{cm/Å})^3 = 1.269 \times 10^{-22}\ \text{cm}^3$

Density = mass/volume = $4.561 \times 10^{-22}\ \text{g}/(1.269 \times 10^{-22}\ \text{cm}^3) = 3.594\ \text{g/cm}^3$

5. Contents of unit cell:
(8 corner Al atoms)(1/8 occupancy) = 1 Al atom
(6 face-centered atoms)(1/2 occupancy) = 3 Al atoms
There are 4 Al atoms per cell

Mass of a unit cell:
(4 Al atoms)(1 mol Al/6.022×10^{23} atoms)(26.98 g/mol Al) = 1.792×10^{-22} g

Volume of a unit cell:
$(4.050\ \text{Å})^3(1 \times 10^{-8}\ \text{cm/Å})^3 = 6.643 \times 10^{-23}\ \text{cm}^3$

Density = mass/volume = $1.792 \times 10^{-22}\ \text{g}/\ 6.643 \times 10^{-23}\ \text{cm}^3 = 2.698\ \text{g/cm}^3$

6. Contents of unit cell:
(8 corner Ca^{2+} ions)(1/8 occupancy) = 1 Ca^{2+} ion
(12 edge Ca^{2+} ions)(1/4 occupancy) = 3 Ca^{2+} ions
There is a total of 4 Ca^{2+} ions per cell
(8 internal F^- ions)(1 occupancy) = 8 F^- ions

Mass of a unit cell, in atomic mass units (amu):
(4 Ca^{2+} ions)(40.08 amu/Ca^{2+}) + (8 F^- ions))(19.00 amu/F^-) = 312.32 amu

Volume of unit cell:
$(5.46295\ \text{Å})^3(1 \times 10^{-8}\ \text{cm/Å})^3 = 1.63035 \times 10^{-22}\ \text{cm}^3$

Mass of a unit cell, in grams:
$(1.63035 \times 10^{-22}\ \text{cm}^3)(3.1805\ \text{g/cm}^3) = 5.1853 \times 10^{-22}$ g

Division of the amu mass by the gram mass gives Avagadro's number (N)
N = (312.32 amu)/(5.1853×10^{-22} g) = 6.0232×10^{23}

7. Yes, both represent CsCl. Both show the 1:1 ratio of ions and the correct cubic coordination (Chapter 5) of both cations and anions.

8. As examples, we do the Ga(P,As) solid solution system.
Maximum % Ga occurs in GaP (molar mass = 100.69 g/mol):
% Ga = (69.72 g/100.69 g/mol)(100) = 69.24%

Minimum % Ga occurs in GaAs (molar mass = 144.64 g/mol):
% Ga = (69.72 g/144.64 g/mol)(100) = 48.20%

9. Contents of top unit cell:
(8 corner Ti^{4+} ions)(1/8 occupancy) = 1 Ti^{4+} ion
(12 edge O^{2-} ions)(1/4 occupancy) = 3 O^{2-} ions
(1 internal Ca^{2+} ion)(1 occupancy) = 1 Ca^{2+} ion

Contents of bottom unit cell:
(8 corner Ca^{2+} ions)(1/8 occupancy) = 1 Ca^{2+} ions
(6 face-shared O^{2-} ions)(1/2 occupancy) = 3 O^{2-} ions
(1 internal Ti^{4+} ion)(1 occupancy) = 1 Ti^{4+} ion

Both representations lead to the empirical formula $CaTiO_3$ and the same spatial relationships of atoms. They represent identical structures.

10. The molar mass of $KCl_{0.30}Br_{0.70}$ is calculated as follows:
(1)(39.0983 g K/mol) + (0.30)(35.4527 g Cl/mol) + (0.70)(79.904 g Br/mol) = 105.667 g

The percent by mass of bromide ions in $KCl_{0.30}Br_{0.70}$ is given by
(56 g Br)(100)/(105.667 g $KCl_{0.30}Br_{0.70}$) = 53% Br

Therefore, the mass of Br in a 1.00-g sample of $KCl_{0.30}Br_{0.70}$ is
(1.00 g $KCl_{0.30}Br_{0.70}$) (53 g Br)/(100 g $KCl_{0.30}Br_{0.70}$) = 0.53 g Br

The amount of AgBr that can be produced from 0.53 g Br is calculated to be

(0.53 g Br)(1 mol Br/79.904 g Br) = 6.6×10^{-3} mol Br

$$(6.6 \times 10^{-3} \text{ mol Br}) \frac{1 \text{ mol AgBr}}{1 \text{ mol Br}} \frac{187.77 \text{ g AgBr}}{1 \text{ mol AgBr}} = 1.2 \text{ g AgBr}$$

11. a. volume of unit cell = $(4.29 \times 10^{-8} \text{ cm})^3 = 7.90 \times 10^{-23} \text{ cm}^3$
mass of unit cell = $(7.90 \times 10^{-23} \text{ cm}^3)(5.66 \text{ g/cm}^3) = 4.47 \times 10^{-22}$ g

b. Let x = the number of oxygen atoms in a unit cell
Then $0.932x$ = the number of iron atoms in a unit cell
$(0.932x)$(55.847 g/mol)/(6.022×10^{23} ions per mol) +
(x)(16.00 g/mol)/(6.022×10^{23} ions per mol) = 4.47×10^{-22} g
(1.13×10^{-22} g)(x) = 4.47×10^{-22} g
$x = 3.96$ = number of oxygen atoms per unit cell
(0.932)(3.96) = 3.69 iron atoms per unit cell

c. (3.96/4.00)(100) = 99% of oxygen sites occupied
(3.69/4.00)(100) = 92% of iron sites occupied

12. Assuming 100 g TiO_x
(70.90 g Ti)(1 mol Ti/47.88 g) = 1.481 mol Ti
(29.10 g O)(1 mol O/16.00 g) = 1.819 mol O
Molar ratios: Ti: (1.481/1.481) = 1.000; O: (1.819/1.481) = 1.228
Empirical formula = $TiO_{1.228}$

13. We do isolation of Cd as an example: What mass of $Cd_{0.010}Zn_{0.99}S$ must be processed to isolate 1,000 kg of cadmium?
molar mass $Cd_{0.010}Zn_{0.99}S$ = 98 g/mol
(1.000×10^3 kg Cd)(1,000 g/kg)(1 mol Cd/112.41 g) = 8.896×10^3 mol Cd

$$8.896 \times 10^3 \text{ mol Cd} \times \frac{1 \text{ mol } Cd_{0.010}Zn_{0.99}S}{0.010 \text{ mol Cd}} \times \frac{98 \text{ g}}{1 \text{ mol } Cd_{0.010}Zn_{0.99}S}$$

$= 8.7 \times 10^7$ g $Cd_{0.010}Zn_{0.99}S$

14. Molar mass of $CdS_{0.29}Se_{0.71}$ = 1$\underline{7}$8 g
% Cd = 63%, % S = 5.2%, % Se = 32%

Molar mass of $Na_{0.24}WO_3$ = 2$\underline{3}$7.37 g
% Na = 2.3%, % W = 77.5%, % O = 20.2%

15. $x = 0.54$ and $y = 0.63$

16. For example, if $x = 0.50$
$Ca_5(PO_4)_3OH + 0.50\ NaF \rightarrow Ca_5(PO_4)_3(OH)_{0.50}F_{0.50} + 0.50\ NaOH$
$(1.00\ g\ Ca_5(PO_4)_3OH)(1\ mol/\ 502.32\ g) = 1.99 \times 10^{-3}\ mol\ Ca_5(PO_4)_3OH$
$(1.99 \times 10^{-3}\ mol\ Ca_5(PO_4)_3OH)(0.50\ mol\ NaF/1\ mol\ Ca_5(PO_4)_3OH) =$
$9.\underline{9}5 \times 10^{-4}\ mol\ NaF$
$(9.\underline{9}5 \times 10^{-4}\ mol\ NaF)(41.99\ g/\ 1\ mol\ NaF) = 4.2 \times 10^{-2}\ g\ NaF$

17. For example, if $x = 0.50$
$Si_{0.50}Ge_{0.50} + O_2 \rightarrow 0.50\ SiO_2 + 0.50\ GeO_2$
$(0.5356\ g\ Si_{0.50}Ge_{0.50})(1\ mol/5\underline{0}.35\ g) = 1.\underline{0}64 \times 10^{-2}\ mol\ Si_{0.50}Ge_{0.50}$
$(1.\underline{0}64 \times 10^{-2}\ mol\ Si_{0.50}Ge_{0.50})(0.50\ mol\ SiO_2/1\ mol\ Si_{0.50}Ge_{0.50}) =$
$5.\underline{3}18 \times 10^{-3}\ mol\ SiO_2$
$(5.\underline{3}18 \times 10^{-3}\ mol\ SiO_2)(60.09\ g/1\ mol\ SiO_2) = 0.32\ g\ SiO_2$
$(1.\underline{0}64 \times 10^{-2}\ mol\ Si_{0.50}Ge_{0.50})(0.50\ mol\ GeO_2/1\ mol\ Si_{0.50}Ge_{0.50}) =$
$5.\underline{3}18 \times 10^{-3}\ mol\ GeO_2$
$(5.\underline{3}18 \times 10^{-3}\ mol\ GeO_2)(104.61\ g/1\ mol\ GeO_2) = 0.56\ g\ GeO_2$
Total mass of oxides = 0.88 g

18. For example, if $x = 0.50$
$2\ Ca_5(PO_4)_3(OH)_{0.50}F_{0.50} + 7H_2SO_4 \rightarrow 3Ca(H_2PO_4)_2 + 7CaSO_4 + H_2O + HF$
$(1.000 \times 10^3\ kg\ Ca_5(PO_4)_3(OH)_{0.50}F_{0.50})(1 \times 10^3\ g/kg)(1\ mol/5\underline{0}3.31\ g) =$
$1.\underline{9}87 \times 10^3\ mol\ Ca_5(PO_4)_3(OH)_{0.50}F_{0.50}$
$(1.\underline{9}87 \times 10^3\ mol\ Ca_5(PO_4)_3(OH)_{0.50}F_{0.50})\ x$
$(1\ mol\ HF/2\ mol\ Ca_5(PO_4)_3(OH)_{0.50}F_{0.50})(20.01\ g/mol\ HF) =$
$2.0 \times 10^4\ g\ HF$

19. For example, if $x = 0.50$
$CdS_{0.50}Se_{0.50} + 2\ HCl \rightarrow CdCl_2 + 0.50\ H_2S + 0.50\ H_2Se$
$(0.5356\ g\ CdS_{0.50}Se_{0.50})(1\ mol\ /1\underline{6}8\ g) = 3.\underline{1}9 \times 10^{-3}\ mol\ CdS_{0.50}Se_{0.50}$
$(3.\underline{1}9 \times 10^{-3}\ mol\ CdS_{0.50}Se_{0.50})(1\ mol\ CdCl_2/1\ mol\ CdS_{0.50}Se_{0.50}) =$
$3.\underline{1}9 \times 10^{-3}\ mol\ CdCl_2$
$(3.\underline{1}9 \times 10^{-3}\ mol\ CdCl_2)(183.3\ g/1\ mol) = 0.58\ g\ CdCl_2$

20. For example, if $x = 0.50$
$CdS_{0.50}Se_{0.50} + 2\ HCl \rightarrow CdCl_2 + 0.50\ H_2S + 0.50\ H_2Se$
$(0.5356\ g\ CdS_{0.50}Se_{0.50})(1\ mol\ /1\underline{6}8\ g) = 3.\underline{1}9 \times 10^{-3}\ mol\ CdS_{0.50}Se_{0.50}$
$(3.\underline{1}9 \times 10^{-3}\ mol\ CdS_{0.50}Se_{0.50})(0.50\ mol\ H_2S/1\ mol\ CdS_{0.50}Se_{0.50}) =$
$1.\underline{6}0 \times 10^{-3}\ mol\ H_2S$

$PV = nRT$
$V = nRT/P = (1.\underline{6}0 \times 10^{-3}\ mol\ H_2S)(0.0821\ L{\cdot}atm/mol{\cdot}K)(273\ K)/(1\ atm)$
$V = 0.036\ L$

21. liquid H_2: $(1\ cm^3)(0.070\ g/cm^3) = 0.070\ g\ H_2$

For LiH:
volume of unit cell = $(4.085\ Å)^3(1 \times 10^{-8}\ cm/1\ Å)^3 = 6.817 \times 10^{-23}\ cm^3$
cell contents: (8 corner Li^+ ions)(1/8 occupancy) = 1 Li^+ ion
(6 face-centered Li^+ ions)(1/2 occupancy) = 3 Li^+ ions
(12 H^- edge-shared ions)(1/4 occupancy) = 3 H^- ions
(1 internal H^- ion)(1 occupancy) = 1 H^- ion

mass of unit cell

$= (4\ H^-\ \text{ions})\dfrac{1.008\ g}{\text{mol H}}\dfrac{1\ \text{mol}\ H^-}{6.022 \times 10^{23}\ \text{ions}} + (4\ Li^+\ \text{ions})\dfrac{6.941\ g}{\text{mol Li}}\dfrac{1\ \text{mol}\ Li^+}{6.022 \times 10^{23}\ \text{ions}}$

$= 5.280 \times 10^{-23}\ g$

$$\text{density of LiH} = \text{mass/volume} = \frac{5.280 \times 10^{-23}\ g}{0.817 \times 10^{-23}\ cm^3} = 0.7745\ g/cm^3$$

mass of hydrogen in 1 cm^3 LiH

(1 cm^3 LiH)(0.7745 g/cm^3)(1.008 g H/7.949 g LiH) = 0.09821 g H

∴ there is more hydrogen by mass in 1 cm^3 of LiH than in 1 cm^3 of H_2

22. $KCl + RbCl \rightarrow 2\ K_{0.50}Rb_{0.50}Cl$

(10.0 g $K_{0.50}Rb_{0.50}Cl$)(1 mol/9$\underline{7}$.7 g) = 0.1$\underline{0}$2 mol $K_{0.50}Rb_{0.50}Cl$

$(0.1\underline{0}2\ \text{mol}\ K_{0.50}Rb_{0.50}Cl)\dfrac{1\ \text{mol KCl}}{2\ \text{mol}\ K_{0.50}Rb_{0.50}Cl} = 5.\underline{1}0 \times 10^{-2}\ \text{mol KCl}$

$(5.\underline{1}0 \times 10^{-2}\ \text{mol KCl})(74.55\ g/1\ \text{mol KCl}) = 3.8\ g\ KCl$

$(0.1\underline{0}2\ \text{mol}\ K_{0.50}Rb_{0.50}Cl)\dfrac{1\ \text{mol RbCl}}{2\ \text{mol}\ K_{0.50}Rb_{0.50}Cl} = 5.\underline{1}0 \times 10^{-2}\ \text{mol RbCl}$

$(5.\underline{1}0 \times 10^{-2}\ \text{mol RbCl})(120.92\ g/1\ \text{mol RbCl}) = 6.2\ g\ RbCl$

23. a. salt b. molcule c. alloy d. molcule e. salt

24. Ba and O; extended ionic solid
Cu and Ni; extended metallic solid
N and O; discrete molcule
Na and F; extended ionic solid

25. Ni and Cu have the same structure (fcc) and similar atomic radii. The same is true for Mo and V, which are bcc metals. These pairs are expected to most easily substitute for one another and form solid solutions, Ni_xCu_{1-x} and Mo_xV_{1-x}.

26. a.

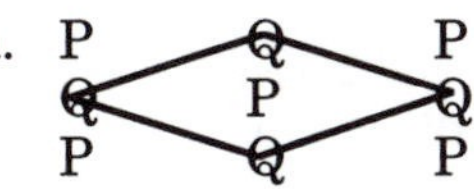

b)

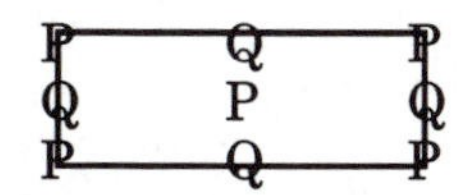

a.
P: (1)(1 occupancy) = 1 P
Q: (4)(1/4 occupancy) = 1 Q

PQ

b.
P: (4)(1/4 occupancy) = 1 P
(1)(1 occupancy) = 1 P
Q: (4)(1/2 occupancy) = 2 Q
P_2Q_2 = PQ

27. a. three in the unit cell, but nine others in neighboring cells also exist, giving a total of twelve.
b. only one in the unit cell, but seven more exist in neighboring cells, giving a total of eight.

28. six

29. Cs and Cl

30. $GaAs_{0.75}P_{0.25}$ molar mass = 133.65 g

31. $CdS_{0.10}Se_{0.90}$ molar mass = 187 g

32. Elemental analysis would not be able to distinguish between $Cu_{0.5}Ni_{0.5}$ and an equimolar homogeneous mixture of Cu and Ni. The magnetic properties of elemental Cu and Ni are quite different. Nickel is ferromagnetic (Chapter 2), and could be separated from a mixture with copper by the use of a magnet. The Cu and Ni in a $Cu_{0.5}Ni_{0.5}$ solid solution cannot be separated by a magnet. Thus magnetic properties could distinguish the two samples.

Chapter 4

1. $c = \nu\lambda$
   $(2.998 \times 10^8 \text{ m/s}) = (\nu)(1.54 \text{ Å})(1 \times 10^{-10} \text{ m/Å})$
   $\nu = 1.95 \times 10^{18} \text{ s}^{-1}$

2. a. $(2)(d)(\sin\theta) = n\lambda$
   $(2)(4.123 \text{ Å})(\sin\theta) = (2)(1.54 \text{ Å})$
   $\theta = 22.0°$
   b. $\theta = 15.8°$ c. $\theta = 27.4°$ d. $\theta = 21.0°$ e. $\theta = 25.9°$

3. $\sin\theta = n\lambda/(2d)$ For $n = 1$ and $\lambda = 1.54$ Å, $\sin\theta = 0.770$ Å$/d$
   a. $\sin\theta = (0.77\text{Å}/1.79 \text{ Å}) = 0.430$ $\theta = 25.5°$
   b. $\theta = 15.9°$ c. $\theta = 17.2°$ d. $\theta = 20.9°$ e. $\theta = 18.1°$

4. $n = 1$ $\lambda = 1.54$ Å $d = (1)(1.54 \text{ Å})/[(2)(\sin\theta)] = 0.770 \text{ Å}/(\sin\theta)$
   a. $d = 1.79$ Å b. $d = 2.81$ Å c. $d = 2.60$ Å d. $d = 2.16$ Å e. $d = 2.47$ Å

5. $c = \nu\lambda$
   $(2.998 \times 10^8 \text{ m/s}) = (\nu)(670 \text{ nm})(1 \times 10^{-9} \text{ m/nm})$
   $\nu = 4.47 \times 10^{14} \text{ s}^{-1}$

7. a. vertical spots b. horizontal spots c. square pattern of spots

8. a. The smaller square array gives the larger square diffraction pattern.
   b. The pattern on the left gives a square pattern; the array on the right gives a rectangle with the long distance between spots in the horizontal direction.
   c. Both arrays give rectangular diffraction patterns. The array on the left will give a diffraction pattern with the longer edge of the rectangle in the vertical direction. The array on the right will produce a diffraction pattern with the longer edge of the rectangle in the horizontal direction.
   d. An array of lines gives a series of dots in the diffraction pattern. The horizontal pattern of lines will give a diffraction pattern consisting of a set of spots running in the vertical direction, and the vertical array of lines gives a diffraction pattern consisting of a set of spots in the horizontal direction.

9. a. The diffraction pattern with fewer spots is due to array b which has an extra dot in the middle of each square.
   b. Light diffracted by the added dots is out of phase with the light diffracted by the other dots of the array, creating destructive interference that removes every other spot from the diffraction pattern.
   c. The unit cell of Figure 9a can be considered to be a rectangle, with its longer sides in the vertical direction. Long distances in dot arrays become short in the diffraction pattern (reciprocal lattice effect) so the diffraction pattern will also be a rectangle, with the short edge in the vertical direction. The unit cell of Figure 9b can be viewed as a slightly squashed, tilted parallelogram. The repeat unit in the diffraction pattern will also be a tilted parallelogram.

10. Assume a d-spacing of 1.0×10^{-4} m (1.0×10^{6} Å) and compare ϕ values for first-order diffraction using the equation for Fraunhofer diffraction:

IR: $d \sin \phi = n\lambda$	UV: $d \sin \phi = n\lambda$
$(1.0 \times 10^{6}$ Å$) \sin \phi = (1)(8{,}000$ Å$)$	$(1.0 \times 10^{6}$ Å$) \sin \phi = (1)(3{,}000$ Å$)$
$\sin \phi = 8.0 \times 10^{-3}$	$\sin \phi = 3.0 \times 10^{-3}$
$\phi = 0.46°$	$\phi = 0.17°$

The angle and therefore the size of the diffraction pattern increases with the wavelength of light used. UV radiation would lead to a small diffraction pattern, and IR radiation would yield a large diffraction pattern.

11. In order for diffraction to occur, the wavelength of incident radiation must be approximately the same as the spacing of the atoms. Atomic separations are typically a few angstroms. X-rays, with wavelength of about 1 Å, are a far better match to interatomic spacings than visible light ($\lambda \sim 4{,}000$ Å to 7,000 Å).

12. The STM can only collect two-dimensional information. The penetrating nature of X-rays and subsequent diffraction off various planes of atoms adds information in the third dimension.

13. Referring to Figure 4.1, ϕ can be calculated from
   $\tan \phi = X/L = (3.8 \text{ cm}/6.0 \times 10^{2} \text{ cm}) = 6.3 \times 10^{-3}$
   $\phi = 0.36°$ (compare this angle to those calculated in problem 4.10)
   $(d)(\sin \phi) = n\lambda$
   $(d)(\sin 0.36°) = (1)(6.33 \times 10^{-7} \text{ m})$ $d = 1.00 \times 10^{-4}$ m

14. For a given order of diffraction (n value), as d becomes smaller, $\sin \theta$ must become larger to satisfy the Bragg equation: $2d(\sin \theta) = n\lambda$. Thus, as $\sin \theta$ and θ increase, the observed spacing of the spots in a diffraction pattern increases.

15. a. In moving from sc to bcc, one half of the spots in the diffraction pattern disappear. This disappearance is due to complete destructive interference that occurs with some of the X-ray waves. The destructive interference is caused by the presence of the scattering center in the body of the bcc cell. In moving from bcc to fcc, one half of the spots again disappear, as a result of destructive interference that occurs from the scattering centers located on the faces of the cell.

   b. We do CsCl. The diffraction pattern of CsCl should reflect both the positions of the ions in the cell, as well as their relative sizes. Because Cs ions and Cl ions are different, the destructive interference due to the Cs ion in the center of the cell is *incomplete*. The CsCl diffraction pattern is predicted to have spots at the same locations as the sc pattern of question 15a, but the intensities of half of the spots will be less than the corresponding spots in the sc diffraction pattern. The spots of reduced intensity will be at the same locations as the spots that completely disappeared in the diffraction pattern of the bcc structure of 15a.

Chapter 5

1. a. $V = (4/3)\pi r^3$ $V_{atom} = (4/3)(\pi)(0.50 \text{ Å})^3 = 0.52 \text{ Å}^3$
   b. In a simple cubic structure, the corner atoms of the unit cell touch. Therefore, the length of the unit cell will equal $2r$. In this problem, the length of the cell = $2r$ = 1.00 Å

c. V_{cell} = (1.00 Å)^3 = 1.00 Å^3

d. Volume of atoms in cell:
(8 corner atoms)(1/8 occupancy) = volume of one atom = 0.52 Å^3.
Packing efficiency = volume of atoms in cell/V_{cell} = $\frac{0.52 \text{ Å}^3}{1.00 \text{ Å}^3} \times 100 = 52\%$

2. a. $V = (4/3)\pi r^3$ $V_{atom} = (4/3)(\pi)(0.50 \text{ Å})^3 = 0.52 \text{ Å}^3$

b. In a bcc structure, the corner atoms do not touch each other. The atoms touch along the body-diagonal; the body-diagonal has a length equal to $4r$ or $2d$. If the bcc cell has an edge length of a, the face-diagonals have a length given by the Pythagorean theorem to be $(2)^{0.5}a$. The body-diagonal of the cell constitutes the hypotenuse of a right triangle; the face-diagonal and cell edge are the other two sides. Thus we have

$$[(2)^{0.5}a]^2 + a^2 = (2d)^2$$
$$3a^2 = 4d^2$$
$$a^2 = (4/3)d^2$$
$$a = 1.155d$$

c. $V = (a)^3 = [(1.155)(1.00 \text{ Å})]^3 = 1.54 \text{ Å}^3$

d. Atoms in unit cell:
(8 corner atoms)(1/8 occupancy) = 1 atom
(1 body-centered atom)(1 occupancy) = 1 atom
or a total of 2 atoms per unit cell.
Volume of atoms per unit cell = (2 atoms)(0.52 Å^3/atom) = 1.04 Å^3
Volume of unit cell = 1.54 Å^3
Packing efficiency = (100)(1.04 Å^3)/1.54 Å^3 = 68%

3. a. $V = (4/3)\pi r^3$
$V_{atom} = (4/3)(\pi)(0.500 \text{ Å})^3 = 0.524 \text{ Å}^3$

b.

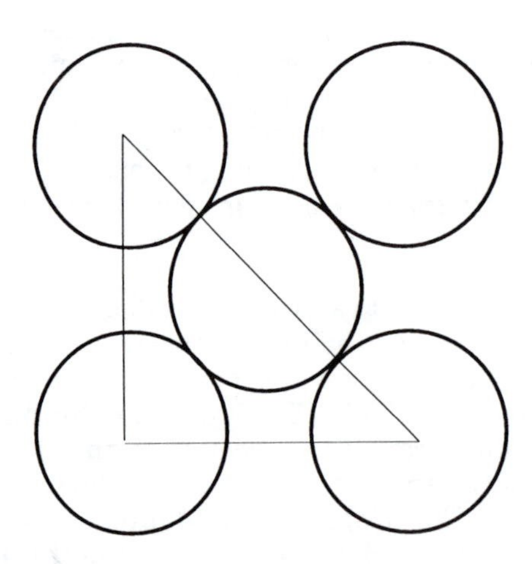

The diagram represents one face of a fcc system. As shown, a right triangle can be drawn with a hypotenuse of length $4r$ or 2.00 Å, and the other sides equal in length to that of the cell edge, a. This information allows calculation of the cell edge length by the Pythagorean theorem:

$$a^2 + a^2 = (4r)^2 = (2.00 \text{ Å})^2$$
$$2a^2 = 4.00 \text{ Å}^2$$
$$a = 1.41 \text{ Å}$$

c. V_{cell} = (1.41 Å)^3 = 2.80 Å^3

d. Atoms in unit cell:
(8 corner atoms)(1/8 occupancy) = 1 atom
(6 face-shared atoms)(1/2 occupancy) = 3 atoms
or a total of 4 atoms per unit cell.

V_{atoms} = (0.524 Å^3/atom)(4 atoms) = 2.08 Å^3

Packing efficiency = (2.08 Å^3)(100)/(2.80 Å^3) = 74.3%

4. d.

5. The solid may be described as two interpenetrating sc arrays of ions. Thus, both figures accurately describe the CsCl structure.

6. LiH adopts the NaCl structure with a cell edge length of 4.085 Å. A generalized view of one face of the NaCl structure is

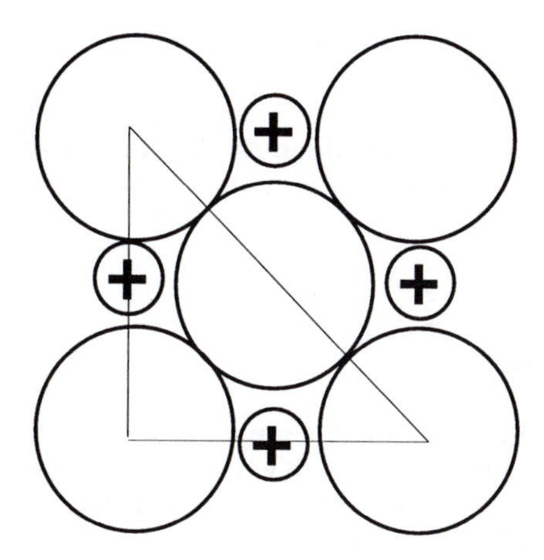

A right triangle can be drawn on the cell face with a hypotenuse of length $4r$, and the other sides equal in length a to that of the cell edge (4.085 Å). This information allows calculation of the ionic radius of H^-:

$$(4r)^2 = (4.085\ \text{Å})^2 + (4.085\ \text{Å})^2$$
$$r = 1.443\ \text{Å}$$

7. For example, Ba (bcc) cell length = 5.025 Å.
V_{cell} = (5.025 Å)^3(1 × 10^{-8} cm/1 Å)^3 = 1.269 × 10^{-22} cm^3

The bcc structure contains the following Ba atoms:
(8 corner atoms)(1/8 occupancy) = 1 atom
(1 atom within cell)(1 occupancy) = 1 atom
or a total of 2 atoms per unit cell.

(2 Ba atoms)(1 mol Ba/6.022 × 10^{23} atoms)(137.33 g/mol Ba) = 4.561 × 10^{-22} g

Density = mass/volume = (4.561 × 10^{-22} g)/(1.269 × 10^{-22} cm^3) = 3.594 g/cm^3

8. Two tetrahedral holes exist per ccp sulfur atom and half of them are full. Therefore there is one zinc atom per sulfur atom, so the empirical formula of the compound is ZnS.

9. Two tetrahedral holes exist per ccp sulfide ion, and each of them contains a lithium ion. Therefore the empirical formula is Li_2S.

10. One octahedral hole exists per ccp chloride ion and all of the octahedral holes contain silver ions. The empirical formula of the compound is AgCl.

11. There is one cubic hole per iodide ion in a simple cubic array of iodide ions. All of the cubic holes contain a thallium ion. The empirical formula is TlI.

12. The cubic array of F^- yields
(8 corner ions)(1/8 occupancy) = 1 F^- ion
One cubic hole exists per unit cell. Half of the cubic holes are filled with Ba^{2+} ions and the other half are empty. Therefore, on average, there is one half Ba^{2+} ion per unit cell. There is one F^- ion per unit cell. The formula of the compound is $Ba_{0.5}F$ or BaF_2.

14. For example, aluminum adopts the fcc structure with a cell edge length of 4.050 Å. The diagram presented in the solution to problem 3 of this chapter represents one face of the fcc array of aluminum atoms. A right triangle can be drawn on the cell face with a hypotenuse of length $4r$, and the other sides equal in length to that of the cell edge (4.050 Å). This information allows calculation of the atomic radius of Al:

$$(4r)^2 = (4.050)^2 + (4.050)^2$$
$$16r^2 = 32.805 \text{ Å}^2$$
$$r = 1.432 \text{ Å}$$

15. For example, carbon. Diamond has a unit cell edge length of 3.5668 Å. If the unit cell is subdivided into eight smaller cubes of equal volume, each edge of the small cubic cells has a length of (3.5668 Å)/2 = 1.7834 Å. The length x of the face-diagonal of the mini-cell is then given by

$$(1.7834 \text{ Å})^2 + (1.7834 \text{ Å})^2 = x^2 \qquad x = 2.5221 \text{ Å}$$

The length y of the body-diagonal of the small cubic cell is calculated to be

$$(1.7834 \text{ Å})^2 + (2.5221 \text{ Å})^2 = y^2 \qquad y = 3.0889 \text{ Å}$$

The body-diagonal has a length equal to $4r$. Thus, $r = 0.77223$ Å.

16. For example, CaO. Cell edge length = 4.8105 Å. A generalized view of one face of the NaCl structure is shown in problem 6 of this chapter.

A right triangle can be drawn on the cell face with a hypotenuse of length (4)(oxide radius), and the other sides equal in length to that of the cell edge (4.8105 Å). This information allows calculation of the ionic radius of O^{2-}:

$$(4r)^2 = (4.8105 \text{ Å})^2 + (4.8105 \text{ Å})^2$$
$$16r^2 = 46.281 \text{ Å}^2$$
$$r = 1.7008 \text{ Å}$$

The length of the CaO cell edge is the sum of two oxide radii plus two calcium ion radii:

$$4.8105\ \text{Å} = (2)(O^{2-}\ \text{radius}) + (2)(Ca^{2+}\ \text{radius})$$
$$4.8105\ \text{Å} = (2)(1.7008\ \text{Å}) + (2)(Ca^{2+}\ \text{radius})$$
$$Ca^{2+}\ \text{radius} = 0.70445\ \text{Å}$$

17. The plot of unit cell volume versus anion volume was prepared from the following data:

Anion	Anion radius (Å)	Anion volume (Å^3), $V = (4/3)\,\pi\, r^3$
F^-	1.19	7.06
Cl^-	1.67	19.5
I^-	2.06	36.6

Salt	Cell Length (Å)	Cell Volume (Å^3)
KF	5.347	152.9
KCl	6.2929	249.20
KBr	6.6000	287.50
KI	7.0656	352.73

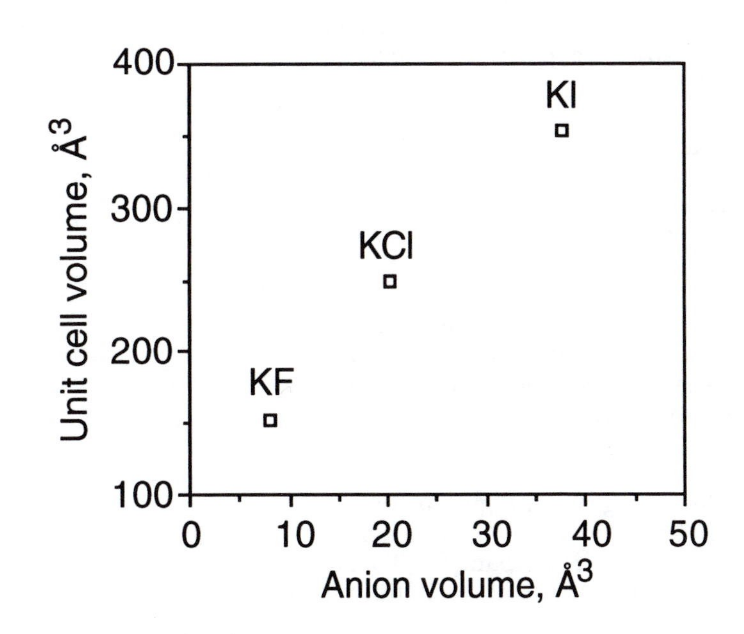

The volume of KBr (287.50 Å^3) and this plot allow one to estimate the volume of the Br^- anion (31 Å^3), and to calculate an ionic radius for Br^- of 1.9 Å.

18. undistorted cubic holes: Po
undistorted octahedral holes: Al, Co, Pb
undistorted tetrahedral holes: Al, Co, Pb

19.

Type	Anisotropic	Isotropic
Ionic	CdI_2	NaCl
Covalent	graphite	diamond

20. Aluminum adopts an efficiently packed fcc structure, but silicon adopts the less efficiently packed diamond structure: The aluminum atoms pack with far less empty space between atoms than silicon, overcoming the difference in size and atomic weight.

21. Although the two-dimensional arrangement of atoms in individual planes of hcp and ccp structures are identical, the stacking pattern or three-dimensional sequencing of layers is different, and therefore the structures are different.

22. The radius ratio rule may be used to predict the coordination number and type of holes favored by ionic compounds, but not of solids with strong directional covalent bonding like diamond and graphite.

23. Ratio = 1.00; radius ratio rule predicts 8-coordinate. See No. 22.

24. The cell shown in Figure 5.15 cannot account for the stoichiometry of $CdCl_2$:
(8 corner chloride ions)(1/8 occupancy) = 1 chloride ion
(6 face-centered chloride ions)(1/2 occupancy) = 3 chloride ions
(6 edge-shared cadmium ions)(1/4 occupancy) = 1.5 cadmium ions
= $Cd_{1.5}Cl_4$ or Cd_3Cl_8

25. a. Referring to the z-level diagram of Figure 5.16:
(8 corner arsenic ions)(1/8 occupancy) = 1 arsenic ion
(1 arsenic ion within cell)(1 occupancy) = 1 arsenic ion
(2 nickel ions within cell)(1 occupancy) = 2 nickel ions
= Ni_2As_2 or NiAs

b. (8 corner chloride ions)(1/8 occupancy) = 1 chloride ion
(6 face-centered chloride ions)(1/2 occupancy) = 3 chloride ions
(12 edge-shared sodium ions)(1/4 occupancy) = 3 sodium ions
(1 sodium ion within cell)(1 occupancy) = 1 sodium ion
= Na_4Cl_4 or NaCl

c. Referring to the z-level diagram of Figure 5.17A:
(8 corner iodide ions)(1/8 occupancy) = 1 iodide ion
(1 iodide ion within cell)(1 occupancy) = 1 iodide ion
(1 cadmium ion within cell)(1 occupancy) = 1 cadmium ion
= CdI_2

d. Referring to the z-level diagram of Figure 5.19:
(8 corner oxide ions)(1/8 occupancy) = 1 oxide ion
(6 face-centered oxide ions)(1/2 occupancy) = 3 oxide ions
(8 lithium ions within cell)(1 occupancy) = 8 lithium ions
= Li_8O_4 or Li_2O

e. .Referring to the z-level diagram of Figure 5.20:
(8 corner fluoride ions)(1/8 occupancy) = 1 fluoride ion
(6 face-centered fluoride ions)(1/2 occupancy) = 3 fluoride ions
(12 edge-shared fluoride ions)(1/4 occupancy) = 3 fluoride ions
(1 fluoride ion within cell)(1 occupancy) = 1 fluoride ion
(4 calcium ions within cell)(1 occupancy) = 4 calcium ions
= Ca_4F_8 or CaF_2

26. As shown in Figure 5.17B, a single crystal of CdI_2 is expected to cleave easily between adjacent layers of iodide ions. Along this plane there is no electrostatic attraction between Cd^{2+} and I^- ions.

27. The z-level diagram of Figure 5.23B represents the placement of holes for a unit cell containing two hexagonal close packed atoms:
z = 0 and 1.0: (8 corner atoms)(1/8 occupancy) = 1 atom
z = 0.5: (1 atom within cell)(1 occupancy) = 1 atom

This cell contains four T_d holes (or two holes per close packed atom):
$z = 0.375$ and 0.625: (8 T_d holes)(1/4 occupancy) = 2 T_d holes
$z = 0.125$ and 0.875: (2 T_d holes within cell)(1 occupancy) = 2 T_d holes

The cell also contains two O_h holes (or one hole per close packed atom):
$z = 0.25$ and 0.75: (2 O_h holes within cell)(1 occupancy) = 2 O_h holes

In summary, a unit cell containing two hcp atoms contains four T_d holes and two O_h holes. Thus, for every atom there is one O_h hole and two T_d holes.

28. Figure 5.13A shows a ccp array of four chloride ions:
(8 corner chloride ions)(1/8 occupancy) = 1 chloride ion
(6 face-centered chloride ions)(1/2 occupancy) = 3 chloride ions
The sodium cations are located in the octahedral holes of the lattice:
(12 edge-shared sodium ions)(1/4 occupancy) = 3 sodium ions
(1 sodium ion within cell)(1 occupancy) = 1 sodium ion
for a total of four octahedral holes (one per close-packed ion).

The unit cell shown in Figure 5.19 may also be described as a combination of eight small cubes. Each of these eight small cubes has four oxide ions that form a tetrahedral hole. Each of the tetrahedral holes is fully contained within the unit cell, which has a total of four oxide ions. Thus there are eight tetrahedral holes in the unit cell, or two per close-packed oxide ion.

29. There are two tetrahedral holes and one octahedral hole per C_{60} unit in the fcc structure (*see* problem 28). If these holes are all filled with potassium ions, the formula of the salt will be K_3C_{60}.

30.

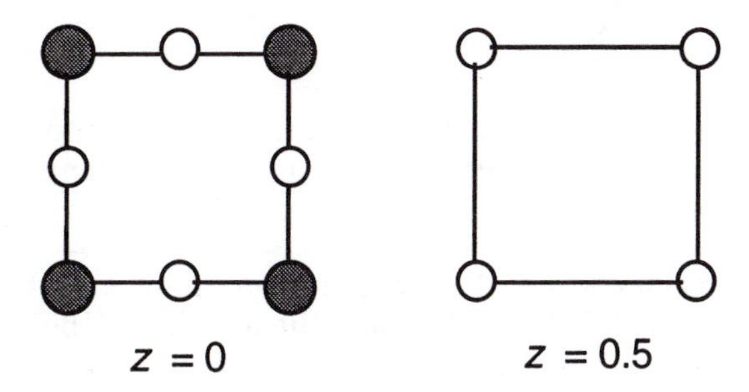

31. (8 corner copper ions)(1/8 occupancy) = 1 copper ion
(8 edge-shared copper ions)(1/4 occupancy) = 2 copper ions
(12 edge-shared oxide ions)(1/4 occupancy) = 3 oxide ions
(8 face-shared oxide ions)(1/2 occupancy) = 4 oxide ions
(2 barium ions within cell)(1 occupancy) = 2 barium ions
(1 yttrium ion within cell)(1 occupancy) = 1 yttrium ion
= $YBa_2Cu_3O_7$

32. For example, Figure 5.31D:
(2 silicon atoms within cell)(1 occupancy) = 2 silicon atoms
(5 oxygen atoms within cell)(1 occupancy) = 5 oxygen atoms
(2 corner-shared oxygen atoms)(1/2 occupancy) = 1 oxygen atom
= $Si_2O_6^{4-}$ or SiO_3^{2-}

Figure 5.32:
(2 silicon atoms within cell)(1 occupancy) = 2 silicon atoms
(4 face-shared silicon atoms)(1/2 occupancy) = 2 silicon atoms
(8 oxygen atoms within cell)(1 occupancy) = 8 oxygen atoms
(6 face-shared oxygen atoms)(1/2 occupancy) = 3 oxygen atom
= $Si_4O_{11}^{6-}$

Figure 5.33:
(1 silicon atom within cell)(1 occupancy) = 1 silicon atom
(4 edge-shared silicon atoms)(1/4 occupancy) = 1 silicon atom
(4 oxygen atoms within cell)(1 occupancy) = 4 oxygen atoms
(4 edge-shared oxygen atoms)(1/4 occupancy) = 1 oxygen atom
= $Si_2O_5^{2-}$

33. a. $Si_4O_{11}^{6-} \rightarrow AlSi_3O_{11}^{7-}$
b. $Si_4O_{11}^{6-} \rightarrow Al_2Si_2O_{11}^{8-}$

34. a. six
b. $[Fe^{3+}]_4[Fe(CN)_6^{4-}]_3$
c. The salt does not have a 1:1 cation:anion stoichiometry and therefore cannot have a perfect NaCl structure. Rather, it exists with a NaCl-like structure with periodic anion vacancies.

35.

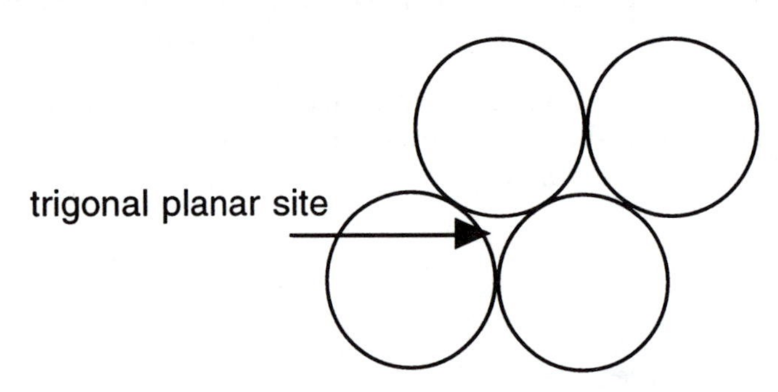

36. Two-thirds of the octahedral holes are filled with aluminum ions, and one octahedral hole exists per O^{2-} ion. Therefore, the formula of the compound is $Al_{2/3}O$ or Al_2O_3.

37. One-fourth of the tetrahedral holes are filled with mercury ions, and there are two tetrahedral holes per iodide ion. Therefore the formula of the compound is $Hg_{1/2}I$ or HgI_2.

38. One-eighth of the tetrahedral holes are filled with tin ions, and there are two tetrahedral holes per iodide ion. Therefore the formula of the compound is $Sn_{1/4}I$ or SnI_4.

39. One-third of the tetrahedral holes are filled with aluminum ions, and there are two tetrahedral holes per selenium ion. Therefore the formula of the compound is $Al_{2/3}Se$ or Al_2Se_3.

40. a. LiI adopts the NaCl structure:

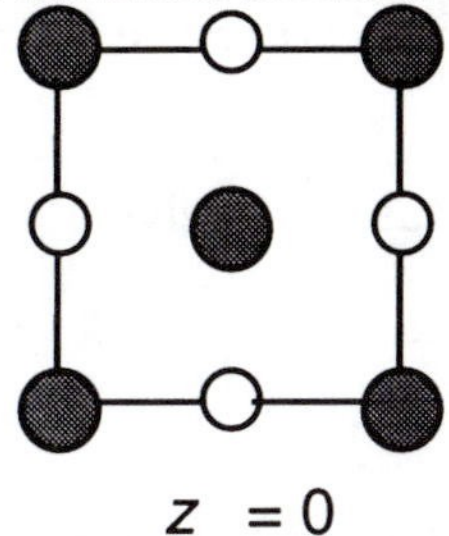

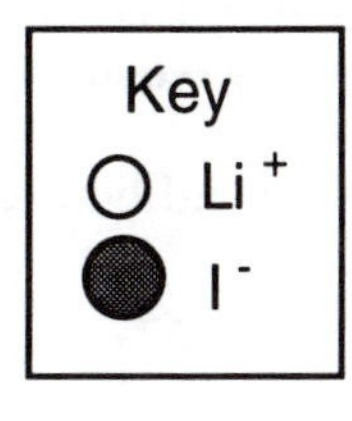

b. The LiI cell edge length = 6.000 Å. A generalized, space-filling view of one face of the LiI structure is shown in the solution to problem 6 of this chapter.

A right triangle can be inscribed on the cell face with a hypotenuse of length (4)(I radius), and the other sides equal the length of the cell edge (6.000 Å). This information allows calculation of the ionic radius r of I^-:

$$(4r)^2 = (6.000\ \text{Å})^2 + (6.000\ \text{Å})^2$$
$$16r^2 = 72.000\ \text{Å}^2$$
$$r = 2.121\ \text{Å}$$

c. The length of the LiI cell edge is the sum of two iodide radii plus two lithium ion radii:

6.000 Å = (2)(I^- radius) + (2)(Li^+ radius)
6.000 Å = (2)(2.121 Å) + (2)(Li^+ radius)
Li^+ radius = 0.8790 Å

d. For example, LiCl also adopts the NaCl structure with cell edge length = 5.12954 Å. The chloride ionic radius can be calculated as in part c as follows:

5.12954 Å = (2)(Cl^- radius) + (2)(Li^+ radius)
5.12954 Å = (2)(Cl^- radius) + (2)(0.8790 Å)
Cl^- radius = 1.686 Å

Other results calculated in this manner are tabulated as follows:

Ion	Ionic Radius (Å)	Ion	Ionic Radius (Å)
Li^+	0.879	F^-	1.130
Na^+	1.134	Cl^-	1.686
K^+	1.460	Br^-	1.872
		I^-	2.121

41. Layers A and B of MoS_2 in Figure 5.26 will slip relative to one another because of the weak van der Waals forces that hold these layers together.

42. The sheets of sp^2-hybridized carbon atoms of graphite will slip relative to one another because of the weak van der Waals forces that hold these layers together.

43. a. The cation positions are the same. The centers of the C_2^{2-} units of CaC_2 match the Cl^- positions of NaCl.
b. The arsenic atoms in MgAgAs and the sulfur atoms in ZnS both pack in fcc structures. The Mg (or Ag) atoms in MgAgAs and the Zn atoms in ZnS fill half of the tetrahedral holes. In the MgAgAs structure the Ag atoms (or Mg atoms) occupy the remaining tetrahedral hole positions.

c. Both CdI_2 and NiAs are based on hcp anions. The cations in these solids are located in O_h holes formed by the anions. In the CdI_2 structure, alternating planes of cations are vacant.

44. A bcc metal has eight nearest neighbors, and a fcc metal has twelve.

Chapter 6

2. Iron is body-centered cubic with cell edge length = 2.8606 Å, and unit cell volume of $(2.8606\ \text{Å})^3(1 \times 10^{-8}\ \text{cm/Å})^3 = 2.3408 \times 10^{-23}\ \text{cm}^3$. Each unit cell contains (8 corner atoms)(1/8 occupancy) = 1 iron atom
(1 body-centered atom)(1 occupancy) = 1 iron atom
= two iron atoms per unit cell.

The mass of the two atoms is
(2 Fe atoms)(1 mol Fe/6.022×10^{23} atoms)(55.85 g/mol Fe) = 1.855×10^{-22} g

Density = mass/volume = $(1.855 \times 10^{-22}\ \text{g})/(2.3408 \times 10^{-23}\ \text{cm}^3) = 7.925\ \text{g/cm}^3$

If 0.10% of the iron sites were vacant, the mass of the unit cell would equal $(0.999)(1.855 \times 10^{-22}\ \text{g}) = 1.8\underline{5}3 \times 10^{-22}\ \text{g}$

$$\text{density of the iron} = \text{mass/volume} = \frac{1.8\underline{5}3 \times 0^{-22}\ \text{g}}{2.3408 \times 0^{-23}\ \text{cm}^3} = 7.92\ \text{g/cm}^3$$

1 cm^3 of defect free iron would weigh: $(1\ \text{cm}^3)(7.925\ \text{g/cm}^3) = 7.925$ g
1 cm^3 of iron with 0.10% vacancies would weigh essentially the same amount: $(1\ \text{cm}^3)(7.92\ \text{g/cm}^3) = 7.92$ g

3. STM images the surface of a sample, and defects are frequently observed at grain boundaries like those at the surface of a solid. In addition, STM provides atom-by-atom images that might indicate the presence of a point defect. X-ray crystallography is a technique that cannot distinguish a single atom, but rather examines the bulk of a sample. The diffraction phenomenon depends on many planes of repeating atoms and yields averaged data based on a large number of these atoms.

4. $(1.00 \times 10^{-3}$ g NaCl) (1 mol NaCl/58.44 g) = 1.71×10^{-5} mol NaCl
$(1.71 \times 10^{-5}$ mol NaCl)(6.022×10^{23} NaCl formula units/mol NaCl)
= 1.03×10^{19} NaCl formula units
$(1.03 \times 10^{19}$ NaCl formula units)(1 defect/1×10^{15} NaCl formula units)
= 1.03×10^{4} defects (or missing pairs of ions)

5. Figure 6.17 and lattice information from Appendix 5.6 can be used to generate the following approximations:

Compound	Edge Length (Å)	λ_{max} (nm)
NaI	6.4728	~ 600
LiI	6.000	~ 500

7. Each substitution of a Na^+ ion by a Mg^{2+} ion places one unit of positive charge in excess of electrical neutrality in the solid. Answers a–c all maintain or exacerbate the charge imbalance imposed by the dopant. Only d offers a method of establishing charge balance by removing Na^+ ions equal in number to those substituted for the dopant.

8. Rather than dragging its entire body over a surface, a caterpillar moves by displacing only a portion of its body at any given instant. This is similar to the movement of a dislocation, seen in Figure 6.10A.

9. Aisles fill in the gaps that would naturally occur between columns of seats as you move away from the stage or field.

10. Refer to Figure 6.17 for the necessary data.

Compound	Unit Cell Length (Å)	F-center Abs. Max. (nm)
KCl	6.3	560
RbCl	6.6	620

As indicated in the text of this chapter, F-centers in $K_{1-x}Rb_xCl$ solid solutions have absorption maxima that lie between pure KCl and pure RbCl. In a $K_{0.5}Rb_{0.5}Cl$ solid solution, the expected parameters would be approximately 6.4 Å and 590 nm.

11. In the solid solution only one absorption peak would be expected, at about 590 nm (*see* problem 10). In the two-component physical mixture you would expect two peaks, at about 560 and 620 nm.

12. In an hcp metal there is only one set of slip planes; the slip planes are located between the ABAB... stacking planes. This feature can be contrasted to ccp metals in which close-packed slip planes exist in four different sets.

Chapter 7

1. a. x =0.75 and y = 0.40.

2. c. Referring to Figure 7.16, complementary AZ pairs are indicated. Cadmium (Group 12) and tellurium (group 16) can form cadmium telluride, a solid that is structurally similar to α-Sn with the same number of valence electrons.

3. a. A substance with a band gap in the infrared region of the electromagnetic spectrum will absorb light throughout the visible portion of the electromagnetic spectrum and appear black.

4. d.

5. d. When the reducing agent fills the lower energy band with electrons, the sample will become an insulator with a large band gap.

6. b. Metals are characterized by increasing resistivity with increasing temperature; or alternatively, by increasing conductivity with decreasing temperature.

7. b.

8. One liter of 1.0 M NaOH will contain 2.0 mols of carriers (ions).
(2.0 mols carriers)(6.022×10^{23} carriers/mol) = 1.2×10^{24} carriers
carrier conc. = 1.2×10^{24} carriers/1.0×10^3 cm^3 = 1.2×10^{21} carriers/cm^3

9. a. $R = \rho L/A$
$A = \rho L/R$
$A = [(1.67 \times 10^{-6}$ ohm-cm)(150 m)(100 cm/m)]/(150 ohm)
$A = 1.67 \times 10^{-4}$ cm^2

b. $A = \pi r^2$ $\quad r = 7.29 \times 10^{-3}$ cm, $d = 1.46 \times 10^{-2}$ cm

10. Referring to Figure 7.18, a band gap energy of 2.4 eV will give the material a yellow or yellow–orange color.

11. If red light is to be transmitted (not absorbed), then the material should begin to absorb in the orange region, at about 2.1 eV.

12. The color emitted from an LED when a voltage is applied depends on the magnitude of the band gap. Approximate values for band-gap energies and corresponding emission colors are given in the first column of Figure 7.18. Referring to Figure 7.19 then allows determination of approximate values for x in GaP_xAs_{1-x}.

Color	Band Gap Energy (eV)	x
Red	1.9	0.5
Orange	2.1	0.7
Yellow	2.2	0.9
Blue	2.8	—

A blue LED cannot be made from this family, because none of these solids has a band gap large enough to correspond to blue light.

13. $x = 0.6$; $y = 0.3$ The gallium and indium atoms occupy positions equivalent to zinc in zinc blende; the arsenic and phosphorus atoms occupy positions equivalent to sulfur in zinc blende. The 1:1 stoichiometry of Zn:S dictates that the sum of the Ga and In atoms must equal the sum of the As and P atoms present in the solid. The As/P atoms form a fcc lattice with Ga/In atoms in half the tetrahedral holes. Fractional stoichiometries correspond to probabilities for finding a particular atom at a site (*see* Chapter 3).

14. The 2-eV solar cell should be on top as this material absorbs higher energy light but transmits low energy light to be absorbed by the 1-eV cell. Any high energy light passing through the first cell will also be absorbed by the 1-eV cell.

15. a. The solid should be a metallic conductor based on the partially filled higher energy band.
b. The solid will become an insulator (large band gap) or semiconductor (small band gap) if an oxidizing agent is used to remove all the electrons from the higher energy band.

16. First, calculate the frequency and energy that corresponds to 7000 Å:
$\nu = (3.00 \times 10^8 \text{ m s}^{-1})/(7000 \times 10^{-10} \text{ m}) = 4.29 \times 10^{14} \text{ s}^{-1}$
$E = (4.29 \times 10^{14} \text{ s}^{-1})(6.6262 \times 10^{-34} \text{ J·s}) = 2.84 \times 10^{-19} \text{ J}$
$E = (2.84 \times 10^{-19} \text{ J})(1 \text{ eV}/1.6022 \times 10^{-19} \text{ J}) = 1.77 \text{ eV}$

Second, calculate the value of x in GaP_xAs_{1-x}. The problem states that the band gap increases linearly with x, when $0 \leq x \leq 0.45$. When $x = 0$, the band gap is 1.4 eV; when $x = 0.45$, the band gap is 2.0 eV. The slope m of the line is determined by these two points:

$$m = (2.0 - 1.4)/(0.45 - 0) = 1.33$$

The equation of the line determined by the two data points in the problem is

$$y = 1.33x + 1.4$$

The desired band gap is 1.77 eV. Substituting this value into the equation for the line gives a value for x:

$$1.77 = 1.33x + 1.4$$
$$x = 0.28$$

Thus, the solid solution that is predicted to give 7000 Å is $GaP_{0.28}As_{0.72}$. This problem can also be solved graphically, of course.

17. InSb and CdTe are isoelectronic with α-Sn. α-Sn is metallic with a small band gap (less than 0.1 eV). Ionic character increases with the difference in electronegativity between atoms A and Z. As ionic character increases, so does E_g. The expected trend in band-gap energies is Sn < InSb < CdTe.

18. BN is isoelectronic with diamond and is also extremely hard.

19. Phosphorus is isovalent with arsenic, so substitution of As with P does not change the valence count. Cadmium has one less valence electron than gallium, but tin has one additional electron than gallium. Substituting one half of the Ga atoms with Cd and the other half with Sn leads to the same valence count. We therefore have $Cd_{0.5}Sn_{0.5}P$, or $CdSnP_2$.

20. GaP = zinc blende structure:

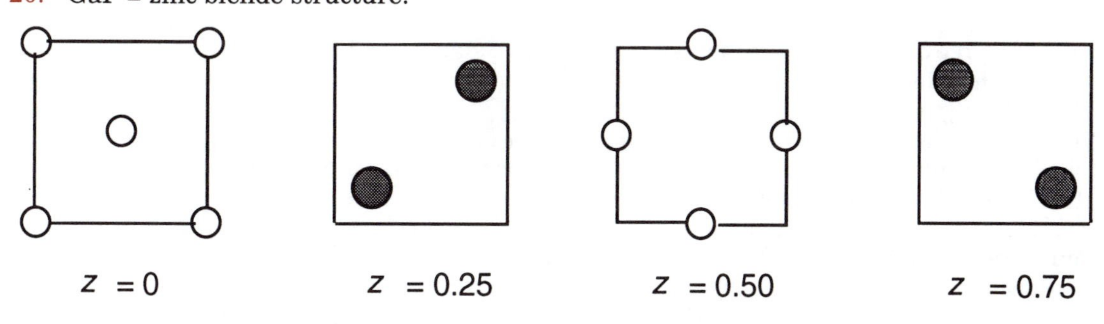

Si = diamond structure:

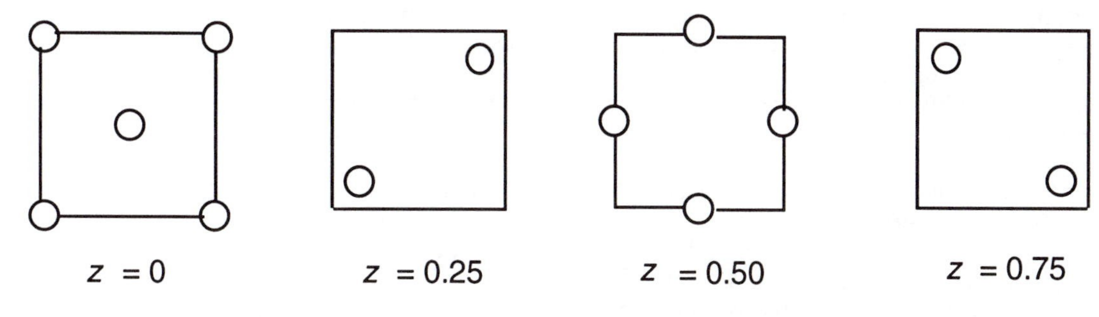

21. The partially filled valence band in sodium is exactly half filled because each sodium atom contributes a half-filled 3s atomic orbital to the formation of the valence band.

22. The conductivity of a semiconductor depends on its carrier concentration. At higher temperatures more thermal energy is available, and more electrons can be promoted across the band gap into the conduction band, where they contribute to electrical conductivity.

23. Metals are good electrical conductors at room temperature, but semiconductors generally are not. At liquid helium temperature, metals are excellent conductors (or superconductors), and semiconductors are insulators due to the lack of thermal energy necessary to promote electrons across the band gap.

24. Mg: Mg serves as an example of a metal. The Mg 3s band is filled but overlaps the Mg 3p band to create a partially filled band.

    Si: Si serves as an example of a semiconductor. At relatively low temperatures few electrons would be in the conduction band, whereas at higher temperatures a greater fraction of the electrons are promoted from the valence band to the conduction band; thus partially filled bands are generated.

    NaCl: NaCl serves as an example of an ionic insulator. Because of the large band gap in NaCl (*see* Table 7.4), this insulator contains bands that are either full or empty of electrons.

25. Oxidation of graphite removes electrons from the previously full valence band, leaving it partially filled and thereby giving it metallic properties. Bromine and sodium react to yield NaBr, an ionic insulator with a large band gap.

26. b. sp^2, three electrons per carbon are used to form sigma bonds.
    c. a nonhybridized 2p orbital; one electron per carbon
    d. half-full, metal
    e. polyacetylene would either be a semiconductor or an insulator, depending on the magnitude of the band gap. In fact, polyacetylene has a band gap of 1.9 eV, which makes it a semiconductor. It has a small electrical conductivity, similar to other common semiconductors, in the range of 10^{-9} $ohm^{-1}\ cm^{-1}$ (cis form) to 10^{-5} $ohm^{-1}\ cm^{-1}$ (trans form).
    f. Reaction with an oxidizing agent such as Br_2 causes electrons to be removed from the valence band, thereby giving a partially filled band. This greatly increases the conductivity of polyacetylene. If polyacetylene reacts with a reducing agent such as Li, electrons would be added to the conduction band, again giving a partially filled band. This also greatly increases polyacetylene's conductivity. In fact, by reacting *trans*-polyacetylene with either an oxidizing agent or a reducing agent, conductivities as high as 10^3 $ohm^{-1}\ cm^{-1}$ have been recorded, which are similar to the conductivities of many metals.

27. a. (8 corner Se atoms)(1/8 corner atom per cell) = 1 Se atom
    (10 face-centered Se atoms)(1/2 face atom per cell) = 5 Se atoms
    (1 Se atom within structure)(1 atom/cell) = 1 Se atom
    (4 edge-shared Se atoms)(1/4 atom/cell) = 1 Se atom
    Total Se atoms shown in structure = 8

    (4 Cu atoms within structure)(1 atom/cell) = 4 Cu atoms
    (4 In atoms within structure)(1 atom/cell) = 4 In atoms

    Formula of compound: $Cu_4In_4Se_8$ or $CuInSe_2$

b. Comparison of the chalcopyrite structure to that of zinc blende (*see* Figure 5.21) shows that chalcopyrite has the same structure as zinc blende, with In atoms replacing half of the Zn atoms and Cu atoms replacing the other half of the zinc atoms. The unit cell of the chalcopyrite structure contains more planes of atoms along the *z* axis than does the zinc blende structure, because of the alternating positions occupied by the In and Cu atoms in the structure.

c. An indium atom has one additional valence electron compared to a zinc atom, and a copper atom contains one less valence electron than a zinc atom. Selenium is isovalent with sulfur. In the formula ZnS we therefore can replace every Zn atom with one half of an atom of In and one half of an atom of Cu, giving $Cu_{0.5}In_{0.5}Se$, or $CuInSe_2$.

28. Absorption begins at the band-gap energy. For the physical mixture, expect absorption at the band gap of GaAs. For the solid solution expect the onset of absorption midway between the band gaps of GaAs and GaP. For emission, expect two bands from the physical mixture at the two band-gap energies. In the solid solution expect emission at its band gap energy, midway between these two energies.

29. The most bonding orbital occurs with p orbitals overlapping with no sign changes (no nodes between orbitals). The most antibonding orbital occurs with sign mismatch at each overlap.

30. See Demonstration 7.11 and pursuant discussion.

31. The band shifts from about 840 to 800 nm, or by ~ 0.074 eV.

Chapter 8

1. Ge has a smaller band gap (0.68 eV) than Si (1.11 eV); thus, a sample of Ge will have more electron–hole pairs at room temperature, as more electrons are able to be promoted across the smaller band gap. Similarly, Si (1.11 eV) will have more electron–hole pairs than GaAs (1.43 eV).

2. a. The value of p for Si is approximately 10^{16} cm^{-3} from the figure; therefore, $p = n = 10^{16}$ cm^{-3}
$K = pn = (10^{16} \text{ cm}^{-3})^2 = 10^{32} \text{ cm}^{-6}$

b. The negative slope of all data in Figure 8.5 indicates that autoionization is endothermic.

c. Figure 8.5 shows that at 20 °C, water has about 10^{14} H^+ ions/cm^3. Gallium arsenide has a hole concentration of about 10^6 h^+/cm^3 at 20 °C. Comparison of these values indicates that there are roughly 10^8 times more positively charged carriers in pure water than in undoped gallium arsenide.

3. a. Indium-doped silicon is a p-type semiconductor with the following band diagram (*see* Figure 8.9C):

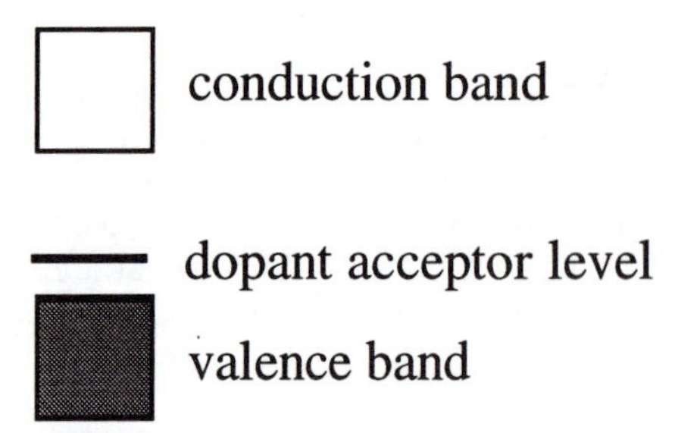

The Fermi level of In-doped silicon will initially be near the valence band, and will move toward the conduction band as P is added to the solid. This increase in energy results from mobile electrons (the result of P doping) neutralizing free holes generated by the original In dopants (*see* Figure s 8.9 and 8.11).

When the added P is about the same concentration as the original In dopant concentration, the Fermi level is near the center of the band gap (*see* Figure 8.9B). Addition of still more P causes the Fermi level to move progressively closer to the conduction band and yields a n-type semiconductor (*see* Figure 8.9D).

b. Antimony-doped silicon is a n-type semiconductor with the following band diagram (*see* Figure 8.9D):

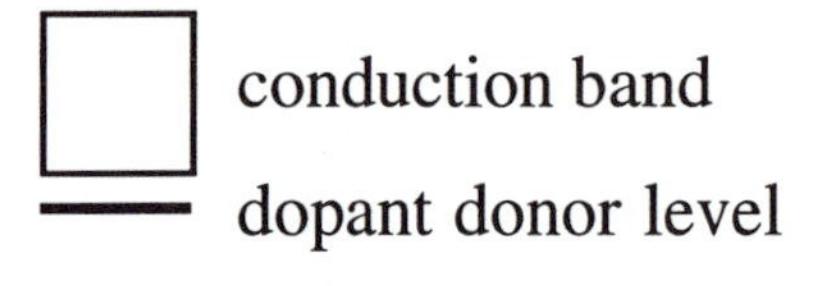

The Fermi level of Sb-doped silicon will be near the conduction band, and will move toward the valence band as Al is added to the solid. This behavior results from mobile holes (the result of Al doping) neutralizing free electrons generated by the original Sb dopants (see Figure 8.11). When the added Al is about the same concentration as the original Sb concentration, the Fermi level is near the center of the band gap. Addition of still more Al causes the Fermi level to move progressively closer to the valence band and yields a p-type semiconductor.

4. a. Half cell reactions and standard potentials for the cell in Figure 8.12 are as follows:

cathode: $2\,H^+_{(aq)} + 2\,e^- \leftrightarrows H_{2(g)}$; $E° = 0.00$ V

anode: $H_{2(g)} + 2\,OH^-_{(aq)} \leftrightarrows 2\,H_2O_{(l)} + 2\,e^-$; $E° = 0.83$ V

overall: $2H^+_{(aq)} + 2\,OH^-_{(aq)} \leftrightarrows 2\,H_2O_{(l)}$

$E_{cell} = (0.00\text{ V}) + 0.83\text{ V} = 0.83\text{ V}$

b. $[H^+]$ in base compartment:

$$[H^+][OH^-] = 1.0 \times 10^{-14}$$
$$[H^+](1.0\text{ M}) = 1.0 \times 10^{-14}$$
$$[H^+] = 1.0 \times 10^{-14}\text{ M}$$

As given by equation 8.14;

$E_{cell} = -\{(2.3)(RT)/(F)\}\log\{[H^+_{base}]/[H^+_{acid}]\}$

$$E_{cell} = -\frac{(2.3)(8.314\ J/K)(298\ K)}{(9.65 \times 10^4\ C)}\log\{\frac{1.0 \times 10^{-14}\ M}{1.0\ M}\}$$

$E_{cell} = 0.83$ V

5. a. A silver atom contains one less valence electron than a cadmium atom. The valence-band hole concentration is increased in the solid, making the sample a p-type semiconductor.

   b. A bromine atom contains one more valence electron than a sulfur atom. The conduction-band electron concentration is increased in the solid, making the sample an n-type semiconductor.

6. a. $K = pn = 6 \times 10^{26}$ cm^{-6}
   $p = n = K^{1/2} = (6 \times 10^{26}\ cm^{-6})^{1/2} = 2 \times 10^{13}$ cm^{-3}
   Germanium doped with Group 13 elements like aluminum will be p-type with $p > n$. Germanium doped with Group 15 elements like phosphorus will be n-type with $n > p$.

   b. $K = pn$ $\quad 6 \times 10^{26}\ cm^{-6} = p(3 \times 10^{18}\ cm^{-3})$ $\quad p = 2 \times 10^{8}$ cm^{-3}

   c.

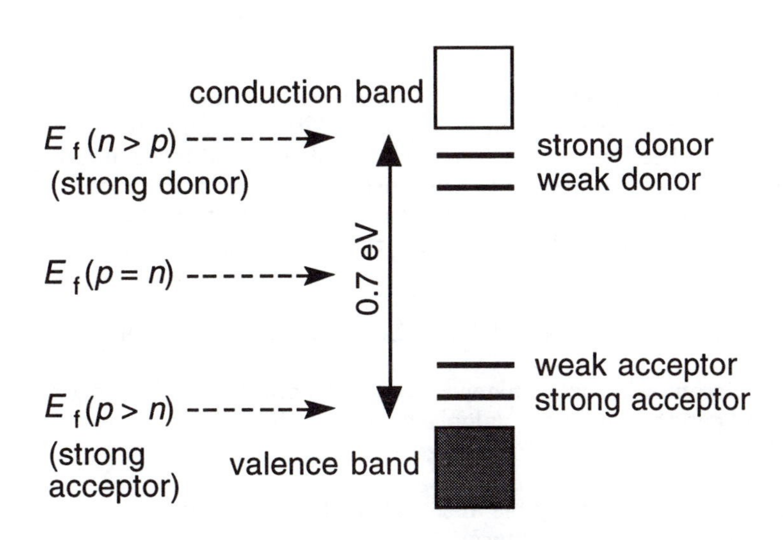

   d. The smaller K value for Si (2×10^{20} cm^{-6}) relative to that for Ge (6×10^{26} cm^{-6}) reflects the larger band gap of Si (1.11 eV) versus Ge (0.68 eV). The trend in band gaps is discussed in Chapter 7.

   e. The band diagram for a germanium semiconductor doped with indium or antimony is given as

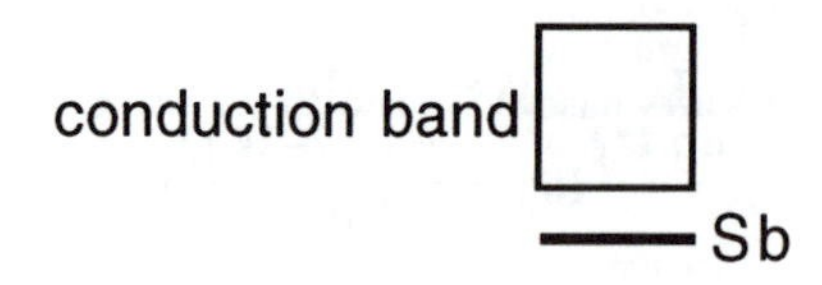

Indium has one fewer valence electron than germanium. When doped into Ge, In acts as an acceptor of Ge valence-band electrons, creating localized In^- ions and additional valence band mobile holes:

$$In \rightarrow In^- + h^+$$

Antimony has one more valence electron than germanium. When doped into Ge, Sb acts as a donor of electrons to the Ge conduction band, creating localized Sb cations and additional mobile conduction band electrons:

$$Sb \rightarrow Sb^+ + e^-$$

f. The Fermi level for In-doped Ge will initially be near the valence band, as In is an acceptor and the semiconductor is p-type. Addition of Sb (a donor) to In-doped Ge will cause the Fermi level to move toward the conduction band. When the concentration of Sb is equal to the concentration of In, the Fermi level will be near the center of the band gap. As the concentration of Sb exceeds that of In, the Fermi level will continue to rise and approach the conduction band.

This process is similar to a neutralization reaction in aqueous solution.

g. Gold is a weak acceptor (ionization energy = 0.2 eV) and should be placed about midway between the top of the valence band and the middle of the band gap (0.2 eV above the valence band edge).

7. a. Gallium arsenide exhibits negligible absorbance of light below approximately 1.4 eV, its band-gap energy. Light higher in energy than 1.4 eV is strongly absorbed by GaAs, and corresponds to transitions between filled valence-band orbitals and unfilled conduction-band orbitals.

b. Substitution of Se for As generates mobile conduction-band electrons and immobile Se^+, forming the n-type portion of the junction. Substitution of Zn for Ga generates mobile valence-band holes and immobile Zn^-, forming the p-type portion of the junction. The p–n junction is defined in this case by the diffusion of Zn into the n-type solid, as described in the chapter. Gallium arsenide is isoelectronic with germanium.

c. For example, consider photons with energies of 1.0, 1.5, and 2.0 eV. Only photons with energy equal to or greater than the band-gap energy will be absorbed to produce valence holes and conduction band electrons. The 1.0-eV photon will not be absorbed by the solar cell and does not produce electricity.

Both the 1.5- and 2.0-eV photons will be absorbed, and both will promote one valence electron into the conduction band leading to a contribution to the electric current. However, the energetic cost of these two events differs by 0.5 eV. The 1.5-eV photon was more efficient because it successfully promoted a valence band electron at a lower energetic cost than the 2.0-eV photon. The remainder of the energy of the 2.0-eV photon produces heat.

d. The solar spectrum shows that the sun has substantial output in the 1.5–2.1-eV range. These are energies that can be absorbed by GaAs with a band gap of 1.4 eV.

8. The semiconductivity of pure Si relies on an absorption of energy from an external source to promote electrons from the valence band to the conduction band. At low temperatures fewer electrons are promoted to the conduction band and conductivity drops dramatically. Doped Si can create conducting holes and electrons in the manner described for pure Si, but also by the promotion of electrons into or out of the dopant energy levels, which lie near the band edges for strong donors and acceptors. Less energy is needed for this process than for promotion of electrons across the band gap, and the semiconductivity of doped Si is thus much less temperature-dependent than it is for pure Si.

9. a.

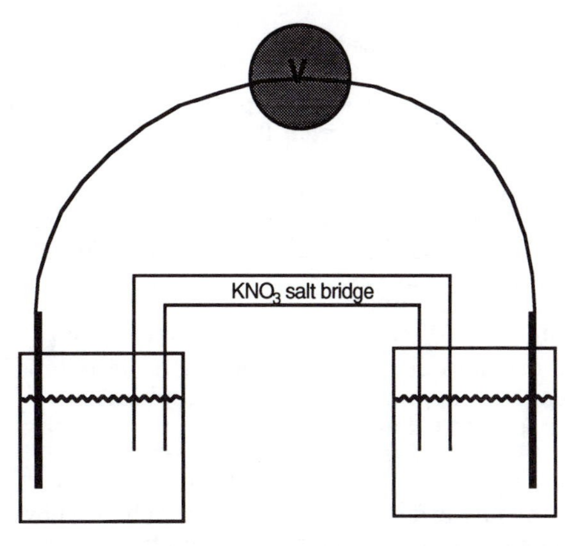

b. cathode: $2\,Ag^{+}_{(aq)} + 2\,e^{-} \leftrightarrows 2\,Ag_{(s)}$ $E° = 0.80$ V

anode: $Mg_{(s)} \leftrightarrows Mg^{2+}_{(aq)} + 2\,e^{-}$ $E° = 2.37$ V

overall: $2\,Ag^{+}_{(aq)} + Mg_{(s)} \leftrightarrows 2\,Ag_{(s)} + Mg^{2+}_{(aq)}$

$E_{cell} = 0.80\text{ V} + 2.37\text{ V} = 3.17\text{ V}$

c.

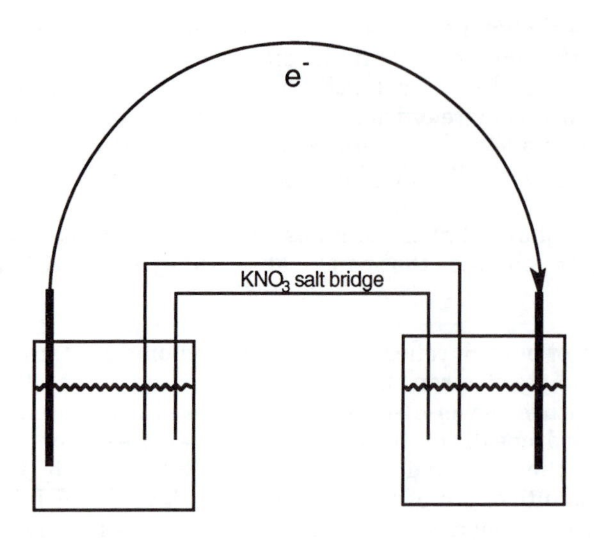

$Mg_{(s)} \leftrightarrows Mg^{2+}_{(aq)} + 2e^-$
(anode reaction)

$Ag^+_{(aq)} + e^- \leftrightarrows Ag_{(s)}$
(cathode reaction)

d. The LED will glow only when forward biased; thus, the electrodes must be attached properly. Attach the LED to the electrochemical cell with the n-type portion connected to the anode (Mg electrode), and p-type portion connected to the cathode (Ag electrode), the direction of current flow.

e. InP may be doped with Group 12 acceptors like Cd or Ag to form the p-type region, as these atoms contain fewer valence electrons than In. Donors such as S and Se from Group 16 can be used to create the n-type region, as these atoms contain more valence electrons than P.

f. InP will emit light close to its band-gap energy which is the threshold energy for light absorption by a semiconductor. The absorption spectrum of InP shows the onset of absorbance at approximately 1.3 eV, and this is approximately where the emitted light is found.

g. The LED will remain lit so long as the electrochemical cell delivers a sufficient voltage to maintain the forward bias. The potential delivered by the cell will decrease over time in accord with the Nernst equation.

10. The ionization reaction $D \rightarrow D^+ + e^-$, where D is a weak donor, is analogous to writing

$$H_2O + NH_3 \leftrightarrows NH_4^+ + OH^-$$

for the weak base NH_3.
The ionization reaction $D + h^+ \rightarrow D^+$ is analogous to writing

$$NH_3 + H^+ \leftrightarrows NH_4^+$$

The former is an Arrhenius-like formulation of the basicity of NH_3 (NH_3 increases the OH^- concentration), whereas the latter is a Brφnsted-like formulation (NH_3 is a proton acceptor that reduces the concentration of H^+).

11. The energy associated with a given transition corresponds to the difference in energy between the two quantized levels involved in the process, or

$$\Delta E = |E_{n_2} - E_{n_1}|$$

The energies of the quantized levels of the hydrogen atom can be calculated using equation 7 of Chapter 8. If a value of 12 is used for the relative dielectric constant, ε_r, and a value of $0.3m_e$ is used for the effective mass of the electron, the equation can be used to approximate the energies of the quantized levels of phosphorus-doped silicon.

The energy levels are calculated as follows:

$E_{joules} = (-e^4)(0.3\,m)/(8\varepsilon_o{}^2\varepsilon_r{}^2h^2n^2)$, where
$e = 1.602 \times 10^{-19}$ C $\quad m = 9.109 \times 10^{-31}$ kg $\quad \varepsilon_o = 8.85 \times 10^{-12}$ C^2 N^{-1} m^{-2}
$\varepsilon_r = 12 \quad h = 6.626 \times 10^{-34}$ kg m^2 s^{-1}

This reduces to $E_{joules} = -4.54 \times 10^{-21}$ kg m^2 s^{-2}/n^2, or -4.54×10^{-21} J/n^2
For $n = 1$, $E = -4.54 \times 10^{-21}$J For $n = 2$, $E = -1.14 \times 10^{-21}$ J
For $n = 3$, $E = -5.05 \times 10^{-22}$ J For $n = 4$, $E = -2.84 \times 10^{-22}$ J

a. $n = 4$ to $n = 3$, $E = 2.21 \times 10^{-22}$ J
b. $n = 4$ to $n = 1$, $E = 4.26 \times 10^{-21}$ J
c. $n = 3$ to $n = 2$, $E = 6.35 \times 10^{-22}$ J

These energies correspond to wavenumber values of
a. 11.1 cm^{-1} b. 214 cm^{-1} c. 31.9 cm^{-1}

These transitions occur in the far-IR or microwave portion of the electromagnetic spectrum.

d. The ambient thermal energy is given by kT. At 4.2 K this corresponds to 5.8×10^{-23} J or 3.6×10^{-4} eV.

12. b. 13.
d. 14. A zinc atom contains two fewer valence electrons than a germanium atom. Substitution of Zn for Ge generates mobile valence-band holes and immobile Zn^{2-}. There are two energy levels within the band gap, one corresponding to $Zn \rightarrow Zn^- + h^+$ and one to $Zn^- \rightarrow h^+ + Zn^{2-}$. This is analogous to a diprotic acid.

15. Referring to equation 17 in this chapter, the voltage that could be obtained from the p–n junction can be determined as follows:

E (in volts) $= -(2.3RT/F)[\log([e^-]_{p\text{-type side}}/[e^-]_{n\text{-type side}})]$
$K = pn$
$[e^-]_{p\text{-type side}} = K/p = 10^{20}$ cm^{-6}/10^{17} cm^{-3} = 10^3 cm^{-3}

$$E = -\frac{(2.3)(8.314\ \text{J/K})(298\ \text{K})}{(9.65 \times 10^4\text{C})} \log \left\{\frac{10^3\ \text{cm}^{-3}}{10^{17}\ \text{cm}^{-3}}\right\}$$

$E = 0.83$ V

16. a. The energy of one photon of 670-nm radiation is given by

$E = hc/\lambda = \frac{(6.626 \times 10^{-34}\ J{\cdot}s)(3.00 \times 10^{8}\ m\ s^{-1})}{(670 \times 0^{-9}\ m)} = 2.96 \times 10^{-19}$ J/photon

The energy of 1 mol of 670-nm photons therefore is
$(2.96 \times 10^{-19}$ J/photon)$(6.02 \times 10^{23}$ photons/mol photons) =
= 1.79×10^{5} J/mol photons

5 mW corresponds to 5×10^{-3} J/s.
$(5 \times 10^{-3}$ J/s)/$(1.79 \times 10^{5}$ J/mol photons) = 3×10^{-8} mols photons/s.

b. $(3 \times 10^{-8}$ mols photons/s)(1 mol e^-/1 mol photons)(96,500C/mol e^-) =
= $3 \times 10^{-3}\ C\ s^{-1} = 3$ mA

Chapter 9

1. d. 2. d. 3. d.

4. a. Assuming idealized oxygen stoichiometry, one method of preparation is
$Y_2O_{3(s)} + 4\ BaCO_{3(s)} + 6\ CuO_{(s)} + 0.5\ O_{2(g)} \rightarrow 2\ YBa_2Cu_3O_{7(s)} + 4\ CO_{2(g)}$
b. (12 g $YBa_2Cu_3O_7$)(1 mol /666.19 g) = 1.8×10^{-2} mol $YBa_2Cu_3O_7$
$(1.8 \times 10^{-2}$ mol $YBa_2Cu_3O_7$)(1 mol Y_2O_3/(2 mol $YBa_2Cu_3O_7$) =
9.0×10^{-3} mol Y_2O_3
$(9.0 \times 10^{-3}$ mol Y_2O_3)(225.81 g/1 mol Y_2O_3) = 2.03 g Y_2O_3
$(1.8 \times 10^{-2}$ mol $YBa_2Cu_3O_7$)(4 mol $BaCO_3$/2 mol $YBa_2Cu_3O_7$) =
3.6×10^{-2} mol $BaCO_3$ = 7.1 g $BaCO_3$
$(1.8 \times 10^{-2}$ mol $YBa_2Cu_3O_7$)(3 mol CuO/1 mol $YBa_2Cu_3O_7$) =
5.4×10^{-2} mol CuO = 4.3 g CuO

5. Compounds like Cu_2HgI_4 could be incorporated into the design of control circuits of refrigerators. When the interior of the refrigerator is cold, the insulating form of the compound would be present; this form would prevent the compressor from functioning. When the temperature inside the refrigerator rises, Cu_2HgI_4 (or a similar material) becomes conducting, and turns on the refrigeration unit, which will run until the compound cools and becomes an insulator again.

6. Thermochromic compounds could be incorporated into simple warning devices for tools such as drill bits that can fail under stresses of high temperature and pressure. At normal or safe operating temperatures the warning device is one particular color; when the temperature rises to an unsafe or undesirable temperature, a color change occurs, warning the operator of a potential safety problem.

7. I^-: (8 corner iodide ions)(1/8 occupancy) = 1 iodide ion
(10 face-shared iodide ions)(1/2 occupancy) = 5 iodide ions
(1 iodide ion within cell)(1 occupancy) = 1 iodide ion
(4 edge-shared iodide ions)(1/4 occupancy) = 1 iodide ion
Cu^+: (4 cuprous ions within cell)(1 occupancy) = 4 cuprous ions
Hg^{2+}: (2 mercuric ions within cell)(1 occupancy) = 2 mercuric ions
= $Cu_4Hg_2I_8$ or Cu_2HgI_4

8. a. white tin $\leftrightarrows$ gray tin
$\Delta G^\circ = (1\ \text{mol})(\Delta G^\circ_f\ \text{gray tin}) - (1\ \text{mol})(\Delta G^\circ_f\ \text{white tin})$
$\Delta G^\circ = (1\ \text{mol})(0.13\ \text{kJ/mol}) - (1\ \text{mol})(0\ \text{kJ/mol}) = 0.13\ \text{kJ}$
The reaction is spontaneous right to left at 298 K.
b. $\Delta G = \Delta H - T\Delta S$
$0 = [(1\ \text{mol})(-2.09\ \text{kJ/mol}) - (1\ \text{mol})(0\ \text{kJ/mol})] - (T)[(1\ \text{mol})(0.044\ \text{kJ/mol·K}) - (1\ \text{mol})(0.052\ \text{kJ/mol·K})]$
$(T)[(1\ \text{mol})(0.044\ \text{kJ/mol·K}) - (1\ \text{mol})(0.052\ \text{kJ/mol·K})] = -2.09\ \text{kJ}$
$T = 260\ \text{K}$

9. Vaporization of water requires the disruption of strong intermolcular forces (hydrogen bonding) and therefore requires a great deal of energy in order to convert the condensed, liquid phase to the disordered gaseous phase. A solid-state phase change like that exhibited by NiTi involves only slight shifts in atomic positions.

10. The hcp and ccp structures both have a 74% packing efficiency (Table 5.1). Application of pressure would not be expected to convert one form to the other.

11. If a pellet of 1-2-3 is cooled to below its critical temperature (easily done with liquid nitrogen), the superconducting sample can be placed over a magnet to levitate the superconductor. Use tongs, tweezers, or gloves to handle the cold 1-2-3 sample.

12. The interaction between the paramagnetic or ferromagnetic solid and the magnet depicted in Figure 2.10 demonstrates that paramagnetic and ferromagnetic materials are attracted towards magnetic fields; the result is that the balance measures a lower apparent weight for the sample. The superconductor, in contrast, will repel an external magnetic field (Meissner effect). The 1-2-3 sample should have an increased apparent weight as a result.

13. Two samples of 1-2-3 superconductor, cooled below their critical temperatures, could form the "bread" of a sandwich that has a magnet suspended in between. (*see* Jacob, A. T.; Pechmann, C. I.; Ellis, A. B. *J. Chem. Educ.* **1988**, *65*, 1094.)

14. The stoichiometry determination follows the procedure worked in the solution to exercise 7. In the Cu_2HgI_4 unit cell, the vacancies and divalent cations (Hg^{2+}) occupy the same planes ($z = 0.37$ and 0.87), whereas in the Ag_2HgI_4 unit cell the vacancies and the monovalent cations (Ag^+) occupy the same planes ($z = 0.13$ and 0.63).

15. Iodine: 5p orbitals; mercury: 6s

16. The yellow color of the low-temperature phase corresponds to a higher energy band gap than the orange color of the high-temperature phase (Figure 7.18). If a change in band widths is the reason for the smaller band gap at higher temperature, then the bands must have widened at higher temperature so as to shrink the energetic separation between the two band edges. A similar color shift and explanation can be invoked for the Cu_2HgI_4 with the red (low-temperature) to red–brown (high-temperature) color change also consistent with larger band widths and a smaller band gap at higher temperature.

17. Because the same disordered cubic structures occur for the high-temperature phases of Ag_2HgI_4 and Cu_2HgI_4, it might be possible to prepare solid

solutions of composition $Ag_{2x}Cu_{2-2x}HgI_4$. However, because of the large difference in ionic radii for the two monovalent cations, only limited substitution of one monovalent cation for the other might be possible.

18. For example, for $Bi_2Ca_2Sr_2Cu_3O_{10}$, the oxidation number of Cu is +2.

19. Electricity can be used to resistively heat the sculpture, thereby transforming the memory metal from its low-temperature to its high-temperature shape.

20. From Figure 9.10, at 0 °C, a sample of Ni_xTi_{1-x} with $x = 0.51$ would be in the high-temperature phase, whereas a sample with $x = 0.49$ would be in the low-temperature phase.

21. The consistent applied stress over this processing cycle favors a particular orientation of variants in the low-temperature phase. *See* Wayman, C. M. *MRS Bull.* **!993**, *XVIII*, 49–56.

22. This is an exothermic reaction, the heat from which can be used to drive the martensite-to-austenite conversion of the unreacted NiTi. One of the reaction products is nickel bromide. Further details can be found in Gisser, K. R. C.; Philipp, D. M.; Ellis, A. B. *Chem. Mater.* **1992**, *4*, 700.

23. A sample could be prepared that is coiled or spring-shaped in the hig-temperature form and linear in the low-temperature form. Insert the cold linear sample into the artery, and when it warms to body temperature it assumes the coiled shape that can help keep the artery opened.

24. The rock salt structure has the lower density or the greater molar volume.

Chapter 10

1. In solution, Pb^{2+} and I^- ions can easily diffuse together to produce PbI_2. In the solid, this diffusion is far slower.

2. Grind the mixture, then heat it.

3. Co will oxidize to Co_3O_4 in air. Thus no Co will be available to reduce the Co_3O_4 initially present.

4. a. $Ti_2O_3(s) + V_2O_5(s) \rightarrow 2\ TiVO_4(s)$
   b. $Pb(s) + PbO(s) + Nb_2O_5(s) \rightarrow 2\ PbNbO_3(s)$
   c. $Ti(s) + 2S(s) \rightarrow TiS_2(s)$
   d. $2CrCl_3(s) + H_2(g) \rightarrow 2\ CrCl_2(s) + 2\ HCl(g)$
   e. $MnCl_2(aq) + 2\ NaOH(aq) \rightarrow Mn(OH)_2(s) + 2NaCl(aq)$
   $Mn(OH)_2(s) \rightarrow MnO(s) + H_2O(g)$
   f. $SiCl_4(l) + 4H_2O \rightarrow Si(OH)_4(s) + 4HCl$
   $Si(OH)_4\ (s) \rightarrow SiO_2(s) + 2H_2O$

5. V^{5+} is a weak oxidizing agent and will not oxidize chloride ion. Iodide ion is easier to oxidize.

6. The Li^+ is small enough to diffuse into the V_2O_5 structure with little or no change in the structure. The Cs^+ is rather large.

7. Decomposition of $La[Fe(C_2O_4)_3]$ produces an oxide with the La^{3+} and Fe^{3+} ions mixed on an atomic scale. Such mixing is not present in a mixture of $La_2(CO_3)_3$ and Fe_2O_3.

8. All three samples would give the same elemental analysis, because they all have the same relative numbers of Al, Ga, and As atoms. However, they would have distinguishable X-ray diffraction patterns: The physical mixture would have two sets of peaks corresponding to pure AlAs and pure GaAs. The solid solution would have one set of peaks corresponding to pure $Al_{0.5}Ga_{0.5}As$. The layered solid grown by CVD methods would have a diffraction pattern that reflects the periodic arrangement of 20 atomic layers (10 layers of AlAs followed by 10 layers of GaAs).

9. There are countless possibilities. Structures like the following layer sequences would all correspond to 1:1 A:Z stoichiometry:

```
AAAAAAAAAA        AAAAAAAAAA        AAAAAAAAAA
ZZZZZZZZZZ        AAAAAAAAAA        AAAAAAAAAA
AAAAAAAAAA        ZZZZZZZZZZ        AAAAAAAAAA
ZZZZZZZZZZ        ZZZZZZZZZZ        ZZZZZZZZZZ
                                    ZZZZZZZZZZ
                                    ZZZZZZZZZZ
```

All of these structures would have different unit cells (different repeat distances) and thus different X-ray diffraction signatures.

10. One-eighth of the tetrahedral holes are occupied by Mg^{2+} ions and one-half of the octahedral holes are occupied by Al^{3+} ions.

11. a. The large band-gap material is A and thus corresponds to $Al_{0.3}Ga_{0.7}As$; the smaller band-gap material is B and is GaAs.

b. Both conduction band electrons and valence band holes will be at lower energy if they are at the band edge in region B (recall, Chapter 8, that holes "rise" to reach lower energy).

c. i. absorption in region B (photon energy is between the band-gap energies of the two materials).

ii. absorption in neither region (photon energy is below the band-gap energies of the two materials).

iii. absorption in both regions (photon energy is above the band-gap energies of both materials).

12. *See* the polymer sketch in Experiment 15 on silica gel. Because Ge is isovalent with Si, it could replace it in the structure. Fluorine might replace oxygen in the structure, in which case its tendency to form only a single bond may terminate the growth of the polymer at that position.

13. a. $0.5\ Ni(s) + 0.5\ Al(s) \rightarrow Ni_{0.5}Al_{0.5}(s)$
b. Dividing the enthalpy of formation by the specific heat capacity gives an estimated temperature increase of nearly 2400 °C.

14. For example, platinum and palladium have nearly identical unit cell dimensions and would be good candidates for epitaxial growth.

15. Isoelectronic semiconductors like GaAs and ZnSe with nearly identical unit cell dimensions are good candidates for epitaxial growth as are Ge and GaAs.

Index

Numbers in **boldface type** refer to Experiments 1 through 15. The word "demo" after an entry refers to demonstrations, and the word "lab" refers to suggested laboratory activities.

F

Q

R

X

Y

Z